D1440957

Multiply Bonded Main Group Metals and Metalloids

This is a paperback edition of *Advances in Organometallic Chemistry*, Volume 39.

Multiply Bonded Main Group Metals and Metalloids

EDITED BY

ROBERT WEST

DEPARTMENT OF CHEMISTRY
UNIVERSITY OF WISCONSIN
MADISON, WISCONSIN

F. GORDON A. STONE

DEPARTMENT OF CHEMISTRY
BAYLOR UNIVERSITY
WACO, TEXAS

ACADEMIC PRESS
San Diego New York Boston
London Sydney Tokyo Toronto

This book is printed on acid-free paper. ♾

Academic Press, Inc.
A Division of Harcourt Brace & Company
525 B Street, Suite 1900, San Diego, California 92101-4495

United Kingdom Edition published by
Academic Press Limited
24-28 Oval Road, London NW1 7DX

International Standard Serial Number: 0065-3055

International Standard Book Number: 0-12-744740-7

PRINTED IN THE UNITED STATES OF AMERICA
96 97 98 99 00 01 EB 9 8 7 6 5 4 3 2 1

Contents

Silicon–Phosphorus and Silicon–Arsenic Multiple Bonds

MATTHIAS DRIESS

Chemistry of Stable Disilenes

RENJI OKAZAKI and ROBERT WEST

Stable Doubly Bonded Compounds of Germanium and Tin

K. M. BAINES and W. G. STIBBS

Diheteroferrocenes and Related Derivatives of the Group 15 Elements: Arsenic, Antimony, and Bismuth

ARTHUR J. ASHE III and SALEEM AL-AHMAD

Boron–Carbon Multiple Bonds

JOHN J. EISCH

Contributors

Numbers in parentheses indicate the pages on which the authors' contributions begin.

SALEEM AL-AHMAD (325), Department of Chemistry, The University of Michigan, Ann Arbor, Michigan 48109

ARTHUR J. ASHE III (325), Department of Chemistry, The University of Michigan, Ann Arbor, Michigan 48109

K. M. BAINES (275), Department of Chemistry, University of Western Ontario, London, Ontario, Canada N6A 5B7

ADRIAN G. BROOK (71), Lash Miller Chemical Laboratories, University of Toronto, Toronto, Canada M5S 1A1

MICHAEL A. BROOK (71), Department of Chemistry, McMaster University, Hamilton, Ontario, Canada L8S 4R8

PENELOPE J. BROTHERS (1), Department of Chemistry, University of Auckland, Auckland, New Zealand

MATTHIAS DRIESS (193), Anorganisch-chemisches Institut der Universtät Heidelberg, D-69120 Heidelberg, Germany

JOHN J. EISCH (355), Department of Chemistry, State University of New York at Binghamton, Binghamton, New York 13902

INA HEMME (159), Institute of Inorganic Chemistry, University of Goettingen, D-37077 Goettingen, Germany

UWE KLINGEBIEL (159), Institute of Inorganic Chemistry, University of Goettingen, D-37077 Goettingen, Germany

RENJI OKAZAKI (231), Department of Chemistry, Graduate School of Science, The University of Tokyo, Hongo, Tokyo 113, Japan

PHILIP P. POWER (1), Department of Chemistry, University of California, Davis, Davis, California 95616

W. G. STIBBS (275), Department of Chemistry, University of Western Ontario, London, Ontario, Canada N6A 5B7

ROBERT WEST (231), Department of Chemistry, University of Wisconsin, Madison, Wisconsin 53706

Preface

This volume of *Advances in Organometallic Chemistry* is devoted entirely to the chemistry of multiply bonded main group metals and metalloids.

Persistent efforts to prepare stable multiply bonded compounds of main group metals during the early years of this century were uniformly unsuccessful, eventually leading to the textbook *double-bond rule*: "Elements outside the first row of the periodic table do not form stable multiple bonds." By the late 1960's, however, evidence was beginning to accumulate that multiply bonded species of heavier main group elements, especially silicon, were present as intermediates in thermal or photochemical reactions. And, in 1976, Si=C compounds were isolated in argon matrices at cryogenic temperatures.

Major advances took place in 1981, when the synthesis of stable compounds containing Si=C, Si=Si, and P≡P multiple bonds was announced. These developments led to a rapid flowering of this field of research, and in the intervening 14 years, the chemistry of multiply bonded main group metals has become an active and fascinating area. Stable multiple-bond compounds are now known for Si, Ge, and Sn in group 14; for B, Al, Ga, and In in group 13; and for As, Sb, and Bi, as well as phosphorus in group 15.

In this volume, the current state of knowledge of many of these compounds is surveyed. It is our hope that this book will be of use to workers in the field for many years to come.

Robert West
F. Gordon A. Stone

Multiple Bonding Involving the Heavier Main Group 3 Elements Al, Ga, In, and Tl

PENELOPE J. BROTHERS

Department of Chemistry, University of Auckland
Private Bag 92019, Auckland, New Zealand

PHILIP P. POWER

Department of Chemistry, University of California
Davis, California

I
INTRODUCTION

A. *Factors Favoring the Stability of Multiple Bonds to the Heavier Main Group 3 Elements*

The chemistry of compounds that involve the heavier main group elements in multiple bonding has undergone very rapid and extensive development in the last two decades. Much of this impressive advance has occurred in the main group 4[1] and 5[2] elements. In sharp contrast, multiple bonding in the neighboring heavier main group 3 elements has received little attention until recently. It is probable that one of the major factors that has inhibited growth in this area has been the close association of the chemistry of the elements of the boron group with the concept of electron deficiency—a characteristic that generally precludes the existence of multiple bonds. In spite of this viewpoint, an extensive number of boron compounds have multiple bonds either to boron itself[3] or to a range of other elements. Some of these—many B—N compounds for example—have π-bonding, which has been known and studied for many years.[4] For other moieties, such as B—C (in acyclic compounds),[5] B—P,[6] and B—As,[7] π-bonding has become well established fairly recently. Furthermore, there have been numerous instances of short B—B bonds in various solid-state metal boride structures,[8] whereas formal B=B double bonding in acyclic boron compounds was reported in molecular compounds only within the past few years.[9]

These significant developments in boron chemistry are only now beginning to find parallels in compounds of the heavier elements Al, Ga, In, and Tl. A major reason for these recent advances has been the recognition that the heavier main group 3 elements are quite large[10] (e.g., covalent radius of aluminum = 1.3 Å; cf. radii B = 0.85 Å, Si = 1.17 Å, P = 1.11 Å) and require sufficient steric protection to ensure that they remain low-coordinate (i.e., coordination number of 3 or less), thereby preventing decomposition (which usually occurs via disproportionation) and allowing the unused orbitals to interact and form multiple bonds. This requirement has often been difficult to satisfy not only because of their large size but also because of the very strong tendency of Al, Ga, In, and Tl centers to form complexes with Lewis bases such as ethers, amines, or phosphines. This essentially prevents the observation of classical *p–p*, π-bonding for these Lewis acid–base complexes.

The ready formation of such complexes has also meant that much of the inorganic and organometallic chemistry of these elements has had to be carried out in the absence of donor solvents.[11] Thus, the methods used for the synthesis of many aluminum and gallium organometallic compounds are often quite different from those employed for the neighboring elements silicon and germanium. For example, in the synthesis of aluminum or gallium alkyls and aryls, Grignard reagents are not generally used since their preparation almost invariably involves the use of an ether donor solvent which may complex to the metal center.[12–14] Likewise, many organolithium compounds (e.g., LiMe or LiPh) are difficult to employ owing to their insolubility in nondonor solvents. In addition, there is a tendency to form metallate complexes of the type $LiMR_4$ or $LiMClR_3$ (M = Al or Ga).[11,13] However, many of the problems outlined above may be overcome by the use of suitably large substituents. Thus, the use of the $—CH(SiMe_3)_2$ ligand allows the ready synthesis of $[(Me_3Si)_2CH]_2Al$-$Cl(OEt_2)$ (which has a synthetically useful and reactive Al—Cl moiety) via the reaction of $LiCH(SiMe_3)_2$ with $AlCl_3$ in ether.[15] The ether is not tenaciously held by the aluminum and may easily be removed under reduced pressure. It is dimeric in freezing benzene and presumably in the solid state.[15] More recently, extensive investigations by a number of groups[16–20] have resulted in the isolation of several compounds involving the elements Al, Ga, or In in unusual bonding, coordination numbers, and/or oxidation states,[16,17] e.g., R_2MMR_2 (R = $CH(SiMe_3)_2$, M = Al,[18] Ga,[19] or In[20]). In addition, many sterically hindered species with the synthetically useful metal–halogen moiety have been synthesized. These include $(t\text{-Bu})_2MCl$ (M = Al[21] or Ga[22]); $Trip_2MX$ (M = Al, X = Br; Ga, X = Cl or Br[23]; Trip = 2,4,6-i-$Pr_3C_6H_2$); Mes*MCl_2 (M = Ga or In; Mes* = 2,4,6-t-$Bu_3C_6H_2$[23,24]); Mes*MBr_2 (M = Al,[23] In[24]); Mes^*_2MCl (M = Ga[25] or In[26]). These compounds, along with the previously mentioned $[(Me_3Si)_2CH]_2AlCl$ species, represent the key starting materials for several new species of compounds described in this review.

The second major consideration in the formation of multiple bonds involving Al, Ga, In, and Tl is the provision of sufficient electron density for their formation. As already noted, these elements do not have sufficient numbers of electrons in their valence shells to occupy all the valence orbitals. The normal result of this is a strong tendency toward association or Lewis acid–base complex formation—characteristics that often define the chemistry of these elements. Even if these two processes are blocked by the introduction of sterically hindering substituents, there remains the problem of providing sufficient electron density to generate the multiple bond itself. This problem is currently approached in two major ways. First, one of the substituents at the main group 3 element can have one or more lone pairs which may behave as electron donors to an empty p-orbital on the main group 3 element. Second, electrons may be added to the main group 3 element p-orbitals by reduction. Again, both of these methods of formation of a π-bond (particularly the former) to afford multiple bonding have been established in boron chemistry.[3–5,27] Ways of applying these methods to the heavier congeners have been found only within the past few years.

B. π-Bonding

Prior to discussing multiple bonding in heavier main group 3 compounds, a brief review of the nature of multiple bonds is in order. In the context of the lighter main group elements C, N, and O, multiple bonds are usually considered to be a combination of σ- and π-bonds. The σ-bond involves a conventional, positive-wavefunction orbital overlap along the bond axis. It is worth noting, however, that many possible single bonds (i.e., σ-bonds) involving the heavier main group elements are currently unknown or poorly documented. For example, the first well-characterized molecular

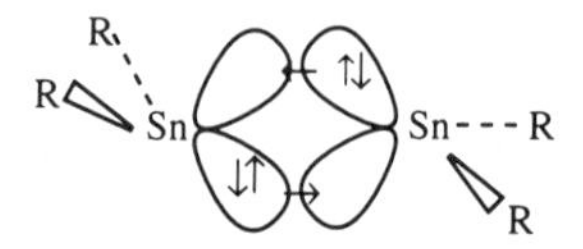

FIG. 1. Schematic drawing[31] of the proposed double bond in $[Sn\{CH(SiMe_3)_2\}_2]_2$.

compound involving an Al—Al bond was reported as recently as 1988.[18] In the case of π-bonds different orbital-overlap models are applicable. First, there is the conventional π-bond that involves side-on overlap of p-orbitals. It is this type of bonding that this review primarily addresses. In addition, there are also π-bonding models that involve the participation of d-orbitals or σ^*-orbitals of the main group element. For many main group elements there is little evidence to substantiate significant d-orbital involvement in π-overlap owing to the large difference in energy between them and the valence s- and p-orbitals.[28] The involvement of σ^*-orbitals in π-bonding has been invoked to account for short metal–ligand bonds and wide angles at the ligand in certain four-coordinate aluminum aryloxide derivatives.[29] These experimental data have been supported by calculations.[30] However, it is possible to argue that the short bond lengths are the result of ionic contributions owing to the high electronegativity difference between aluminum and oxygen. In addition, the ionic characteristics of the Al—O bond may imply that the normal directionalized bonds associated with covalent bonding are of reduced importance. This may mean that the bond angles are more easily distorted without weakening the strength of the bond itself.

One other significant model for multiple bonding among heavier main group elements has also been proposed. This is the donor–acceptor model proposed by Lappert to account for the unique bonding characteristics of $Sn[CH(SiMe)_2]_2$ in the dimeric form,[31] as illustrated by Fig. 1. This dimeric compound, which is a tin analog of a substituted ethene, has pyramidal rather than planar coordination at tin. The nonplanarity of the geometry at the main group 4 atom has also been observed in related silicon and germanium compounds.[1,32] The discovery of such compounds has shown that the classical σ/π-model of the double bond in carbon, nitrogen, or oxygen compounds does not necessarily apply to the heavier elements.

C. *Significance of Bond Lengths*

The primary method for the assessment of the extent of multiple bonding in compounds involves the use of structural data. Shorter bond lengths are usually taken to imply stronger bonds. A bond length that is shorter than the sum of the covalent radii of the bonded atoms, however, does not necessarily imply the presence of formal multiple bonds. It has been known for many years that bonds between atoms of different electronegativities (i.e., bonds with partial ionic character) are generally shorter than those predicted by the sum of the covalent radii.[33] This has been taken as evidence for a resonance contribution to the bonding wavefunction from its ionic form. Moreover, the more ionic the bond, the greater this contribution should be. Indeed, experimentally, it is generally found that the greater the difference in electronegativity between the bonding atoms, the greater the deviation from the predicted bond distance based on covalent radii. For example, the predicted lengths for Al—O, Al—N, and Al—C bonds are 1.96, 2.00, and 2.07 Å, respectively.[10] Experimentally, most terminal Al—O, Al—N, and Al—C bond lengths are in the ranges 1.70–1.78 Å, 1.78–1.85 Å, and 1.95–2.05 Å (*vide infra*). Thus, the differences between predicted and experimental values are 0.18–0.26 Å for Al—O, 0.11–0.22 Å for Al—N, and 0.02–0.12 Å for Al—C bonds. These data demonstrate that greater discrepancies are observed for the more ionic bonds. This strongly implies the existence of a significant ionic contribution to bond strength. Further evidence for the relative weakness of π-bonding in Al—O or Al—N species comes from the fact that conventional π-bonding involving *p–p* overlap should be stronger in Al—N than in Al—O compounds. Yet, the largest differences between predicted and measured distances occur in Al—O rather than Al—N bonds.

A number of empirical methods exist for the adjustment of covalent bond lengths for ionic effects.[34,35] These are based primarily on formulas that involve the sum of the covalent radii corrected by a factor that is dependent on the electronegativity difference between the atoms. In many instances, quite good agreement is obtained between the predicted and experimental values, as shown by the listing in Table I.

Two other important factors may influence observed bond distances: the coordination numbers of the bonded atoms and the size of the substituent ligands. In the case of main group 3 multiply bonded species the primary concern is with the abnormally low (i.e., 2 or 3) coordination number of the main group 3 atom, which usually necessitates a lower than normal estimate of its radius. Thus, the use of half the interatomic distance found, for example, in metallic aluminum (i.e., 1.43 Å; each Al is 12-coordinate with an Al—Al distance of 2.86 Å[36]), frequently gives predicted bond lengths that are too long for most aluminum compounds. It is usually found that a radius of 1.3 Å is more realistic and in much closer agreement with the 1.32–1.33 Å value estimated from the Al—Al distances in the recently reported compounds R_2AlAlR_2 (R = $CH(SiMe_3)_2$[18] or Trip[37]).

TABLE I

ESTIMATED, EMPIRICALLY CORRECTED, AND EXPERIMENTAL M—E SINGLE BOND LENGTHS FOR MONOMERIC $R_2M—ER'_2$ OR $R_2M—ER'$ COMPOUNDS

Bond type	Σ Covalent radii[a,b] (Å)	ΔEN[a]	Empirically corrected bond lengths (Å) Schomaker and Stevenson[c]	Empirically corrected bond lengths (Å) Blom and Haaland[d]	Observed bond lengths (Å)
Al—N	2.00	1.60	1.85	1.79	1.78–1.88
Ga—N	1.95	1.25	1.84	1.80	1.82–1.94
In—N	2.10	1.59	1.96	2.03	2.05–2.20
Al—P	2.40	0.59	2.35	2.36	2.34
Ga—P	2.35	0.24	2.33	2.27	2.26–2.39
In—P	2.50	0.57	2.45	2.51	2.57–2.61
Ga-As	2.47	0.38	2.44	2.37	2.40–2.47
Al—O	1.96	2.03	1.78	1.72	1.65–1.71
Ga—O	1.91	1.68	1.76	1.74	1.78–1.82
Al—S	2.32	0.97	2.23	2.17	2.19
Ga—S	2.27	0.62	2.21	2.18	2.21–2.27
In—S	2.42	0.95	2.34	2.42	2.40
Ga—Se	2.42	0.66	2.36	2.31	2.32
In—Se	2.57	0.99	2.48	2.54	2.51
Ga-Te	2.60	0.19	2.58	2.58	2.54

[a] Covalent radii and electronegativity values from Ref. 10. ΔEN is the difference in electronegativity between the two elements using the Allred–Rochow scale.

[b] Covalent radii: O, 0.66 Å (Ref. 96); N, 0.70 (see the reference Pestana, D.C., Power, P. P. *Inorg. Chem.* **1991,** 30, 528 for further discussion). The radius for In was determined from the average In—In bond length in R_2InInR_2 compounds (Refs. 20, 46–48).

[c] Correction given by the formula $r_{A-B} = r_A + r_B - 0.09\,|EN_A - EN_B|$ (Ref. 34).

[d] Correction given by the formula $r_{A-B} = r_A + r_B - 0.085\,|EN_A - EN_B|^{1.4}$ (Ref. 35). Slightly different values of the covalent radii were used in these estimates.

Since low coordination numbers are generally achieved by the use of large substituents, the size of these substituents may effect a lengthening in the observed bond length through steric crowding. This phenomenon can be illustrated by a comparison of intermetallic bond distances in the series of compounds R_2MMR_2 (M = Al, Ga, In: R = $CH(SiMe_3)_2$ or Trip), where it can be seen that the M—M distances are consistently longer in the more crowded alkyl derivatives (*vide infra*).

II

HOMONUCLEAR MULTIPLE BONDING INVOLVING THE HEAVIER MAIN GROUP 3 ELEMENTS

The new developments in homonuclear bonding described in this section for Al → Tl have precedent in the synthesis of molecular species with boron–boron multiple bonds. In essence, it has been shown that tetraorganodiboron compounds[38–40] can undergo two successive one-electron reductions[9,41] to give multiple B—B bonds as shown by Eq. (1).

$$R_2B{-}BR_2 \xrightarrow{1e^-} [R_2B \overset{\cdot}{\text{---}} BR_2]^- \xrightarrow{1e^-} [R_2B{=}BR_2]^{2-} \qquad (1)$$

Tetraorganodiboron species, which are normally unstable toward disproportionation, can be stabilized through the use of large substituents[38,39] with bulky aryl derivatives exhibiting significantly higher stability than the currently known purely alkyl analogs.[40] A one-electron reduction was demonstrated in the case of the tetraalkyl species which were characterized by EPR spectroscopy in solution.[41] The EPR studies indicated that the unpaired electron was located in a π- rather than a σ-type orbital. Furthermore, the coupling to each boron nucleus was observed to be equivalent. In other words, the EPR data were consistent with the formation of a one-electron π-bond between the boron centers. Unfortunately, this species has not yet been characterized structurally, so the B—B distance is unavailable. The structure of such a species is obviously most desirable since it would confirm the existence of a shortened B—B bond. Structures have been obtained, however, for the doubly reduced species $[R_2BBR_2]^{2-}$ which may have either aromatic or amide substituents.[9,27] These exhibit an approximately planar core geometry (Fig. 2) as well as shortened B—B distances (~1.63 Å vs. 1.71 Å in the neutral precursors). The shortening of 0.08 Å is not as great as that seen in the isoelectronic alkenes owing, most probably, to the large coulombic repulsion in the doubly reduced diboron species.

These reductions suggested that it should be possible to carry out similar reactions for the heavier element congeners R_2MMR_2 (M = Al, Ga, In, or Tl). The neutral tetraorganodimetallane precursors for such reactions have only become available relatively recently—the first tetraorganodimetallane derivatives were reported in 1988—although species containing Ga—Ga[42] or In—In[43] interactions have been known for some years in compounds such as the base-stabilized $Ga_2X_4 \cdot 2L$ (X = halide; L = variety of donors including X^-) and related species. The first example of a tetraorganodialane was synthesized by the reduction of the sterically hindered

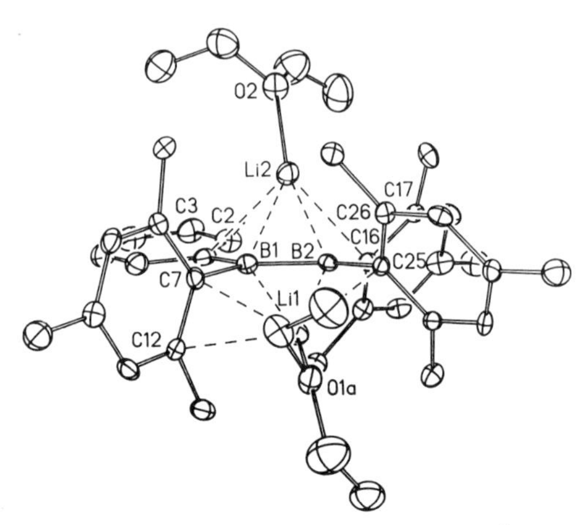

FIG. 2. Thermal ellipsoidal drawing of $[Mes_2BB(Ph)Mes]^{2-}$ which contains formal boron–boron double bonds. [Reprinted with permission from Moezzi *et al.*[9a] Copyright 1992 American Chemical Society.]

$[(Me_3Si)_2CH]_2AlCl$ precursor with potassium in hydrocarbon solvent (Eq. 2).[18]

$$2\,R_2AlCl \xrightarrow[\text{hexane, 25°C}]{2K} R_2AlAlR_2 + 2\,KCl \qquad (2)$$

This remarkable compound is stabilized by the presence of bulky substituents which prevent disproportionation and decomposition. The Al—Al distance is 2.660(1) Å and the torsion angle between the coordination planes at the aluminum atoms is near 0°. Subsequently, it was shown that the gallium (Ga—Ga = 2.541(1) Å[19]) and indium (In—In = 2.828(1)Å[20]) analogs could also be synthesized. More recently, the related Tl—Tl bonded species $\{(Me_3Si)_3Si\}_2TlTl\{Si(SiMe_3)_3\}_2$ (Tl—Tl distance = 2.9142(5) Å) has been reported.[44]

The successful use of aryl substituents in the stabilization of the tetraorganodiboron compounds, together with the ease of their subsequent reduction to the dianions,[9,27] has prompted the synthesis of tetraaryldialuminum, -gallium, and -indium analogs for the purpose of further reduction. This was accomplished with use of the Trip substituents, and these derivatives displayed somewhat shorter M—M distances (Al—Al = 2.647(3) Å,[37] Ga—Ga = 2.515(3) Å,[45] In—In = 2.775(2) Å[46]) than those observed in the alkyls, presumably because of the slightly smaller size of the Trip group. A number of other interesting related Ga—Ga and In—In bonded species have also been reported; these include $In_2[2,4,6\text{-}(CF_3)_3C_6H_2]_4$[47] and the amido derivatives $[MN(t\text{-Bu})SiMe_2\{N(t\text{-Bu})\}_2SiMe_2N(t\text{-Bu})]_2$

(M = Ga or In[48a]) and the recently reported $(TMP)_2GaGa(TMP)_2$[48b] (TMP = 2,2,6,6-tetramethylpiperidine) $(Me_3SiC)_2B_4H_4GaGaB_4H_4(CSiMe_3)_2$[48c] and (t-Bu)NCHCH(t-Bu)NGaGaN(t-Bu)CHCHN(t-Bu)[48d]. The structure of the indium compound [InN(t-Bu)$SiMe_3${N(t-Bu)}$_2SiMe_2$N(t-Bu)]$_2$ shows that the In—In distance is 2.768(1) Å with the two indium atoms being each coordinated by three nitrogens. In the case of $In_2[2,4,6\text{-}(CF_3)_3C_6H_2)]_4$, stabilization is thought to be a consequence of close In · · · F interactions involving fluorines from the ortho CF_3 groups. The In—In bond length (2.744(2) Å] is considerably shorter than that seen in $In_2[CH(SiMe_3)_2]_4$ and is almost identical to the one observed in In_2Trip_4. The gallium species (t-Bu)NCHCH(t-Bu)NGaGaN(t-Bu)CHCHN(t-Bu)[48d] is of significant interest since it has a very short Ga–Ga bond of 2.333(1) Å which is thought to be a consequence of reduced steric crowding. Also worthy of mention is the fact that several very interesting cluster species involving M—M bonds, which have the general formula $(MR)_n^{m-}$ (M = Al, Ga, or In: $n \geq 4$; m = 0, 1, or 2[49]) have been reported. These species are distinguished by coordination numbers of 4 or higher at the metal. Furthermore, they are formally electron-deficient and, in consequence, their formal metal–metal bond orders are less than 1. Their metal–metal distances are thus generally longer than the single bond lengths cited above.

Reduction of R_2MMR_2 (M = Al; R = $CH(SiMe_3)_2$ or Trip; M = Ga; R = Trip) to give the anion $[R_2MMR_2]^-$ was reported essentially simultaneously by three different groups. The isolation of the following compounds were described: $[Li(TMEDA)_2]^+[R_2AlAlR_2]^-$[50] (Fig. 3), $[K(DME)_3]^+[R_2AlAlR_2]^-$[51] (TMEDA = N,N,N',N'-tetramethylethylenediamine; DME = 1,2-dimethoxyethane; R = $CH(SiMe_3)_2$), and $[Li(12\text{-crown-}4)_2]^+[Trip_2GaGaTrip_2]^-$[45] (Fig. 4). The X-ray crystal structures of the lithium salts have been determined, and the synthesis and structure of the aluminum analog of the last species, $[Li(TMEDA)_2]^+[Trip_2AlAlTrip_2]^-$, was reported separately.[37] Structural data for these compounds and their neutral precursors are summarized in Table II. The most striking feature of the data is the degree of shortening in the M—M bonding produced by the one-electron reduction. The torsion angles between the metal coordination planes close to a very small value (in the case of the two aryl derivatives) and the M—C distances are increased upon reduction. The shorter M—M bond may be attributed to the creation of a one-electron π-bond (bond order $\frac{1}{2}$) between the metals. In the case of the aluminum alkyl, the shortening is ~0.13 Å. For both the aluminum and gallium aryl derivatives, however, the shortening is more than 0.17 Å. Thus, significantly shorter M—M distances are observed in both the unreduced and reduced aryl derivatives in comparison to the alkyl analog. The Al—Al (2.470(2) Å) and Ga—Ga

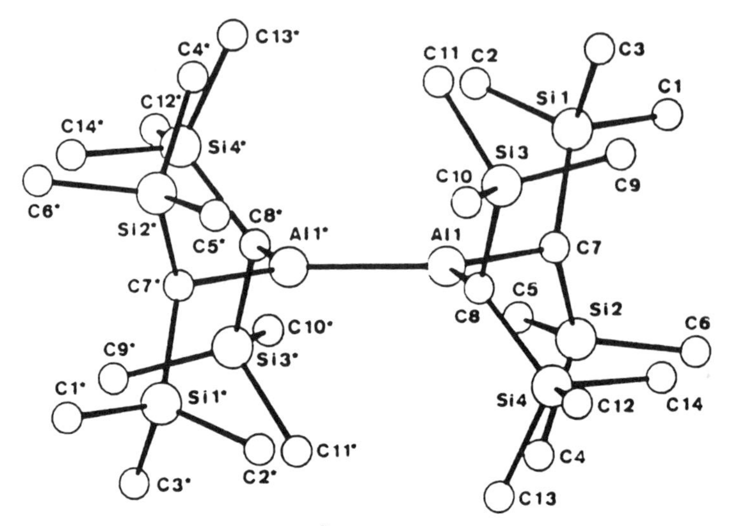

FIG. 3. Thermal ellipsoidal drawing[50] of $[R_2AlAlR_2]^-$ ($R = CH(SiMe_3)_2$). Important structural details are given in Table II.

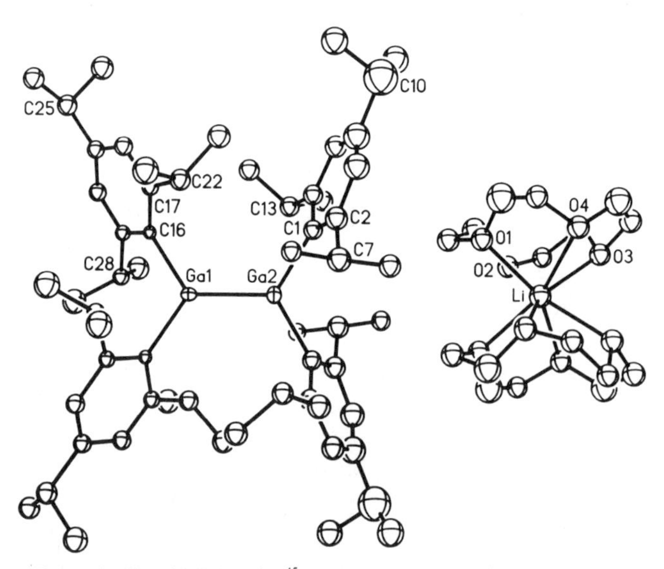

FIG. 4. Thermal ellipsoidal drawing[45] of $[Li(12\text{-crown-}4)_2]^+$ $[Trip_2GaGaTrip_2]^-$. Important bond distances and angles for this compound, and its aluminum analog, are given in Table II.

TABLE II

METAL–METAL BOND LENGTHS AND TORSION ANGLES IN STRUCTURALLY CHARACTERIZED TETRAORGANODIMETALLANES AND THEIR MONOREDUCED ANALOGS

Compound	M—M (Å)	M—C (Å)	Torsion angle[a] (deg.)	Reference
Aluminum				
$[(Me_3Si)_2CH]_2AlAl[CH(SiMe_3)_2]_2$	2.660(1)	1.982(3)	~0	18
$Trip_2AlAlTrip_2$	2.647(3)	1.996(3)	44.8	37
Gallium				
$[(Me_3Si)_2CH]_2GaGa[CH(SiMe_3)_2]_2$	2.541(1)	1.995(5)	~0	19
$Trip_2GaGaTrip_2$	2.515(3)	2.008(7)	43.8	45
$(TMP)_2GaGa(TMP)_2$	2.525(1)	1.901(4)[b]	31	48b
$(Me_3SiC)_2B_4H_4GaGaB_4H_4(CSiMe_3)_2$	2.340(2)	—	—	48c
(*t*-Bu)NCHCH(*t*-Bu)NGaGaN(*t*-Bu)CHCHN(*t*-Bu)	2.333(1)	1.836(4)[b]	90	48d
Indium				
$[(Me_3Si)_2CH]_2InIn[CH(SiMe_3)_2]_2$	2.828(1)	2.194(5)	~0	20
$Trip_2InInTrip_2$	2.775(2)	2.184(7)	47.8	46
$[2,4,6\text{-}(CF_3)_3C_6H_2]_2InIn[C_6H_2\text{-}2,4,6,(CF_3)_2]_2$	2.744(2)	2.21(1)	85.9(5)	47
$(t\text{-}BuN)\{SiMe_2(t\text{-}Bu)N\}_2InIn\{N(t\text{-}Bu)SiMe_2\}_2N(t\text{-}Bu)$	2.768(1)	2.12(1)→2.29(8)[c]	—	48a
Thallium				
$[(Me_3Si)_3Si]_2TlTl[Si(SiMe_3)_3]_2$	2.914(5)	2.675(2)[d]	78.1	42
Reduced species				
$[(Me_3Si)_2CH]_2Al\text{—}Al[CH(SiMe_3)_2]_2^-$	2.53(1)	2.040(5)	0	50
$Trip_2Al\text{—}AlTrip_2^-$	2.470(2)	2.021(1)	1.4	37
$Trip_2Ga\text{—}GaTrip_2^-$	2.343(2)	2.038(10)	15.5	45
$Ga_3\{C_6H_3\text{-}2,6\text{-}Mes_2\}_3^{2-}$	2.441(1)	2.037(3)Å	—	48e

[a] Angle between the perpendiculars to the $M\text{—}C_2$ coordination planes.
[b] Ga—N distances.
[c] In—N distances.
[d] Tl—Si distance.

(2.343(2) Å) bond lengths in the reduced aryl derivatives are, at present, among the shortest Al—Al and Ga—Ga bonds to have been reported.

The creation of a π-bond between the metals and the location of the unpaired electron in a π-orbital are supported by EPR and electronic absorption spectra.[37,45,50,51] For the EPR data the signals have a g-value very close to 2. The hyperfine interactions with the metal nuclei are low: ~10 G for the Al species and 35–40 G for the gallium species. Furthermore, there is equal coupling to the two metal nuclei in each ion. Both the aluminum and gallium radical anions are strongly colored. Their absorption spectra are characterized by several medium-intensity bands in the visible region in addition to an intense band in the near-IR region. For the aryl compounds, $[Trip_2MMTrip_2]^-$ (M = Al or Ga)[37,45] the maxima observed were at 742 nm (Al) and 840 nm (Ga). For $[R_2AlAlR_2]^-$, the longest wavelength band reported was a shoulder at 650 nm.[51] If it is assumed that the maxima observed for the aryl derivatives correspond to a π–π^* transition, the approximate π-bond energy may be estimated. For the Al and Ga compounds, the wavelengths 742 nm and 840 nm correspond respectively to energies of 38.5 and 34 kcal mol^{-1}, thereby suggesting estimates of less than half of these values for the strength of the π-bond. Thus, the values for the Al and Ga species are <19 and <17 kcal mol^{-1}, respectively. These are somewhat less than the energies of π-bonds reported for Si=Si[1] and P=P[2] compounds.

The extent of the shortening in the Al and Ga ions, is somewhat surprising in view of data for the doubly reduced boron species $[Mes_2BB(Ph)Mes]^{2-}$ (Mes = 2,4,6-trimethylphenyl)[9a] whose B—B distance of 1.636(11) Å is only 0.08 Å shorter than that in the unreduced precursor $Mes_2BB(Ph)Me$. In this dianion the B—B π-orbital is, in effect, fully occupied by two electrons; yet the extent of the shortening is less than one-half that observed for the heavier aluminum and gallium congeners in which only one electron occupies the π-orbital. Part of the explanation for the large apparent difference lies in the existence of a large coulombic repulsion between the two negative charges in the case of the boron dianion. A further reason relates to the probable charge distribution within the unreduced and reduced heavier element derivatives is illustrated in Fig. 5. In the neutral precursor there is a charge separation across the M—C bond

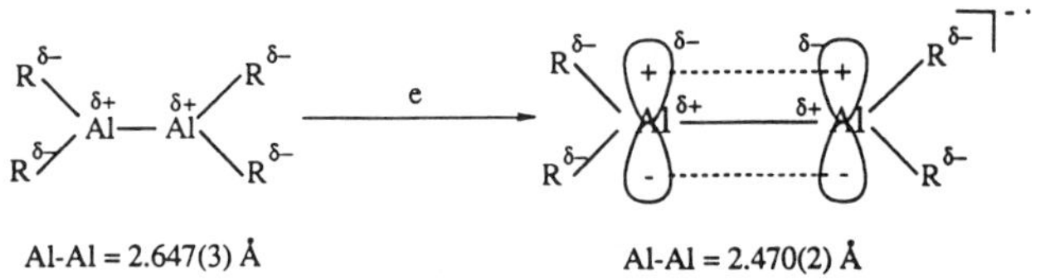

Al-Al = 2.647(3) Å   Al-Al = 2.470(2) Å

FIG. 5. Schematic drawing showing the charge distribution in R_2AlAlR_2 (R = Trip) species.

owing to the more electropositive character of the metal in comparison to the carbon substituents. This δ^+–δ^- attraction results in an ionic contribution to the M—C bond which affords an M—C bond distance shorter than that predicted from the sum of the atomic radii. By the same token, there is a δ^+–δ^+ repulsion across the M—M bond which makes it slightly longer than that predicted by the sum of the radii. Upon the reduction of this system to give a singly reduced anion, however, the negative charge is added primarily to the M—M π-system. This, in addition to creating the half π-bond, tends to offset the δ^+–δ^+ repulsion between the metals. Furthermore, the δ^+–δ^- attractive contribution to the M—C bond is diminished. Thus, the shortening in the M—M bond is somewhat larger than expected owing to the simultaneous reduction in the M—M δ^+–δ^+ repulsion as well as the creation of the π-bond. The lengthening in the M—C bonds seen in the reduced species is also consistent with the above picture. It is notable that the changes in M—M bond lengths are opposite to what is expected on steric grounds since steric interference is presumably greatest in the reduced species which have small torsion angles between the M—M planes. It could be argued that the lengthening observed in the M—C bonds in the reduced species is indeed due to the increase in steric repulsion. The structures of other sterically hindered Al and Ga species do not bear out this argument. In particular, the Ga—C distance in the crowded species $GaTrip_3$[23] is similar to those observed in the reduced $[Trip_2GaGaTrip_2]^-$ ion.

Attempts to doubly reduce the R_2MMR_2 to give $[R_2MMR_2]^{2-}$ species have not been successful to date. One possible explanation is that the second electron which is added occupies an M—M σ^* orbital which weakens the M—M bond sufficiently to cause decomposition. Nonetheless, double reduction can occur in a Ga_3 ring system, as exemplified by the recently reported species $Na_2Ga_3(C_6H_3\text{-}2,6\text{-}Mes_2)_3$.[48e] In this interesting compound the Ga—Ga bond length in the formally aromatic Ga_3^{2-} ring is 2.441(1) Å, which is not notably shortened in comparison to a Ga—Ga single bond.

III

HETERONUCLEAR MULTIPLE BONDING INVOLVING THE HEAVIER MAIN GROUP 3 ELEMENTS

A. *Main Group 3–4 Multiple Bonding*

Compounds that involve multiple bonding between main group 3 and 4 elements have received very little attention to date. It has been suggested for many years, however, that formally single bonds such as B—C[52] and

Al—C[13a] may involve a hyperconjugative interaction as illustrated schematically by

$$\mathrm{Al{-}C(H)_3 \longleftrightarrow \overset{-}{Al}{=}C \;\; H^+}$$

Other types of hyperconjugative interactions are also thought to exist. A recent example involves the compound R_2AlAlR_2 (R = $—CH(SiMe_3)_2$) whose planar structure has been accounted for in terms of a hyperconjugative interaction involving the orbitals of the C—Si bond.[16,18]

A compound for which there is spectroscopic and structural evidence for more conventional *p–p* π-bonding involving main group 3 and 4 elements was not reported until recently.[53a] Treatment of the stable phosphinocarbene $(R_2N)_2PC(SiMe_3)$ with MMe_3 (M = Al, Ga, or In) results in the product depicted by

$$\mathrm{(R_2N)_2P{-}\ddot{C}{-}SiMe_3 \xrightarrow{MMe_3} (R_2N)_2(Me)P{=}C(SiMe_3)(MMe_2)}$$

(M = Al, Ga, or In; R = Cy)

The ^1H and ^{13}C NMR data for the three metal derivatives were similar. An X-ray crystallographic structure for the gallium derivative (Fig. 6) indicated a shortened Ga—C (carbene) distance of 1.935(6) Å, whereas the Ga—C(Me) distances had normal single bond values, 1.994(8) and 1.988(8) Å. Thus, the structure may be visualized as being composed of the two canonical forms

$$\mathrm{(R_2N)_2(Me)P{=}C(SiMe_3)(GaMe_2) \longleftrightarrow (R_2N)_2(Me)P^{+}{-}C(SiMe_3){=}Ga^{-}(Me)_2}$$

one of which has significant Ga—C multiple bond character. There is, however, a rather large torsion angle (36.7°) between the perpendiculars to the coordination planes at gallium and carbon, so the π-interaction in this molecule is probably significantly weaker than a full-fledged π-bond.

More recent results[53b–d] have shown that three-coordinate gallium can be incorporated in potentially delocalized systems such as Mes* $\overline{\mathrm{GaCMe(CMe)_2C}}\mathrm{Me}$ but no significant shortening of the ring Ga—C bonds was observed. Nonetheless the ^{13}C NMR spectrum of the gallatabenzene

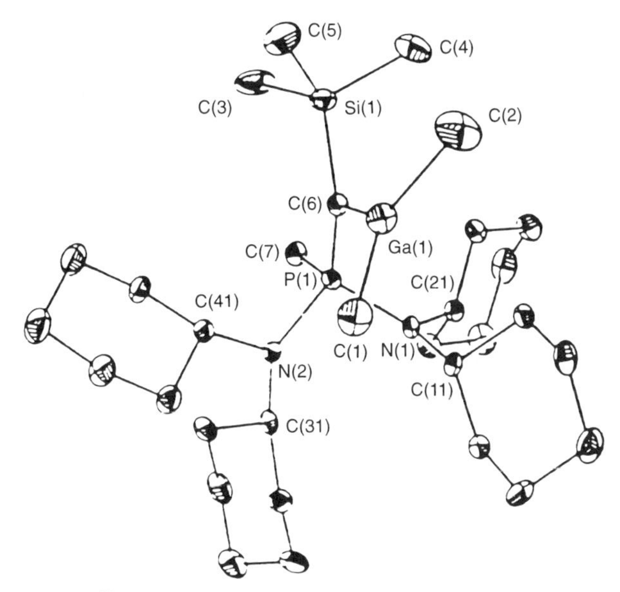

FIG. 6. Drawing[53a] of $(Cy_2N)_2(Me)PC(SiMe_3)GaMe_2$, which is thought to contain a significant component of Ga—C multiple bonding.

ion $[Mes^*GaC_5H_5]^-$ is consistent with significant delocalization of the ring π-electrons.[53d]

B. *Main Group 3–5 and 3–6 Multiple Bonding*

1. *Introduction*

Unassociated compounds of the form $R_2M—ER'_2$ have the potential for π-bonding between the main group 3 element (M) and the main group 5 element (E) through overlap of a lone pair on E with the empty *p*-orbital on M. In this case M must have a coordination number of 3 (or less), so that the required *p*-orbital is in fact empty and available to form a π-bond. Similarly, three coordination is required at the group 5 center. Such π-bonding is well established in molecules containing boron–nitrogen bonds[4] and has recently been demonstrated in the boron–phosphorus[6] and boron–arsenic[7] congeners. For the heavier main group 3 elements, associated compounds $(R_2M—ER'_2)_n$ are well-established and have been detailed in several reviews.[4b,54] However, most of these feature four coordination at both M and E and thus do not have the potential for π-bonding (except possibly through hyperconjugation involving the σ-bonding framework). Monomeric species (n = 1, M = Al–Tl, E = N–As) are fewer in number and the bulk of the examples have been reported since 1990. Compounds

FIG. 7. Alignment of p-orbitals in R_2MER_2': (a) torsion angle 0°, maximum π-overlap; (b) torsion angle 90°, minimum π-overlap.

with main group 3–6 bonds, $R_2M—ER'$, have one fewer substituent than their group 5 counterparts, and structurally characterized, monomeric examples of these that contain three-coordinate heavier group 3 elements are even rarer.[55]

A major structural indicator for the presence of a π-interaction is a significant shortening of the M—E bond relative either to a standard single bond length or to the sum of the covalent radii. Such comparisons are not as simple as they may seem, for reasons outlined in the introduction. In essence, differences in electronegativity between M and E can be quite large and lead to bond shortening through an ionic contribution to the bond strength.[33] Furthermore, the unusually low coordination number at the main group 3 element may decrease its effective covalent radius. The second major structural indication of π-bonding concerns the correct p-orbital alignment. The π-overlap will be maximized where both M and E display trigonal-planar geometry and the normals to the planes at M and at E are parallel or nearly so, as illustrated by Fig. 7a (in Fig. 7b overlap is essentially zero). Formation of a π-bond will result in some energy gain, but this will be favored only if the energy costs in achieving the correct geometry for π-bonding, which can be significant, do not exceed this stabilization. The chemistry of the heavier group 3 elements is dominated by the tendency to form dimers or more highly associated structures, especially when the substituents are heteroatom donors. Three coordination is achieved by the use of bulky substituents at either M or E or at both. If the substituents are sufficiently bulky, steric barriers to bond shortening can arise since the low torsion angle between the trigonal planes required for optimum π-overlap forces the bulky groups into the same plane. Trigonal-planar geometry at the group 5 atom also has an energy cost that is related to the barrier to inversion in the pyramidal ground state ER_3 molecule. This barrier is much higher for the heavier elements—for example, ~35 kcal mol^{-1} and ~44 kcal mol^{-1} for PH_3 and AsH_3, respectively, compared to 5–6 kcal mol^{-1} for ammonia.[56] These factors are reflected in the structures of $R_2B—NR_2'$ species, in which planar geometry at nitrogen is almost always observed, and in $R_2B—PR_2'$ compounds, in which the geometry at phosphorus ranges from planar through increasing degrees of pyramidicity depending on substituent bulk

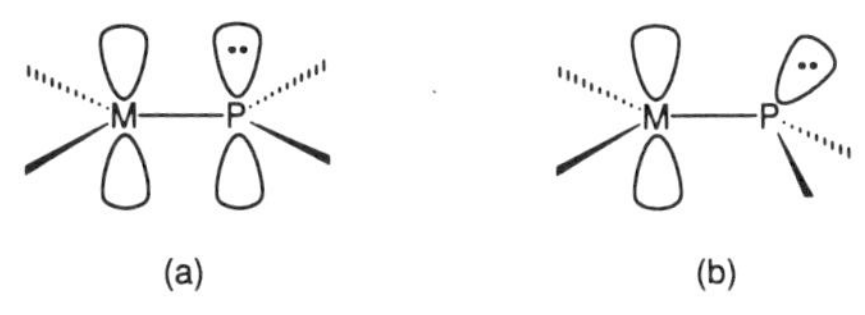

FIG. 8. Planar (a) and pyramidal (b) geometry at phosphorus in $R_2MPR'_2$.

and electronegativity,[4,6] as shown in Fig. 8. Inversion barriers are lowered when electropositive or very bulky substituents are present. It is worth bearing in mind that the R_2M— fragment is itself an electropositive σ-donor with an empty *p*-orbital. The π-bonding interaction with this *p*-orbital may thus act as a mechanism to lower the inversion barrier through delocalization (i.e., conjugation) of the lone pair on E in the transition state.[57]

Shortening of the M—E bond arising from differences in electronegativity between M and E was mentioned above and in the introduction. For reference, Table I gives a selection of M— E single bond distances calculated from the simple sum of covalent radii[10] together with estimated bond distances arising from the application of two different empirical electrostatic corrections, using the formulas developed either by Shomaker and Stevenson[34] or by Blom and Haaland.[35]

If a π-bond is present, it should result in a barrier to rotation around the M—E bond, which may in some cases be observed by variable-temperature (VT) NMR. However, an observed barrier to rotation may be due to steric factors associated with the presence of very bulky substituents rather than to the presence of a π-bond, or may relate to rotation of the substituent groups on M or E rather than around the M—E bond itself.

Another complication arises from the fact that in main group 3–6 species the dynamic process involving the M–E bond may occur by two different mechanisms as illustrated by transition states (i) and (ii):

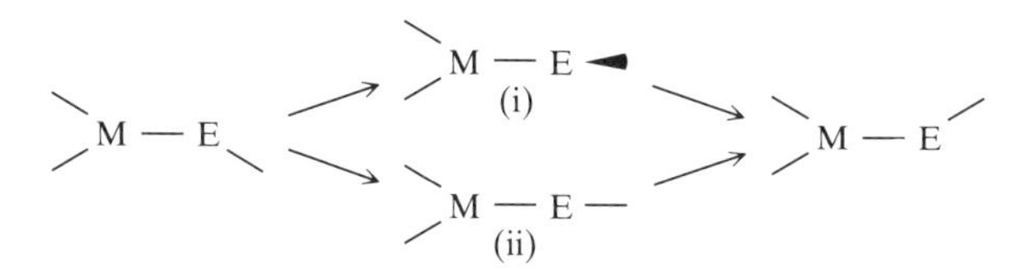

In the transition state (i) the coordination planes at M and E are perpendicular to each other. In (ii) E is linearly coordinated and the "rotation" takes place through a "linear inversion" at E which allows E to remain π-bonded to M throughout the dynamic process. Recent calculations[58] have shown that pathway (ii) is of significantly lower energy than (i) for the M = B and E = O system but is of much higher energy where M = B and E = S. It may be that, for M—OR species (M = Al, Ga, or In), apparent rotation around the M—O bond takes place by a linear inversion since it is much easier to achieve linear coordination for O than for S, Se, or Te. Rotation around the M—E bond in R_2M—ER′ derivatives (M = Al, Ga, or In; E = S, Se, or Te) most probably occurs exclusively through pathway (i), throughout which E retains a bent geometry. It is thus clear that caution should be exercised when interpreting VT NMR results; such results can probably most safely be used to estimate an upper limit for the strength of a putative π-interaction.

Synthetic strategies for the preparations of the three-coordinate compounds discussed in the following sections will not be addressed, and can be obtained from the references. To summarize, the bulk of the compounds have been prepared by salt elimination reactions upon combining R_nMX_{3-n} (X = halide) with $3 - n$ equivalents of $LiER'_2$ (E = N, P, or As) or $3 - n$ equivalents of LiER′ (E = S or Se). Monomeric aluminum and gallium aryloxide compounds are usually prepared by hydrogen or alkane elimination from the reaction of an aryl alcohol with a metal hydride or alkyl, respectively. Similarly, amine or alkane elimination may be employed to produce indium thiolates and selenolates, even though the salt elimination method was suitable for the gallium and aluminum congeners. Special case syntheses of the four- and six-membered ring compounds utilize alkane or phosphine elimination from suitable precursors.

2. *Metal Amide Derivatives*

Structural data for monomeric, three-coordinate aluminum, gallium, indium, and thallium amides, including the M—N bond length and the torsion angle between the normals to the planes at the metal and at nitrogen, are given in Table III.[59–70] Mono-, di-, and triamides have been structurally characterized for both aluminum and gallium, while for indium, only mono- and trisubstituted examples currently exist, and thallium is limited to just two examples—the triamide $Tl\{N(SiMe_3)_2\}_3$[59a,61a] and the Tl(I) compound $[TlN(SiMe_3)Dipp]_4$ (Dipp = 2,6-di(isopropyl)phenyl).[61b] Several species are also known that probably have three coordinate metals, e.g., MeAl $\{N(SiMe_3)_2\}_2$[60c] or $Me_2AlN(H)Mes^*$,[60d] but are not structurally characterized. All of the structurally characterized compounds show essentially planar geometry at both the metal and nitrogen centers, with the exception of the weakly π-associated, indium pyrrolyl complex, $Me_2In(NC_4Me_4)$,[70] in which the sum of the angles at indium is close to 349°. The planar geometry at the nitrogens in these compounds reflects the low inversion barrier at this atom. Many of the metal and nitrogen centers in the com-

TABLE III

METAL–NITROGEN BOND LENGTHS AND TORSION ANGLES BETWEEN THE METAL AND NITROGEN COORDINATION PLANES FOR THREE-COORDINATE ALUMINUM, GALLIUM, INDIUM, AND THALLIUM AMIDES

Compound	M—N (Å)	Torsion angle[a] (deg.)	Reference
Aluminum			
Monoamides			
$Trip_2AlN(H)Dipp$	1.784(3)	5.5	65
t-$Bu_2AlNMes_2$	1.823(4)	49.5	64
t-$Bu_2AlN(Dipp)SiPh_3$	1.834(3)	16.1	64
t-$Bu_2AlN(1\text{-}Ad)SiPh_3$	1.849(4) (av)	86.3 (av)	64
t-$Bu_2AlN(SiPh_3)_2$	1.879(4) (av)	64.3 (av)	64
Diamides			
$\{MeAlNDipp\}_3$	1.782(4)	0	66a,73a
$Cp^*\{(Me_3Si)_2N\}AlN(\mu\text{-}AlCp^*)$-$\{\mu\text{-}AlN(SiMe_3)_2\}NAlCp^*_2$	1.807(b)[b]	0	66b
	1.793(t)[b]	—	
$(Mes^*AlNPh)_2$	1.824(3)	0	66c
$MesAl\{N(SiMe_3)_2\}_2$	1.804(2)	49.7	60b
	1.809(2)	44.5	
$Mes^*Al\{N(H)Ph\}_2$	1.794(4)	16.9 (av)	66c
Triamides			
$Al\{N(SiMe_3)_2\}_3$	1.78(2)	50	59b
$Al\{N(i\text{-}Pr)_2\}_3$	1.791(4) (av)	36.6 (av)	60b
	1.794(4) (av)	38.3 (av)	
	1.801(4) (av)	75.5 (av)	
Gallium			
Monoamides			
$Mes^*Ga(Cl)N(H)Ph$	1.832(10) (av)	2.4 (av)	60b
$Trip_2GaN(H)Dipp$	1.847(12) (av)	9.0	65
$Mes^*_2GaN(H)Ph$	1.874(4)	6.7	60b
$Trip_2GaNPh_2$	1.878(7)	0	65
t-$Bu_2GaN(t\text{-}Bu)SiPh_3$	1.906(5)	88.7	65
t-$Bu_2GaN(1\text{-}Ad)SiPh_3$	1.924(2)	71.8	65
$Et_2GaN(t\text{-}Bu)BMes_2$	1.937(3)	69.7	68
Diamides			
$PhOGa(TMP)_2$	1.818(3)	—	67a
	1.849(3)	—	67a
$Mes^*Ga\{N(H)Ph\}_2$	1.837(8)	7.3	60b
$ClGa\{N(SiMe_3)_2\}_2$	1.834(4)	40.5	60b
	1.844(4)	49.5	

(continues)

TABLE III—(*Continued*)

Compound	M—N (Å)	Torsion angle[a] (deg.)	Reference
$PhGa(TMP)_2$	1.883(2)	—	67a
$(Me_3Si)_3SiGa(TMP)_2$	1.908(3)	79.2	67b
	1.913(2)	64.3	
$(TMP)_2GaGa(TMP)_2$	1.901(4)	53	48b
Triamides			
$Ga\{N(SiMe_3)_2\}_3$	1.868(1)	48.6	59a,60b,62,60e
	1.870(6)	50	
Indium			
Monoamides			
$t\text{-}Bu_2InN(Dipp)SiPh_3$	2.104(3)	15.5	63
$Et_2In(NC_4H_4)$	2.166(4)	16.1	69
$Me_2In(NC_4Me_4)$	2.197(3)		70
Triamides			
$In\{N(SiMe_3)_2\}_3$	2.049(1)	48.6	59a,62,63
Thallium			
$Tl\{N(SiMe_3)_2\}_3$	2.089(18)	49.1	59a,61a
$[TlN(SiMe_3)Dipp]_4$	2.306(6)	—	61b

[a] Angle between the perpendiculars to the MC_2 and NC_2 coordination planes.
[b] The designations (b) and (t) refer to bridging and terminal amido atoms.

pounds in Table III show angular distortions from ideal trigonal-planar geometry which can be attributed largely to steric effects arising from the bulky substituents. For example, the angles at aluminum and nitrogen in $t\text{-}Bu_2AlNMes_2$[64] deviate from 120° by less than 3° (Fig. 9), while in $Mes^*_2GaN(H)Ph$[60b] the C—Ga—C angle between the two bulky Mes* groups is 135.3(2)°, more than 20° wider than either of the N—Ga—C angles. The compounds that have aryl substituents at the metal show high torsion angles (56° to 89°) between the plane at the metal and the aryl rings, with the very bulky Mes* groups having the highest torsion angles. The exception is $Mes^*_2GaN(H)Ph$[60b] in which one Mes* ring is tilted away from Ga—C vector and has a torsion angle of only 42.7°. The Mes* compounds show short intramolecular contacts between the gallium and the Mes* *ortho-t*-Bu substituents which project above and below the plane at gallium (Fig. 10). All of the compounds are discrete monomeric entities with no close intermolecular contacts, with the exception of the two indium pyrrolyl complexes. The species $Me_2In(NC_4Me_4)$[70] exists as dimeric units with intermolecular contacts (In—C 2.575 Å) between the indium atoms and the pyrrole carbon atoms of the other molecule within each pair.

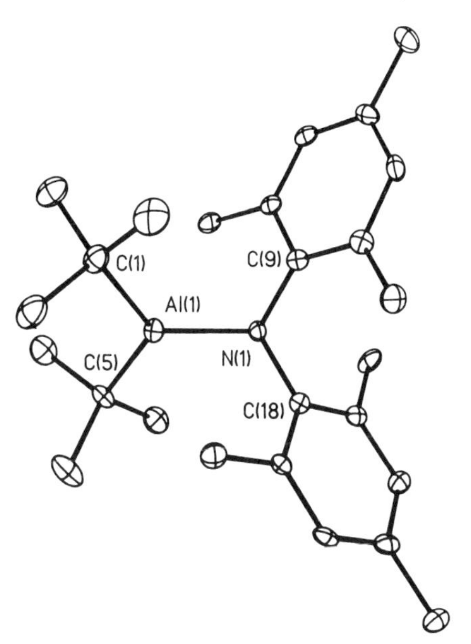

FIG. 9. Thermal ellipsoidal drawing of t-$Bu_2AlNMes_2$. Further details of the structure are given in Table III. [Reprinted with permission from Petrie *et al.*[64] Copyright 1993 American Chemical Society.]

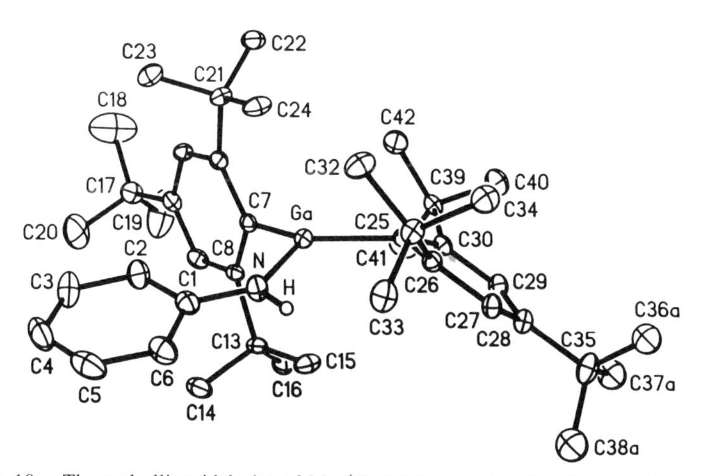

FIG. 10. Thermal ellipsoidal plot of $Mes^{*}_{2}GaNHPh$ showing the alignment between the coordination planes of N and Ga and the distortion in one of the Mes* rings. Further details are given in Table III. [Reprinted with permission from Brothers *et al.*[60b] Copyright 1994 American Chemical Society.]

$Et_2In(NC_4H_4)$[69] has similar contacts (In—C 2.950 Å) and the molecules are stacked as chains in the crystal.

All of the Al—N and Ga—N bond distances are shorter than the simple sum of the covalent radii for the three-coordinate aluminum or gallium and nitrogen centers. More appropriate estimates for the M—N single bond length, which take into account the differences in electronegativity between the metal and nitrogen, can be calculated using the empirical electrostatic corrections discussed above. This gives values in the range 1.79–1.85 Å for Al—N, 1.80–1.84 Å for Ga—N, and 1.96–2.03 Å for In—N (see Table I). As shown in Table III, most of the Al—N and Ga—N bond lengths fall within these predicted single bond ranges. The exceptions occur in the compounds that have the very bulky $SiPh_3$ substituent at nitrogen—namely, t-$Bu_2AlN(SiPh_3)_2$,[64] t-$Bu_2GaN(t$-$Bu)SiPh_3$,[65] and t-$Bu_2GaN(1$-$Ad)SiPh_3$ (1-Ad = 1-adamantanyl)[65]—or substituents (other than gallium) with which the nitrogen lone pair interacts, as in $Et_2GaN(t$-$Bu)BMes_2$.[68] In these examples, any M—N π-interaction that might lead to bond shortening must be weak compared to the unfavorable interactions arising from steric crowding. Large torsion angles are also seen for these complexes, further illustrating the importance of steric effects in many of these very crowded molecules. The fact that $Et_2GaN(t$-$Bu)BMes_2$[68] possesses the longest Ga—N bond length for a monoamide does, however, lend some credence to the existence of some π-interaction in the Ga—N bond, which in this case has been weakened by a competitive B—N π-interaction. By the same token, however, the Ga—N ionic interaction may also be reduced since some of the electron density on nitrogen is delocalized onto boron. The shortest Ga—N bond length (1.818(3) Å[67a] is observed for one of the Ga—N bonds in $PhOGa(TMP)_2$. Presumably, the greater ionic character at Ga (owing to the presence of the phenoxide ligand) contributes to the shortening in this case. For both aluminum and gallium it is the complexes with primary amide substituents NHPh and NHDipp, the least bulky secondary amide NPh_2, and the ring compound $\{MeAlNDipp\}_3$[66] (Fig. 11), which has only one substituent at nitrogen and aluminum, that show both the shortest M—N distances and the smallest torsion angles. In sum, these structural data emphasize the importance of steric effects in determining the M—N bond length.

Trends for the M—N bond lengths are less obvious for the indium and thallium amides, for which there are many fewer examples. Two of the indium amides, t-$Bu_2InN(Dipp)SiPh_3$[63] and $In\{N(SiMe_3)_2\}_3$,[63] have bond lengths that are close to or shorter than the sum of the covalent radii (2.10 Å), but not as short as the single bond length estimated using the electrostatic correction. The two indium pyrrolyl complexes[69,70] have bond lengths somewhat longer than 2.10 Å, but these molecules are both weakly associated with intermolecular In—C(aromatic) contacts. They represent

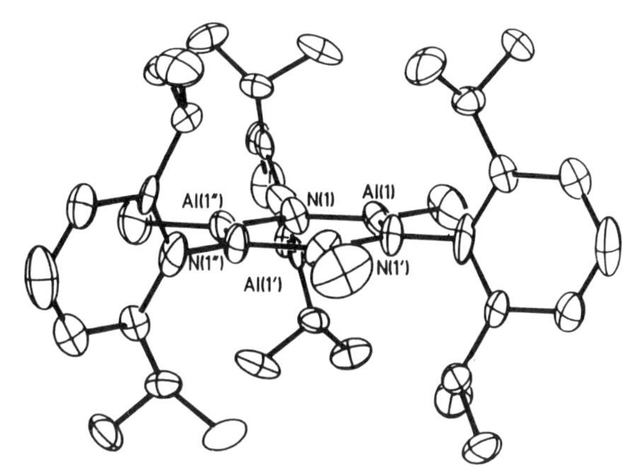

FIG. 11. Thermal ellipsoidal plot[66a] of $(MeAlNDipp)_3$ demonstrating the planarity of the Al_3N_3 core. Further details are given in Table III.

an intermediate stage between the truly monomeric, unassociated amides and those in which the metal centers are coordinated to a heteroatom lone pair or the electron density in a π-system to form dimers or solvated species. In addition to these compounds there is the recently reported weakly associated, Tl(I) tetramer $[TlN(SiMe_3)Dipp]_4$[63] (Tl · · · Tl = 4.06 Å) which has long Tl—N distances averaging 2.306(6) Å. The long Tl—N distance, which does not support significant π-bonding, is presumably a consequence of the larger size of the Tl(I) ion.

If π-bonding is significant in these metal amides, short M—N bonds should occur in the molecules that have low torsion angles. Inspection of the M—N bond lengths and the torsion angles around the M—N bonds for the aluminum and gallium compounds in Table III shows that there is no strong correlation. To a first approximation the shortest and longest M—N bonds are indeed associated with the lowest and highest torsion angles, respectively. However, there are several exceptions; for instance, $Al\{N(SiMe_3)_2\}_3$[59b] has the shortest Al—N bond (1.78 Å) but a torsion angle of 50°. Likewise, the indium analog $In\{N(SiMe_3)_2\}_3$[62,63] has the shortest In—N bond (2.049 Å) and a very similar torsion angle of 48.6°. For the compounds with intermediate bond lengths the torsion angles appear to vary almost randomly. For example, the gallium diamides $Mes^*Ga\{N(H)Ph\}_2$[60b] (Fig. 12) and $ClGa\{N(SiMe_3)_2\}_2$[60b] show very similar Ga—N bond lengths, close to 1.84 Å, while the torsion angles vary from 7° to more than 40°.

In the π-bonding model, diamides and triamides are expected to show longer M—N distances than their monoamido counterparts, as there are two or three nitrogen lone pairs competing for just one empty *p*-orbital

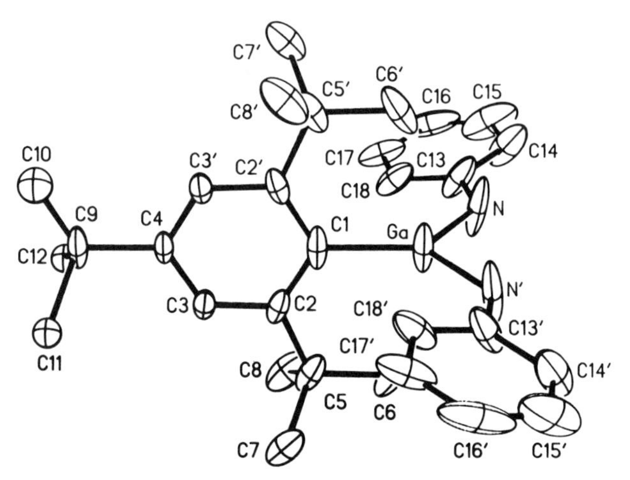

FIG. 12. Thermal ellipsoidal plot of Mes*Ga{N(H)Ph}$_2$. Further details are given in Table III. [Reprinted with permission from Brothers *et al.*[60b] Copyright 1994 American Chemical Society.]

on the metal. This effect is observed in boron–nitrogen compounds, where there is good evidence for significant p–p π-bonding.[4] The diamides usually exhibit longer B—N bonds than do the monoamides, and the triamides show one long and two short B—N bonds, indicating that the π-bonding interaction is confined to the BN_2 framework.[71] However, inspection of the data in Table III indicates that any π-bonding in the aluminum and gallium amides must be of much lesser significance. The M—N bond lengths for the aluminum di- and triamides and the indium triamide lie toward the shorter end of the range of observed bond lengths, while the Ga—N bond lengths in the gallium examples are in the lower to middle part of the range. These results are consistent with the hypothesis that ionic effects rather than π-bonding are the significant influence on structural features. In particular, the bond length data for the aluminum and indium compounds are consistent with bond shortening resulting from a progressive contraction of the metal orbitals upon substitution by more electronegative substituents.[72] Gallium is more electronegative than aluminum; consequently, electronegativity differences between the gallium atom and its substituents are less marked, thereby moderating the ionic resonance effect. Even so, Mes*Ga(Cl)N(H)Ph,[60b] ClGa{N(SiMe$_3$)$_2$}$_2$,[60b] and PhOGa(TMP)$_2$,[67a] with a relatively small, electronegative chloride or phenoxide ligand on the metal in addition to one or two amides, exhibit three of the four shortest Ga—N bond distances.

The ring compound {MeAlNDipp}$_3$[66a,73a] (Fig. 11) is an analog of borazine, $(HBNH)_3$ and its derivatives, which are in turn isoelectronic and isostructural with benzene. While this family of boron–nitrogen rings is large, this compound is the only known example of an aluminum–nitrogen six-membered ring with three coordination at both aluminum and nitrogen. The extent of the π-delocalization in borazine (11.1 kcal mol^{-1}) has been calculated to be about half that of benzene (22.9 kcal mol^{-1}.[73b] Calculations for a homodesmotic reaction sequence for $(HAlNH)_3$ give a value of 1.9 kcal mol^{-1} for the stabilization energy imparted to the ring by the delocalization process. Since this involves six Al—N bonds, these results can be interpreted in terms of a negligibly small Al—N π interaction in $\{MeAlNDipp\}_3$, despite the short Al—N bond length and the optimum 0° torsion angle. The recently reported species $Cp^*\{(Me_3Si)_2N\}AlN(\mu\text{-}Al\text{-}Cp^*)\mu\text{-}AlN(SiMe_3)_2\}NAlCp^*_2$ (Cp^* = η^5-pentamethylcyclopentadienide)[66b] and $(Mes^*AlNPh)_2$[66c] contain a central Al_2N_2 unit. In the former compound the exo-substituents at the ring nitrogens are $—AlCp^*_2$ and $—AlCp^*\{N(SiMe_3)_2\}$ whereas the two ring aluminums are substituted by Cp^* and $—N(SiMe_3)_2$. Some variation in Al—N distances is observed, with the Al—N ring bonds averaging 1.807 Å and the exo-Al—N distances averaging 1.793 Å. In $(Mes^*AlNPh)_2$ the Al-N distance is 1.824(3) Å.[66c]

The energy of the potential π-bond in the other compounds can be probed by VT NMR studies. A barrier to rotation about an M—N bond of ~8 kcal mol^{-1} or greater may be observed by dynamic behavior in the NMR spectra of a compound over the experimentally accessible range −100° to +100°C. Dynamic processes may be attributed either to aryl ring rotation about the M—C or N—C bond or to restricted rotation about the M—N bond. This latter barrier may arise either from the presence of an M—N π-interaction or from steric repulsion by the bulky substituents. A number of the compounds from Table III have been investigated by VT NMR. $t\text{-}Bu_2AlN(1\text{-}Ad)SiPh_3$,[64] $t\text{-}Bu_2GaN(1\text{-}Ad)SiPh_3$,[65] and $t\text{-}Bu_2GaN(t\text{-}Bu)SiPh_3$[65] show no dynamic behavior over the temperature range investigated. The high torsion angles in these molecules are indicative of steric strain, which may be preventing rotation about the M–N bond or π-bond formation. Dynamic processes are observed for $Trip_2GaNPh_2$ (9.5 kcal mol^{-1}),[65] $MesAl\{N(SiMe_3)_2\}_2$ (11.5 kcal mol^{-1}),[60b] $t\text{-}Bu_2InN(Dipp)SiPh_3$ (15.5 kcal mol^{-1}),[65] $t\text{-}Bu_2AlNMes_2$ (15.6 kcal mol^{-1}),[64] and $t\text{-}Bu_2AlN(Dipp)SiPh_3$ (16.6 kcal mol^{-1}).[68] In all of these molecules the observed process is likely to be rotation of an aryl substituent around the M—C or N—C bond or, in the case of $MesAl\{N(SiMe_3)_2\}_2$,[60b] flipping of the coordination planes at nitrogen rather than rotation around the M—N bond. For $t\text{-}Bu_2AlN(Dipp)SiPh_3$[64] a second process is observed; this process has a barrier of 9.9 kcal mol^{-1} and can be attributed to rotation around the Al—N bond. Similar barriers are observed for $Trip_2AlN(H)Dipp$ (9.4 kcal mol^{-1}),[65] $Trip_2GaN(H)Dipp$ (9.7 kcal mol^{-1}),[65] and $Mes^*_2GaN(H)Ph$ (10.1 kcal mol^{-1})[60b]; these are all believed to arise from rotation about the M—N bond, giving an upper limit of about 9–10 kcal mol^{-1} for the strength of the M—N π-bond. These values may be compared to barriers to rotation around B—N π-bonds in the range 5–25 kcal mol^{-1}.[4] Aryl ring rotation, versus rotation around the M—N bond, can often be distinguished by observing which resonances are affected during VT NMR studies. For the amide $Ar_2MN(H)R$, aryl ring rotation should lead to broadening only of the ortho substituents of the Mes, Trip, or Mes* groups, while rotation around the M—N bond should also lead to broadening of the para substituents. The values observed here for aryl ring rotation are consistent with the range of 10–16 kcal mol^{-1} reported for alkoxydiarylboranes and triarylboranes.[74] Restricted rotation about the M—N bond may not be observable by NMR for t-$Bu_2AlNMes_2$[64] and $Trip_2GaNPh_2$[65] due to the higher symmetry in these molecules, which have two identical substituents at the metal and the nitrogen.

Table III also shows three groups of comparable sets of compounds for which the ligands are identical and only the metal changes. The aluminum and gallium pairs $Trip_2MN(H)Dipp$[65] and t-$Bu_2MN(1\text{-}Ad)SiPh_3$[64,65] show a close similarity within each pair, with short M—N bonds and low torsion angles (<10°) for the first pair and longer M—N bonds and high torsion angles (>70°) for the second pair. The difference between the Al—N and Ga—N bond lengths is constant at about 0.06 Å in both pairs. Similarly, the aluminum[64] and indium[63] pair t-$Bu_2MN(Dipp)SiPh_3$ show almost identical torsion angles. These comparisons suggest that steric rather than π-bonding effects are determining the geometry and affecting bond lengths. A series of structures of the triamides $M\{N(SiMe_3)_2\}_3$ is now complete for the four elements aluminum, gallium, indium, and thallium. Apart from the expected lengthening of the M—N bond with increasing atomic number, remarkably similar structural features occur in all four compounds, with N—Si distances in the narrow range 1.75–1.738 Å, Si—N—Si angles that vary from 118° to 122.5°, and torsion angles in the range 48.6°–50°. The high torsion angles indicate that π-bonding is of limited significance in these triamides and that the relatively short M—N bonds must be due to enhanced ionic effects arising from contraction of the metal orbitals.

Overall, the most likely candidates for which π-bonding might be significant are the compounds with short M—N bond lengths, low torsion angles, and less bulky substituents. Compounds from Table III that fall into this category are the monoamides Mes*Ga(Cl)N(H)Ph, $Mes^*_2GaN(H)Ph$, $Trip_2GaN(H)Dipp$, and $Trip_2AlN(H)Dipp$ and the diamide $Mes^*Ga\{N(H)Ph\}_2$. Barriers to rotation around the M—N bond of 9–10 kcal mol^{-1} have been observed for the last three of these, indicating that this is a maximum value for the small π-component of the bonding in these compounds. The lack of a strong correlation between M—N bond length, torsion angle, and the number of amide ligands per metal center suggests that the conformation of these compounds is determined primarily by steric effects. The major contribution to bond shortening is an ionic resonance effect, with π-bonding playing a small but significant role in compounds where steric effects allow the requisite geometry for π-overlap to be achieved.

3. *Metal Phosphide and Arsenide Derivatives*

Three-coordinate aluminum, gallium, and indium phosphides are shown in Table IV,[75–86] and gallium and indium arsenide complexes are given in Table V.[77–79,87–90] For completeness, Tables IV and V include several examples that have been spectroscopically, but not structurally characterized. In addition, data for the interesting Zintl anions $M_2E_4^{6-}$ (M = Al or Ga; E = P or As)[77] are included. A notable feature of Tables IV and V is that there are very few aluminum and indium complexes, and structurally characterized examples are limited to just two aluminum and one indium organophosphide. The bulk of the compounds are gallium phosphide or arsenide complexes. These include simple mono-, di-, and triphosphides and -arsenides; a digallylphosphane; four- and six-membered ring compounds $(RGaPR')_n$, where $n = 2$ or 3; and a cage compound with a Ga_4P_5 core that contains both three- and four-coordinate gallium and phosphorus. The digallylphosphine MesP{Ga(Trip)$_2$}$_2$[78] and the diphosphide *t*-Bu Ga(PHMes*)$_2$ can be viewed as Ga—P analogs of the allyl cation and anion, respectively. In *t*-Bu$_2$GaP(Ph)BMes$_2$[81] and *t*-BuGa{P(Ph) BMes$_2$}$_2$,[81] each phosphorus atom is bonded to both gallium and boron, thus facilitating an intramolecular comparison of gallium–phosphorus and boron–phosphorus π-bonding.

All of the complexes have planar or close to planar geometry at the metal atom. However, in contrast to the metal amides discussed above, in which nitrogen also has a planar configuration, the phosphides and arsenides show pyramidal or flattened pyramidal geometry at the group 5 atom. The sum of the angles at phosphorus range from normal pyramidal ($\Sigma°P = 314°$) to one example of essentially planar geometry, MesP{Ga (Trip)$_2$}$_2$ ($\Sigma°P = 359°$) (Fig. 13).[78] For comparison, $\Sigma°P$ in the free phosphines PPh_3 and $PMes_3$ are 310° and 329°, respectively.[91,92] All the arsenides are close to fully pyramidal, and the range of $\Sigma°As$ is much narrower, from 310° to 330°. This difference between the amides and the phosphides and arsenides reflects the much higher inversion barrier for the heavier elements (>30 kcal mol^{-1}). This has important implications for π-bonding in these compounds. First, overlap of the lone pair on the group 5 atom and the empty *p*-orbital on the metal is much less favorable if the group 5 element has pyramidal geometry. Second, pyramidal coordination increases the steric interactions between the substituents at phosphorus or arsenic and those at the metal, particularly in the configuration where the perpendiculars to the planes at phosphorus or arsenic and the metal are aligned, as required for maximum *p*–*p* orbital overlap. Finally, the higher

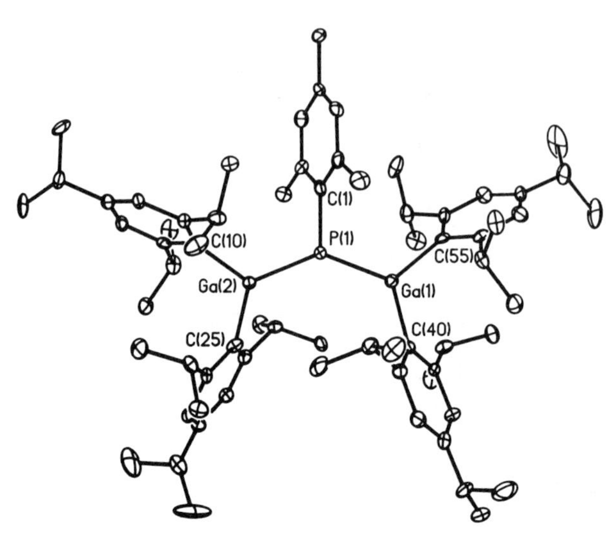

FIG. 13. Thermal ellipsoidal drawing of MesP{Ga(Trip)$_2$}$_2$. Further structural details are given in Table IV. [Reprinted with permission from Petrie and Power.[78] Copyright 1993 American Chemical Society.]

inversion barrier increases the energy cost to achieve the planar or flattened pyramidal geometry at phosphorus or arsenic necessary to form a strong π-bond. The question of pyramidicity at the phosphorus or arsenic atom also has a significant σ-effect arising from the hybridization. An atom in the planar, sp^2-hybridized geometry will have more s-character in its bonds than in the pyramidal, sp^3-hybridized configuration, and this may also lead to some bond shortening.

A number of other factors can affect the inversion barrier at phosphorus or arsenic. Bulky substituents, for example increase steric interactions in the pyramidal arrangement and lead to flattening of the coordination geometry. In addition, very electropositive substituents, which affect the hybridization at the central element and favor configurations with more s-character (as in the sp^2-planar geometry), may also lead to a more planar geometry. Significant π-bonding has been observed in complexes with boron–phosphorus bonds, particularly when the phosphorus atom bears bulky or electropositive substituents.[6] The group 3 atom itself is electropositive and should lower the inverison barrier relative to the free phosphines or arsines. Many of the compounds in Tables IV and V also bear additional electropositive substituents such as $SiMe_3$, $SiPh_3$, or $BMes_2$ on the phosphorus or arsenic atom.

Until very recently, the only example of a quasi-3-coordinate aluminum phosphide was $(Me_3SiCH_2)_2AlPPh_2$,[76] which was reported to exist in solution as a monomer–dimer equilibrium. However, a crystal structure of

TABLE IV

METAL–PHOSPHORUS BOND LENGTHS, TORSION ANGLES BETWEEN THE METAL AND PHOSPHORUS COORDINATION PLANES, SUMS OF ANGLES AT PHOSPHORUS, AND ^{31}P NMR CHEMICAL SHIFTS FOR THREE-COORDINATE ALUMINUM, GALLIUM, AND INDIUM PHOSPHIDES

Compound	M—P (Å)	Torsion angle (deg.)	Σ°P (deg.)	δ (^{31}P NMR)[a]	Reference
Aluminum					
$Trip_2AlP(1\text{-}Ad)SiPh_3$	2.342(2) (av)	47.5 (av)	330.0	−157	75a
$Trip_2AlP(Mes^*)SiPh_3$	—	—	—	−94	75a
$(Me_3SiCH_2)_2AlPPh_2$	—	—	—	—	76
$(Mes^*AlPPh)_3$	2.328(3)	—	329.1	−144.2	75b
$Al_2P_4^{6-}$	2.251(2) (t)[b]	0	360.0	—	77
	2.339(1) (b)[b]				
Gallium					
Monophosphides					
$MesP\{Ga(Trip)_2\}_2$	2.256(3)	14.7	358.8	−58	78
	2.258(3)	15.6			
$t\text{-}Bu_2GaP(Mes^*)SiPh_3$	2.295(3)	3.2	346.4	−119	79
$t\text{-}Bu_2GaP(Trip)SiPh_3$	2.296(1)	1.9	340.5	−146	80
$t\text{-}Bu_2GaP(Ph)BMes_2$	2.319(1)	56.4	344.2		81
$t\text{-}Bu_2GaP(SiMe_3)SiPh_3$	2.358(4)	76.5	326.2	−267	80
$t\text{-}Bu_2GaP(Mes)SiPh_3$	—	—	—	−141	80
Diphosphides					
$t\text{-}BuGa\{P(Ph)BMes_2\}_2$	2.194(2)	—	351.8	—	81
	2.390(2)		347.7		
$(t\text{-}BuGaPMes^*)_2$	2.274(4)	—	336.0	−70	82
$(TriphGaPCy)_3$	2.30 (av)	—	325 (av)	−61	83
$Ga_4(Trip)_3\{P(1\text{-}Ad)\}_4\text{-}$	2.290(4)		317.0		84
$P(H)(1\text{-}Ad)$	2.320(4)		334.0		
$t\text{-}BuGa(PHMes^*)_2$	2.326(4)			−113	82
	2.323(5)				
Triphosphides					
$Ga(PHMes^*)_3$	2.34(1) (av)			−92	85
$Ga(P\text{-}t\text{-}Bu_2)_3$	—	—	—	55	85
$Ga_2P_4^{6-}$	2.247(5)	0	360.0	—	77
	2.358(5)				
Indium					
$In(P\text{-}t\text{-}Bu_2)_3$	2.574(11)	84	315.6	72	86
	2.588(14)	89	319.9		
	2.613(12)	88	326.7		

[a] ^{31}P chemical shift (δ) relative to external 85% H_3PO_4.

[b] The designations (b) and (t) refer to bridging and terminal phosphido atoms.

TABLE V

METAL–ARSENIC BOND LENGTHS, TORSION ANGLES BETWEEN THE METAL AND ARSENIC COORDINATION PLANES, AND SUMS OF THE ANGLES AT ARSENIC FOR THREE-COORDINATE GALLIUM AND INDIUM ARSENIDES

Compound	M—As (Å)	Torsion angle (deg.)	Σ°As (deg.)	Reference
Aluminum				
$(Mes^*AlAsPh)_3$	2.430(5)	—	319.7	75b
$Al_2As_4^{6-}$	2.342(5) (t)	0	360	77
	2.438(5) (b)			
Gallium				
Monoarsenides				
$PhAs\{Ga(Trip)_2\}_2$	2.401(1)	41.7	329.8	78
	2.418(1)	81.3		
$(\eta^1\text{-}Cp^*)_2GaAs(SiMe_3)_2$	2.433(4)	~90	~320	87
$t\text{-}Bu_2GaAs\{CH(SiMe_3)_2\}SiPh_3$	2.459 (av)	70.6 (av)	316 (av)	79
$t\text{-}Bu_2GaAs\text{-}t\text{-}Bu_2$	2.466(3)	90	317.0	88
Triarsenides				
$Ga\{As(SiMe_3)_2\}_3$	2.421(4)	—	101.85	89b
$Ga(AsMes_2)_3$	2.470(1)	52	320.2	
	2.498(1)	58	310.5	
	2.508(1)	86	313.6	89a
$Ga(As\text{-}t\text{-}Bu_2)_3$	—	—	—	90
$Ga_2As_4^{6-}$	2.343(2) (t)[a]	0	360	77
	2.459(3) (b)[a]			
Indium				
$In(As\text{-}t\text{-}Bu_2)_3$	—	—	—	90

[a] The designation (b) and (t) refer to bridging and terminal arsenido ligands.

$Trip_2AlP(1\text{-}Ad)SiPh_3$[75a] (Fig. 14) and spectroscopic characterization of $Trip_2AlP(Mes)SiPh_3$[75a] have now firmly established this class of compounds. The Al—P bond distance (2.342(2) Å) in $Trip_2AlP(1\text{-}Ad)SiPh_3$ is slightly shorter than the sum of the covalent radii (2.4 Å) and very close to the value corrected for ionic effects of 2.34 Å (see Table I). The markedly pyramidal phosphorus center ($\Sigma°P = 330°$) and the high torsion angle (47.5°) are indicative of little π-overlap in this molecule. Similar structural parameters have been observed for $(Mes^*AlPPh)_3$ which has an average Al-P distance of 2.328(3) Å and $\Sigma°P$ of 329.1°.[75b] The arsenic analog is also known[75b] (Al—As = 2.430(5) Å, $\Sigma°As = 319.7°$) and this compound is the only molecular species with bonding between three coordinate alumi-

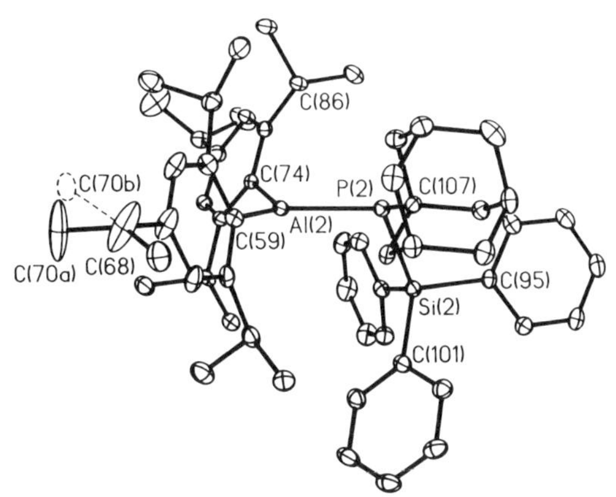

FIG. 14. Thermal ellipsoidal plot of $Trip_2AlP(1\text{-}Ad)SiPh_3$, which is currently the only structurally characterized unassociated aluminum diorganophosphide. [Reprinted with permission from Wehmschulte, Ruhlandt-Senge, and Power.[75a] Copyright 1994 American Chemical Society.]

num and arsenic centers. Apparently, even one or more electropositive aluminum or silicon substituents on the phosphorus atom are unable to lower the high inversion barrier sufficiently for efficient π-overlap to occur. Furthermore, the very bulky substituents at phosphorus together with its pyramidal geometry preclude alignment of the phosphorus lone pair and the *p*-orbital on aluminum. Variable-temperature ^1H and ^{13}C NMR studies of both $Trip_2AlP(Mes)SiPh_3$ and $Trip_2AlP(1\text{-}Ad)SiPh_3$ show dynamic behavior consistent with rotation around the Al—P bond with barriers of 8–9 and 14 kcal mol^{-1}.[75a] Most likely, however, this is not a measure of the π-bond strength because similar barriers are observed in gallium complexes where there are much better structural indications for a π-interaction. The barrier measured in the aluminum phosphides is probably steric in origin. Oddly, a much less pyramidal phosphorus center ($\Sigma°P$ = 351.6°) is found in the "base-stabilized" species $t\text{-}Bu_2Al(OEt_2)P$ (Trip)($SiPh_3$), which has very similar substituents at phosphorus.[80]

Despite being heavier than aluminum, gallium is closer to phosphorus both in size and electronegativity (EN) (see Table I). In fact, phosphorus is only a little more electronegative than gallium (EN values of P and Ga are 2.1 and 1.8);[10] thus the sum of the covalent radii (2.35 Å) is only slightly higher than the value corrected for electrostatic effects (2.33 Å), and ionic contributions to either the Ga—P or Ga—As bonds are expected to be quite small. Structural indications expected for π-bonding in gallium phosphides are Ga—P bond shortening, a low torsion angle, and planarity—or at least flattened pyramidal geometry—at phosphorus. Several of the gallium phosphide structures show both structural and VT NMR evidence for a weak gallium–phosphorus π-interaction. In particular, the gallium monophosphides show a good correlation between the Ga—P bond length and the torsion angle between the planes at phosphorus and at gallium. All of the monophosphides, with the exception of t-$Bu_2GaP(SiMe_3)SiPh_3$,[80] have Ga—P bonds that are shorter than the corrected sum of the covalent radii (2.33 Å, Table I). The digallylphosphane $MesP\{Ga(Trip)_2\}_2$[78] has a very short Ga—P bond of 2.256(3) Å, planar coordination at phosphorus, and small (although not the smallest) torsion angles of ~15°. In fact, bond shortening in the two compounds with the most accurate torsion angles, t-$Bu_2GaP(Mes^*)SiPh_3$ and t-$Bu_2GaP(Trip)SiPh_3$,[79] may be inhibited by their very bulky substituents. Each of the three compounds above has two electropositive substituents at phosphorus (Ga, Ga; Ga, Si; or Ga, B) which is expected to lower the inversion barrier; however, it is interesting to note that t-$Bu_2GaP(SiMe_3)SiPh_3$, which has three electropositive substituents at phosphorus (Ga,Si,Si), has the highest degree of pyramidicity. The compounds with the shortest Ga—P bond lengths (2.32 Å or less) show considerably flattened geometry at phosphorus (Σ°P 340–359°), showing that the degree of planarity at phosphorus also correlates well with the Ga—P bond length and the torsion angle. Of all the monophosphides, t-$Bu_2GaP(SiMe_3)SiPh_3$[80] has the longest Ga—P bond length, the highest torsion angle, and the most pyramidal phosphorus atom and probably has no π-component in the Ga—P bond.

There may also be a σ-contribution to the bond shortening observed in these complexes, as the flattened geometry at phosphorus increases the s-character in the bond to gallium. An estimate of the size of the σ-contribution to bond shortening can be made by comparing the P—C bond lengths and Σ°P for t-$Bu_2GaP(Mes^*)SiPh_3$ (2.259(4) Å, 346.4°)[79,80] (Fig. 15) and its parent free phosphine $HP(Mes^*)SiPh_3$ (2.293(1) Å, 303.9°).[89] The P—C bond length shortens by 0.034 Å on going from the pyramidal free phosphine to the almost planar phosphide. However, the Ga—P bond in the phosphide is 0.063 Å shorter than that in t-$Bu_2GaP(SiMe_3)SiPh_3$, whose high torsion angle and long Ga—P bond can be used as a model for a non-π-bonded gallium phosphide. It appears, therefore, that about half of the Ga—P bond shortening can be attributed to the σ-effect resulting from rehybridization at phosphorus, while the other half may arise from a small π-bonding contribution, bearing in mind that the Ga—P bond in t-$Bu_2GaP(Mes^*)SiPh_3$ might be even shorter if not for the bulky substituents. The very small differencs (0.02 Å) between the P—Si bond lengths and Σ°P (321.4° versus 326.3°) in the phosphine $P(SiMe_3)_2(SiPh_3)$[80] and in t-$Bu_2GaP(SiMe_3)SiPh_3$ are consistent with the lack of a π-interaction in the latter compound.

Variable-temperature ^1H NMR studies of $MesP\{Ga(Trip)_2\}_2$[78] and t-Bu_2-$GaP(Mes^*)SiPh_3$[79,80] show barriers to rotation around the Ga—P bond of

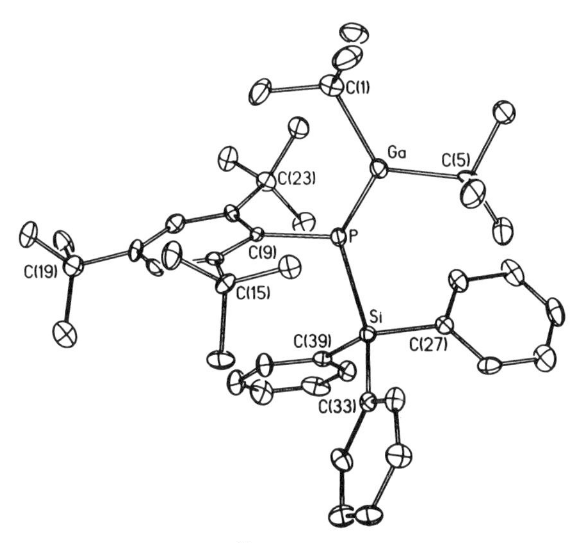

FIG. 15. Thermal ellipsoidal plot[80] of t-$Bu_2GaP(Mes^*)SiPh_3$ showing the flattened nature of phosphorus geometry and the alignment between the gallium p-orbital and the phosphorous lone pair. [Reprinted with permission from Petrie, Ruhlandt-Senge, and Power.[79] Copyright 1992 American Chemical Society.]

10.2 and 12.7 kcal mol^{-1}, respectively. The latter compound has a higher barrier even though the former has the shorter bond length, again reinforcing the view that bond shortening through π-overlap in some of these compounds might be inhibited by steric factors. The related compounds t-$Bu_2GaP(R)SiPh_3$ (R = Trip, $SiMe_3$, or Mes)[80] were also studied, showing dynamic behavior for R = Trip although a barrier could not be measured due to overlapping peaks. No dynamic behavior was observed for R = $SiMe_3$ and Mes as low as −100°C, indicating that any π-interaction in these compounds must be less than ~8 kcal mol^{-1} in magnitude and reinforcing the conclusions from the structural data for R = $SiMe_3$ that indicate negligible π-bonding.

$MesP\{Ga(Trip)_2\}_2$ has an almost planar C_2Ga—P(C)—GaC_2 array and is related to the diboron complex $PhP\{B(Mes)_2\}_2$, which has an exactly planar C_2B—P(C)—BC_2 framework.[93] Both compounds are structural analogs of the allyl cation, and the short B—P bonds (1.871(2) Å) and coplanar configuration in the boron–phosphorus compound are consistent with delocalized B—P—B π-bonding analogous to that in the allyl cation. The structural similarity and short Ga—P bond lengths in $MesP\{Ga(Trip)_2\}_2$ suggest that it also might feature a delocalized Ga—P—Ga π-interaction as illustrated in Fig. 16. A third compound, t-$Bu_2GaP(Ph)BMes_2$ (Fig. 17),[81] is also a structural analog of the allyl cation but with a Ga—P—B framework similar to the Ga—N—B species discussed earlier.[68] This con-

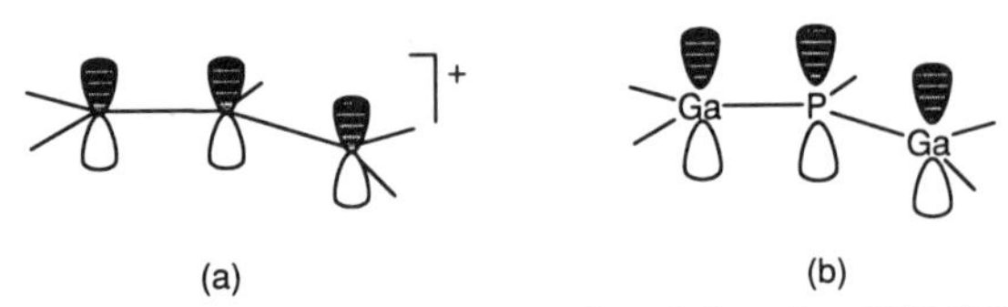

FIG. 16. Orbitals of π-symmetry in (a) the allyl cation $[CH_2CHCH_2]^+$ and (b) $MesP\{Ga(Trip)_2\}_2$.

figuration allows an intramolecular comparison of gallium–phosphorus and boron–phorphorus π-bonding, as the empty p-orbitals on boron and gallium must compete for the single lone pair on phosphorus. In this compound the Ga—P bond distance of 2.319(1) Å and torsion angle of 56.4° between the gallium and phosphorus planes indicate weaker π-bonding than in the related compounds discussed above, while the short B—P bond length (1.838(3) Å) and low torsion angle (>10°) are consistent with the conclusion that the π-bonding in this compound is primarily confined to the B—P bond rather than the Ga—P bond. This comparison can be extended even further in t-BuGa$\{P(Ph)BMes_2\}_2$,[81] which has a B—P—Ga—P—B framework and can be viewed as an analog of the pentadienyl cation. The geometry at the boron centers is planar and both phosphorus centers are considerably flattened, but marked alternation is observed in the bond lengths. The B—P—Ga—P—B array shows a short–long–

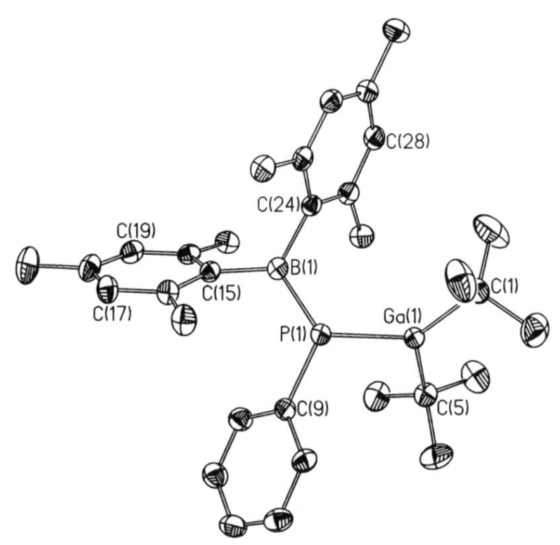

FIG. 17. Thermal ellipsoidal drawing[81] of t-$Bu_2GaP(Ph)BMes_2$ showing the large torsion angle between the planes at Ga and P.

short–long arrangement, giving rise to both very short and very long Ga—P and B—P bond distances within the same molecule. In fact the Ga—P distances of 2.194(2) Å and 2.390(2) Å are the shortest and longest, respectively, for the three-coordinate compounds listed in Table IV. Similarly, there is an extremely short B—P distance of 1.828(6) Å and a relatively long B—P bond of 1.896(7) Å. The π-bonding in this compound, then, is neither symmetrically delocalized, as in $PhP\{B(Mes)_2\}_2$[93] and $MesP\{Ga(Trip)_2\}_2$, nor is it concentrated in the B—P bonds, as in t-$Bu_2GaP(Ph)BMes_2$[81]; rather, there is a diene-like alteration of the bond lengths.

In the same way that the digallylphosphane $MesP\{Ga(Trip)_2\}_2$[78] is an analog of the allyl cation, the diphosphide t-$BuGa(PHMes^*)_2$ is an allyl anion analog.[82] The structural features of the former are consistent with the delocalized π-bond over the Ga—P—Ga framework, giving rise to very short Ga—P bonds although the latter compound shows lengthened Ga—P distances. In terms of a simple molecular orbital picture for the π-orbitals of an allyl-type arrangment composed of gallium and phosphorus atoms, the Ga—P—Ga (allyl cation) system has two π-electrons which occupy a strongly bonding orbital. In the P—Ga—P (allyl anion) analog the additional two electrons must be either nonbonding or weakly π-antibonding, accounting for the longer Ga—P distances. Alternatively, longer distances could arise from increased interelectronic repulsion in the four π-electron anion analog. A parallel situation is observed in the boron–phosphorus allyl cation analog $PhP\{B(Mes)_2\}_2$,[93] this complex has a planar B_2PC_5 framework and B—P bond distances that are ~0.02 Å shorter than those in the allyl anion analog $MesB(PPh_2)_2$, which has pyramidal phosphorus atoms.[6,93] However, the difference in Ga—P bond lengths of ~0.07 Å between the gallium allyl cation and anion analogs is much larger than in the boron–phosphorus compounds.

The compounds listed in Table IV as diphosphides are not readily comparable with each other, as they represent several different classes of compounds. The boryl-substituted diphosphide t-$BuGa\{P(Ph)BMes_2\}_2$, which is a pentadienyl cation analog, was discussed above.[81] $(t$-$BuGaPMes^*)_2$[82] and $(TriphGaPCy)_3$[83] (Triph = 2,4,6-$Ph_3C_6H_2$, Cy = cyclohexyl) are the only examples from the class of rings $(RGaPR')_n$, where n = 2 and 3, respectively. $Ga_4(Trip)_3\{P(1\text{-}Ad)\}_4P(H)(1\text{-}Ad)$[84] contains a three-coordinate P—Ga—P fragment as part of a six-membered Ga_3P_3 ring, which is itself contained in a cage with a Ga_4P_5 framework. The remaining gallium and phosphorus centers in the cage display four-coordination, allowing an internal comparison of Ga—P bonds between three- or four-coordinate atoms. All of these ring compounds have relatively short Ga—P bonds, ranging from 2.274(4) to 2.330(4) Å, yet they all contain pyramidal phosphorus, indicating that a π-interaction is not particularly strong. In the cage compound, the average Ga—P bond length (2.319 Å) involving

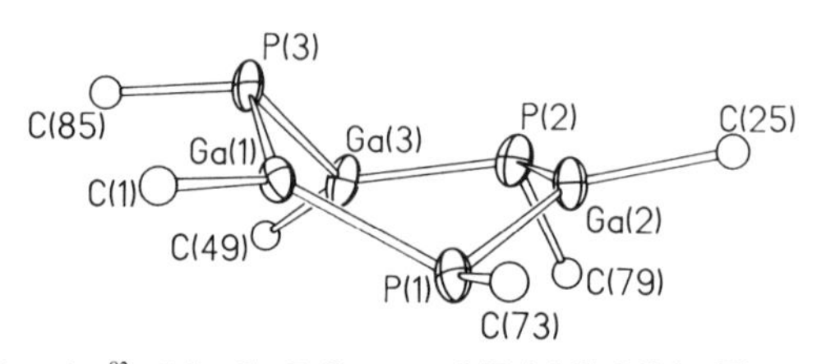

FIG. 18. A drawing[83] of the $Ga_3P_3C_6$ core of $(TriphGaPCy)_3$. The coordination at the P atoms is nonplanar and there is no evidence of delocalization in the potentially aromatic Ga_3P_3 core.

the three-coordinate phosphorus and gallium centers is significantly shorter than that between the four-coordinate atoms (2.426 Å).

Like the aluminum–nitrogen analog $(MeAlNDipp)_3$,[66a,73a] the six-membered ring $(TriphGaPCy)_3$[83] is electronically related to borazine and benzene. None of the phosphorus atoms in $(TriphGaPCy)_3$ (Fig. 18) has a planar configuration, however, and the $\Sigma°P$ ranges from 315.7° to 331.1°. Six different Ga—P bond lengths in the range from 2.279(4) to 2.338(5) Å are observed with the shortest Ga—P bonds corresponding to the most flattened phosphorus geometry. Despite the relatively short Ga—P bond distances, the puckered nature of the ring is evidence that π-delocalization is not significant. Even in $(MeAlNDipp)_3$, which has short Al—N bond lengths and a planar Al_3N_3 ring homodesmotic calculations have shown that π-interaction is negligible.[73b] Very short Ga—P bonds are also found in conjunction with pyramidal phosphorus centers in the four-membered ring $(t\text{-BuGaPMes*})_2$ (Fig. 19). In both the four- and six-membered rings $(RGaPR')_n$ the element : substituent ratio is 1 : 1, in contrast to the simple phosphides $R_2GaPR'_2$ where the ratio is 1 : 2. The shorter bonds in the rings could be attributed simply to the smaller number of bulky substituents. In a similar vein, there is less steric crowding around an ideal four-membered ring whose external angles are 270° (at each ring member with planar coordination) than around a six-membered ring whose corresponding angles are 240°. This, perhaps, explains why $(t\text{-BuGaPMes*})_2$[82] has even shorter Ga—P bonds (2.274(4) Å) than $(TriphGaPCy)_3$.[83] Although the Ga_2P_2 array is planar in $(CyGaPMes*)_2$, the pyramidal phosphorus centers again indicate that π-bonding is not likely to be significant in this compound.

The only simple diphosphide complex in Table IV is $t\text{-BuGa(PHMes*)}_2$,[82] which has Ga—P bond lengths averaging 2.325 Å. This distance is, with one exception, longer than those observed in the monophosphide complexes. This is consistent with a weak but significant gallium–phosphorus π-interaction where competition between two phospho-

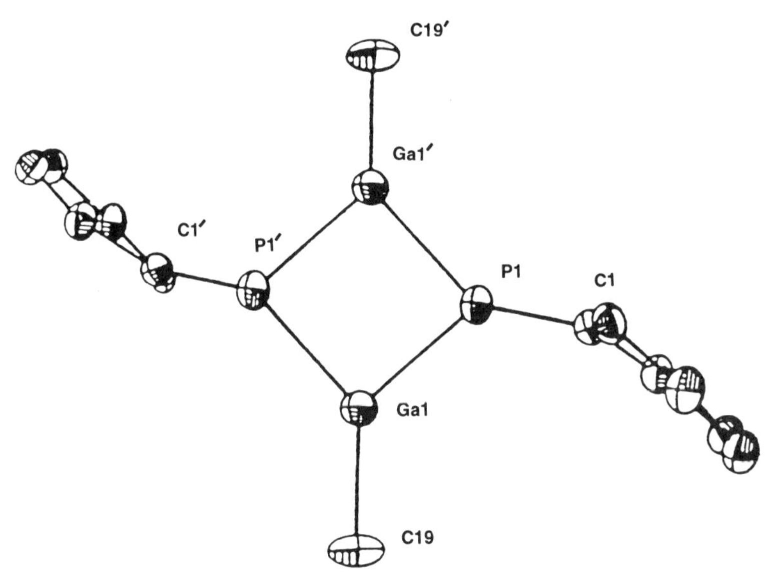

FIG. 19. Illustration of (t-BuGaPMes*)$_2$. The coordination at the phosphorous atom is pyramidal and no delocalization of the phosphorous lone pairs is evident. [Reprinted with permission from Atwood *et al.*[82] Copyright 1991 American Chemical Society.]

rus lone pairs for a single gallium p-orbital results in a longer bond for the diphosphide relative to the monophosphides. This trend continues for the triphosphide Ga(PHMes*)$_3$,[85] which has an average Ga—P bond length of 2.34(1) Å. The hydrogen substituents on the primary phosphides in these compounds prevent accurate determination of the orientation of the phosphorus lone pair or the sum of the angles at phosphorus.

The triphosphide In(P-t-Bu$_2$)$_3$[86] is currently the only well characterized three-coordinate compound with In—P bonds. These bonds average 2.592 Å, and all are significantly longer than the sum of the covalent radii for indium and phosphorus, 2.50 Å. The high torsion angles (>80°) and the pyramidal geometry at phosphorus indicate that π-bonding is probably not significant in this compound. The gallium analog Ga(PHMes*)$_3$[85] has lengthened Ga—P bonds, presumably due to intramolecular competition of the three phosphorus lone pairs for the single gallium p-orbital. Although there are no indium mono- or diphosphides available for comparison, they might be expected to have shorter In—P bonds than does In(P-t-Bu$_2$)$_3$.

The ^{31}P NMR chemical shifts, shown in Table IV, should be a useful indicator of the electronic environment at phosphorus. Almost all the ^{31}P chemical shifts are upfield (negative δ values) in contrast to the boron–phosphorus compounds where downfield shifts (positive δ values) are observed. This can be interpreted in terms of weak gallium–phosphorus π-

interactions in which the phosphorus lone pair resides largely on that atom, thus leading to increased shielding and upfield shifts. Among the compounds in Table IV, only the structurally characterized gallium monophosphides are truly comparable, and these do show a correlation of the structural data with increasingly negative chemical shifts as the gallium–phosphorus π-bond becomes weaker. $t\text{-}Bu_2GaP(SiMe_3)SiPh_3$ shows a very large upfield ^{31}P NMR shift of −267 ppm, which is consistent both with the lack of π-bonding in this compound and with the three electropositive substituents at phosphorus (Ga, Si, Si). Apart from these simple observations, it is difficult to draw any further conclusions at present because the environments around the phosphorus centers in Table IV are too varied and the ^{31}P NMR data show a corresponding lack of clear trends.

The sum of the covalent radii for gallium and arsenic is 2.47 Å[10] and the value corrected for ionic effects is 2.44 Å. The Ga—As bond lengths for the gallium monoarsenides in Table V are clustered around the latter value, with two shorter and two longer bonds. However, unlike the gallium phosphides, the gallium monoarsenides do not have small torsion angles and the arsenic centers are all markedly pyramidal. The pyramidal geometry at the arsenic centers is probably a consequence of the higher inversion barrier at arsenic. The longer Ga—As bonds (relative to Ga—P bonds) reduce the steric influence of the bulky GaR_2 group and also tend to permit a more pyramidal geometry at arsenic. The most flattened arsenic geometry, $\Sigma°$ As = 330°, occurs in $PhAs\{Ga(Trip)_2\}_2$,[78] which also has the shortest Ga—As bonds (2.41(1) Å) (average). However, Ga—As bonds of similar length (2.401(4) Å) have been observed in $As\{GaBr_2(THF)\}_3$ (THF = tetrahydrofuran),[94] which has three-coordinate arsenic but four-coordinate gallium centers, thus precluding π-bonding in this compound. $PhAs\{Ga(Trip)_2\}_2$[78] and $MesP\{Ga(Trip)_2\}_2$[78] are both allyl cation analogs, but the digallylarsane does not show the coplanar $C_2GaP(C)GaC_2$ skeleton observed both in the digallylphosphane and its boron analog, and thus does not exhibit the same kind of weak delocalized π-bonding. In the gallium triarsenide $Ga(AsMes_2)_3$,[89] all three Ga—As bonds, which average 2.492 Å, are longer than the Ga—As bonds in the monoarsenides. This compound represented the first structural characterization of a species featuring bonding between three-coordinate heavier main group 3–5 atoms. Overall, the structural data for the gallium arsenides in Table V indicate simple Ga—As σ-bonding with little or no π-component.

A further interesting feature of the gallium phosphides and arsenides is that the former compounds are colorless whereas the latter range from yellow to orange. Color can arise from $\pi \rightarrow \pi^*$ transitions in main group compounds: for example, in the disilylenes and digermenes $R_2E = ER_2$ (E = Si, Ge) in which the $\pi \rightarrow \pi^*$ transitions occur at lower energy than in the alkene analogs (E = C).[95] However, the yellow to orange color of the gallium arsenides can be attributed to ligand-to-metal charge-transfer transitions which are at lower energy than the corresponding transitions in the colorless gallium phosphides.

Boron–phosphorus and boron–arsenic π-bonds have been shown to be of comparable strength in some types of compounds.[6,7] Thus, despite the higher inversion barrier at arsenic, gallium–arsenic π-bonding is not expected to be inherently weaker than gallium–phosphorus π-bonding. The stabilization realized by both the B—P and B—As π-bonds must be sufficient to overcome the inversion barriers at the group 5 elements. However, gallium–phosphorus π-bonding is weaker than boron–phosphorus π-bonding, but is sufficient to allow flattening of the geometry at the phosphorus center in some cases. Gallium–arsenic π-bonding might also be expected to be weaker than that in the boron–arsenic case, but the higher inversion barrier at arsenic (~10 kcal mol^{-1} higher than phosphorus) is probably just sufficient to preclude attainment of the geometry required to form even weak gallium–arsenic π-bonds in the compounds examined so far.

4. *Metal Alkoxides and Aryloxides*

Monomeric main group 3 alkoxides R_2MOR' have one substituent fewer than the corresponding amides $R_2MNR'_2$. As a result, it is correspondingly more difficult to achieve three-coordination at the metal owing to the reduced degree of steric crowding afforded by the OR′ ligands. Very few such complexes have, in fact, been structurally characterized, and only five aluminum and four gallium examples, as shown in Table VI,[96–101] are currently known. In addition, there are a small number of three-coordinate complexes whose structures have not been determined. In many of the structurally characterized species, long-range contacts (>2.0 Å) are observed between the metal and hydrogen atoms from the alkyl or aryl substituents. The metal centers have very close to planar coordination although the angles between the substituents are distorted from regular trigonal-planar geometry. The slight displacement of the metal from the plane in the Mes* species is toward the closest *ortho-t*-butyl hydrogen atom. The structural parameters given in Table VI include the metal–oxygen bond length, the M—O—C angle, and the angle between the planes at the metal and at oxygen (torsion angle). The torsion angle is a less useful parameter for the group 6 compounds (relative to the group 5 compounds) because the interligand angle at O (S, Se, or Te) and the torsion angle do not unequivocally define the positions of the lone pairs, which can assume a range of relative orientations depending on the hybridization at oxygen.

TABLE VI

METAL–OXYGEN BOND LENGTHS, M—O—C ANGLES, AND TORSION ANGLES FOR THREE-COORDINATE ALUMINUM AND GALLIUM ALKOXIDES AND ARYLOXIDES

Compound	M—O (Å)	M—O—C angle (deg.)	Torsion angle[a] (deg.)	Reference
Aluminum				
Monoaryloxides				
t-$Bu_2AlOMes^*$	1.709 (av)	135.2 (av)	12.8	96
t-$Bu_2Al(O$-2,6-t-Bu_2-4-$MeC_6H_2)$	1.710(2)	129.4(1)	3.0	96
t-$Bu_2Al(O$-2,6-t-$Bu_2C_6H_3)$	—	—	—	96
i-$Bu_2Al(O$-2,6-t-Bu_2-4-$MeC_6H_2)$	—	—	—	97
Bisaryloxides				
$MeAl(O$-2,6-t-Bu_2-4-$MeC_6H_2)_2$	1.685(2)	148.8(2)	—	98
	1.687(2)	140.5(2)	—	
i-$BuAl(O$-2,6-t-$Bu_2C_6H_3)_2$	1.682(1)	157.3(1)	—	99
	1.702(1)	134.7(1)	—	
i-$BuAl(O$-2,6-t-Bu_2-4-$MeC_6H_2)_2$	—	—	—	97
Trisarylkoxides				
$Al(O$-2,6-t-Bu_2-4-$MeC_6H_2)_3$	1.648(7) (av)	177.2 (av)	—	100a
Aluminoxane				
$O[Al\{CH(SiMe_3)_2\}_2]_2$	1.6877(4)	180	0	100b
$O[Al\{CH(SiMe_3)_2\}_2]_2$-$ONMe_3$	1.753(3)	162.3(2)	—	100c
	1.846(3)			
Gallium				
Monoaryloxides				
$O\{Ga(Mes^*)Mn(CO)_5\}_2$	1.784(11)	150.2(5)[b]	—	67c
	1.789(11)			
t-$Bu_2Ga(O$-2,6-t-Bu_2-4-$MeC_6H_2)$	1.821(3)	125.0(2)	1.5	96
$(TMP)_2GaOPh$	1.822(3)	128.6(2)	23.5	
t-$Bu_2GaOCPh_3$	1.831(4)	127.5(3)	~90	96b
Bisalkoxides				
$Mes^*Ga(OSiPh_3)_2$	1.783(8)	140.3(5)[c]	—	101
	1.814(8)	141.9(5)	—	

[a] The torsion angle given is the angle between the O—C vector and the plane at the metal.
[b] Ga—O—Ga angle.
[c] Ga—O—Si angle.

The very large electronegativity difference between aluminum and oxygen means that estimates of an aluminum–oxygen single bond length can be calculated either from ionic radii, giving a range 1.65–1.68 Å (Table I[96]), or from the sum of the covalent radii (1.96 Å) using a substantial electrostatic correction (0.18–0.24 Å) to give an estimate of 1.72–1.78 Å (Table I). For the smaller gallium atom the sum of the covalent radii is 1.91 Å, but gallium is also less electropositive than aluminum so the ionic correction is smaller and the estimated Ga—O single bond length (~1.76 Å) may be longer than the Al—O bond. The Al—O bond lengths in Table VI range from 1.648 to 1.710 Å. These are close to the range predicted from the ionic model. The Ga—O bond lengths are slightly longer than the estimated value. Shorter Ga—O bonds ~1.79 Å are observed in the digalloxane $O\{Ga(Mes^*)Mn(CO)_5\}_2$[67c] and the gallium siloxide Mes*Ga$(OSiPh_3)_2$.[101] The three monosubstituted R_2GaOR' compounds have almost equal Ga—O bond lengths (range 1.821(3)–1.831(4) Å) in spite of large variations in ligand type and torsion angles. The aluminum aryloxides show a correlation that involves decreasing Al—O bond length and increasing Al—O—C bond angle on going from mono- (1.71 Å, 132°) (Fig. 20)[96] to bis- (1.69 Å, 145°) (Fig. 21)[98,99] to trisaryloxides (1.65 Å, 177°) (Fig. 22).[100] These observations parallel those for the aluminum and gallium amides, where decreasing bond length with increasing substitution by electronegative atoms favors an increasingly ionic rather than a π-bonding model. Variable-temperature NMR studies support this result and no barrier to rotation is observed at temperatures as low as −100°C for the first three compounds in Table VI and t-$Bu_2Ga(O$-2,6-t-Bu_2-4-$MeC_6H_2)$, suggesting that any π-interaction is less than 8 kcal mol^{-1} in magnitude. However, it is possible that the M—O—C moiety undergoes a linear inversion without significant disruption of a π-interaction. Thus, the VT NMR study may not give an accurate picture of the strength of the π-

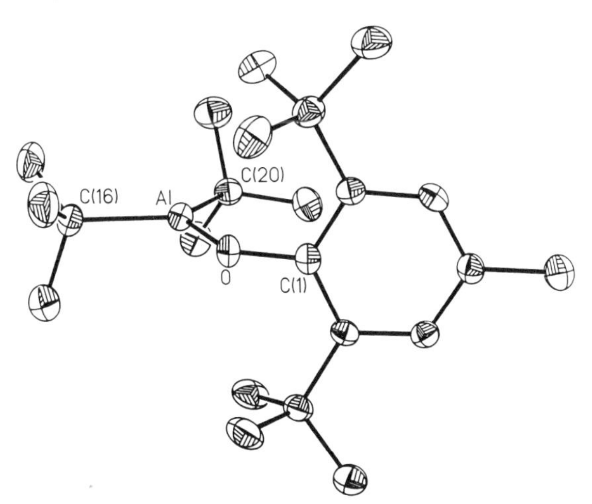

FIG. 20. Drawing of t-$Bu_2AlO(2,6$-t-Bu_2-4-$MeC_6H_2)$ showing the alignment of the coordination planes at Al and O. Further details are given in Table VI. [Reprinted with permission from Petrie, Olmstead, and Power.[96] Copyright 1991 American Chemical Society.]

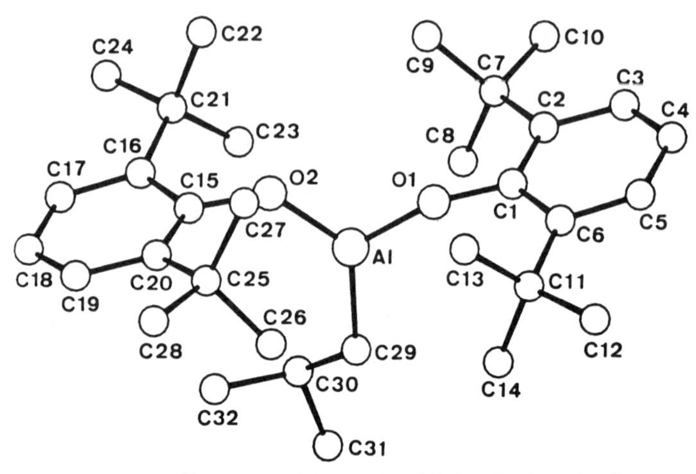

FIG. 21. Illustration[99] of i-BuAl(O-2,6-t-$Bu_2C_6H_3)_2$. Further details are given in Table VI.

interaction that might be present. There is a progressive upfield shift in the ^{27}Al NMR chemical shift in the series of three-coordinate monoaryloxides (~190 ppm), bisaryloxides (~100 ppm), and the single example of a trisaryloxide (~0 ppm).[99,100] However, this chemical shift change is the opposite of what would be expected on purely electronegativity grounds.

Calculations at the HF/3-21G* level on H_2AlOH and H_2AlOMe, which

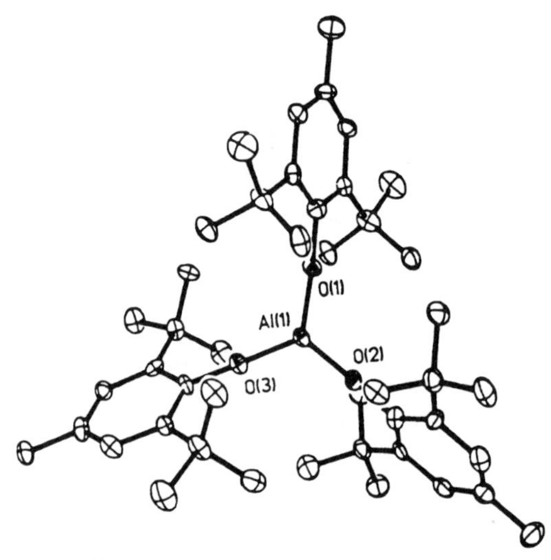

FIG. 22. Illustration[100a] of the structure of Al(O-2,6-t-Bu_2-4-$MeC_6H_2)_3$ showing the almost linear coordination at the oxygen centers.

have been interpreted in favor of Al—O π-bonding, indicate that the interaction is stabilized as the Al—O—C angle increases.[30] However it is not apparent why the average bond angles should increase with the number of aryloxide ligands. Just as the barrier to inversion at nitrogen is low, the angles at oxygen are relatively soft owing to a low linear inversion barrier at this atom. A combination of steric factors, long intramolecular M · · · H contacts, and crystal packing forces may influence the flexible M—O—C angles. However, in the bisaryloxides $MeAl(O\text{-}2,6\text{-}t\text{-}Bu_2\text{-}4\text{-}MeC_6H_2)_2$[96] and $i\text{-}BuAl(O\text{-}2,6\text{-}t\text{-}Bu_2C_6H_3)_2$,[99] it is interesting that the narrowest angle at aluminum is O—Al—O, close to 111° in both compounds, even though the O atoms bear the very bulky substituents. The O—Al—C angles in both compounds are between 120 and 130°. In addition, the planar framework (cf. allene) seen for $O[Al\{CH(SiMe_3)_2\}_2]_2$[100b] does not strongly support the existence of significant Al—O π-bonding.

The simplest view of a metal alkoxide involves an sp^2-hybridized oxygen center which bonds to the metal and the substituent, with one lone pair occupying the trigonal plane. The second lone pair is available for overlap with the empty p-orbital on the metal, requiring that this lone pair and the p-orbital be aligned perpendicular to the planes at the metal and oxygen. However, if the M—O—C angle widens toward 180°, the oxygen approaches sp-hybridization where one lone pair may remain orthogonal to the metal plane as required for p–p π-overlap while the other lone pair can participate in in-plane overlap with orbitals of suitable symmetry. This kind of linear inversion mechanism for M—O π-overlap has also been proposed for both three- and four-coordinate aluminum aryloxides.[29,102]

A large family of four-coordinate compounds having the general formula $R_{3\text{-}n}Al(O\text{-}2,6\text{-}t\text{-}Bu_2\text{-}4\text{-}MeC_6H_2)_n(L)$, where $n = 1$ or 2, R = Me or Et, and L is a neutral oxygen, nitrogen, or phosphorus donor, have been structurally characterized.[29] The Al-O bond lengths (1.713(4)–1.749(5) Å) are short, although they are significantly longer than the values observed for the three-coordinate compounds in Table VI, and the Al—O—C bond angles span the range 140.6(6)–174.8(3)°. Like the three-coordinate compounds, the average Al—O bond length is shorter for the bisaryloxides (1.72 Å) than for the monoaryloxides (1.74 Å). The short Al—O bond lengths and large Al—O—C bond angles are interpreted in terms of an in-plane Al—O π-interaction between the in-plane oxygen lone pair and aluminum-centered σ^* orbitals of the same symmetry which are antibonding with respect to the other substituents on aluminum, particularly the Al—L bond (Fig. 23). The results of extensive studies on these compounds have been summarized in a recent review.[29] The monoaryloxides show an inverse correlation between the Al—O bond length and the Al—O—C bond angle, while the bis aryloxides show no correlation. Significant π-bonding through the in-plane mechanism is belived to be present in both

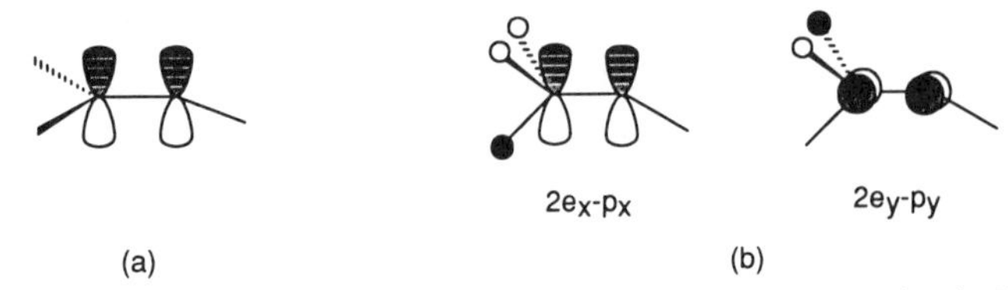

FIG. 23. (a) Alignment of p-orbitals on Al and O for p–p π-overlap in X_2AlOR; (b) alignment of Al—X σ^*-orbitals and p-orbital on O for "in-plane" overlap in the x and y planes in X_3Al—OR (X is a neutral or anionic σ-bonded substituent).

series of compounds, with the lack of correlation between bond lengths and angles in the bisaryloxides being attributed to steric factors. *Ab initio* calculations of the charges on the oxygen atoms are believed not to be in accord with an ionic model for the bonding in either the three- or four-coordinate compounds, although [30] it should be borne in mind that the model compounds used in this theoretical study, H_2AlOH and H_2AlOMe, are greatly simplified in comparison to the structurally characterized compounds.

In summary, the question of aluminum–oxygen and gallium–oxygen π-bonding is not completely resolved at present. The data for the three-coordinate aluminum and gallium aryloxides show patterns that are very similar to those observed for the three-coordinate amides, and the short bond lengths and variations with the number of substituents can be accounted for in terms of an ionic model. This is not surprising in view of the large electronegativity differences, which are maximized for the aluminum and oxygen pair. On the other hand, a second, "in-plane" mode of π-bonding has been proposed for four-coordinate aluminum aryloxides. The proponents of this interaction favor largely covalent bonding between aluminum and oxygen in both three- and four-coordinate compounds. Given the probable importance of ionic effects in main group 3 amides, it seems unlikely that covalent bonding predominates in Al—O bonds, which have the largest difference in electronegativity of any of the main group 3–5 or 3–6 pair combinations (Table I).

5. *Metal Thiolates, Selenolates, and Tellurolates*

Like the three-coordinate aluminum and gallium alkoxides, the corresponding thiolates, selenolates, and tellurolates require very bulky substituents at the chalcogen atom to ensure a low metal coordination number. In addition, the steric requirements associated with the preservation of the low coordination number at the metal are increased by the longer M—S and M—Se bond lengths. The ten examples of low-coordinate, unassociated M—S or M—Se compounds and the three examples of M—Te compounds that have been reported to date are given in Table VII.[103–109] All have been structurally characterized. With the exception of the μ-sulfido or μ-tellurido dialane $E[Al\{CH(SiMe_3)_2\}_2]_2$ (E = S[103a]; E = Te[103b]) the μ-tellurido digallane $Te[Ga\{CH(SiMe_3)_2\}_2]_2$[104], and the gallium tellurolate $\{(Me_3Si)_2CH\}_2GaTeSi(SiMe_3)_3$[108] all of the remaining compounds in Table VII bear the bulky Mes* group on the sulfur or selenium. The M—O—C angles for the alkoxides in Table VI span the range 125–177°, averaging 148°. In contrast the angles at sulfur and selenium for the compounds in Table VII are much narrower and each lies within a narrow range, close to 100° and 93°, respectively. The exceptions are the μ-sulfido or μ-tellurido dialanes $E[Al\{CH(SiMe_3)_2\}_2]_2$ (E = S[103a]; E = Te[103b]) digallane, or the metal atom. $Te[Ga\{CH(SiMe_3)_2\}_2]_2$[104], which have wider

TABLE VII

METAL–SULFUR, –SELENIUM, AND –TELLURIUM BOND LENGTHS, M—E—C ANGLES, AND TORSION ANGLES FOR THREE-COORDINATE ALUMINUM, GALLIUM, AND INDIUM THIOLATES, SELENOLATES AND TELLUROLATES[a], AND RELATED COMPOUNDS

Compound	M—E (Å)	M—E—C angle (deg.)	Torsion angle[b] (deg.)	Reference
Aluminum				
$S[Al\{CH(SiMe_3)_2\}_2]_2$	2.187(4)	117.5(3)[c]	39.9	103a
$BuAl(SMes^*)_2$	2.188(9)	98.5(2.0)	14.8	106
t-$BuAl(SMes^*)_2$	2.196(3)	102.7(3)	19.8	106
$Al(SMes^*)_3$	2.185(2)	100.4(1)	20.7	105
$Te[Al\{CH(SiMe_3)_2\}_2]_2$	2.549(1)	110.40(6)[d]	—	103b
Gallium				
Mes^*_2GaSMe	2.271(2)	102.9(3)	14.4	106
$BuGa(SMes^*)_2$	2.210(11)	99.9(4.5)	4.1, 14.2	106
$Ga(SMes^*)_3$	2.205(1)	100.4(1)	20.5	105
$Ga(SeMes^*)_3$	2.324(1)	93.9(2)	10.3	107
$\{(Me_3Si)_2CH\}_2GaTeSi(SiMe_3)_3$	2.535(1)	117.4(1)[e]	0	108
$Te[Ga\{CH(SiMe_3)_2\}_2]_2$	2.552(4)	109.82(2)[f]	—	104
Indium				
$In(SMes^*)_3$	2.398(3)	98.1(1)	22.0	109
$In(SeMes^*)_3$	2.506(1)	93.7(1)	11.5	109

[a] Where more than one chalcogenolate is present, averaged values are given.

[b] The torsion angle given is the angle between the perpendiculars to the MEC and C_2ME planes and is quoted as an average value where there is more than one chalcogenate present.

[c] Al—S—Al angle.

[d] Al—Te—Al angle.

[e] Ga—Te—Si angle.

[f] Ga—Te—Ga angle.

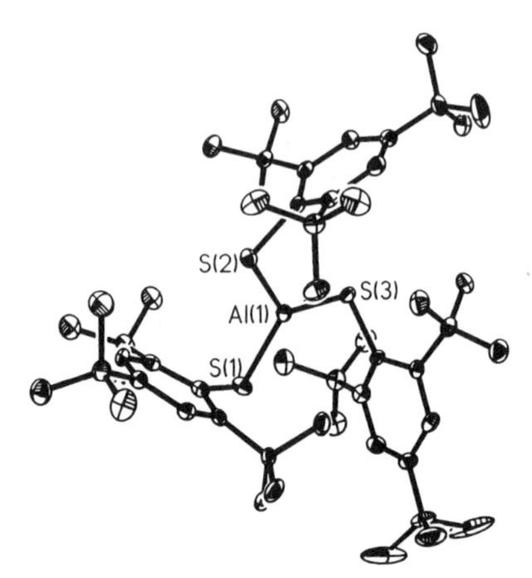

FIG. 24. Illustration of the structure of $Al(SMes^*)_3$. The structures of the gallium and indium analogs are very similar. Further details are given in Table VII. [Reprinted with permission from Ruhlandt-Senge and Power.[105] Copyright 1991 American Chemical Society.]

angles of 117.5(3), 110.40(6)°, and 109.82(2)° at the chalcogenide. The angle at tellurium in $\{(Me_3Si)_2CH\}_2GaTeSi(SiMe_3)_3$[108] is also relatively large, 117.4 (1)°.

The series of compounds $M(SMes^*)_3$ (M = Al, Ga, or In) (Fig. 24) and $M(SeMes^*)_3$ (M = Ga, In) show remarkable structural similarity. Each has a propeller-like arrangement of EMes* groups around a planar coordinated metal with the aromatic rings approximately orthogonal to the ME_3 plane. The aromatic rings are all rotated around the M—E bond so that they lie on the same side of this plane. Slight displacements of the central metal from the ME_3 plane are also toward this side. For the thiolates, each compound has two Mes* groups for which the angle between the S—C vector and the ME_3 plane (torsion angle) is 10–15° while the angle for the third ring is close to 39°. The planes of the Mes* rings are tilted ~12° relative to the S—C vector, and the two rings associated with the small torsion angles are tilted toward the metal while the third is tilted away. There are long-range M···H contacts (>2 Å) involving *ortho-t*-butyl groups on the Mes* rings which are tilted toward the metal. The two small and one larger torsion angles in the thiolates may be a reflection of steric strain, with one of the rings rotating further away from the ME_3 plane to relieve crowding. This phenomenon is not observed in the selenolates, presumably because the longer M-Se bonds help to alleviate the strain. In the selenolates the torsion angles range from 4.9° to 15.5°. There is no apparent correlation between torsion angle and bond length for the thiolates or selenolates.

The aluminum and gallium bisthiolates, like their bisaryloxide counterparts, show marked distortions from trigonal-planar angles at the metal.

The narrowest angle in each case is the S—M—S angle (averaging 108°) despite the fact that the largest steric interactions might be expected between the bulky SMes* groups.

The compounds in Table VII represent the shortest bond lengths reported to date for their respective main group 3–6 element pairs. The bond lengths are all shorter than the sum of the covalent radii by large margins (see Table I) and are shorter than comparable four-coordinate compounds containing each element pair. In spite of this, none of the aluminum and gallium thiolates or selenolates shows strong evidence for π-bonding. However, most of the compounds bear more than one chalcogenolate ligand, which would diminish the effect of any π-interaction in each bond. Even so, the aluminum and gallium bis- and tristhiolates show no significant difference between the average M—S bond lengths for each metal. However, the only unassociated gallium monothiolate, Mes^*_2-GaSMe[106] (Fig. 25), has a Ga—S bond length of 2.271(2) Å, which is significantly longer than that in the bis- or tristhiolates. As in the amide and aryloxide compounds, this observation supports the importance of ionic effects rather than π-overlap in M—S bonding. However, VT NMR studies indicate a barrier to rotation around the Ga—S bond of 10 kcal mol^{-1} in Mes^*_2GaSMe. In addition, the structural distortions in the series $M(EMes^*)_3$ can largely be explained by invoking steric or electrostatic arguments.[105] The compounds do not have geometries in which the planes at the metal and the chalcogen are most favorably aligned for optimum π-overlap. Barriers to rotation around the M—E bonds were not observed at temperatures as low as -90°C for any of the trithiolate or triselenolate derivatives, which places an upper limit of about 8 kcal mol^{-1} on a putative π-interaction. Nonetheless, a barrier to rotation of ~12 kcal mol^{-1} has been observed for $\{(Me_3Si)_2CH\}_2GaTeSi(SiMe_3)_3$. The very small electro-

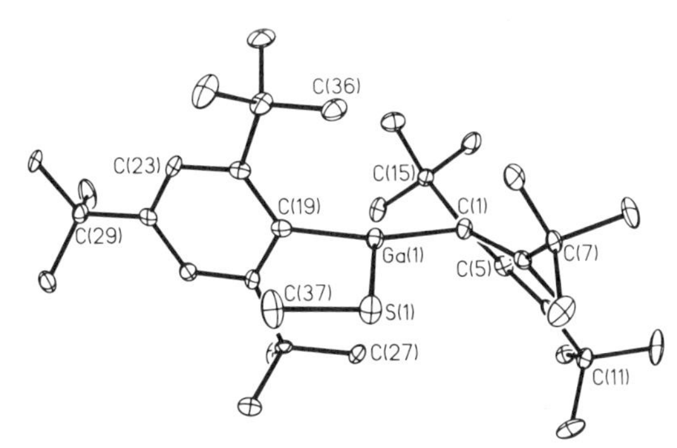

FIG. 25. Illustration of the structure of Mes^*_2GaSMe which shows the alignment of the coordination planes at Ga and S. Further details are given in Table VII. [Reprinted with permission from Wehmschulte, Ruhlandt-Senge, and Power.[106] Copyright 1995 American Chemical Society.]

negativity difference between gallium and tellurium (ΔEN = 0.19, see Table I) and the presence of only one tellurolate ligand in the compound make this an attractive candidate for significant π-bonding. However, $\{(Me_3Si)_2CH\}_2GaTeSi(SiMe_3)_3$ is a rare example of a three-coordinate main group 3 tellurolate, so it is difficult to draw generalized conclusions. Barriers to rotation have also been observed in $E[Al\{CH(SiMe_3)_2\}_2]_2$ (E = S or Te[103a,b]), but these are thought to be due to steric effects rather than significant π-interactions.

C. *Transition Metal Derivatives*

A small number of three-coordinate, heavier main group 3 compounds that have bonds to transition metals have been prepared and structurally characterized over the last two decades. To avoid confusion in this section, the main group 3 element will be denoted by E and the transition metal center, together with its ligands, by ML_n. These compounds, which have the general formula $L_nM—EX_2$ (where X can be a halide or other heteroatom substituent, an alkyl or aryl group, or even another ML_n fragment), are quite well established for boron.[110] Currently, only two structurally characterized three-coordinate aluminum complexes (i.e., $\{Cp^*(C_2H_4)$-$CoAlEt\}_2$[111a] and $(CpNiAlCp^*)_2$[111b]) exist. Transition metal complexes of aluminum are almost entirely limited to adducts of AlX_3 (where X is broadly defined as above), such as $[Ph_3AlFeCp(CO)_2]^-$,[112a,b] or to complexes in which aluminum and the transition metal are connected through bridging hydrides.[13a] Another example contains an $AlMe_2$ fragment bridging an Ir—N bond with the four-coordinate aluminum atom bonded to both the indium and nitrogen atoms.[112]

There is a small selection of three-coordinate compounds containing gallium and thallium and a larger number containing indium, all of which are listed in Table VIII.[67c,111–135] An approximately equal number of four-coordinate compounds have been structurally characterized, but these do not represent all of the M—E element combinations listed in Table VIII. Four-coordinate compounds arise when the relative size of M and E together with the steric bulk of the ligands is insufficient to stabilize three-coordination. L_nMEX_2 compounds can dimerize through coordination of either the ML_n fragment or, more commonly, a lone pair on X (when X = halide) to the main group center E of a second molecule, resulting in four-coordination at E. Alternatively, coordination of a neutral solvent molecule to E, forming $L_nME(L')X_2$, or formation of a metallate complex through addition of an extra anionic ligand to give $[L_nMEX_3]^-$ or $[\{L_nM\}_2EX_2]^-$ (X = halide) has the same effect. The synthetic and structural chemistry of the heavier main group 3 compounds with bonds to transition

TABLE VIII

COMPOUNDS WITH BONDS BETWEEN TRANSITION METALS AND THREE-COORDINATE GALLIUM, INDIUM, AND THALLIUM

Compound	E—M	E—M (av) (Å)	Reference
$\{Cp^*(C_2H_4)CoAlEt\}_2$	Al—Co	2.335(2)	111a
$(CpNiAlCp^*)_2$	Al—Ni	2.278(3)	111b
$Cp(CO)_3WGaMe_2$	Ga—W	2.708(3)	115
$\{Cp(CO)_3W\}_3Ga$	Ga—W	2.739(3)	116
$(CO)_5MnGa\{\mu\text{-}Mn(CO)_4\}_2GaMn(CO)_5$	Ga—Mn	2.445(1)[a]	117
		2.455(1)[b]	
$(CO)_5MnGa(Cl)Mes^*$	Ga—Mn	2.495(4)	67c
$O\{Ga(Mes^*)Mn(CO)_5\}_2$	Ga—Mn	2.533(4)	67c
$(CO)_5ReGa\{\mu\text{-}Re(CO)_4\}_2GaRe(CO)_5$	Ga—Re	2.569(5)[a]	118a
		2.599(4)[b]	
$(Ph_3P)(CO)_4ReGa\{Re(CO)_2(PPh_3)(\mu\text{-}X)\}_2$	Ga—Re	2.494(2)[a]	118b
(X = Br, I)		2.524(3)[b]	118c
$\{Cp(CO)_2Fe\}_2Ga(t\text{-}Bu)$	Ga—Fe	2.411(1)	113
$Cp(CO)_2FeGa(t\text{-}Bu)_2$	Ga—Fe	2.413(1)	113
$\{Cp(CO)_2Fe\}_3Ga$	Ga—Fe	2.444(1q)	114
$\{(OC)_4Co\}_2GaMes^*$	Ga—Co	2.50(1)	67c
$(Cy_2PCH_2CH_2PCy_2)(Me_3CCH_2)Pt$ $Ga(CH_2CMe_3)_2$	Ga—Pt	2.438(1)	119
$\{(CO)_5Cr\}InBr(THF)$	In—Cr	2.554(3)	120
$[\{(CO)_5Cr\}_2InBr]^{2-}$	In—Cr	2.657(2)	121
$\{Cp(CO)_3Cr\}_2In$	In—Cr	2.794(1)	122
$\{Cp(CO)_3Mo\}_3In$	In—Mo	2.878(1)	123
$\{HB(3,5\text{-}Me_2pz)_3\}InW(CO)_5$	In—W	2.783	124
$(CO)_5MnIn\{\mu\text{-}Mn(CO)_4\}_2InMn(CO)_5$	In—Mn	2.596(1)[a]	117
		2.601(1)[b]	
$(CO)_5MnIn\{\mu\text{-}Fe(CO)_4\}_2InMn(CO)_5$	In—Mn	2.635(1)[a]	125
	In—Fe	2.663(1)[b]	
$(CO)_5ReIn\{\mu\text{-}Re(CO)_4\}_2InRe(CO)_5$	In—Re	2.738(1)[a]	117
		2.781(1)[b]	
$\{HB(3,5\text{-}Me_2pz)_3\}InFe(CO)_4$	In—Fe	2.463	124
$[\{(CO)_4Fe\}_3In]^{3-}$	In—Fe	2.633(1)	126
$\{Cp(CO)_2Ru\}_3In$	In—Ru	2.676(1)	127
$\{(CO)_4Co\}_3In$	In—Co	2.594(3)	128
$Ir(InEt_2)H(Et)(PMe_3)_3$	In—Ir	2.601(1)	129
$(Cy_2PCH_2CH_2PCy_2)(Me_3SiCH_2)$- $PtIn(CH_2SiMe_3)_2$	In—Pt	2.6012(2)	
$[\{(CO)_5Cr\}_2Tl]^-$	Tl—Cr	2.639(4)	131
$[\{(CO)_5Cr\}_2TlX]^{2-}$ (X = Cl, Br, I)	Tl—Cr	2.681(3)	121
$\{Cp(CO)_3Cr\}_3Tl$	Tl—Cr	2.871(2)	122
$\{Cp(CO)_3Mo\}_3Tl$[c]	Tl—Mo	2.931(1)	122
$\{Cp(CO)_3Mo\}_3Tl$[d]	Tl—Mo	2.965(3)	132
$[Tl_2Fe_4(CO)_{16}]^{2-}$	Tl—Fe	2.553(5)[e]	133
		2.632(5)[e]	
		3.038(4)[e]	

(*continues*)

TABLE VIII (*continued*)

Compound	E—M	E—M (av) (Å)	Reference
$[Tl_2Fe_6(CO)_{24}]^-$	Tl—Fe	2.754(2)[f] 2.756(2)[f]	134
$[Tl_4Fe_3(CO)_{30}]^{4-}$	Tl—Fe	2.530(7)[e] 2.647(6)[e] 2.712(6)[f] 2.784(6)[f]	133
$[Tl_8Fe_{10}(CO)_{36}]^{5-}$	Tl—Fe	2.551(4)[e] 2.616(4)[e] 2.727(3)[f] 2.803(3)[f]	135

[a] Terminal E—M bond.
[b] Bridging E—M bond.
[c] Crystallized from CH_2Cl_2/hexane.
[d] Crystallized from acetone/pentane.
[e] Bond to three-coordinate thallium.
[f] Bond to four-coordinate thallium.

metals has been reviewed several times, although none of these treatments has considered in depth the question of π-bonding.[13,136–139]

The discussion of the main group 3–5 and 3–6 compounds in the previous sections was limited to examples in which the group 3 element E is three-coordinate, so that an empty *p*-orbital on E is available for overlap with a lone pair on the group 5 or 6 atom. For the same reason, the discussion here will focus on those compounds with three-coordination at gallium, indium, or thallium. In the case of the transition metal derivatives, it is transition metal *d*-electrons that are available to overlap with the empty *p*-orbital on E to form the potential π-bond, as illustrated in Fig. 26.

The first important structural indication for π-bonding is shortening of the M—E bond, although it can be difficult to determine a standard M—E single bond length for each of the element pairs in Table VIII. For various reasons, ionic rather than covalent radii are used in estimating ligand bond lengths for transition elements. In addition, covalent radii are expected

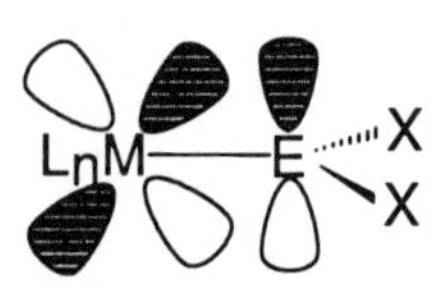

FIG. 26. Schematic drawing of the overlap of transition metal (M) *d*-orbital and the heavier main group element (E) *p*-orbital.

to be greatly affected by the nature of the ligands in the coordination sphere. Four-coordinate compounds are known for some of the M—E combinations, but not always with comparable ligands or substituents, which makes useful comparisons difficult. The second structural indication for π-bonding is the orientation of the *p*-orbital on E (perpendicular to the MEX_2 plane) relative to a filled orbital of π-symmetry on the ML_n fragment. Some of the impetus for studying transition metal complexes with main group 3 substituents derives from the fact that a main group 3 fragment $:ER_2^-$ is isolobal with the carbene moiety $:CR_2$. Alternatively, cationic carbene complexes $[L_nMCR_2]^+$ are isolobal with neutral main group 3 complexes L_nMER_2. Owing to a wealth of theoretical and experimental studies on many different transition metal carbene complexes, the orientation of the ER_2 substituent for a given ML_n fragment can be compared to the orientation of a carbene ligand in the corresponding complex. Even so, a theoretical analysis of the expected orientation of the orbitals for each of the transition metal/ligand combinations listed in Table VIII is not available in many instances.

The formal analogy between carbene (CR_2) and ER_2 complexes is useful; currently, however, there are only eleven compounds in Table VIII that contain organoaluminum-gallium or -indium fragments bound to the transition metal. These are $\{Cp^*(C_2H_4)NiAlEt\}_2$, $\{CpNiAlCp^*\}_2$, $\{Cp(CO)_2Fe\}_2Ga(t\text{-Bu})$, $Cp(CO)_2FeGa(t\text{-Bu})_2$, $Cp(CO)_3WGaMe_2$ (Cp = η^5-cyclopentadienide), $(CO)_5MnGa(Cl)Mes^*$, $O\{Ga(Mes^*)Mn(CO)_5\}_2$, $\{(OC)_4Co\}_2$-$GaMes^*$, $(Cy_2PCH_2CH_2PCy_2)(Me_3CCH_2)PtGa(CH_2CMe_3)_2$, $(Cy_2PCH_2$-$CH_2PCy_2)$ $(Me_3CCH_2)PtIn(CH_2CMe_3)_2$, and $IrH(Et)(InEt_2)(PMe_3)_3$. The last three are the only examples in which the ML_n fragments are relatively electron-rich, owing to the fact that they are late transition elements without strong π-acceptor ligands in the coordination sphere. The remaining complexes in Table VIII contain either three transition metal centers or two ML_n centers with one halide substituents at E, and all of the ML_n groups contain between two and five carbonyl ligands. Few of these compounds are good candidates for M—E π-bonding, owing to the fact that the *d*-electrons on two or three transition metal centers will be competing for the single empty *p*-orbital on E, thereby moderating any π-effects. In addition, the strongly π-accepting CO ligands in the ML_n coordination sphere will compete with the empty *p*-orbital on E for the available *d*-electron density on M, which diminishes the possibility of significant M—E π-bonding. All of the three-coordinate complexes in Table VIII exhibit approximately planar geometry at E, although some angles depart considerably from the ideal trigonal angles of 120°. Nonetheless the two aluminum compounds, both of which feature two AlR fragments bridging two transition metal moieties, are notable for their short aluminum—metal bonds. For example the Al—Co distance in $\{Cp^*(C_2H_4)CoAlEt\}_2$ is over

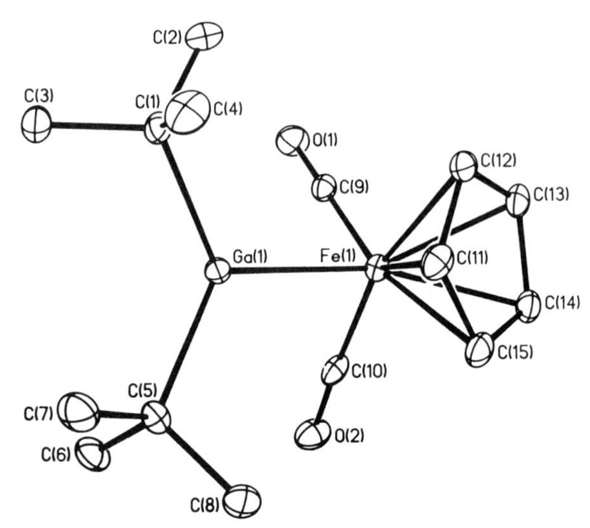

FIG. 27. Thermal ellipsoidal drawing of $Cp(CO)_2FeGa(t\text{-}Bu)_2$. Further details are provided in Table VIII. [Reprinted with permission from St. Denis *et al.*[113] Copyright 1994 American Chemical Society.]

0.16 Å shorter than the Ga—Co bond in $\{(OC)_4Co\}_2GaMes^*$. The Al—Ni distance, 2.278(3)Å, is also very short and the data for the two complexes are consistent with the presence of some multiple bonding. Clearly, further examples will be required before definitive conclusions can be made.

The compounds with Ga—Fe bonds form a series $\{L_nM\}_xER_{3-x}$, where ML_n = $CpFe(CO)_2$, R = *t*-Bu, and x = 1, 2, or 3. The first complex of this series, $Cp(CO)_2FeGa(t\text{-}Bu)_2$ (Fig. 27),[113] should be the best candidate for an M—E π-bond, since the diorganogallium group is bonded to a single iron center. However, the Fe—Ga bond length of 2.413(1) Å is almost identical to that found (2.411(1) Å) for the next member of the series, $\{Cp(CO)_2Fe\}_2Ga(t\text{-}Bu)$. Neither bond length is significantly shorter than those in either the three-coordinate triiron compound $\{Cp(CO)_2Fe\}_3Ga$ (2.444(1) Å) or the four-coordinate gallium compounds $Cp(CO)_2FeGa(\eta_1\text{-}C_5H_4Me)_2(py)$ (2.427(1) Å)[140] and $Cp(CO)_2FeGa(t\text{-}Bu)_2 \cdot \{CpFe(CO)_2\}_2$ (2.441(1) Å).[113] The latter compound contains a $Cp(CO)Fe(\mu\text{-}CO)_2Fe(CO)Cp$ molecule weakly coordinated to gallium through the oxygen atom of one bridging carbonyl ligand. The lack of significant change in the Ga—Fe bond length when the number of iron substituents or the coordination number at gallium is varied suggests a negligible π-component in the Ga—Fe bond. Similarly, π-bonding is probably not a significant feature in the Ga—W compounds $(Cp(CO)_3WGaMe_2$ and $\{Cp(CO)_3W\}_3Ga$, which

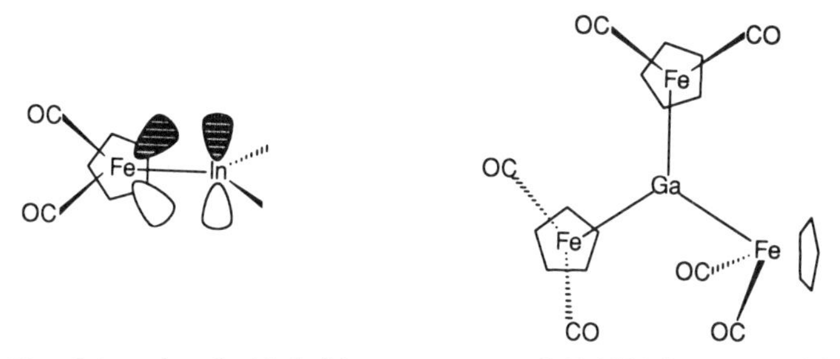

FIG. 28. Orientation of orbital with π-symmetry on $CpFe(CO)_2$ fragment and the orientation of these groups in $\{CpFe(CO)_2\}_3Ga$.

have very similar Ga—W bond lengths of 2.708(3) Å and 2.739(3) Å, respectively.

Molecular orbital arguments can also be used to rationalize the lack of Ga—Fe π-bonding in such molecules as $Cp(CO)_2FeGa(t\text{-Bu})_2$. The filled orbital of π-symmetry on the $CpFe(CO)_2$ fragment is parallel to the cyclopentadienyl ring and is bisected by the plane of symmetry through the $CpFe(CO)_2$ fragment[141]; it is this orbital that is available to form the π-bond in the cationic carbene complexes $[CpFe(CO)_2(CR_2)]^+$ which are isolobal with the gallium compound $Cp(CO)_2FeGa(t\text{-Bu})_2$. In an ideal C_{3h} environment, this π-orbital has a'' symmetry, as does the empty gallium p_z-orbital. The conformation for maximum π-overlap between these two orbitals occurs when the plane bisecting the $CpFe(CO)_2$ fragment is coincident with the $GaFe_3$ plane in $\{Cp(CO)_2Fe\}_3Ga$ (in other words, when the cyclopentadienyl ring is orthogonal to this plane), as illustrated by the schematic drawing of the gallium derivative in Fig. 28. In this complex, one of the $CpFe(CO)_2$ groups does have this conformation, allowing for maximum π-overlap, while the other two are rotated at 90° which should preclude π-overlap. However, there is no significant difference between the three Fe—Ga bond lengths, all of which are within 0.01 Å of the average value (2.444(1) Å).[114] In both alkylgallium complexes $Cp(CO)_2FeGa(t\text{-Bu})_2$ and $\{Cp(CO)_2Fe\}_2Ga(t\text{-Bu})$, the $CpFe(CO)_2$ fragments are oriented at 90° to the conformation required for maximum overlap with the empty p-orbital on gallium, again precluding a significant π-interaction. The related indium–ruthenium compound $\{Cp(CO)_2Ru\}_3In$ also shows no orientational preference for the $CpRu(CO)_2$ groups, of which none has the optimum orientation for π-bonding. The In—Ru bond length in this compound, 2.676(1) Å, is not markedly different from that in the chloro-bridged dimer $[\{Cp(CO)_2Ru\}_2In(\mu\text{-Cl})]_2$, 2.629 Å.[127]

Table VIII also includes a series of gallium, indium, and thallium compounds of the general formula $\{Cp(CO)_3M\}_3E$ (E,M = Ga,W; In,Cr; In,Mo;

Tl,Cr; or Tl,Mo) which have bonds to three chromium, molybdenum, or tungsten groups. All of these complexes are approximately isostructural, having ideal C_{3h} symmetry with an almost completely planar EM_3 core (displacement of E from the M_3 plane <0.06 Å).[122] The cyclopentadienyl ring on each M center is approximately perpendicular to this plane, so that the rings are bisected by the plane. However, one of the two independent crystal structure determinations of the Tl—Mo compound $\{Cp(CO)_3Mo\}_3Tl$ displays a different geometry. One of the Cp rings lies perpendicular to, and is bisected by, the Mo_3 plane, while the other two are approximately parallel to the Mo_3 plane, one above and one below. The displacement of the thallium atom from the Mo_3 plane, 0.59 Å,[132] is much more marked. Given the importance of the orientation of the transition metal fragment with respect to the plane at E in determining the optimum geometry for π-overlap, the polymorphism induced in this compound simply by crystallization from different solvents suggests that π-bonding is not of major importance. Several compounds related to the chromium and molybdenum complexes $\{Cp(CO)_3M\}_3In$ have four-coordinate indium centers: $\{Cp(CO)_3Cr\}_2InCl(py)$, In—Cr = 2.773(1) Å; $[\{Cp(CO)_3Mo\}_2InCl_2]^-$, In—Mo = 2.851(1) Å[123,139]; and $\{Cp(CO)_3Mo\}_2InCl(THF)$, In—Mo = 2.825(1) Å.[139] Here again, comparisons show that the M—E bond lengths in the three- and four-coordinate compounds are very similar. Surprisingly, a much shorter In—Mo bond length of 2.739(1) Å is observed in the four-coordinate indium compound $[\{Cp(CO)_3Mo\}InCl_3]^-$.[123]

Extended Hückel molecular-orbital (EHMO) calculations using structural parameters from the X-ray determination of $\{Cp(CO)_3Mo\}_3In$ (Fig. 29) and ideal symmetry C_{3h} were carried out for this compound in order to investigate the extent of the indium–molybdenum π-bonding.[122] The HOMO of the compound is the $3e'$ molecular orbital, which is In—Mo σ-bonding and involves the $p_{x,y}$-orbitals on indium.[142] However, the In p-orbital is much higher in energy than the molybdenum d-orbital, so the $3e'$ HOMO is largely molybdenum in character (occupancy of the indium

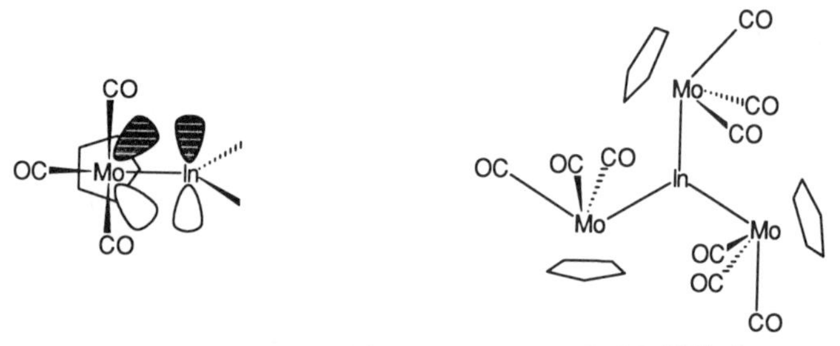

FIG. 29. Orientation of orbital with π-symmetry on $Cp(Mo(CO)_3$ fragment and the orientation of these groups in $\{Cp(Mo(CO)_3\}_3In$.

$p_{x,y}$-orbitals is calculated to be 0.21 electrons). The conformation of the $Cp(CO)_3Mo$ fragments in the complex means that the orbitals of π-symmetry on each molybdenum are aligned with the p_z-orbital on indium and all have a'' symmetry. This allows for a weak indium–molybdenum π-interaction, but, as before, there is a large difference in energy between the indium and molybdenum orbitals and the occupancy of the indium p_z-orbital is 0.12 electrons. Even though the π-orbitals on the molybdenum atoms are correctly oriented for maximum π-overlap with indium, the geometric alignment is poor and the molybdenum π-orbitals are tilted away from the In—Mo vectors by 30° (Fig. 29). These factors, together with the presence of the strongly π-accepting CO ligands on molybdenum and the fact that three molybdenum centers are competing for a single p-orbital on indium, indicate that, although π-bonding is possible on symmetry grounds, the π-interaction is very weak and the indium–molybdenum bonding is largely polar in nature. Chemical evidence for this conclusion comes from the observation that $[CpMo(CO)_3]^-$ groups dissociate from $\{Cp(CO)_3Mo\}_3In$ in polar solvents such as acetone and acetonitrile.[122,123]

A general feature of the gallium, indium, and thallium complexes with bonds to $CpM(CO)_n$ fragments (M = Cr, Mo, W, $n = 3$; M = Fe, Ru, $n = 2$) is that the carbonyl C—O stretching frequencies are low, which indicates considerable negative charge on the transition metal center and increased back-donation into the π^*-orbitals.[113,114,122,127] The opposite effect would be expected if π-bonding to the main group 3 element were significant, in which case competition for d-electron density on M would be expected to result in increased C—O stretching frequencies.

A different mechanism for indium–transition metal π-bonding has been proposed for the indium–iron dimer $[\{CpFe(CO)_2\}_2In(\mu\text{-}Cl)]_2$ which contains four-coordinate indium centers. Short indium–iron bonds (average In—Fe bond lengths 2.556(1) Å) have been attributed to negative hyperconjugation arising through π-overlap of iron d-orbitals of π-symmetry with In—Cl σ^*-orbitals.[143] This proposal is similar to that for Al—O π-bonding in four-coordinate aluminum aryloxides discussed in Section III.B.4. However, as in the aluminum–oxygen case, it is difficult to determine whether short bonds result from this type of π-overlap or from an electrostatic interaction between the electropositive indium center and the CO-substituted iron centers combined with a contraction of the indium orbitals arising from the electronegative chloride ligands.

In the family of compounds $(CO)_5ME\{\mu\text{-}M'(CO)_4\}_2EM(CO)_5$ (E,M,M' = Ga,Mn,Mn; Ga,Re,Re; In,Mn,Mn; In,Mn,Fe; or In,Re,Re), the common structural arrangement consists of a central $E_2M'_2$ four-membered ring with terminal $M(CO)_5$ groups bonded to E (Fig. 30). When M' = Mn or Re, there is a Mn—Mn or Re—Re bond across the ring, but the compound with M' = Fe has two more electrons and no Fe—Fe bond. When M =

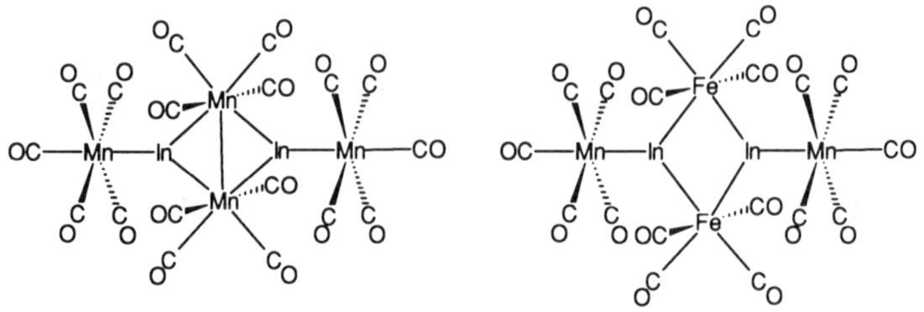

FIG. 30. Structures of $(CO)_5MnIn\{\mu\text{-}Mn(CO)_4\}_2InMn(CO)_5$, and $(CO)_5MnIn\{\mu\text{-}Fe(CO)_4\}_2InMn(CO)_5$.

M′, the terminal and bridging Ga—Mn, In—Mn, or In—Re bond lengths are very similar, with the former ~0.01–0.05 Å shorter than the latter, as shown by the two values given for each compound in Table VIII. Comparisons can be made with four-coordinate In—Mn and In—Re compounds, $\{(CO)_5M\}_2In\{\mu\text{-}X\}_2In\{M(CO)_5\}_2$ (M = Mn, Re; X = Cl, Br, I) for which the average In—Mn and In—Re bond lengths are 2.667(1) Å[144] and 2.801(2) Å,[145] less than 0.1 Å longer than the three-coordinate compounds in Table VIII. $(CO)_5MnIn\{\mu\text{-}Fe(CO)_4\}_2InMn(CO)_5$ and $[\{(CO)_4Fe\}_3In]^{3-}$ have Fe—In bonds to three-coordinate indium of 2.663(1) Å and 2.633(1) Å, respectively, and both of these are similar to or longer than several examples of compounds with Fe—In bonds to four-coordinate indium: $[In(bipy)\{Fe(CO)_4\}_2]^-$ (bipy = 2,2′-dipyridine), 2.549(3) Å[146]; $[In(\mu\text{-}Cl)\{CpFe(CO)_2\}_2]_2$, 2.556(1) Å[143]; $InCl(PMe_2Ph)\{CpFe(CO)_2\}_2$, 2.576(1) Å[143]; and $[InBr_2\{\mu\text{-}Fe(CO)_4\}]_2^{2-}$, 2.668(1) Å.[126] All these data indicate that, again, π-bonding is probably of limited importance in the managanese-, iron-, and rhenium-bridged compounds as well as in the indium triiron compound.

An alternative view of the bonding in these compounds has been put forward, noting that $Mn(CO)_4$ is isolobal with CH and $InMn(CO)_5$ is isolobal with CH^+. Thus the central In_2Mn_2 ring in $(CO)_5MnIn\{\mu\text{-}Mn(CO)_4\}_2InMn(CO)_5$ is analogous to $C_4H_4^{2+}$, which would be expected to contain a delocalized orbital of π-symmetry. On the other hand, the related compound $(CO)_5MnIn\{\mu\text{-}Fe(CO)_4\}_2InMn(CO)_5$ contains two additional electrons in the central In_2Fe_2 ring which would then be comparable to C_4H_4, cyclobutadiene.[137]

Two other related compounds containing Ga—Re bonds to three-coordinate gallium are $(Ph_3P)(CO)_4ReGa\{Re(CO)_2(PPh_3)(\mu\text{-}X)\}_2$ (X = Br, I), which has a central $GaRe_2$ ring (Fig. 31). Each rhenium is substituted by a PPh_3 ligand which lies trans to the Ga—Re bond. It is interesting that the terminal and bridging Ga—Re bond lengths, which are 2.494(2) and 2.524(3) Å, respectively, for X = I, are shorter than the comparable bond lengths of 2.569(5) and 2.599(4) Å in $(CO)_5ReGa\{\mu\text{-}Re(CO)_4\}_2GaRe(CO)_5$.

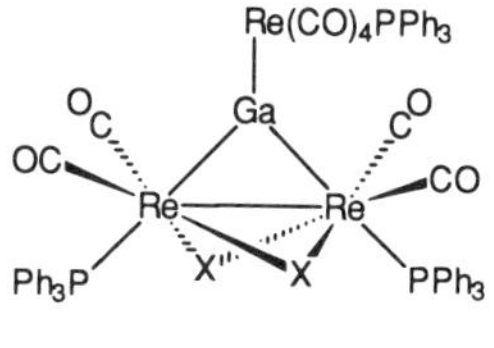

(X = Br, I)

FIG. 31. Structure of $(Ph_3P)(CO)_4ReGa\{Re(CO)_2(PPh_3)_2(\mu\text{-}X)\}_2$ (X = Br, I).

Replacement of the strongly π-accepting CO ligand trans to the gallium substituent by the more electron-rich PPh_3 ligand seems to strengthen the Ga—Re bond, which is exactly what would be expected if π-bonding effects were significant.

The indium–iron and indium–cobalt complexes $[\{(CO)_4Fe\}_3In]^{3-}$ and $\{(CO)_4Co\}_3In$ are isoelectronic and have similar In—M bond distances of 2.633(1) and 2.594(3) Å, respectively. As outlined above, the In—Fe distance is comparable to that in several compounds containing four-coordinate indium centers. There are structural data for a selection of compounds containing bonds between four-coordinate indium and cobalt, ranging from 2.542(3) Å in $InCl_2(PPh_3)\{Co(CO)_4\}$[147] to 2.705(1) Å in $[InI\{Co(CO)_4\}_3]^-$.[148] In both $[\{(CO)_4Fe\}_3In]^{3-}$ and $\{(CO)_4Co\}_3In$, as in the four-coordinate iron and cobalt complexes, the indium atom occupies the axial position in a trigonal-bipyramidal geometry about the iron or cobalt atom. In complexes related to $Fe(CO)_4L$ or $CoX(CO)_3L$, it is well known that π-bonding to L is maximized when this ligand occupies an equatorial site rather than the axial site observed in the complexes described here.

Although the 16-electron fragment $Cr(CO)_5$ is isolobal with $Fe(CO)_4$, there is evidence of significant bond shortening in indium–chromium and thallium–chromium complexes containing the $Cr(CO)_5$ group. The discussion earlier in this section established that $\{Cp(CO)_3Cr\}_3In$ involves negligible π-bonding, so its In—Cr bond length of 2.794(1) Å is probably a good estimate for an In—Cr single bond. The In—Cr bonds in $[\{(CO)_5Cr\}_2InBr]^{2-}$ and $\{(CO)_5Cr\}InBr(THF)$ are notably shorter, at 2.657(2) and 2.554(3) Å, respectively. Similarly, if $\{Cp(CO)_3Cr\}_3Tl$ with a Tl—Cr bond length of 2.871(2) Å is used as the benchmark for a single bond, then $[\{(CO)_5Cr\}_2\text{-}TlX]^{2-}$ (X = Cl, Br, I) and $[\{(CO)_5Cr\}_2Tl]^-$ also have much shorter bonds, 2.681(3) (av) and 2.639(4) Å, respectively. The indium and thallium complexes $[\{(CO)_5Cr\}_2EX]^{2-}$ are considered to be the first main group 3 members of the "inidene" class $[L_nME(R)ML_n]^m$ (E = In, Tl, $m = -2$; E = Ge, Pb, $m = -1$; E = As, Sb, $m = 0$),[149] for which an M—E π-interaction is well established.[131] Similarly, the two-coordinate thallium complex $[\{(CO)_5Cr\}_2Tl]^-$ is an example of the related cumulene complexes, $[L_nM\text{=}E\text{=}ML_n]^{m+1}$, which are known for E = Ge, Pb, As, and Sb.[121] The proposed oxidation state of thallium in $[\{(CO)_5Cr\}_2Tl]^-$ is Tl^{-I}, acting as a four-electron donor isolobal with Pb^0 and Sb^I. The marked bond shortening in the indium and thallium inidene and cumulene complexes and close isoelectronic relationship with the main group 4 and 5 analogs, for which there is good evidence for π-bonding, may indicate that π-bonding also features in the main group 3 compounds. The further shortening of the In—Cr bond, upon formal replacement of a $[Cr(CO)_5]^{2-}$ group in $[\{(CO)_5Cr\}_2InBr]^{2-}$ by a neutral THF ligand to give $\{(CO)_5Cr\}InBr(THF)$, is also evidence for multiple bonding. In the latter compound, In—Cr π-bonding is localized in just one bond, and one $Cr(CO)_5$ fragment with its competing CO ligands has been lost. However, on going from main group 5 to 4 to 3, the elements become increasingly electropositive, thereby raising the energy of the valence orbitals and increasing the energy mismatch with the *d*-orbitals on the transition metal fragment. On these grounds significant π-overlap becomes less likely along the series, even though the compounds are all formally isoelectronic. Alternatively, $\{(CO)_5Cr\}InBr(THF)$ can be viewed as involving an In^I fragment, InBr(THF), acting as a donor toward a $\{(CO)_5Cr\}$ fragment.

The cumulene analog $[\{(CO)_5Cr\}_2Tl]^-$ is isoelectronic with $[(CO)_4FeTl\text{-}Fe(CO)_4]^-$, which is unknown as a monomer but can be isolated in its dimeric form as $[(CO)_4FeTl\{\mu\text{-}Fe(CO)_4\}_2TlFe(CO)_4]^{2-}$, shown in Table VIII as $[Tl_2Fe_4(CO)_{16}]^{2-}$ [133] (compound (a) in Fig. 32). This compound has a short terminal Tl—Fe bond length of 2.553(5) Å and considerable asymmetry in the Tl_2Fe_2 ring, with one Tl—Fe bond of 2.632(5) Å and the other much longer at 3.038(4) Å, indicating that the interaction between monomers is weak. The terminal $Fe(CO)_4$ groups have the Tl atom in the axial position, which is not as favorable as the equatorial position for π-bonding, indicating that the monomer might not be a candidate for cumulene-type π-bonding. Interestingly, the conformation of the $Cr(CO)_5$ fragments in the chromium inidene complexes is not significant since maximum π-overlap in $Cr(CO)_5L$ complexes is possible for any conformation about the Cr—L bond. The thallium–iron dimer $[Tl_2Fe_4(CO)_{16}]^{2-}$ is isoelectronic with $(CO)_5MnIn\{\mu\text{-}Fe(CO)_4\}_2InMn(CO)_5$,[125] through the replacement of Tl with In and a terminal $[Fe(CO)_4]^-$ group with an isoelectronic $Mn(CO)_5$ moiety. In contrast to the asymmetric Tl_2Fe_2 ring in the former compound, the In_2Fe_2 ring in the latter is completely symmetric.

A series of thallium–iron carbonyl cluster compounds is also provided in Table VIII. The apparently complex structures can be derived from $(CO)_3Fe(\mu\text{-}CO)_3Fe(CO)_3$ (compound (b) in Fig. 32) by successive replacement of bridging μ-CO groups by isoelectronic $\mu\text{-}[TlFe(CO)_4]^-$ groups. In each resulting molecule, one of the Tl—$Fe(CO)_4$ fragments bridges through a four-membered Tl_2Fe_2 ring, analogous to that in $[Tl_2Fe_4$

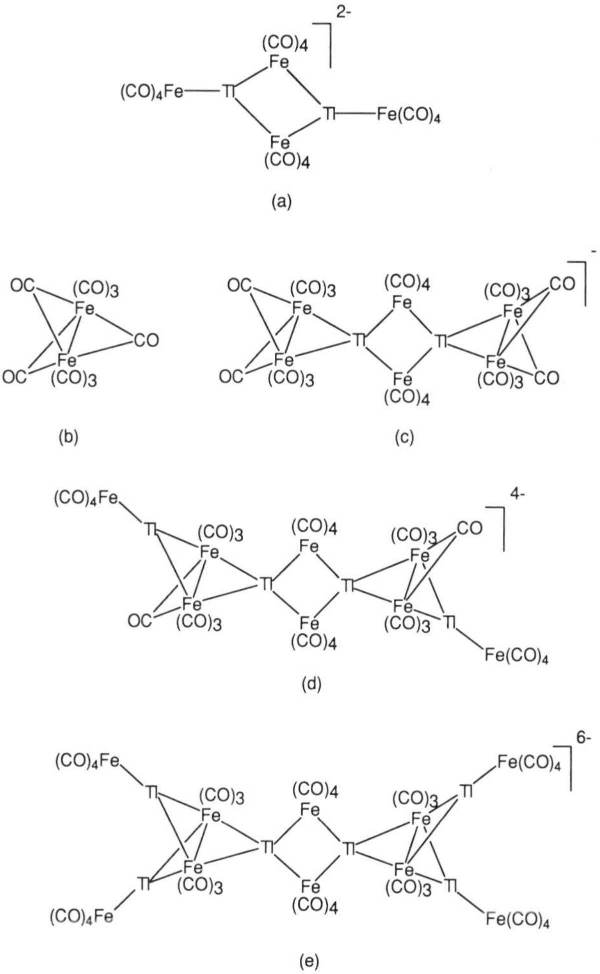

FIG. 32. Schematic illustration of the thallium–iron cluster compounds: (a) $[Tl_2Fe_4(CO)_{16}]^{2-}$, (b) $Fe_2(CO)_9$, (c) $[Tl_2Fe_6(CO)_{24}]^{-}$, (d) $[Tl_4Fe_8(CO)_{30}]^{4-}$, (e) $[Tl_6Fe_{10}(CO)_{36}]^{6-}$.

$(CO)_{16}]^{2-}$, to afford a dimer. The resulting structures (Fig. 32) have the following formulas: $[Tl_2Fe_6(CO)_{24}]^{-}$ [134] (c), $[Tl_4Fe_8(CO)_{30}]^{4-}$ [133] (d), and $[Tl_6Fe_{10}(CO)_{36}]^{6-}$ [135] (e). Whereas the Tl_2Fe_2 ring in the Tl_2Fe_4 dimer is very asymmetric, the corresponding rings in the higher clusters possess much greater symmetry. The last two clusters contain both three- and four-coordinate thallium atoms, but all show a pattern with short Tl—Fe bonds between three-coordinate thallium and the terminal $Fe(CO)_4$ groups (2.55 Å) and μ_3-$Fe(CO)_3$ groups (2.63 Å) and longer bonds to the four-coordinate thallium atoms (>2.7 Å). However, all the terminal $Fe(CO)_4$ groups contain the thallium atom in the less favorable (with respect to π-bonding) axial position. The structural features of the clusters can be rationalized in terms of electron deficiency, multicenter bonding, and possible Tl · · · Tl bonds[133] rather than in terms of π-interactions involving the three-coordinate thallium centers. Representative monomeric four-coordinate thallium–iron complexes $[Tl(LL)\{Fe(CO)_4\}_2]^-$, where LL is a neutral bidentate nitrogen-donor ligand, have Tl—Fe bonds in the range from 2.568(2) to 2.611(4) Å, in the same range as that of the three-coordinate Tl—Fe bonds in the clusters.[150]

Two examples of compounds that are four-coordinate at indium, $\{HB(3,5\text{-}Me_2pz)_3\}InFe(CO)_4$ and $\{HB(3,5\text{-}Me_2Pz)_3\}InW(CO)_5$ (pz = pyrazolyl) are also included in Table VIII.[124] Each contains the tridentate tris(3,5-dimethylpyrazolyl)borate ligand at indium and either an isolobal $Fe(CO)_4$ or $W(CO)_5$ fragment. The $\{(CO)_5Cr\}InBr(THF)$ complex can be considered to contain a "formal" In=Cr double bond by replacement of two Br^- ligands in $InBr_3$ by $Cr(CO)_5^{2-}$. Likewise, $\{HB(3,5\text{-}Me_2pz)_3\}In$-$Fe(CO)_4$ and $\{HB(3,5\text{-}Me_2pz)_3\}InW(CO)_5$ can be considered to contain formal In=Fe and In=W double bonds. The compounds were prepared by the reaction of $\{HB(3,5\text{-}Me_2pz)_3\}InCl_2$ with $Fe(CO)_4^{2-}$ or $W(CO)_5^{2-}$, respectively. Structural evidence for this formalism is certainly evident, as $\{HB(3,5\text{-}ME_2pz)_3\}InFe(CO)_4$ has the shortest known In—Fe bond length of 2.463 Å. There are no other structurally characterized compounds with In—W bonds for comparison, but the In—W bond length of 2.783 Å in $\{HB(3,5\text{-}Me_2pz)_3\}InW(CO)_5$ is shorter than the In—Mo bond length in $\{Cp(CO)_3Mo\}_3In$ (2.878(1) Å). Since the pyrazolylborate ligand is much more electron-rich than the carbonylmetallate ligands found in comparable complexes, bond shortening would be expected if π-bonding were significant. However, as in all other compounds containing an $InFe(CO)_4$ group, the indium atom in $\{HB(3,5\text{-}Me_2pz)_3\}InFe(CO)_4$ occupies the axial position, even though this is less favored for π-bonding. The $Fe(CO)_4$ C—O stretching frequencies are consistent with the view that the $HB(3,5\text{-}Me_2pz)_3In^I$ fragment acts as a donor to Fe^0 in $Fe(CO)_4$.

Finally, there are three examples in Table VIII of electron-rich late transition metal complexes containing dialkylgallium or -indium ligands: $(Cy_2PCH_2CH_2PCy_2)(Me_3CCH_2)PtGa(CH_2CMe_3)_2$,[119] $(Cy_2PCH_2CH_2PCy_2)$-$(Me_3SiCH_2)PtIn(CH_2SiMe_3)_2$,[130] and $Ir(InEt_2)H(Et)(PMe_3)_3$.[129] It is difficult to evaluate the bond lengths in these complexes since each E—M element pair is the only structurally characterized example. The $InEt_2$ group in the iridium complex is trans to a PMe_3 ligand and is staggered with respect to the four equatorial ligands, although in octahedral d^6 ML_6 complexes there is no orientational preference for π-bonding. In the square-planar platinum complexes the plane at gallium is rotated by 57° relative to the plane at platinum, while in the indium complex the torsion angle appears to be even larger. The GaR_2 or InR_2 groups are expected to lie in the plane (torsion angle of 0°) for maximum π-overlap. However, the very bulky $Cy_2PCH_2CH_2PCy_2$ ligands necessary for stabilization of the complex result in a very hindered environment, so it may be that steric effects determine the torsion angle. In each of the complexes the gallium or indium ligand exhibits a trans influence that results in a lengthening of the trans M—P bond by 0.03 to 0.06 Å relative to the cis M—P bond. Overall, for these complexes, it is difficult to estimate from either bond length or conformational data whether π-bonding is significant. The lack of electron-withdrawing groups on either the M or E fragments and the presence of only one M—E bond in each molecule would seem to make the prospect more attractive than in the carbonylmetallate complexes.

To summarize, most of the complexes in Table VIII are of the form $E(ML_n)_3$, where ML_n is a carbonylmetallate fragment. The thallium–iron clusters can be included in this grouping. The strongly electron-withdrawing carbonyl ligands on the transition metal fragment, and the presence of three M—E bonds in each molecule, render any M—E π-bonding negligible in these complexes. This is consistent with the lack of correlation of M—E bond length with either the orientation of the ML_n fragment or the coordination number (3 or 4) at E. Even the alkylgallium iron and tungsten complexes show little evidence for π-bonding. Spectroscopic and chemical evidence for all of these complexes is consistent with largely polar M—E bonding.

The inidene and cumulene analogs $[\{(CO)_5Cr\}_2MX]^{2-}$ (M = In, Tl; X = Cl, Br, I), $[\{(CO)_5Cr\}_2Tl]^-$, and $\{(CO)_5Cr\}InBr(THF)$ display significant bond shortening, as do the four-coordinate trispyrazolylborate complexes $\{HB(3,5\text{-}Me_2pz)_3\}InML_n$ ($ML_n = Fe(CO)_4$, $W(CO)_5$). However, it is much more difficult to separate electrostatic contributions to bond shortening for transition metal–main group 3 derivatives than it is for the group 3–5 and 3–6 complexes discussed earlier. There is likely to be a large energy difference between the frontier orbitals on the main group 3 and transition metal fragments which would limit effective π-overlap. Where the orientation of the transition metal moiety is important for maximum π-overlap, the main group 3 fragment is usually found to be in the less favorable coordination site or conformation. The same is true for the iridium and platinum alkylindium and -gallium complexes, which are relatively electron-rich and contain only one M—E bond. Overall, while transition metal–main group 3 element π-bonding is possible in several complexes, unequivocal evidence for π-bonding has been established in none.

Very little theoretical work has been carried out on any of the transition metal–main group 3 bonded compounds, and this is clearly an area that could provide useful insight into the question of E—M π-bonding. Analogs of many of the structural types that appear in Table VIII are known in which E is a main group 4 or 5 atom, and a systematic study across main groups 3, 4, and 5 would also be useful. For example, Te═Mn multiple bonding is well established in the complex $Te\{CpMn(CO)_2\}_3$,[151] yet there is little evidence that it is important in the isoelectronic and isolobal compound $[In\{Fe(CO)_4\}_3]^{3-}$. Variable-temperature NMR studies, which have been useful in determining barriers to rotation in the main group 3–5 and 3–6 compounds, have not featured strongly in studies of the transition metal derivatives and might also prove useful.

The field of transition metal–main group 3 compounds is now maturing, and reports of new element combinations and unusual new structural types are now appearing regularly. An interesting recent example is the unprecedented indium–manganese compound $[(\mu_5\text{-In})\{Mn(CO)_4\}_5]^{2-}$; this compound contains an indium atom in an ideal pentagonal-planar environment within an $\{Mn(CO)_4\}_5$ ring, which is itself isolobal to C_5H_5.[152] The average In—Mn distance of 2.644(2) Å is not especially short but delocalized bonding of some kind is undoubtedly important in this novel compound. Theoretical studies of the bonding in the transition metal–main group 3 compounds clearly lag behind the developments in the synthetic and structural chemistry.

IV

CONCLUSIONS

The preceding sections have shown that π-bonding involving the heavier main group 3 elements may exist in several classes of compound. The π-interactions are weak, however, and in the compounds studied to date the π-bond energies rarely exceed 15 kcal mol^{-1}. The bond shortenings and absorption spectra suggest that the homonuclear π-bonds in the radical species $[R_2MMR_2]^-$ (M = Al or Ga; R = $CH(SiMe_3)_2$ or Trip) are probably the strongest currently known π-bonds and may have values as high as 20 kcal mol^{-1}. These energy values may be compared with those in disilenes (~30 kcal mol^{-1})[1] and diphosphenes (~25 kcal mol^{-1}).[2] Presumably, the main reason for the relative weakness of the homonuclear π-bond strengths in the heavier main group elements is the larger size of these elements. The electropositive characteristics of the elements probably also play a significant role; and, with the exception of the homonuclear bonded species, the compounds discussed in this review feature the main group 3 element bound to atoms that are often quite electronegative. Thus, both the σ- and π-bonding in the compounds always have polar character, sometimes markedly so, as exemplified by the Al—O bond which has a difference of about 2 units on the electronegativity scale. To generalize, orbital overlap, essential for π-bonding, is often reduced by the large size and electronegativity differences between main group 3 elements and their substituents.

It should be possible, however, to strengthen π-bonding in several ways. One method, already proven in the case of B—P and B—As species, is to increase the charge density on the π-donor atom. Thus, in compounds such as $[R_2BER']^-$ (E = P or As), where there is a formal negative charge on the P or As atom, short B—E bonds and alignment of the coordination planes at boron and phosphorus or arsenic are observed. The key element in this strategy is that sufficient shortening of the bond is induced by electrostatic means so that stronger π-overlap can occur. Dynamic ^1H NMR measurements have shown that rotation barriers as high as 23 kcal mol^{-1} are present in the B—P and B—As bonds.[6,7] It is probable that if similar compounds, e.g., the ions $[R_2MPR']^-$ (M = Al, Ga, or In), were synthesized, they would be expected to have much stronger π-bonds than the, apparently, weak interactions observed in such molecules as the neutral gallium species $R_2GaPR'_2$.[79] Another approach would be to use electron-releasing groups on the π-donor atoms and strongly electron-withdrawing groups on the main group 3 element. Relevant indium derivatives of electron-withdrawing groups such as $2,4,6\text{-}(CF_3)_3C_6H_2$ have already been documented.[47] This substituent, or a related one, may play a key role in facilitating π-electron delocalization in new, heavier, main group 3 derivatives. Electron-donating alkyl groups are already in general use as substituents at the electron-donating center.

The references show that much of the work described in this review has been carried out since the late 1980s. The field of multiple bonding involving the heavier main group 3 elements is thus very young. Significant new developments may be anticipated for this rapidly evolving area.

Acknowledgments

P.J.B. is grateful to the University of Auckland for the award of sabbatical leave in 1993. The authors are grateful to Professor D. L. Reger for providing material in advance of publication. P.P.P. thanks the NSF and PRF for financial support.

Note added in proof. This review was originally submitted in October 1994. Revisions, that include important developments up to August 31, 1995, were added at the proof stage. Only brief discussions of these could be given for technical reasons.

References

(1) (a) West, R. *Agnew. Chem.*, Int. Ed. Engl. **1987,** *26*, 1201. (b) Raabe, G.; Michl, J. *Chem. Rev.* **1985,** *85*, 419. (c) Brook, A. G. In *Heteroatom Chemistry*; (Block, E., Ed.; VCH: Weinheim, 1990; Chapter 6, p. 105. (d) Barrau, J.; Escudié, J.; Satgé, J. *Chem. Rev.* **1990,** *90*, 1.

(2) (a) *Multiple Bonds and Low Coordination in Phosphorus Chemistry*; Regitz, M.; Scherer, O.J., Eds.; Thieme Verlag: Stuttgart, 1990. (b) Cowley, A. H.; Norman, N. C. *Prog. Inorg. Chem.* **1986,** *34*, 1. (c) Weber, L. *Chem. Rev.* **1992,** *92*, 1839.
(3) *Comprehensive Organometallic Chemistry*; Wilkinson, G.; Stone, F. G. A.; Abel, E. W., Eds.; Pergamon: Oxford, 1982; Vol. 1: (a) Morris, J. H. Chapter 5.2, p. 311. (b) Herberich, G. E. Chapter 5.3, p. 381. (c) Robinson, W. T.; Grimes, R. N. *Inorg. Chem.* **1975,** *14*, 3056. (d) Herberich, G. E.; Hessner, B.; Hostalek, M. *Angew. Chem.*, Int. Ed. Engl. **1986,** *25*, 642. (e) Herberich, G. E.; Hessner, B.; Hostalek, M. *J. Organomet, Chem.* **1988,** *355*, 473. (f) Davis, J. H.; Sinn, E.; Grimes, R. M. *J. Am. Chem. Soc.* **1989,** *111*, 4784. (g) Dirschl, F.; Hanecker, E.; Nöth, H.; Rattay, W.; Wagner, W. *Z. Naturforsch.*, B: Anorg. Chem., Org. Chem. **1986,** *41B*, 32. (h) van Bonn, K.-M.; Schreyer, P.; Paetzold, P.; Boese, R. *Chem. Ber.* **1988,** *121*, 1045. (i) Meyer, H.; Schmidt-Lukasch, G.; Baum, G.; Massa, W., Berndt, A. *Z. Naturforsch.*, B: Chem. Sci. **1988,** *43B*, 801.
(4) (a) Niedenzu, K.; Dawson, J. W. *Boron-Nitrogen Compounds*; Springer-Verlag: Berlin, 1964. (b) Lappert, M. F.; Power, P. P.; Sanger, A. R.; Srivastava, R. C. *Metal and Metalloid Amides*; Ellis Horwood: Chichester, England, 1979. (c) Barfield, P. A.; Lappert, M. F.; Lee, J. *Proc. Chem. Soc. London* **1961,** *421*. *J. Chem. Soc. A* **1968,** 554. (d) Ryschekewitsch, G. E.; Brey, W.S.; Saji, A. *J. Am. Chem. Soc.* **1961,** *83*, 1010. (e) Neilson, R.; Wells, R. L. *Inorg. Chem.* **1977,** *16*, 7. (f) Kölle, P.; Nöth, H. *Chem. Rev.* **1985,** *85*, 399. (g) Paetzold, P. *Adv. Inorg. Chem.* **1987,** *31*, 123.
(5) (a) Wilson, J. W.; *J. Organomet. Chem.* **1980,** *185*, 297. (b) Glaser, B.; Nöth, H. *Angew. Chem.*, Int. Ed. Engl. **1985,** *24*, 416. (c) Olmstead, M. M., Power, P. P., Weese, K. J., Doedens, R. J. *J. Am. Chem. Soc.* **1987,** *109*, 2541. (d) Berndt, A. *Angew, Chem.*, Int. Ed. Engl. **1993,** *32*, 985.
(6) (a) Power, P. P. *Angew. Chem.*, Int. Ed. Engl. **1990,** *29*, 449. (b) Paine, R. T.; Nöth, H. *Chem. Rev.* **1995,** *95*, 343.
(7) Petrie, M. A.; Olmstead, M. M.; Hope, H.; Bartlett, R. A.; Power, P.P. *J. Am. Chem. Soc.* **1993,** *115*, 3221.
(8) Wells, A. F. *Structural Inorganic Chemistry*, 5th ed.; Clarendon: Oxford, 1984; p. 1052.
(9) (a) Moezzi, A.; Olmstead, M. M.; Power, P. P. *J. Am. Chem. Soc.* **1992,** *114*, 2715. (b) Moezzi, A.; Bartlett, R. A.; Power, P. P. *Angew. Chem.*, Int. Ed. Engl. **1991,** *31*, 108.
(10) Covalent radii and electronegativity values (Allred-Rochow) are abstracted from Huheey, J. E.; Keiter, E. A.; Keiter R. L. *Inorganic Chemistry*, 4th ed.; Harper & Row: New York, 1993; pp. 187 and 292.
(11) Mole, T.; Jeffery, E. A. *Organoaluminum Compounds*; Elsevier: New York, 1972.
(12) Eisch, J. J. *Organometallic Synthesis*, Academic Press: New York, 1981; Vol. 2.
(13) *Comprehensive Organometallic Chemistry;* Wilkinson, G.; Stone, F.G.A.; Abel, E. W., Eds.; Pergamon: Oxford, 1982; Vol. 1: (a) Eisch, J.J. Chapter 6, p. 555. (b) Tuck, D. G.; Chapter 7, p. 683. (c) Kurosawa, H. Chapter 8, p. 725.
(14) (a) *Chemistry of Aluminum, Gallium, Indium and Thallium;* Downs, A. J., Ed.; Blackie-Chapman-Hall: London, 1993. (b) *Coordination Chemistry of Aluminum*; Robinson, G. H., Ed.; VCH: Weinheim, 1993.
(15) Al-Hashimi, S.; Smith, J. D. *J. Organomet. Chem.* **1978,** *153*, 253.
(16) Uhl, W. *Angew. Chem.*, Int. Ed. Engl. **1993,** *32*, 1386.
(17) (a) Dohmeier, C.; Mocker, M.; Schnöckel, H.; Lotz, A.; Schneider U.; Ahlrichs, R. *Angew. Chem.*, Int. Ed. Engl. **1993,** *32*, 1428, (b) Loos, D.; Schnöckel, H. *J. Organomet. Chem.* **1993,** *463*, 37.
(18) Uhl, W. *Z. Naturforsch.*, B; Chem. Sci. **1988,** *43B*, 1113.

(19) Uhl, W.; Layh, M.; Hildenbrand, T. *J. Organomet Chem.* **1989,** *364*, 289.
(20) Uhl, W.; Layh, M.; Hiller, W. *J. Organomet. Chem.* **1989,** *368*, 139.
(21) Uhl, W.; Wagner, J.; Fenske, D.; Baum, G. *Z. Anorg. Allg. Chem.* **1992,** *612*, 25.
(22) Cleaver, W. M.; Barron, A. R. *Chemtronics* **1989,** *4*, 146.
(23) Petrie, M. A.; Power, P. P.; Dias, H. V. R.; Ruhlandt-Senge, K.; Waggoner, K. M.; Wehmschulte, R. J. *Organometallics* **1993,** *12*, 1086.
(24) Schultz, S.; Pusch, S.; Pohl, E.; Dielkus, S.; Herbst-Irmer, R.; Roesky, H. W. *Inorg. Chem.* **1993,** *32*, 3343.
(25) Meller, A.; Pusch, S.; Pohl, E.; Häming, L.; Herbst-Irmer, R. *Chem. Ber.* **1993,** *126*, 2255.
(26) Rahbarnoohi, H.; Heeg, M. J.; Oliver, J. P. *Organometallics* **1994,** *13*, 2123.
(27) Power, P. P. *Inorg. Chim. Acta* **1992,** *198*, 443.
(28) Luke, B. T.; Pople, J. A.; Krogh-Jespersen, M.-B.; Apeloig, Y.; Chandrasekhar, J.; Schleyer, P. von R. *J. Am. Chem. Soc.* **1986,** *108*, 260.
(29) Healy, M. D.; Power, M. B.; Barron, A. R. *Coord. Chem. Rev.* **1994,** *130*, 63.
(30) Barron, A. R.; Dobbs, K. D.; Francl, M. M. *J. Am. Chem. Soc.* **1991,** *113*, 39.
(31) (a) Goldberg, D. E.; Harris, D. H.; Lappert, M. F. *J. Chem. Soc.*, Chem. Commun. **1976,** 261. (b) Davidson, P. J.; Harris, D. H.; Lappert, M. F. *J. Chem. Soc.*, Dalton Trans. **1976,** 2268.
(32) Hitchcock, P. B.; Lappert, M. F.; Miles, S. J.; Thorne, A. J. *J. Chem. Soc.*, Chem. Commun. **1984,** 480.
(33) Pauling, L. *The Nature of the Chemical Bond*, 3rd ed.; Cornell University Press: Ithaca, NY, 1960.
(34) Schomaker, V.; Stevenson, D. P. *J. Am. Chem. Soc.* **1941,** *63*, 37.
(35) Blom, R.; Haaland, A. *J. Mol. Struct.* **1985,** *129*, 1.
(36) Wells, A. F. "*Structural Inorganic Chemistry*, 5th ed.; Clarendon: Oxford, 1984; p. 1278.
(37) Wehmschulte, R. J.; Ruhlandt-Senge, K.; Olmstead, M. M.; Hope, H.; Sturgeon, B. E.; Power, P. P. *Inorg. Chem.* **1993,** *32*, 2983.
(38) Biffar, W.; Nöth, H.; Pommerening, H. *Angew. Chem.*, Int. Ed. Engl. **1980,** *19*, 56.
(39) Schlüter, K.; Berndt, A. *Angew. Chem.*, Int. Ed. Engl. **1980,** *19*, 57.
(40) Moezzi, A.; Olmstead, M. M.; Bartlett, R. A.; Power, P. P. *Organometallics* **1992,** *11*, 2383.
(41) (a) Klusik, H.; Berndt. A. *Angew, Chem.*, Int. Ed. Engl. **1981,** *20*, 870. (b) Klusik, H.; Berndt, A. *J. Organomet. Chem.* **1981,** *22*, C25.
(42) Beamish, J. C.; Small R. W. H.; Worrall, I. J. *Inorg. Chem.* **1979,** *18*, 220.
(43) Tuck, D. G. *Chem. Soc. Rev.* **1993,** 269.
(44) Henkel, S.; Klinkhammer, K. W.; Schwarz, W. *Angew. Chem.*, Int. Ed. Engl. **1994,** *33*, 681.
(45) He, X.; Bartlett, R. A.; Olmstead, M. M.; Ruhlandt-Senge, K.; Sturgeon, B. E.; Power, P. P. *Angew. Chem.*, Int. Ed. Engl. **1993,** *32*, 717.
(46) Brothers, P. J.; Olmstead, M. M.; Power, P. P. unpublished results, 1993.
(47) Schluter, R. D.; Cowley, A. H.; Atwood, D. A.; Jones, R. A.; Bond, M. R.; Carrano, C. J. *J. Am. Chem. Soc.* **1993,** *115*, 2070.
(48) (a) Veith, M.; Goffing, F.; Becker S.; Huch, V. *J. Organomet. Chem.* **1991,** *406*, 105. (b) Linti, G.; Frey, R.; Schmidt, M. *Z. Naturforsch.*, B: Chem. Sci. **1994,** *49B*, 958. (c) Saxena, A. K., Zhang, M.; Maguire, J. A.; Hosmane, N. S.; Cowley, A. H. *Angew. Chem.* Int. Ed. Engl. **1995,** *34*, 332. (d) Brown, D. S.; Decken, A.; Cowley, A. H. *J. Am. Chem. Soc.* **1995,** *117*, 5421. (e) Li, X-W.; Pennington, W. T.; Robinson, G. H. *J. Am. Chem. Soc.* **1995,** *117*, 7578.
(49) (a) Dohmeier, C.; Robl, C.; Tacke M.; Schnöckel, H. *Angew. Chem.*, Int. Ed. Engl.

1991, *30*, 564. (b) Hiller, W.; Klinkhammer, K.-W.; Uhl, W.; Wagner, J. *Angew. Chem.*, Int. Ed. Engl. **1991,** *30*, 179. (c) Uhl, W.; Hiller, W.; Layh, M.; Schwarz, W. *Angew. Chem.*, Int. Ed. Engl. **1992,** *31*, 1364. (d) Dohmeier, C.; Mocker, M.; Schnöckel, H.; Lötz, A.; Schneider, U.; Ahlrichs, R. *Angew. Chem.*, Int. Ed. Engl. **1993,** *32*, 1428. (e) Beachley, O. T.; Churchill, M. R.; Fettinger, J. C.; Pazik, J. C.; Victoriano, L. *J. Am. Chem. Soc.* **1986,** *108*, 4666. (f) Schluter, R. D.; Cowley, A. H.; Atwood, D. A.; Jones, R. A.; Atwood, J. L. *J. Coord. Chem.* **1993,** *30*, 25.

(50) Pluta, C.; Pörschke K.-R.; Kruger, C.; Hildenbrand, K. *Angew. Chem.*, Int. Ed. Engl. **1993,** *32*, 388.

(51) Uhl, W.; Vester, A.; Kaim, W.; Poppe, J. *J. Organomet. Chem.* **1993,** *454, 9*.

(52) Boesel, R.; Bläser, D.; Niederprum, N.; Nüsse, M.; Brett, W. A.; Schleyer, P. von R.; Bühl, M.; van E. Hommes, N. J. *Angew. Chem.*, Int. Ed. Engl. **1992,** *31*, 314.

(53) (a) Cowley, A. H.; Gabbai, F. P.; Carrano, C. J.; Mokry, L. M.; Bond, M. R.; Bertrand, G. *Angew. Chem.*, Int. Ed. Engl. **1994,** *33*, 578. (b) Cowley, A. M.; Gabbai, F. P.; Decken, A. *Angew. Chem.* Int. Ed. Engl. **1994,** *33*, 1370. (c) Decken, A.; Gabbai, F. P.; Cowley, A. M. *Inorg. Chem.* **1995,** *34*, 3853. (d) Ashe, A. J.; Al-Ahmad, S.; Kampf, J. W. *Angew. Chem.* Int. Ed. Engl. **1995,** *34*, 1357.

(54) (a) Schauer, S. J.; Robinson, G. H. *J. Coord. Chem.* **1993,** *30*, 197 (b) Robinson, G. H., Ed. *Coordination Chemistry of Aluminum;* VCH: Weinheim, 1993; Chapter 2, p. 57. (c) Taylor, M. J.; Brothers, P. J. In *Chemistry of Aluminum, Gallium, Indium and Thallium; Downs, A. J.*, Ed.; Blackie-Chapman-Hall: London, 1993; Chapter 3, pp. 169–211. (d) Cowley, A. H.; Jones, R. A. *Angew. Chem.*, Int. Ed. Engl. **1989,** *28*, 1208. (e) Wells, R. L. *Coord. Chem. Rev.* **1992,** *112*, 273.

(55) (a) Oliver, J. P.; Kumar, R. *Polyhedron* **1990,** *9*, 409. (b) Oliver, J. P.; Kumar, R.; Taghiof, M., In *Coordination Chemistry of Aluminum*; Robinson, G. H., Ed.; VCH: Weinheim, 1993; Chapter 5, p. 167. (c) Pasynkiewicz, S. *Polyhedron* **1990,** *9*, 429.

(56) (a) Lehn, J. M.; Munsch, B. *Mol. Phys.* **1972,** *23*, 91. (b) Spiro, V.; Civis, S.; Ebert, M.; Daniels, V. *J. Mol. Spectrosc.* **1986,** *119*, 426.

(57) Lambert, J. B. *Top. Stereochem.* **1971,** *6*, 19.

(58) Ashby, M. T.; Shestawy, N. A. *Organometallics* **1994,** *13*, 236.

(59) (a) synthesis: Bürger, H.; Cichon, J.; Goetze U.; Wannagat U., Wismar, H. J. *J. Organomet. Chem.* **1971,** *33*, 1. (b) structure: Sheldrick, G. M.; Sheldrick, W. S. *J. Chem. Soc. A.* **1969**, 2279.

(60) (a) synthesis: Ruff, J. K. *J. Am. Chem. Soc.* **1961,** *83*, 2835. (b) structure: Brothers, P. J.; Wehmschulte, R. J.; Olmstead, M. M.; Ruhlandt-Senge, K.; Parkin, S. R.; Power, P. P. *Organometallics* **1994,** *13*, 2792. (c) Wiberg, N.; Baumeister, W.; Zahn, P. *J. Organomet. Chem.* **1972,** *36*, 267. (d) Hitchcock, P. B.; Jasim, J. A.; Lappert, M. F.; Williams, H. D. *Polyhedron* **1990,** *9*, 245. (e) Atwood, D. A.; Atwood, V. O.; Cowley, A. H.; Jones, R. A.; Atwood, J. L.; Bott, S. *Inorg. Chem.* **1994,** *33*, 3251.

(61) (a) Allman, R.; Henke, W.; Krommes; P.; Lorberth, J. *J. Organomet. Chem.* **1978,** *162*, 183. (b) Walzsada, S. D.; Belgardt, T.; Noltemeyer, M.; Roesky, H. W. *Angew. Chem.* Int. Ed. Engl. **1994,** *33*, 1351.

(62) The compounds $M\{N(SiMe_3)_2\}_3$ (M = Ga, In) were stated in a review to be isomorphous to $Fe\{N(SiMe_3)_2\}_3$, with Ga-N and In-N bond lengths of 1.857(8) and 2.057(12) Å: Eller, P. G.; Bradley, D.C.; Hursthouse, M. B.; Meek, D. *Coord. Chem. Rev.* **1977,** *24*, 1.

(63) Petrie, M. A.; Ruhlandt-Senge, K.; Hope, H.; Power, P. P. *Bull. Soc. Chim. Fr.* **1993,** *130*, 851.

(64) Petrie, M. A.; Ruhlandt-Senge, K.; Power, P. P. *Inorg. Chem.* **1993,** *32*, 1135.

(65) Petrie, M. A.; Ruhlandt-Senge, K.; Wehmschulte, R. J.; He, X.; Olmstead, M. M.; Power, P. P. *Inorg. Chem.* **1993,** *32*, 2557.

(66) (a) Waggoner, K. M.; Hope, H.; Power, P. P. *Angew. Chem.*, Int. Ed. Engl. **1988,** *27*, 1699. (b) Schultz, S.; Häming, L.; Herbst-Irmer, R.; Roesky, H. W.; Sheldrick, G. *Angew. Chem.*, Int. Ed. Engl. **1994,** *33*, 969. (c) Wehmschulte, R. J.; Power, P. P., unpublished results.
(67) (a) Linti, G.; Frey, R.; Polborn, K. *Chem. Ber.* **1994,** *127*, 1387. (b) Frey, R.; Linti, G.; Polborn, K. *Chem. Ber.* **1994,** *127*, 101. (c) Cowley, A. H.; Decken, A.; Olazabal, C.; Norman, N. C. *Inorg. Chem.* **1994,** *33*, 3435.
(68) Linti, G. *J. Organomet. Chem.* **1994,** *465*, 79.
(69) Porchia, M.; Benetello, F.; Brianese, N.; Rosetto, G.; Zanella, P.; Bombieri, P. *J. Organomet. Chem.* **1992,** *424,* 1.
(70) Tödtmann, J.; Schwarz, W.; Weidlein, J.; Haaland, A. *Z. Naturforsch., B: Chem. Sci.* **1993,** *48B*, 1437.
(71) Nöth, H.; Staudigl, R.; Storch, W. *Chem. Ber.* **1981,** *114*, 3204.
(72). Haaland, A. In *Coordination Chemistry of Aluminum*; Robinson, G. H., Ed.; VCH: Weinheim, 1993; Chapter 1.
(73) (a) Waggoner, K.; Power, P. P. *J. Am. Chem. Soc.* **1991,** *113*, 3385. (b) Fink, W. H.; Richards, J. C. *J. Am. Chem. Soc.* **1991,** *113*, 3393.
(74) (a) Finocchiaro, P.; Gust, D.; Mislow, K. *J. Am. Chem. Soc.* **1973,** *95*, 7029. (b) Blount, J. F.; Finocchiaro, P.; Gust, D.; Mislow, K. *J. Am. Chem. Soc.* **1973,** *95*, 7019.
(75) (a) Wehmschulte, R. J.; Ruhlandt-Senge, K.; Power, P. P. *Inorg. Chem.* **1994,** *33*, 3205. (b) Wehmschulte, R. J.; Power, P. P., unpublished results.
(76) Beachley, O. T.; Tessier-Youngs, C. *Organometallics* **1983,** *2*, 796.
(77) Somer, M.; Thiery, D.; Peters, K.; Walz, L.; Hartweg, M.; Popp, T.; von Schnering, H. G. *Z. Naturforsch., B: Chem. Sci.* **1991,** *46B*, 789.
(78) Petrie, M. A.; Power, P. P. *Inorg. Chem.* **1993,** *32*, 1309.
(79) Petrie, M. A.; Ruhlandt-Senge, K.; Power, P. P. *Inorg. Chem.* **1992,** *31*, 4038.
(80) Petrie, M. A.; Power, P. P. *J. Chem. Soc.*, Dalton, Trans. **1993,** 1737.
(81) Ellison, J. J.; Ruhlandt-Senge, K.; Power, P. P., unpublished results, 1991.
(82) Atwood, D. A.; Cowley, A. H.; Jones, R. A.; Mardones, M. A. *J. Am. Chem. Soc.* **1991,** *113*, 7050.
(83) Hope, H.; Pestana, D. C.; Power, P. P. *Angew. Chem.*, Int. Ed. Engl. **1991,** *30*, 691.
(84) Waggoner, K. M.; Parkin, S.; Pestana, D. C.; Hope, H.; Power, P. P. *J. Am. Chem. Soc.* **1991,** *113*, 3597.
(85) Arif, A. M.; Benac, B. L.; Cowley, A. H.; Geerts, R.; Jones, R. A.; Kidd, K. B.; Power, J. M.; Schwab, S. T. *J. Chem. Soc.*, Chem. Commun. **1986,** 1543.
(86) Alcock, N. W.; Degnan, I. A.; Wallbridge, M. G. H.; Powell, H. R.; McPartlin, M. *J. Organomet. Chem.* **1989,** *361,* C33.
(87) Byrne, E. K.; Parkany, L.; Theopold, K. H. *Science* **1988,** *241*, 332.
(88) Higa, K. T.; George, C. *Organometallics* **1990,** *9*, 275.
(89) (a) Pitt, C. G.; Higa, K. T.; McPhail, A. T.; Wells, R. L. *Inorg. Chem.* **1986,** *25*, 2483. (b) Wells, R. L.; Self, M. F.; Baldwin, R. A.; White, P. S. *J. Coord. Chem.* **1994,** *33*, 279.
(90) Arif, A. M.; Benac, B. L.; Cowley, A. H.; Jones, R. A.; Kidd, K. B.; Nunn, C. M. *New J. Chem.* **1988,** *12*, 553.
(91) Daly, J. J.; Zuerick, S. A. *Z. Kristallogr. Kristallgeom., Kristallphys., Kristallchem.* **1963,** *118*, 322.
(92) Blount, J. F.; Maryanoff, C. A.; Mislow, K. *Tetrahedron Lett.* **1975,** *11*, 913.
(93) Bartlett, R. A.; Dias, H. V. R.; Power, P. P. *Inorg. Chem.* **1988,** *27*, 3319.
(94) Wells, R. L.; Shafieezad, S.; McPhail, A. T.; Pitt, C. G. *J. Chem. Soc.*, Chem. Commun **1987,** 1823.
(95) West, R. *Angew. Chem.*, Int. Ed. Engl. **1987,** *26*, 1201.

(96) (a) Petrie, M. A.; Olmstead, M. M.; Power, P. P. *J. Am. Chem. Soc.* **1991,** *113*, 8704. (b) Cleaver, W. M.; Barron, A. R. *Organometallics* **1993,** *12*, 1001.
(97) Skowronska-Ptasinska, M.; Starowieyski, K. B.; Pasynkiewicz, S.; Carewska, M. J. *J. Organomet. Chem.* **1978,** *160*, 403.
(98) Shreve, A. P.; Mulhaupt, R.; Fultz, W.; Calabrese, J.; Robbins, W.; Ittel, S. D. *Organometallics* **1988,** *7*, 409.
(99) Benn, R.; Janssen, E.; Lehmkuhl, H.; Rufinska, A.; Angermund, K.; Betz, P.; Goddard, R.; Krüger, C. *J. Organomet. Chem.* **1991,** *411*, 37.
(100) (a) Healy, M. D.; Barron, A. R. *Angew. Chem.*, Int. Ed. Engl. **1992,** *31*, 921. (b) Uhl, W.; Koch, M.; Pohl, S.; Saak, W.; Hiller, W.; Heckel, M. *Z. Naturforsch* **1995,** *50b*, 635 (c) Uhl, W.; Koch, M.; Hiller, W.; Heckel, M. *Angew. Chem.*, Int. Ed. Engl. **1995,** *34*, 989.
(101) Brothers, P. J.; Ruhlandt-Senge, K.; Power, P. P., unpublished results, 1993.
(102) Healy, M. D.; Wierda, D. A.; Barron, A. R. *Organometallics* **1988,** *7*, 2543.
(103) (a) Uhl, W.; Vester. A.; W. Hiller, *J. Organomet. Chem.* **1993,** *443*, 9. (b) Uhl, W.; Schütz, U. *Z. Naturforsch. B: Chem. Sci.* **1994,** *49B*, 931.
(104) Uhl, W.; Schütz, U.; Hiller, W.; Heckel, M. *Organometallics* **1995,** *14*, 1073.
(105) Ruhlandt-Senge, K.; Power, P. P. *Inorg. Chem.* **1991,** *30*, 2633.
(106) Wehmschulte, R. J.; Ruhlandt-Senge, K.; Power, P. P. *Inorg. Chem.* **1995,** *34*, 2593.
(107) Ruhlandt-Senge, K.; Power, P. P. *Inorg. Chem.* **1991,** *30*, 3863.
(108) Uhl, W.; Layh, M.; Becker, G.; Klinkhammer, K. W.; Hildenbrand, T. *Chem. Ber.* **1992,** *125*, 1547.
(109) Ruhlandt-Senge, K.; Power, P. P. *Inorg. Chem.* **1993,** *32*, 3478.
(110) (a) Burlitch, J. M.; Leonowicz, M. E.; Petersen, R. B.; Hughes, R. E. *Inorg. Chem.* **1979,** *18*, 1097. (b) Gilbert, K. B.; Bocock, S. K.; Shore, S. G. In *Comprehensive Organometallic Chemistry* Wilkinson, G.; Stone, F. G. A.; Abel, E. W., Eds.; Oxford: Oxford University Press, 1982; Vol. 6, p. 880. (c) Hartwig, J. F.; Huber, S. *J. Am. Chem. Soc.* **1993,** *115*, 4908.
(111) (a) Schneider, J. J.; Krüger, C.; Nolte, M.; Abraham, I.; Ertel, T. S.; Bertagnolli, H. *Angew. Chem.* Int. Ed. Engl. **1994,** *33*, 2435. (b) Dohmeier, C.; Krautscheid, H.; Schnöckel, H. *Angew. Chem.* Int. Ed. Engl. **1994,** *33*, 2482.
(112) Fryzuk, M. D.; McManus, N. T.; Rettig, S. J.; White, G. S. *Angew. Chem.*, Int. Ed. Engl. **1990,** *29*, 73.
(113) He, X.; Bartlett, R. A.; Power, P. P. *Organometallics* **1994,** *13*, 548.
(114) Campbell, R. M.; Clarkson, L. M.; Clegg, W.; Hockless, D. C. R.; Pickett, N. L.; Norman, N. C. *Chem. Ber.* **1992,** *125*, 55.
(115) St. Denis, J. N.; Butler, W.; Glick, M. D.; Oliver, J. P. *J. Organomet. Chem.* **1977,** *129*, 1.
(116) Conway, A. J.; Hitchcock, P B.; Smith, J. D. *J. Chem. Soc., Dalton Trans.* **1975,** 1944.
(117) Preut, H.; Haupt, H.-J. *Chem. Ber.* **1974,** *107*, 2860.
(118) (a) Haupt, H.-J.; Flörke, U.; Preut, H. *Acta Crystallogr.*, Sect. C: Cryst. Struct. Commun. **1986,** *C42*, 665 (b) Flörke, U.; Balsaa, P.; Haupt, H.-J. *Acta Crystallogr.*, **1986,** *C42*, 275. (c) Haupt, H.-J.; Balsaa, P.; Flörke, U. *Z. Anorg. Allg. Chem.* **1988,** *557*, 69.
(119) Fischer, R. A.; Kaesz, H. D.; Khan, S. I.; Müller, H.-J. *Inorg. Chem.* **1990,** *29*, 1601.
(120) Behrens, H.; Moll, M.; Sixtus, E.; Thiele, G. *Z. Naturforsch.*, B: Anorg. Chem., Org. Chem. **1977,** *32B*, 1109.
(121) Curnow, O. J.; Schiemenz, B.; Huttner, G.; Zsolnai, L. *J. Organomet. Chem.* **1993,** *459*, 17.

(122) Clarkson, L. M.; Clegg, W.; Hockless, D. C. R.; Norman, N. C.; Marder, T. B. *J. Chem. Soc.*, Dalton Trans. **1991,** 2229.
(123) Clarkson, L. M.; Clegg, W.; Norman, N. C.; Tucker, A. J.; Webster, P. M. *Inorg. Chem.* **1988,** *27*, 2653.
(124) Reger, D. L.; Mason, S. S.; Rheingold, A. L.; Haggerty, B. S.; Arnold, F. P., *Organometallics,* **1994,** *13*, 5049.
(125) Preut, H.; Haupt, H.-J. *Acta Crystallogr.*, Sect. B: Struct. Crystallogr. Cryst. Chem. **1979,** *B35*, 2191.
(126) Albano, V. G.; Cané, M.; Iapalucci, M. C.; Longoni, G.; Monari, M. *J. Organomet. Chem.* **1991,** *407*, C9.
(127) Calabrese, J. C.; Clarkson, L. M.; Marder, T. B.; Norman, N. C.; Taylor, N. J. *J. Chem. Soc. Dalton Trans.* **1992,** 3525.
(128) Robinson, W. R.; Schussler, D. P. *Inorg. Chem.* **1973,** *12*, 848.,
(129) Thorn, D. L.; Harlow, R. L. *J. Am. Chem. Soc.* **1989,** *111*, 2575.
(130) Fischer, R. A.; Behm, J. *J. Organomet. Chem.* **1991,** 413, C10.
(131) Schiemenz, B.; Huttner, G. *Angew. Chem.*, Int. Ed. Engl. **1993,** *32*, 1772.
(132) Rajaram, J.; Ibers, J. A. *Inorg. Chem.* **1973,** *12*, 1313.
(133) Whitmire, K. H.; Cassidy, J. M.; Rheingold, A. L.; Ryan, R. R. *Inorg. Chem.* **1988,** *27*, 1347.
(134) Cassidy, J. M.; Whitmire, K. H. *Inorg. Chem.* **1989,** *28*, 1432.
(135) Whitmire, K. H.; Ryan, R. R.; Wasserman, H. J.; Albright, T. A.; Kang, S.-K. *J. Am. Chem. Soc.* **1986,** *108*, 6831.
(136) Taylor, M. J. *Metal-to-Metal Bonded States of the Main Group Elements;* Academic Press: London, 1975.
(137) Compton, N. A.; Errington, R. J.; Norman, N. C. *Adv. Organomet. Chem.* **1990,** *31*, 91.
(138) Whitmire, K. H. *J. Coord. Chem.* **1988,** *17*, 95.
(139) Clarkson, L. M.; Clegg, W.; Hockless, D. C. R.; Norman, N. C.; Farrugia, L. J.; Bott, S. G.; Atwood J. L. *J. Chem. Soc.*, Dalton Trans. **1991,** 2241.
(140) Green, M. L. H.; Mountford, P.; Smout, G. J.; Speel, S. R. *Polyhedron* **1990,** *9*, 2763.
(141) Schilling, B. E. R.; Hoffmann, R.; Lichtenberger, R. *J. Am. Chem. Soc.* **1979,** *101*, 585.
(142) Kubacek, P.; Hoffmann, R.; Havlas, Z. *Organometallics,* **1982,** *1*, 180.
(143) Clarkson, L. M.; Norman, N. C.; Farrugia, L. J. *Organometallics,* **1991,** *10*, 1286.
(144) Haupt, H.-J.; Wolfes, W.; Preut, H. *Inorg. Chem.* **1976,** *15*, 2920.
(145) Haupt, H.-J.; Preut, H.; Wolfes, W. *Z. Anorg. Allg. Chem.* **1979,** *448*, 93.
(146) Cassidy, J. M.; Whitmire, K. H. *Acta Crystallogr.*, Sect. C: Cryst. Struct. Commun. **1990,** *C46*, 1781.
(147) Clarkson, L.M.; McKrudden, K.; Norman, N. C.; Farrugia, L. J. *Polyhedron* **1990,** *9*, 2533.
(148) Clarkson, L. M.; Farrugia, L. J.; Norman, N. C. *Acta Crystallogr.*, Sect. C: Cryst. Struct. Commun. **1991,** *C47*, 2525.
(149) (a) Huttner, G. *Pure Appl. Chem.* **1986,** *58*, 585. (b) Huttner, G.; Evertz, K. *Acc. Chem. Res.* **1986,** *28*, 1496.
(150) Cassidy, J. M.; Whitmire, K. H. *Inorg. Chem.* **1989,** *28*, 1435.
(151) Herberhold, M.; Reiner, D.; Neugebauer, D. *Angew. Chem.*, Int. Ed. Engl. **1983,** *22*, 59.
(152) Schollenberger, M.; Nuber, B.; Zeigler, M. L. *Angew. Chem.*, Int. Ed. Engl. **1993,** *31*, 350.

The Chemistry of Silenes

ADRIAN G. BROOK

Lash Miller Chemical Laboratories
University of Toronto
Toronto M5S 1A1, Canada

MICHAEL A. BROOK

Department of Chemistry
McMaster University
Hamilton, L8S 4M8, Canada

I

HISTORICAL BACKGROUND

Alkenes have long been recognized as among the most important functional-group classes of organic chemistry; and with the development of organosilicon chemistry—particularly the pioneering work of Frederick Stanley Kipping whose efforts through the period 1898–1939 laid the foundation for the field—it was natural that chemists would seek the organosilicon analogs of these species, namely the silenes $R_2Si{=}CR'_2$ and the disilenes, $R_2Si{=}SiR'_2$. The search was long and unsuccessful, culminating with the conclusion that the silicon–carbon double bond was too weak to allow the synthesis of a silene.

This conclusion was shattered in 1967 when Gusel'nikov and Flowers[1,2] announced results clearly suggesting that a transient silene $Me_2Si{=}CH_2$ was formed during the thermolysis of 1,1-dimethyl-1-silacyclobutane at about 450°C. This result inspired renewed efforts to investigate these species, and the literature between 1967 and about 1980 contains much evidence for the formation of transient silenes and descriptions of their chemical behavior. In 1981 the preparation of the first solid, relatively stable silene $(Me_3Si)_2Si{=}C(OSiMe_3)Ad$ was reported by Brook *et al.*[3–5] Coincidentally, the preparation of the first solid, relatively stable disilene was announced by West at the same meeting.[3] Since that time much effort has been devoted to establishing the chemistry of silenes, and several other relatively stable solid compounds have been prepared and characterized, including crystal structures. In addition, literally hundreds of other less stable silenes have been prepared by a number of very different routes, and much is now known about the physical and chemical properties of the silicon–carbon double bond. This review describes the highlights, particularly of work done in the last ten years.

Considerable literature is currently available about these species. The review by Raabe and Michl in 1985 on "Multiple Bonding to Silicon"[6] is certainly the most detailed and complete reference to the topic to that date. Subsequently a number of other reviews that provide additional details of more limited areas of the field have been written by Brook and Baines,[7] Cowley and Norman,[8] Gordon,[9] Shklover *et al.*,[10] Grev,[11] and Lickiss.[12] In addition, much information is to be found in *The Chemistry of Organic Silicon Compounds* edited by Patai and Rappoport,[13] especially in chapters 2, 3, 8, 15, and 17.

II

TYPES AND SOURCES OF SILENES

A. *Introduction*

Silenes with a wide range of structures have been synthesized over the years, from the simplest possible compound, $H_2Si{=}CH_2$, which has a transient existence, to species that contain four complex groups attached to the ends of the silicon–carbon double bond, e.g., $(Me_3Si)MesSi{=}C$

(OSiMe$_3$)Ad (Mes = 2,4,6-trimethylphenyl, Ad = adamantyl, $C_{10}H_{15}$), which is stable in solution at room temperature indefinitely and at 100°C for several hours. Most silenes are acyclic molecules, although examples of a silene in a five- and in an eight-membered ring were reported (see below). As the number and bulk of the substituents increase, so generally does the stability of the silene; a number of tetrasubstituted silenes survive for long periods in solution or as solids (due to kinetic, rather than thermodynamic, stability) and can be handled easily by the standard methods of organic chemistry using vacuum-line techniques. However, because silenes are very reactive molecules, it is essential to avoid the presence of oxygen or protic solvents, for example, alcohols or other relatively acidic or polar species including such chlorinated reagents as chloroform, carbon tetrachloride, or benzyl chloride, which will react with the double bond. Most silenes generated in the absence of trapping reagents react intermolecularly, usually dimerizing in a head-to-tail manner, although some undergo only head-to-head dimerization. Dimerization is avoided when rather bulky groups are attached to at least one end (usually the carbon end) of the silicon–carbon double bond.

B. *Preparation of Silenes*

Over the years a number of general synthetic pathways to silenes have been developed; these have been employed by research groups to generate families of silenes closely related in structure. It seems useful to list the most common pathways, together with references, in Table I[14–91] and then to comment on some of them in detail.

1. *Thermolysis or Photolysis of Silacyclobutanes or Silacyclobutenes*

The first evidence for the formation of silenes came from the thermolysis of silacyclobutanes, which resulted in a retro-[2+2] process leading to the silene $Me_2Si{=}CH_2$ and ethylene[1]:

$$\text{Me}_2\text{Si}(\text{CH}_2)_3 \text{ (silacyclobutane)} \xrightarrow[h\nu]{\Delta \text{ or}} \text{Me}_2\text{Si=CH}_2 + \text{CH}_2\text{=CH}_2 \qquad (1)$$

Generally, only simple silenes having small groups (H, Me, $CH_2{=}CH$) are obtained as transient species from the thermolysis of silacyclobutanes. In part this is due to the high temperatures (usually above 450°C) required for the ring cleavage. Substitution on the carbon atom adjacent to silicon in the ring can lead to carbon-substituted silenes. 1,3-Disilacyclobutanes do not readily revert to silenes under thermal conditions, but examples

TABLE I

COMMON SYNTHETIC ROUTES TO SILENES

Generalized silene structure	Method of synthesis	Common substituents	Reference
$RR'Si{=}CH_2$	(1) Thermolysis or photolysis of silacyclobutanes	R, R′ = H, Me, $CH_2{=}CH$, $PhCH_2$, Ar	14–18
$Me_2Si{=}C(SiMe_3)R$	(2) Photolysis of silyldiazoalkanes, e.g., $Me_3SiSiMe_2CN_2R$	R = H, Me, COMe, CO_2Et, CO-*t*-Bu, COAd, CO-*p*-Tol	19–27
$R_2Si{=}C(SiMe_3)_2$	(3a) Salt elimination from $R_2SiX{-}C(SiMe_3)_2Li$	R = Me, Ph	28–32
$Me_2Si{=}C(SiMe_3)(SiMe(t\text{-}Bu)_2)$	(3b) Salt elimination from $(t\text{-}Bu)_2SiX{-}C(SiMe_3)_2Li$		33,34
$Cl_2Si{=}C(R)CH_2CMe_3$	(4a) Addition of *t*-BuLi to $RR'SiCl{-}CR''{=}CH_2$	R, R′ = Me, Ph, $-(CH_2)_3-$, Me_3SiO, $(Me_3Si)_2N$, etc. R″ = H, $SiMe_3$, $SiMe_2OCMe_3$	35–49
$RR'Si{=}C(R'')CH_2CMe_3$	(4b) Addition of *t*-BuLi to $Cl_3Si{-}CR{=}CH_2$	R = H, Ph, $SiMe_3$, $SiMe_2OCMe_3$	35,36,48,50–53
$(Me_3Si)_2Si{=}CRR'$	(5) Polysilyllithium reagents + C=O $(Me_3Si)_3SiLi + RR'C{=}O$	R, R′ = Me, Et, *i*-Pr, cyclohexanone, adamantanone,	54–59
$(Me_3Si)_2Si{=}CRR'$	(6) Polysilylacylsilanes + RLi $(Me_3Si)_3SiCOR + R'Li$	R = Me, Ph; R′ = Me, Ph, *o*-Tol, *p*-xylyl	60,61
$(Me_3Si)_2Si{=}C(OLi)R'$	(6a) Silenolates: $(Me_3Si)_3COR + Et_3GeLi$ or $R_3'SiLi$	R = CMe_3, Ad, Mes, *o*-Tol	62,63
$(Me_3Si)_2Si{=}CHR$	(7) Polysilylcarbinols + base $(Me_3Si)_3SiCHOHR$ + NaH or MeLi	R = CMe_3, Mes	57,64
$RR'Si{=}CHCH_2SiMe_3$	(8) Photolysis of vinylsilanes $Me_3SiSiRR'CH{=}CH_2$ and $h\nu$	R = Me, Et, Ph	65–69
$R_2Si{=}C{=}C(R)SiMe_3$	(9) Photolysis or thermolysis of alkynylsilane $Me_3SiSiR_2C{\equiv}CR'$ $h\nu$ or Δ	R′ = Ph, Mes	70–72
	(10) Photolysis of arylpolysilanes, e.g., $Me_3SiSiRR'Ar + h\nu$, Ar = Ph	R, R′ = Me, Ph, $SiMe_3$ Ar = Ph, C_6H_4R'', Np	66,73–84
$(Me_3Si)_2Si{=}C(OSiMe_3)R$	(11c) Photolysis of $(Me_3Si)_3SiCOR$	R = Me, Et, *i*-Pr, CMe_3, Ph, Mes, Ad, CEt_3, BCO	85–87
$(Me_3Si)RSi{=}C(OSiMe_3)R'$	(11b) Photolysis of $(Me_3Si)_2RSiCOR'$	R = Me, *t*-Bu, Mes R′ = Ad, Mes	87–89
$R_2Si{=}C(OSiMe_3)Ad$	(11a) Photolysis of Me_3SiSiR_2COAd	R = Ph, Mes	90
$(Me_3Si)RSi{=}C(SiMe_3)R'$	(12) Silenes + Grignard reagents $(Me_3Si)_2Si{=}C(OSiMe_3)R' + RMgX$	R = Me, CD_3, Et, Ph, $PhCH_2$ R′ = Ph, Ad, CMe_3	91

are known of their photochemical cleavage back to silenes.[17] On the other hand, virtually all of the 1,2-disilacyclobutanes investigated revert to silenes under very mild conditions, even below room temperature.[92,93]

While the decomposition of silacyclobutanes as a source of silenes has continued to be studied in the last two decades, the interest has largely focused on mechanisms and kinetic parameters. However, a few reports are listed in Table I of the presumed formation of silenes having previously unpublished substitution patterns, prepared either thermally or photochemically from four-membered ring compounds containing silicon. Two cases of particular interest involve the apparent formation of bis-silenes. Very low-pressure pyrolysis of 1,4-bis(1-methyl-1-silacyclobutyl)benzene[94] apparently formed the bis-silene **1**, as shown in Eq. (2), which formed a high-molecular-weight polymer under conditions of chemical vapor deposition.

$$\text{(Me)Si(C}_3\text{H}_6\text{)}-\text{C}_6\text{H}_4-\text{Si(Me)(C}_3\text{H}_6\text{)} \xrightarrow{\Delta} \underset{\mathbf{1}}{CH_2{=}SiMe{-}C_6H_4{-}SiMe{=}CH_2} \longrightarrow \text{polymer} \quad (2)$$

The second example of a bis-silene involved the thermolysis of the benzo-1,2-disilacyclobutane **2**, which formed the bis-silene **3** that subsequently dimerized in the unusual manner shown in Eq. (3).[95,96]

$$\underset{\mathbf{2}}{C_6H_4(SiEt_2{-}SiEt_2)} \xrightarrow{\Delta} \underset{\mathbf{3}}{C_6H_4({=}SiEt_2)({=}SiEt_2)} \longrightarrow C_6H_4(SiEt_2SiEt_2SiEt_2)(SiEt_2)C_6H_4 \quad (3)$$

By contrast, when **2** was photolyzed, homolysis of the silicon–silicon bond evidently occurred, forming a diradical centered on the two silicon atoms.[97] These were said to undergo disproportionation leading to a monosilene (see also Section IV.B.3).

2. *Photolysis of Silyldiazoalkanes*

The photolysis of disilyldiazoalkanes $R_3SiSiR_2C(N_2)R'$ leads initially to disilylcarbenes $R_3SiSiR_2(R')C$:, which spontaneously undergo 1,2-silyl rearrangements yielding the isomeric silenes $R_2Si{=}C(SiR_3)R'$. A specific example is given in Eq. (4).

$$Me_3SiSiMe_2C(N_2)COOEt \xrightarrow[\Delta]{h\nu\ \text{or}} Me_3SiSiMe_2\ddot{C}COOEt \longrightarrow Me_2Si{=}C(SiMe_3)(COOEt) \quad (4)$$

This route has led to a number of acyclic silenes, many of which have two methyl groups attached to the sp^2-hybridized silicon atom. This is an artifact of synthetic convenience rather than necessity, being the result of using the readily available Me_3SiMe_2SiCl during synthesis of the starting materials: differently substituted chlorodisilanes that are also available could be used in principle. Silabenzenes and silafulvenes have also been produced by this route (see Section II.B). Table I lists references to the syntheses of silenes reported since 1985; these silenes were formed by the photolysis of disilyldiazoalkanes bearing a wide variety of groups, e.g., R, COR, and COOR attached to the diazo carbon atom.

None of the silenes produced was actually isolated as such, in part because no effort was made to isolate the species directly, but most commonly because the silene formed either reacted intramolecularly or else reacted with a trapping agent present during its formation. The most "relatively stable" silenes produced by this method, based on current knowledge, were probably $(Me_3Si)_2Si{=}C(SiMe_3)_2$[19] and $Me_2Si{=}C(SiMe_3)_2$,[98] the latter being one of the silenes prepared earlier by Wiberg by another route,[28] where it was observed by NMR spectroscopy in solution. In some cases the silyldiazoalkanes were photolyzed at 77 K in 3-methylpentane (3-MP) glass, under nonreactive conditions, thus allowing the measurement of their ultraviolet absorption maxima (see Section III.C.2).[20]

3. *Salt Elimination Forming $R_2Si{=}C(SiMe_3)(SiR_3)$*

In a series of papers in the 1980s Wiberg established optimal conditions for metalating a C—Br bond adjacent to a halosilane. Both temperature and solvent were important factors governing the ease of elimination of the LiX salt leading to the silene (Eq. 5). Two silenes of particular importance were prepared: $Me_2Si{=}C(SiMe_3)_2$,[28] used in many studies of the reactions and reactivity of silenes (see Section IV), and $Me_2Si{=}C(SiMe_3)(SiMe(t\text{-}Bu)_2)$[29,33,34,98a] whose crystal structure was obtained.[29]

$$Me_2\underset{F}{Si}-\underset{H}{C}(SiMe_3)_2 \xrightarrow{t\text{-BuLi}} Me_2\underset{F}{Si}-\underset{Li}{C}(SiMe_3)_2 \longrightarrow Me_2Si{=}C(SiMe_3)_2 \quad (5)$$

The former silene is not stable at room temperature, owing to its tendency to dimerize in the absence of other reactants, but it can be "stored" as its [2+2] cycloadduct with *N*-trimethylsilylbenzophenonimine, from which it is cleanly and easily liberated on mild thermolysis.[99] The latter silene arose from the loss of LiX from the compound $(t\text{-}Bu)_2SiX\text{-}CLi(SiMe_3)_2$, which subsequently underwent a spontaneous 1,3-silicon-to-silicon methyl shift.

Since that time, a few additional silenes or silaallenes have been prepared by this method: references are given in Table I.

4. *Addition of* t-*BuLi to Vinylchlorosilanes*

a. *RR'ClSiCR═CH₂*. The addition of *t*-butyllithium to the carbon–carbon double bond of vinylhalosilanes, followed by loss of lithium halide, thus leading to silenes, was first described by Jones in 1977.[49] Typical examples are shown in Eqs. (6) and (6a):

$$PhMeSi(Cl)\text{-}CH{=}CH_2 + t\text{-}BuLi \longrightarrow (Ph)(Me)Si{=}C(H)CH_2CMe_3 \qquad (6)$$

Auner prepared rather more complex silenes from vinylsilacyclobutanes in which the sp^2-hybridized silicon atom was part of a four-membered ring.[38,43] While the silenes were neither stable nor detectable spectroscopically because of their short lifetimes, subsequent studies showed that, in common with most other silenes, they were readily trapped by trimethylmethoxysilane, various dienes with which they underwent [2+4], [2+2], and ene reactions, and other reagents. Some reactions were shown to be stereospecific (see Section IV.D.1). In the absence of a trapping reagent they almost universally underwent head-to-tail dimerization leading to substituted 1,3-disilacyclobutanes. Recently, the silene $Mes_2Si{=}CHCH_2CMe_3$, the first stable silene bearing only one R group on the sp^2-hybridized carbon of the double bond, was isolated as a solid, mp 152°C, and fully characterized spectroscopically.[99a]

b. *$Cl_3Si—CR{=}CH_2$ and $Cl_2R'Si—CR{=}CH_2$*. In recent years Auner has prepared a wide variety of silenes by the addition of *t*-butyllithium to vinyltrichlorosilane or related vinylchlorosilanes having different substituents either on silicon or on the alpha carbon atom of the vinyl group; e.g.,

$$Cl_3SiC(SiMe_3){=}CH_2 + t\text{-}BuLi \longrightarrow Cl_2Si{=}C(SiMe_3)CH_2CMe_3 \qquad (6a)$$

References to the silenes so prepared since 1985 are listed in Table I. None of these silenes was stable, most undergoing head-to-tail dimerization to give 1,3-disilacyclobutanes in the absence of trapping reagents. Some interesting spontaneous silene-to-silene rearrangements were observed,[52] which will be described in Section IV.E.

5. *Polysilyllithium Reagents + Ketones*

Relatively recently, new routes to silenes have been developed that involve the reactions of organometallic reagents with carbonyl compounds, followed by spontaneous rearrangement and elimination of trimethylsilanolate from the initial adducts. Oehme[54–56] observed that the addition of silylmetallic reagents such as $(Me_3Si)_3SiLi$ to ketones (or aldehydes in a couple of cases) led to adducts that eliminated trimethylsilanolate ion Me_3SiO^-, spontaneously forming a silene, as shown in Eq. (7):

$$(Me_3Si)_3SiLi + Me_2C{=}O \longrightarrow (Me_3Si)_3Si{-}C(O^-)Me_2 \longrightarrow (Me_3Si)_2Si{=}CMe_2 + Me_3SiO^- \quad (7)$$

To date, only $(Me_3Si)_3SiLi$ and $(Me_3Si)_3SiMgBr$ have been used in this reaction, but there is no obvious reason preventing the use of other silylmetallic reagents. In the cases studied to date, the initially formed silene reacted further, either with excess silylmetallic reagent or with other added reagents, in some cases even forming head-to-head silene dimer.[58] References to the silenes formed by this route are listed in Table I.

6. *Polysilylacylsilanes + RLi*

The same general results as those described in Section II.A.5 were obtained from the addition of organolithium reagents to polysilylacylsilanes such as $(Me_3Si)_3SiCOR$ (R = Me, Ph, Ad), namely, that the initial adduct spontaneously eliminated trimethylsilanolate ion yielding a silene (Eq. 8). Other acylsilanes such as $(Me_3Si)_2R'SiCOR$ and perhaps $Me_3SiR'_2SiCOR$ could also be employed in this reaction. To date, the scope of the organometallic reagents used has been limited to MeLi[60] and ArLi,[61] in which loss of silanolate ion yields a silene that is either trapped or dimerized. References for the silenes prepared by this route are listed in Table I.

$$(Me_3Si)_3SiCOPh + MeLi \longrightarrow (Me_3Si)_3Si{-}C(O^-)(Ph)Me \longrightarrow (Me_3Si)_2Si{=}C(Ph)Me + Me_3SiO^- \quad (8)$$

Silenolates. When Et_3GeLi[62] is employed as the organometallic reagent in the reaction with the acylsilanes $(Me_3Si)_3SiCOR$, the germyl nucleophile attacks one of the trimethylsilyl groups attached to the central silicon atom rather than the carbonyl group; this results in the loss of a trimethylsilyl group from the polysilylacylsilane in the form of $Me_3SiGeEt_3$ and yields a lithioenolate **4** as shown in Eq. (9). When an excess of the germyllithium reagent was employed in one case, the initially formed enolate subsequently underwent further complex reactions which were attributed to the [2+2] reaction of a silene with a disilene formed in the course of the overall reaction.[100] References to the known silenolates are given in Table I.

$$Et_3GeLi + (Me_3Si)_3SiCOCMe_3 \longrightarrow Et_3GeSiMe_3 + (Me_3Si)_2Si \overset{-}{=\!=} \overset{O\ Li^+}{C} - CMe_3 \quad \mathbf{4} \tag{9}$$

When silyllithium reagents such as $(Me_3Si)_3SiLi$ or $PhMe_2SiLi$[63] were used in the reaction with polysilylacylsilanes, the reaction products were also silenolates resulting from the loss of a Me_3Si group from the polysilylacylsilane.[63,100] In one case the silenolates were characterized by Diels–Alder reaction with a diene, or by *O*-silylation with triethylchlorosilane, which yielded the triethylsilyl derivative of the enolate.[63]

7. *Polysilylcarbinols + Base*

Another variant of the above-mentioned routes to silenes involved treatment of the carbinols $(Me_3Si)_3SiC(OH)RR'$, formed from the addition of organometallic reagents R′Li to polysilylacylsilanes, with bases such as NaH[64] or MeLi,[57,64] leading to the formation of alkoxides. These alkoxides spontaneously lost trimethylsilanolate ion, yielding silenes: references for these reactions are listed in Table I.

$$(Me_3Si)_3SiC(OH)Me_2 + NaH \longrightarrow Me_3Si\text{-}Si(SiMe_3)_2\text{-}C(O^-)Me_2 \longrightarrow (Me_3Si)_2Si{=}CMe_2 + Me_3SiO^- \tag{10}$$

8. *Photolysis of Di- or Polysilylalkenes*

Early studies by Sakurai[101] demonstrated that vinyldisilanes readily rearrange via 1,3-silyl migration when photolyzed, thereby forming silenes, as shown in Eq. (11), which could be trapped by conventional methods. More complex silenes have been prepared more recently, particularly by Ishikawa *et al.*,[66,67,69] and several references to these studies are given in Table I.

$$EtMe_2SiSiMePhCH{=}CH_2 \xrightarrow{h\nu} PhMeSi{=}CHCH_2SiMe_2Et \tag{11}$$

9. *Photolysis or Thermolysis of Di- or Polysilylalkynes*

Kumada and colleagues found that the 1,3-silyl rearrangements from silicon to carbon occurred during the photolysis of disilylalkynes, giving mixtures of silaallenes and silacyclopropenes,[102] as illustrated by Eq. (12):

$$Me_3SiSiMe_2C{\equiv}CPh \xrightarrow{h\nu} Me_2Si{=}C{=}CPh(SiMe_3) + Me_3Si\text{—}\overline{\quad}\text{—}Ph \; (\text{ring: } SiMe_2) \tag{12}$$

Research has continued on the photolysis and thermolysis of such systems in recent years, particularly by Ishikawa. The silenes formed, most of which were trapped as confirmation of their formation, are referenced in Table I.

10. *Photolysis of Di- or Polysilylaromatic Compounds*

The early investigation by Kumada and colleagues[103] of aryldisilane and polysilane thermolysis or photolysis indicated the formation of compounds that appeared to be silenes, as shown in Eq. (13). Thus they underwent ene reactions with acetone or isobutene to give the anticipated adducts, although the silene acted as the ene component rather than as the enophile component, which is the normal behavior of the more conventional silenes with these same reagents. Also, they did not react with methanol in a simple 1,2-addition across the silicon–carbon double bond—only 1,4- and 1,6-addition of methanol to the conjugated system was observed. Moreover, they did not undergo the usual [2+4] and other cycloaddition reactions with dienes or add methoxytrimethylsilane across the silicon–carbon double bond, nor did they undergo [2+2] cycloadditions with carbonyl compounds. Thus, they definitely seemed to be somewhat unusual silenes. Extensive studies of a wide variety of these types of silenes in recent years by Ishikawa *et al.* and others have clearly established these species as silenes; evidence has now been found that, on occasion, they will undergo dimerization[78] and also [2+2] cycloaddition with acetone.[80] References to the new silenes prepared by this route are listed in Table I.

$$Me\text{-}C_6H_4\text{-}SiMe_2SiMe_3 \xrightarrow{h\nu} Me\text{-}C_6H_4(H)(SiMe_3){=}SiMe_2 \xrightarrow{CH_2=CMe_2} \longrightarrow Me\text{-}C_6H_3(SiMe_3)\text{-}SiMe_2CH_2CMe_3 \tag{13}$$

11. *Photolysis of Di- and Polysilylacylsilanes*

a. *Silenes from* $(Me_3Si)_3SiCOR$. Historically, the first evidence for the existence of relatively stable silenes came from the photolysis or thermolysis of members of the family of acylsilanes with the structure $(Me_3Si)_3SiCOR$.[92] Photolysis was the preferred method because of the mild conditions which could be used. When R = *t*-Bu, NMR techniques revealed a temperature-sensitive equilibrium between the silene $(Me_3Si)_2Si{=}C(OSiMe_3)t\text{-}Bu$ and its head-to-head dimer.[92,93] When R was adamantyl, no dimer was formed—almost certainly because of severe steric interactions; rather, the pure silene was isolated, and its crystal structure was obtained, thereby confirming the structure.[5,104]

Subsequently, other members of the family of silenes $(Me_3Si)_2Si{=}C(OSiMe_3)R$ have been prepared, where R = Me, Et, *i*-Pr, CH_2Ph, bicyclooctyl, CEt_3, 1-methylcyclohexyl, and Mes. The first four silenes listed were not stable in inert solvents, and hence were not observable by NMR spectroscopy, since they rapidly reacted intermolecularly to give linear and/or cyclic head-to-head dimers.[87] This is illustrated in Eq. (14) for the benzyl compound where the initially formed silene **5** yielded the cyclic head-to-head dimer **6** as well as the linear head-to-head dimer **7**. The latter four silenes were all relatively stable and were characterized by NMR spectroscopy.[105]

$(Me_3Si)_3SiCOCH_2Ph$ —hv, > 360 nm→ $(Me_3Si)_2Si{=}C(OSiMe_3)CH_2Ph$ (**5**) →

Me3Si OSiMe3
Me3Si–Si——CH2Ph
Me3Si–Si——CH2Ph
Me3Si OSiMe3
6

\+

Me3Si OSiMe3
Me3Si–Si——=CHPh
H
Me3Si–Si——CH2Ph
Me3Si OSiMe3
7

(14)

b. *Silenes from* $(Me_3Si)_2R'SiCOR$. Silenes of the family $(Me_3Si)R'Si{=}C(OSiMe_3)Ad$ have also been made from the acylsilanes $(Me_3Si)_2R'SiCOR$. When R′ = Me or Ph, the silenes were not directly observed,[87] having undergone other reactions, particularly head-to-head dimerization. When R′ = *t*-Bu, only a single geometric isomer of a relatively stable silene[87] was initially reported. However, subsequent studies[106] suggested that both *E*- and *Z*-isomers were present, based on trapping by various alkynes: such reactions have been shown to be both regio- and stereospecific in other cases.[88] When R = Mes, a mixture of relatively stable geometric isomers, *E*- and *Z*-$(Me_3Si)MesSi{=}C(OSiMe_3)Ad$ was obtained and characterized.[88] One of the isomers could be selectively destroyed thermally, leading to the isolation of the other pure isomer.[106]

c. *Silenes from Acyldisilanes* Me_3SiSiR_2COR'. Two new (unstable) silenes of the family $R_2Si{=}C(OSiMe_3)R'$ were prepared by photolysis of acyldisilanes, namely $Ph_2Si{=}C(OSiMe_3)Ad$ and $Mes_2Si{=}C(OSiMe_3)Ad$. These underwent reactions in solution too rapidly to be detectable by NMR spectroscopy.[90]

12. *Silenes from the Reaction of Silenes with RMgX*

Closely related chemistry leading to differently substituted silenes was found from the reaction of a number of silenes with Grignard reagents.[91] The experimental methodology involved rapid addition of a Grignard reagent to a cold solution of a polysilylacylsilane, followed immediately by photolysis. In almost all cases the rate of photoconversion of the polysilylacylsilane **8** to its silene **9** (involving 1,3-silyl migration from silicon to oxygen) was more rapid than the rate of addition of the Grignard reagent to the acylsilane. Rearrangement of the initially formed carbanion **10** to the isomeric oxyanion **11** followed by elimination of trimethylsilanolate ion led to the formation of the new silene **12** (see Scheme 1) by a process analogous to a Peterson reaction. This silene subsequently either

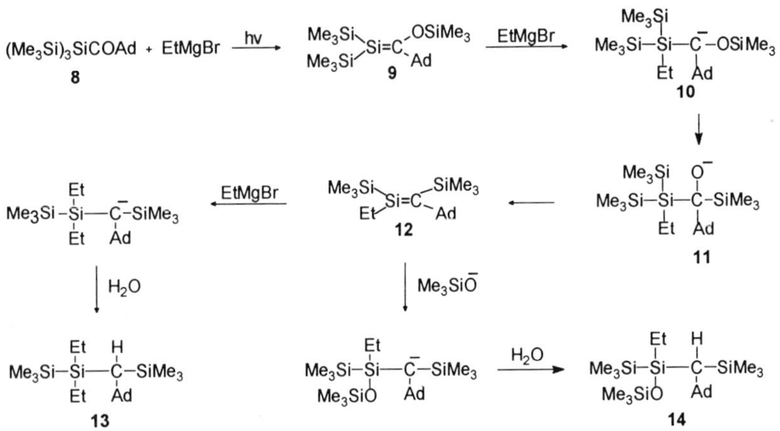

$(Me_3Si)_3SiCOAd$ (**8**) + EtMgBr —hν→ $(Me_3Si)_2Si{=}C(OSiMe_3)Ad$ (**9**) —EtMgBr→ $(Me_3Si)_2(Et)Si{-}C^{-}(Ad)OSiMe_3$ (**10**)

10 → $(Me_3Si)_2(Et)Si{-}C(O^{-})(Ad)SiMe_3$ (**11**) → $(Me_3Si)(Et)Si{=}C(SiMe_3)Ad$ (**12**)

12 —EtMgBr→ $Me_3Si{-}Si(Et)_2{-}C^{-}(Ad)SiMe_3$ —H_2O→ $Me_3Si{-}Si(Et)_2{-}CH(Ad)SiMe_3$ (**13**)

12 —Me_3SiO^{-}→ $Me_3Si{-}Si(Et)(OSiMe_3){-}C^{-}(Ad)SiMe_3$ —H_2O→ $Me_3Si{-}Si(Et)(OSiMe_3){-}CH(Ad)SiMe_3$ (**14**)

SCHEME 1

TABLE II

SILENES FROM THE REACTIONS OF SILENES WITH GRIGNARD REAGENTS[91]

$Me_3Si(Me)Si{=}C(SiMe_3)CMe_3$	$Me_3Si(Me_3SiO)Si{=}C(SiMe_3)Ad$
$Me_3Si(Me)Si{=}C(SiMe_3)Ph$	$PhMeSi{=}C(SiMe_3)Ad$
$Me_3Si(CD_3)Si{=}C(SiMe_3)Ad$	$PhEtSi{=}C(SiMe_3)Ad$
$Me_3Si(Me)Si{=}C(SiMe_3)Ad$	$Me_3Si(Ph)Si{=}C(SiMe_3)Ad$
$Me_3Si(Et)Si{=}C(SiMe_3)Ad$	$Me_3Si(PhCH_2)Si{=}C(SiMe_3)Ad$

reacted further with excess Grignard reagent to give the product **13** after workup, or it added the trimethylsilanolate ion (Me_3SiO^-) formed during the course of the reaction to give compound **14** after workup. Interception and trapping of the silenes by other added reagents should be possible. The new silenes prepared by this route are listed in Table II.

13. *Preparation of Silaaromatics from Silyldiazoalkanes*

It is possible to utilize the thermal or photochemical decomposition of silyldiazoalkanes to prepare cyclic conjugated silenes. Ando found that a six-membered cyclic silyldiazoalkane **15** underwent ring contraction to the silafulvene **16** which was trapped with various reagents.[107] Alternatively, the diazomethylsilacyclopentadiene **17** underwent ring expansion to give the highly substituted silabenzene **18**, accompanied by a 1-ethylidene-1-silacyclopentadiene derivative **19**, as shown in Eq. (15):

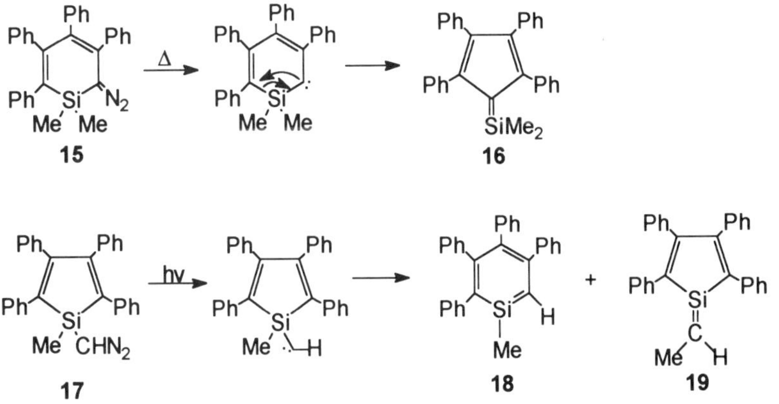

Ph Ph Ph Ph Si N2 Me Me **15** —Δ→ Ph Ph Ph Ph Si Me Me → Ph Ph Ph Ph SiMe2 **16**

(15)

Ph Ph Ph Ph Si Me CHN2 **17** —hν→ Ph Ph Ph Ph Si Me H → Ph Ph Ph Ph Si H Me **18** + Ph Ph Ph Ph Si C Me H **19**

Closely related studies by Märkl using a 2-diazo-1-silacyclohexadiene such as **20** led in some cases to highly substituted silabenzenes,[108,109] one of which, **21**, was "stable" at $-100°C$, based on its ^{29}Si NMR spectrum (Eq. 16).

20 $\xrightarrow{h\nu}$ → **21** (16)

C. *Miscellaneous Photochemical Routes to Silenes*

A few other routes to new silenes involving photochemical processes have been reported, most of which have the structure $RR'Si{=}CH_2$[110–112] and are listed in Table I. Photolysis of 1,1-dimethyl-2-phenyl-1-silacyclobut-2-ene **22** in a glass at 77 K led to the siladiene **23**, which absorbed at 338 nm,[113] as shown in Eq. (17):

$$\underset{\mathbf{22}}{\text{Me}_2\text{Si(CPh=CH-CH}_2\text{)}} \xrightarrow[\text{77K, glass}]{h\nu} \underset{\mathbf{23}}{\text{Me}_2\text{Si=C(Ph)CH=CH}_2} \tag{17}$$

Appropriately substituted silylenes have been shown to undergo rearrangement with ring expansion leading to silenes. Thus Fink[114] irradiated the trisilane **24** in a 3-MP glass forming the cyclopropenyl-substituted silylene **25**, which rearranged to the highly substituted silacyclobutadiene **26** (see Eq. 18). The adduct of this compound with methoxytrimethylsilane was shown by an X-ray crystal structure to have formed from a *syn* addition to the silicon–carbon double bond. Either one or two diastereomers arising from the addition of ethanol to the silicon–carbon double bond were formed, depending on the reaction conditions.

24 $\xrightarrow[\text{254 nm}]{h\nu}$ **25** $\xrightarrow[\text{> 400 nm}]{h\nu}$ **26** (18)

D. *Miscellaneous Thermal Routes to Silenes*

The trapping of silacyclopentadienes has also been reported recently.[115] Using the pyrolysis of **27**, or photolysis or pyrolysis of **28**, the formation of the silylene **29** was inferred. Further photolysis or thermolysis converted the silylene into silene **30**, which could be photochemically isomer-

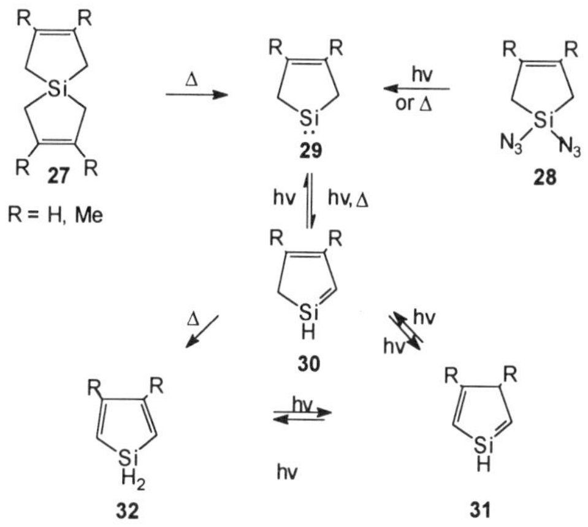

SCHEME 2

ized to silene **31**. Each silene ultimately was converted into the silole **32** (Scheme 2).

A few routes to new silenes, usually involving flash vacuum pyrolysis at high temperatures, have been reported. Silenes were proposed as the result of the thermal expulsion of trimethylmethoxysilane, or a similar volatile fragment, from the starting material; but frequently, proof that the silenes proposed to account for the observed products were in fact formed was not provided.[116–119] The other thermal route employed was the retro-Diels–Alder regeneration of a silene from an adduct with an aromatic compound—often a 9,10-anthracene or 1,4-naphthalene adduct or, in some cases, a 1,4-benzene adduct, as illustrated in Eq. (19).[120]

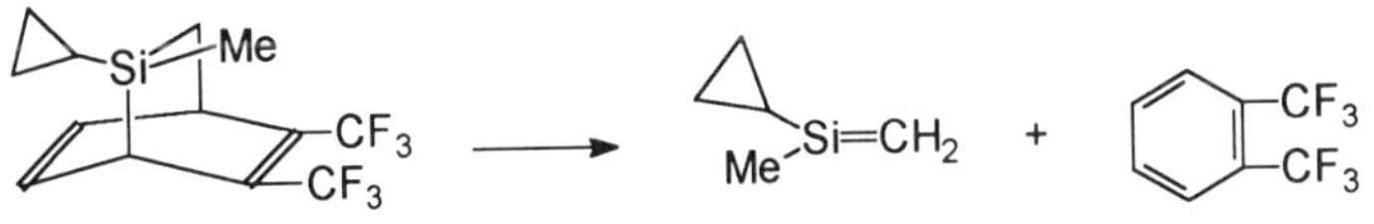

(19)

E. *Silene–Transition Metal Complexes*

Transition metals have been used to trap and stabilize many different types of reactive intermediates, such as carbenes. Reactive silicon intermediates have only recently yielded to this approach. In the case of alkenes, for instance, transition metal complexes are generally made by exposing the alkene to a transition metal bearing suitable leaving groups (e.g., carbonyl). Unlike carbon-based intermediates, however, silicon-based analogs have been very difficult to prepare until recently. Unless the silene bears very bulky groups, it is insufficiently stable to form transition metal complexes before decomposing. However, the presence of bulky groups precludes complexation of the silene with the metal.

Silene–transition metal complexes were proposed by Pannell[121] for some iron and tungsten systems, and such species were observed spectroscopically by Wrighton.[122,123] Thus intermediates such as **33** have been proposed in the preparation of carbosilane polymers from hydrosilanes,[124] both as intermediates in the isotope scrambling observed to occur in similar ruthenium hydride systems[125,126] and in the S_N2' addition of alkyllithium species to chlorovinylsilanes.[47]

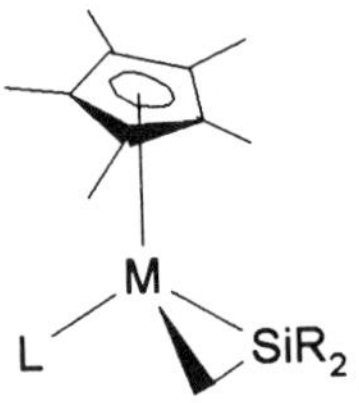

M
L SiR$_2$ **33**

Rearrangements of metal silylgermylalkanes take place within the coordination sphere of the metal. Thus, η^5-$C_5H_5(CO)_2Fe$—CH_2*SiMe*$_2$$GeMe_3$ photochemically rearranges to η^5-$C_5H_5(CO)_2Fe$—*SiMe*$_2$$CH_2GeMe_3$, presumably via the intermediacy of an Fe–silene complex. In contrast, the related tungsten derivative η^5-$C_5H_5(CO)_3WCH_2SiMe_2GeMe_3$ extrudes $Me_2Si{=}CH_2$, as shown in Eq. (20): the silene was trapped by addition of *t*-BuOH.[127]

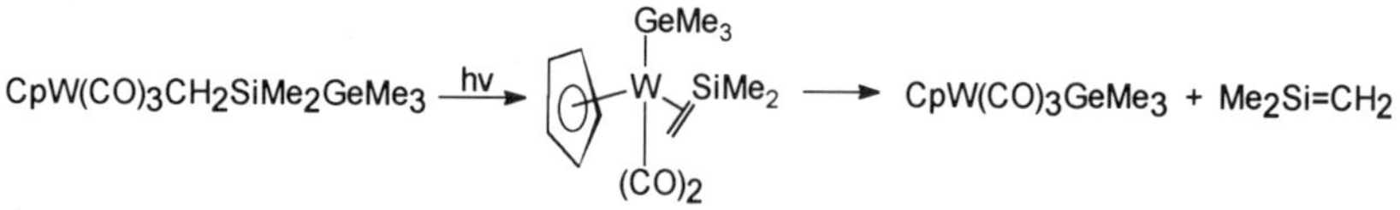

$$CpW(CO)_3CH_2SiMe_2GeMe_3 \xrightarrow{h\nu} CpW(GeMe_3)(CO)_2(\eta^2\text{-}CH_2{=}SiMe_2) \longrightarrow CpW(CO)_3GeMe_3 + Me_2Si{=}CH_2 \quad (20)$$

More recently, silene–transition metal complexes have been prepared by forming the silene in the coordination sphere of the metal; a review has been written on the subject.[12] The Grignard reaction of a haloalkylsilane with a group 8 transition metal leads to a species that can undergo β-elimination to give the silene. Tilley has shown this to be true with Ru and Ir (see Scheme 3).[128,129] The stability of the complexes is dependent upon the groups on silicon; phenyl groups lead to more stable species than do methyl groups.

The crystal structure data suggest some sp^2 character for the silicon and carbon atoms of the double bond, but there is a high degree of π-

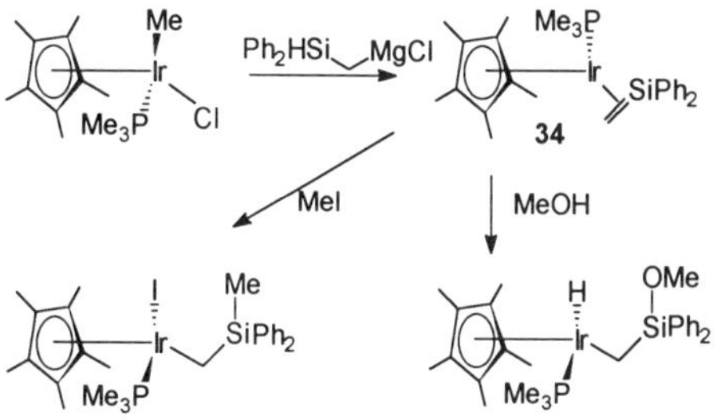

SCHEME 3

backbonding. The bond lengths are somewhat shorter than a simple single Si—C bond, for example, 1.810 Å in **34** compared to the normal single bond lengths of 1.87–1.91 Å (see Section III.B.1). The π-bonding between Ir and the silene leads to a short Ir—Si bond (2.317 Å). The Ir compound **34** undergoes reactions with MeI or MeOH that lead to clean rupture of the Ir—Si bond, giving compounds **35** and **36**, respectively.

Berry has used a related reaction to prepare a W derivative, as shown in Scheme 4.[130] NMR data in this case are similarly consistent with the metal interacting primarily with the Si—C π-orbital.

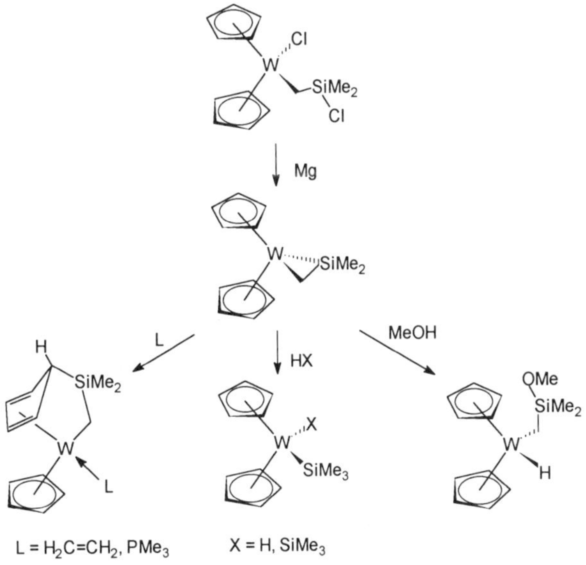

L = $H_2C{=}CH_2$, PMe_3 X = H, $SiMe_3$

SCHEME 4

An intriguing example of a stable aromatic silacyclopentadiene complex has recently been reported by Tilley *et al.*[130a] The sisyl ($Si(SiMe_3)_3$)-substituted silole was prepared in the first two steps by coupling of the dibromosilole (dibromosilacyclopentadiene) with sisyllithium followed by reduction. After complexation with $Cp^*Ru(OMe)_2$, removal of the methoxy groups (as Me_3SiOMe) and exchange of the counterion to the poorly coordinating tetraphenylborate anion, a stable ruthenium hydride complex was isolated. The crystal structure located the bridging hydrogen, and indicated that both rings were flat. Additional evidence that the ring silicon atom had considerable sp^2 character came from the small differences (0.07 Å) of the C—C bond lengths α–β and β–γ to silicon. Deprotonation with sisyllithium led to the free ruthenium compound, whose aromaticity was confirmed principally by NMR evidence. The ^{13}C NMR shifts of the two different carbon atoms are 73.09 and 88.53 ppm (compared with 86.02 for the Cp* ligand). The progression of shifts in the ^{29}Si NMR spectra of the various compounds is also consistent with the identification of the free complex as an aromatic compound.

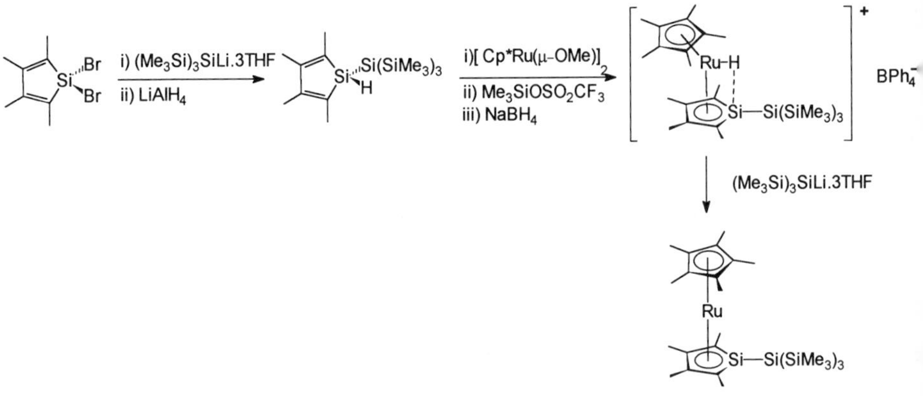

Ishikawa has proposed the formation of various complexes of nickel with silenes, and also with the silicon–carbon double bond component of silaallenes. It was suggested that when polysilylacetylenes were treated with nickel phosphine reagents such as $NiCl_2 \cdot (PEt_3)_2$ or $Ni(PEt_3)_4$, the resulting nickel–silene complexes of silaallenes participated in reactions that formed products different from those formed in the absence of added nickel.[72,131–133b] In addition, silacyclopropenes[133] and vinylsilanes[46,134–136] were also assumed to form nickel complexes in the course of rearrangement or reaction. As one example of the behavior of the nickel complexes, the treatment of the polysilylacetylene **37** or its photoisomer **38** with a nickel phosphine is said to lead to a relatively stable nickelasilacyclobutene

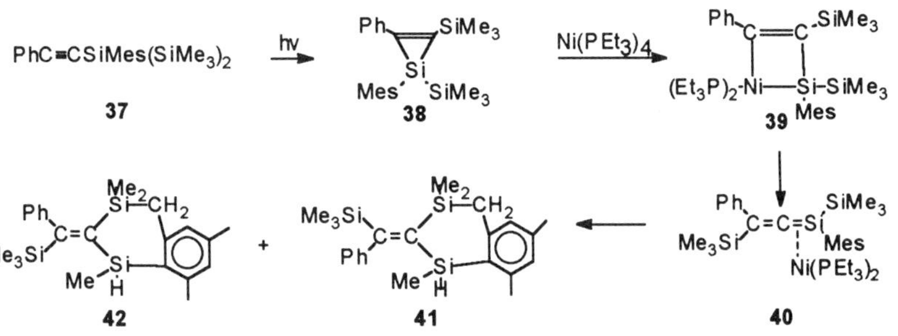

SCHEME 5

39, which isomerizes to the undetectable silaallene–nickel complex **40**, giving the pair of rearranged isomers **41** and **42** obtained in 47 and 41% yields, respectively, as shown in Scheme 5.[72,132]

Finally, Ishikawa found that heating the acylsilane $(Me_3Si)_3SiCOAd$ to 120°C in the presence of $Ni(PEt_3)_4$ and C_6D_6 gave rise to a product that involved the addition of a C—D bond from the deuterobenzene across the double bond of the silene $(Me_3Si)_2Si{=}C(OSiMe_3)Ad$, forming $(Me_3Si)_2Ph{-}CD(OSiMe_3)Ad$.[137] Under the thermal conditions employed, the formation of the silene was expected,[92] but the addition of an aromatic C—H or C—D bond to a silicon–carbon double bond was unprecedented since the addition does not occur in the absence of the nickel complex. It was suggested that an intermediate complex was formed between the silicon–carbon double bond and $Ni(PEt_3)_3$, thus facilitating the addition.

Trimethylenemethane is a special type of alkene that does not exist as the free compound. Various synthetic equivalents to the synthon **43** shown below have been reported. Trost, in particular, has exploited these compounds in 1,3-dipolar cycloaddition reactions.[138,139] A metal-bound, isolated trimethylenemethane species was recently reported by Ando (Scheme 6). It resulted from the complexation of an *exo*-methylenesilacyclopropene with group 8 carbonyls (Fe, Ru).[140,140a] The structure was proved by X-ray crystal structure analysis. ^{29}Si NMR data were consistent with the η^4-structure shown.

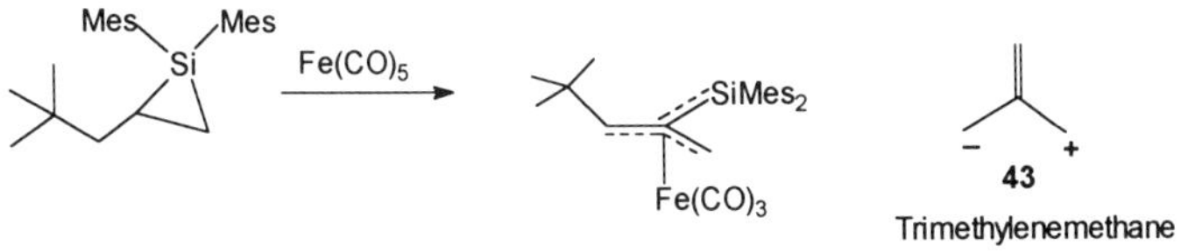

SCHEME 6

F. *Silene–Solvent Complexes*

The stable silene $Me_2Si{=}C(SiMe_3(SiMe(t\text{-}Bu)_2)$ was first reported as a stable complex with THF,[34] and its crystal structure showed the length of the silicon–carbon double bond as 1.747 Å. Subsequently, it was possible to remove the THF and isolate the uncomplexed silene, which had a noticeably shorter Si=C bond length of 1.702 Å.[29] Further investigation showed that stable complexes of this or closely related silenes with trimethyl- or ethyldimethylamine, pyridine, and fluoride ion were also readily formed and moderately stable.[31,141]

No similar complexes have been observed with other stable silenes. Thus, silenes of the family $(Me_3Si)_2Si{=}C(OSiMe_3)R$ show no evidence of complex formation with THF or Me_3N, undoubtedly because of their lower electrophilicity.

G. *Kinetics of Silene Formation*

In the course of many studies concerning silenes, a number of measurements of the rates of various reactions have been determined. In many cases activation parameters have also been obtained. These data are summarized briefly below, but the interested reader is advised to consult the original papers for full details.

Auner, Davidson, and Ijaki-Maghsoodi have examined the kinetics and mechanism of formation of silenes from the pyrolysis (772–847 K) of diallyldimethylsilane.[142] Two competing processes were shown to be involved in the pyrolysis: a radical-forming homolysis and a silene-forming (1,1-dimethylsilabutadiene) retroene reaction. The primary reaction pathway was shown to be the retroene (Scheme 7). The decomposition of diallyldimethylsilane was shown to be first-order with $\log A = 10.7 \pm 0.1$ s^{-1} and $E_{act} = 192 \pm 2$ kJ mol^{-1}; those for the retroene reaction forming the siladiene and propene were $\log A = 11.2 \pm 0.1$ and $E_{act} = 199 \pm 2$ kJ mol^{-1}. The activation parameters for the retroene reaction were shown to be comparable to the analogous reaction with the all-carbon species 1,6-heptadiene.

The rate of pyrolysis of 1,3-disilacyclobutanes **44** (see Table III) was shown by the same authors to be dependent upon the substituents on

Si H Δ Si + Si

SCHEME 7

TABLE III

KINETIC DATA FOR THE PYROLYTIC DECOMPOSITION OF 1,3-DISILACYCLOBUTANES[44]

Compound	Temperature of pyrolysis (°C)	log A (s^{-1})	E ($kJ\ mol^{-1}$)	$k_{550°C}$ (s^{-1})	k_{rel}
 Si Si	664–746	1434	296	4.1×10^{-5}	1
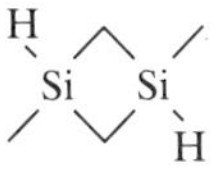 H Si Si H	616–663	13.5	255	2.07×10^{-3}	50
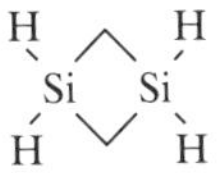 H H Si Si H H	516–573	13.3	230	0.05	1220

silicon.[143] Silene production was shown by trapping studies with Me_3 SiOMe and HCl. The kinetic parameters for the pyrolyses of three species are given in Table III. The decomposition pathways of the hydrogen-bearing compounds were shown to be more complex than those of the tetramethyl compound.

Conlin and co-workers have also studied the fragmentation of a siletane (silacyclobutane). In this case, both the *E*- and *Z*-isomers of 1,1,2,3-tetramethylsilane **45** were prepared and thermolyzed (Scheme 8).[144] Both *E*- and *Z*-isomers of **45** led to the same products in slightly different ratios: the major products were propene with silene **46**, and *E*- and *Z*-2-butenes with silene **47**. Silene formation was inferred from detection of the disilacyclobutane products. During these processes, the stereochemical integrity of the compounds was largely preserved.

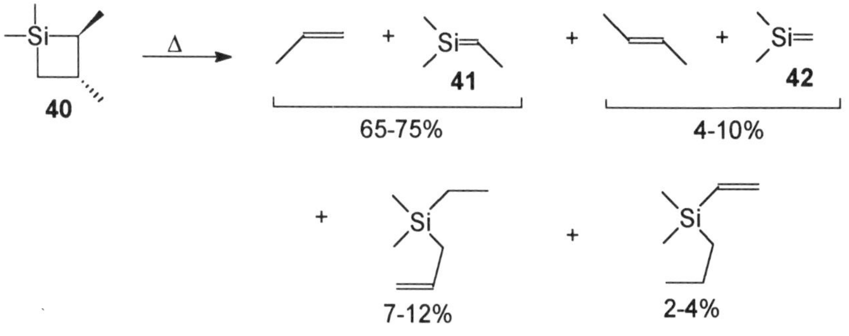

Si **40** Δ → + Si **41** (65-75%) + + Si **42** (4-10%)

+ Si (7-12%) + Si (2-4%)

SCHEME 8

This work has been compared with analogous cyclobutane thermolytic decompositions. The siletanes were found to fragment more readily than the cyclobutanes. Although fragmentation via propene formation (most substituted C—C bond) was favored in both classes of compounds, it was more dominant with the siletanes. These effects are apparent from the kinetic data in Table IV.[144–147]

Conlin[148] also studied the pyrolysis of 1-methyl-1-silacyclobutane in the presence of excess butadiene at various temperatures where the decomposition followed first-order kinetics and where the silene isomerized to the isomeric silylene prior to reacting with the butadiene. The value for the preexponential factor A for the silene-to-silylene isomerization was found to be $9.6 \pm 0.2\ s^{-1}$ and the E_{act} for the isomerization was 30.4 kcal mol^{-1} with $\Delta H^{\ddagger} = 28.9 \pm 0.7$ kcal mol^{-1} and $\Delta S^{\ddagger} = -18.5 \pm 0.9$ cal mol^{-1} deg. More recently, the photochemical ring opening of 1,1-dimethyl-2-phenylcyclobut-3-ene and its recyclization was studied. The E_{act} for cyclization was 9.4 kcal mol^{-1}.[113]

Pola[18] used laser flash photolysis sensitized by SF_6 to study the decomposition of 1-methyl-1-vinylsilacyclobutane which yielded 1-methyl-1-vinylsilene. The reaction followed first-order kinetics.

TABLE IV

KINETIC DATA FOR THE PYROLYTIC DECOMPOSITION OF 1,3-DISILACYCLOBUTANES

Compound	log A (s^{-1})	E (kJ mol^{-1})	$k_{398.2°C}$ (s^{-1})	k_{rel} (ref.)
 Si	15.64	62.6	0.20×10^{-4}	1 (145)
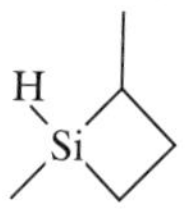 H Si	15.45	60.6	0.53×10^{-4}	2.7 (146)
 Si	16.39	63.6	0.62×10^{-4}	3.1 (147)
 Si			7.00×10^{-4}	35 (144)
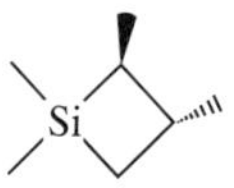 Si			1.97×10^{-4}	10 (144)

H. *Experimental Handling of Silenes*

The manner in which the experimentalist can produce and handle silenes depends greatly upon the structure and stability of the particular silene involved. Relatively stable silenes produced from the polysilylacylsilanes $(Me_3Si)_3SiCOR$ can easily be formed by photolysis at room temperature or lower using wavelengths of >360 nm and handled in solution in normal ways prior to adding the trapping or other reagents in the dark. Also, since few reagents absorb at these wavelengths, photolysis of a mixture of the acylsilane and trapping reagent normally gives identical results to the dark reactions. The stable silene $Me_2Si{=}C(SiMe_3)(SiMe(t\text{-}Bu)_2)$ can also be made and stored prior to treatment with a reactant in solution under mild conditions. The other important silene developed by Wiberg, $Me_2Si{=}C(SiMe_3)_2$, while not stable at room temperature, is readily liberated by mild thermolysis of its cycloadduct with *N*-trimethylsilylbenzophenonimine (see Section IV.C.5).[99]

The many silenes described by Jones and Auner that are derived from addition of *t*-butyllithium to a vinylchlorosilane are much less easy to prepare and manipulate. Since these species are not stable at room or even lower temperatures, it is necessary to work at low temperatures and to add the organolithium reagent to a mixture of the vinylchlorosilane and the trapping reagent. Thus the possibility that, at least in part, the organolithium reagent may react directly with the trapping reagent is difficult to avoid—a phenomenon that may explain the complex mixtures of products and unconsumed starting materials frequently obtained. Silenes that are formed by thermal retro-Diels–Alder or retroene and similar reactions, by the cleavage of silacyclobutanes, or by the thermal rearrangement (or photochemical rearrangement at 254 nm) of aryldi- and polysilanes or other thermal reactions are also difficult to handle under mild conditions; and the problems associated with the efficiency of their trapping, or with the possibility that other processes may intervene prior to trapping, are very difficult to avoid.

III

PHYSICAL PROPERTIES OF SILENES

A. *General Background*

By 1985 much was known about the geometry and bonding of a wide variety of silenes. The majority of these compounds were very short-lived; hence most of this knowledge came from *ab initio* and other kinds of calculations or from IR or UV spectroscopy performed in glasses at very low temperatures. Only two relatively stable silenes were known for which crystal structure data were available.

Calculations or experimental data indicated the following: simple silenes are planar; the rotational barrier involving the double bond is about 40 kcal mol^{-1}; the stretching frequency of the significantly polarized double bond, $Si^{\delta+}{=}C^{\delta-}$ is in the neighborhood of 1000 cm^{-1}, but depends on the substituents present at the ends of the double bond; and the compounds absorb in the ultraviolet region 245–260 nm, depending on the substituents. The length of the silicon–carbon double bond of the simpler silenes was calculated to be about 1.69–1.70 Å.

The data obtained from the stable solid silenes were consistent with these findings. The Wiberg silene $Me_2Si{=}C(SiMe_3)(SiMe(t\text{-}Bu)_2)$[29] was shown to be nearly planar and the length of the silicon–carbon double bond was 1.702 Å, in close agreement with the calculated value, while the Brook silene $(Me_3Si)_2Si{=}C(OSiMe_3)Ad$ had a bond length of 1.764 Å.[5,104] The increased length of the bond in this silene could be satisfactorily accounted for by the effects of the electron-releasing Me_3Si substituents, and particularly the $OSiMe_3$ group on the ends of the Si=C bond.[149] The fact that the molecule was twisted by about 16° could be explained by the steric interactions of the bulky groups on the ends of the double bond: these large groups also inhibited the molecule from dimerizing. Calculations have shown that small twist angles in silenes have a minimal effect on the bond strength.[11]

B. *New Developments*

1. *Bond Lengths*

While no further crystal structures of silenes have been reported in the last decade, the structures of several complexes of the silene $Me_2Si{=}C(SiMe_3)(SiMe(t\text{-}Bu)_2))$ with weak Lewis bases have been determined. The structure of the THF adduct has been described.[34] It was found that the THF coordinated with the sp^2-hybridized silicon atom of the silene, distorting it from planar to near-tetrahedral, and causing an increase in bond length to 1.747 Å from 1.702 Å for the uncomplexed silene.[149a,b] Other complexes of this or closely related silenes with $EtMe_2N$, pyridine, and F^- ion have also been reported[141,150]; it was observed that the length of the double bond increased with increasing strength of the Lewis base, i.e., THF $<$ R_3N $<$ C_5H_5N $<$ F^-. Thus, the closely related but unstable silene $Me_2Si{=}C(SiMe_3)_2$ formed a stable complex with trimethylamine at room temperature[150] and the silene $Me_2Si{=}C(SiMe_2Ph)_2$ formed a complex with $EtMe_2N$ in which the length of the silicon–carbon double bond was 1.761 Å. The fluoride ion complex $FMe_2Si—C^-(SiMe_3)(SiMe(t\text{-}Bu)_2)$ $[Li(12\text{-crown-}4)_2]^+$ had a silicon–carbon bond length of 1.777 Å.[141]

2. *Thermal Stability of Geometric Isomers*

There has been much interest in the relative stabilities of silene geometric isomers, i.e., in the magnitude of the rotational barrier to isomer interconversion. In carbon chemistry the strength of the π bond is about 65 kcal mol^{-1}, whereas with the silicon–carbon double bond it is believed the height of the barrier to *cis/trans* isomerization is about 38 kcal mol^{-1}. [See Raabe and Michl[6] for a detailed review of the available data.] A variety of experiments bear this out.[151] From the results of the liberation of the isomers of $PhMeSi{=}CHCH_2CMe_3$ from its anthracene adduct, Jones[151a] estimated the value to be 43 ± 6 kcal mol^{-1}, which is consistent with previous calculations. Brook[5] heated the silene $(Me_3Si)_2Si{=}C(OSi$ $Me_3)Ad$ to 60°C without finding evidence that the nonequivalent Me_3Si groups were interconverting, observing as well that the remarkably stable silenes $(Me_3Si)MesSi{=}C(OSiMe_3)Ad$ could be heated to 100–120°C without evidence that the geometric isomers interconvert during thermal decomposition.[88]

C. *Spectroscopy*

1. *NMR Spectroscopy*

a. *^{29}Si Spectroscopy.* The chemical shifts of the sp^2 hybridized silicon atoms in silenes are found significantly downfield from the positions normally observed for sp^3-hybridized silicon atoms (e.g., 0 ppm for TMS), as might be expected from comparison with the carbon analogs, alkenes and alkanes. However, the data available are severely limited by the number of relatively stable silenes known. The shifts depend markedly on the nature of the substituents attached to the silicon atom. The available data are listed in Table V. Thus, in the Brook family of silenes $(Me_3Si)_2$ $Si{=}C(OSiMe_3)R$, the signals are usually found in the range 40–54 ppm, as a result of shielding of the sp^2-hybridized silicon atom by the attached trimethylsilyl groups. At the other extreme, silenes having two methyl groups, such as $Me_2Si{=}C(Ad)(SiMe(t\text{-}Bu)OSiMe_3)$ or the Wiberg silene $Me_2Si{=}C(SiMe_3)(SiMe(t\text{-}Bu)_2)$, absorb at 126.5 and 144.2 ppm, respectively. When one trimethylsilyl group is replaced by a mesityl group, little effect on the chemical shifts was observed; however, the recently isolated silene $Mes_2Si{=}CHCH_2CMe_3$ with two aryl groups on the sp^2-hybridized

TABLE V

^{29}Si AND ^{13}C NMR DATA FOR SILENES

Structure	δ of sp^2 Si Atom (ppm)	δ of sp^2 C Atom (ppm)	Reference
$(Me_3Si)_2Si{=}C(OSiMe_3)R$ R = Me_3C, Ad, 1-Me-1-Cyclohexyl, BCO[a]	41.4–43.5	212–214	105
$(Me_3Si)_2Si{=}C(OSiMe_3)CEt_3$	54.3	207	105
$Me_3Si(t\text{-}Bu)Si{=}C(OSiMe_3)Ad$	73.7	195.6	87
E-$(Me_3Si)MesSi{=}C(OSiMe_3)Ad$	41.8	195.7	88
Z-$(Me_3Si)MesSi{=}C(OSiMe_3)Ad$	44.0	191.6	88
$Me_2Si{=}C(Ad)(SiMe(t\text{-}Bu)OSiMe_3)$	126.5	118.1	87
$Me_2Si{=}C(CEt_3)(SiMePhOSiMe_3)$	[b]	110.6	87
$Me_2Si{=}C(SiMe_3)(SiMe(t\text{-}Bu)_2)$	144.2	77.2	149b
$Mes_2Si{=}CHCH_2CMe_3$	77.6	110.4	99a
$Me_3N \cdot Me_2Si{=}C(SiMe_3)_2$	36.9	56.5	150
$Me_3N \cdot Me_2Si{=}C(SiMe_3)(SiMe(t\text{-}Bu)_2)$	34.7	[b]	141
CMe_3 H H Me_3Si Si $SiMe_3$ CMe_3	26.8	126–127	108

[a] BCO = [2.2.2]Bicyclooctyl; [b] not observed.

silicon resonated at 77.6 ppm.[99a] In comparison, it is known that the sp^2-hybridized silicon atoms of disilenes resonate near 63 ppm; for example, the signal for tetramesityldisilene occurs at 63.6 ppm.[152] Few other silenes are stable enough to have been examined by NMR methods. Märkl reported that the chemical shift for a highly substituted silabenzene at −110°C occurred at 26.8 ppm.[108] The silicon atom in this compound may have been complexed with THF in the solvent, resulting in an upfield shift from its true position.

The chemical shifts of sp^3-hybridized silicon atoms, such as a trimethylsilyl group attached to the sp^2-hybridized silicon in silenes, are normally found from −10 to −18 ppm, as compared to the position at −9.8 ppm for the trimethylsilyl groups of $(Me_3Si)_4Si$ (the central silicon atom of this compound resonates at −135.5 ppm). On the other hand, a trimethylsiloxy group attached to carbon, either sp^2- or sp^3-hybridized, normally absorbs in the range from +5 to +20 ppm.

A few coupling constants have been observed for silenes.[105] The $^1J_{Si=C}$ coupling constants had values of 83.5–85.0 Hz, as compared to values of 47–48 Hz for the coupling constants of sp^3-hybridized silicon to attached methyl groups; and the values of $^1J_{Si-Si}$ coupling constants between sp^2-hybridized silicon and Me_3Si groups fell in the range 70–73 Hz.[105]

While not specifically related to sp^2-hybridized silicon, Auner[51] found that the chemical shift of the silicon atom in the products derived from reactions of the silenes $Cl_2Si{=}CRCH_2CMe_3$ with dienes was a useful diagnostic tool. Thus, most [2+4] cycloadducts, namely silacyclohexenes, had silicon NMR absorptions in the range 23–29 ppm, whereas their [2+2] isomers absorbed at higher field, namely in the range 13–18 ppm. Ene products derived from these reactions absorbed in the range 11–12 ppm.

b. *^{13}C Spectroscopy.* It has been possible to observe the influence of sp^2-hybridized silicon atoms (and other substituents) on the chemical shifts of the attached sp^2-hybridized carbon atoms. For most alkenes the observed chemical shifts fall in the range 100–150 ppm. For silenes the range is much larger, from 56 to 214 ppm. For the family of silenes $(Me_3Si)_2Si{=}C(OSiMe_3)R$, the sp^2-hybridized carbon atom is observed in the range 207–214 ppm when groups such as *t*-Bu, Ad, CEt_3, etc., are attached. The large downfield shift undoubtedly reflects the deshielding influence of the attached trimethylsiloxy group At the other extreme, the sp^2-hybridized carbon atom of the Wiberg silene $Me_2Si{=}C(SiMe_3)(SiMe(t\text{-}Bu)_2)$ resonated at 77 ppm, reflecting the shielding influence of the attached silyl groups, and the silene $Mes_2Si{=}CHCH_2CMe_3$ resonated at 110.4 ppm.[99a] The sp^2-hybridized carbon of the silene complex $Me_2Si{=}C(SiMe_3)_2 \cdot NMe_3$ resonated at 56.5 ppm,[150] reflecting the strong effect of the Lewis base on the sp^2-hybridized silicon atom. The known values of the sp^2-hybridized carbon chemical shifts in silenes are listed in Table V along with the silicon resonances.

2. *Ultraviolet and Visible Spectroscopy*

Details of the ultraviolet absorption maxima for simple silenes, silaaromatics, and for some relatively stable silenes are known and have been summarized.[6] The simplest silenes absorb in the region 245–260 nm, with unknown extinction coefficients; but as the substituents become increasingly complex, the λ_{max} values of the silenes increase until, with the silene $(Me_3Si)_2Si{=}C(OSiMe_3)Ad$, the absorption occurs at 340 nm[5] with an extinction coefficient of about 7400, consistent with a π–π^* transition. A few further studies of interest are summarized below.

Several papers have described the UV spectra of some relatively recently reported silenes.[19,20,25] These materials, bearing two methyl groups on sp^2-hybridized silicon and a variety of groups (H, Me, Ph, CO_2Me, CO_2Et, COAd, $SiMe_3$) on sp^2-hybridized carbon, were prepared by the photoinduced decomposition of α-diazosilanes and isolated in an argon matrix at 77 K, under which conditions their UV absorption was obtained (see Section II.B.2). As the parent silene is sequentially substituted with $SiMe_3$ groups, a bathochromic shift is observed; also, the addition of conjugating groups similarly leads to a red shift (see Table VI).

An elegant study of the UV–visible and IR spectra of silacyclopentadienes and details of the photochemical interconversion of the isomeric species **48**, **49**, **50**, and **51** (Eq. 21) has been presented[115] and supported by *ab initio* calculations. Gaspar[116] had postulated, based on thermal experiments claiming that silole (silacyclopentadiene) **51** was formed, that cyclopentenylidene **48** rearranged via the isosilole **49** to 1-silacyclopentadiene **51**. This speculation was confirmed by Michl and colleagues,[115] who established—through synthesis, trapping, matrix isolation, and photochemical isomerization—the complex interrelationship between the three species and the alternative isosilole **50** both for the unsubstituted and the 3,4-dimethyl-substituted series of compounds. The compounds were characterized spectroscopically. The conversion of **48** to **49** presumably involved a 1,2-H shift due to silylene insertion in the C—H bond, and the other rearrangements can be explained by 1,3- or 1,5-sigmatropic rearrangements. The relationships are summarized in Eq. (21).

48 $\underset{h\nu}{\overset{h\nu,\ \Delta}{\rightleftharpoons}}$ **49** $\xrightarrow{\Delta}$ **51**; **49** $\underset{h\nu}{\overset{h\nu}{\rightleftharpoons}}$ **50** $\underset{h\nu}{\overset{h\nu}{\rightleftharpoons}}$ **51** (21)

R = H, Me

For R = H the silole **51** had a λ_{max} at 278 nm while the silylene **48** absorbed at 250 and 480 nm. The isomeric silenes **49** and **50** absorbed at 296 and 270 nm, respectively. The UV absorptions for these species have been calculated and assigned, and their IR spectra have also been obtained. When R = Me, there was little change in the λ_{max}, the four species absorbing at 280, 255 and 480, 312, and 274 nm, respectively.

TABLE VI

ULTRAVIOLET ABSORPTION MAXIMA OF VARIOUS SILENES

Silene	λ_{max} (nm)
$Me_2Si{=}CHMe$	255
$Me_2Si{=}CHSiMe_3$	265
$Me_2Si{=}C(Me)SiMe_3$	274
$Me_2Si{=}C(SiMe_3)_2$	278
$Me_2Si{=}C(Ph)CO_2Me$	280
$Me_2Si{=}C(SiMe_3)C{=}O$ adamantyl	284
$Me_2Si{=}C(SiMe_3)CO_2Et$	288

Using nanosecond laser flash photolysis techniques, Leigh[80] observed transient absorption spectra which he attributed to the silenes derived from photolysis of various methylphenyldisilylbenzenes. Thus the silenes **52**, **53**, and **54** were found to absorb at 425, 460, and 490 nm, respectively, in isooctane, and **55** was also found to absorb at 490 nm.[75] In other studies, the silene $Ph_2Si{=}CH_2$ derived by laser flash photolysis was found to absorb at 323 nm.[111]

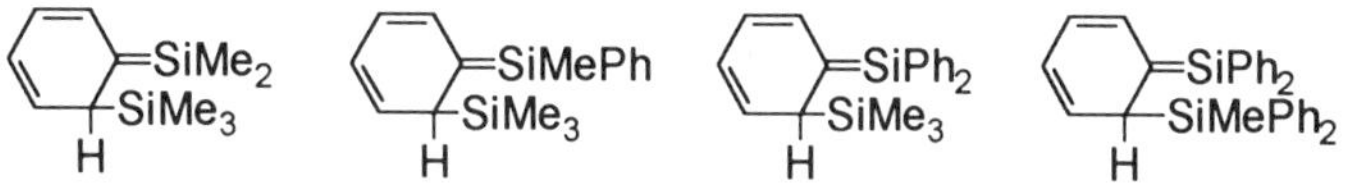

=SiMe$_2$, SiMe$_3$, H — **52**; =SiMePh, SiMe$_3$, H — **53**; =SiPh$_2$, SiMe$_3$, H — **54**; =SiPh$_2$, SiMePh$_2$, H — **55**

These species, which had a lifetime of only a few microseconds, decayed following first-order kinetics.

Conlin[113] reported that the λ_{max} for the siladiene $Me_2Si{=}C(Ph)CH{=}CH_2$ occurred at 338 nm in 3-MP glass.

3. *Infrared Spectroscopy*

Considerable information is available about the infrared absorption of silenes. This information was obtained in large part from experimental studies of simple species such as $H_2Si{=}CH_2$ and $Me_2Si{=}CH_2$ in argon matrices at about 10 K or from calculations, and in small part from room-temperature studies of a few relatively stable silenes in solution.

Bands in the region 950–1100 cm^{-1} have been attributed to the Si=C stretching frequency of the simple silenes, the frequency reported depending on the experimental method, the methodology employed in the calculations, and the substituents present on the ends of the silicon–carbon double bonds: there is generally good agreement between these observations and the appropriate calculations. The material has been well summarized.[6,153]

Similar observations have been made on a few silaaromatics, such as silabenzene and 1-silatoluene, but it is not known which of the many absorption bands observed are associated with the silicon–carbon framework. The same is true for the stable silenes such as $(Me_3Si)_2Si{=}C(OSi Me_3)Ad$. Both in solution and the solid state, a number of bands were observed in the region 1300–930 cm^{-1}, but it is not known with certainty which are associated with the Si=C bond.

In a detailed study of the IR transitions of matrix-isolated 1-methylsilene and its isomer dimethylsilylene the structures of these species[154] have been clearly defined.

4. *Other Spectroscopic Methods*

In a pair of nicely titled papers "Silicon–Carbon Double Bonds: Theory Takes a Round," Gutowsky *et al.*[155,156] noted the controversy that arose when an electron-diffraction determination[157] of the Si=C bond length in $Me_2Si{=}CH_2$ showed it to be 1.83 Å. This value, determined before the crystal structures of the solid silenes $(Me_3Si)_2Si{=}C(OSiMe_3)Ad$ and $Me_2Si{=}C(SiMe_3(SiMe(t\text{-}Bu)_2)$ had been reported, was much longer than theory had predicted. The experimentally measured microwave spectrum of dimethylsilene derived by thermolysis of 1,1-dimethylsilacyclobutane revealed *b*-dipole rotational transitions, and calculations based on these data led to details of the shape of the molecule and to a Si=C bond length of 1.692 Å, consistent with values obtained from *ab initio* calculations.[158]

D. *Computations*

Much effort has been expended on calculating some of the important physical properties of silenes. The early work to 1985 is well summarized by Raabe and Michl[6]; more recent work is given in great detail, particularly work on the calculations of geometries, vibration frequencies, and rotational energies of $H_2Si{=}CH_2$, and determination of the relative energies of this silene and its isomers Me(H)Si: and $H_3Si(H)C$:.[153] Two recent calculations werc made on the relative energies of silaethylene (silene = $H_2Si{=}CH_2$) and methylsilylene (Me(H)Si:). Goddard *et al.*[159] estimated that silene was about 10 kcal mol^{-1} more stable than the silylene, whereas Schaefer *et al.*,[160] using somewhat different methods, found the energy difference to be about 4 kcal mol^{-1}, close to previous theoretical estimates.

Several further calculations of various properties of the silicon–carbon double bond have been reported. As in the previous decade the strength of the silene bond and the reasons for it have continued to attract attention.

There appears to be some consensus that the reactivity of the silene π-bond is a manifestation primarily of the high polarization of the bond, which promotes kinetic reactivity. Recent investigations have focused on the distinction between bond strength and the barrier to rotation around the Si=C bond. In ethylene, the two are the same. Owing to the proclivity of silicon and other heavier heteroatoms toward pyramidalization, however, a distinction must be made between the planar (**56**) and pyramidalized (**57**) biradical intermediates (shown in Scheme 9); the latter is found to be more stable than the former.[161] Using the intrinsic reaction coordinate (IRC: minimum-energy pathway connecting reactant to products via the transition state) to calculate the rotational barrier in silaethylene ($H_2Si{=}CH_2$), the rotational energy barrier through the pyramidalized silicon triplet biradical was found to be 37 kcal mol^{-1}, which is in good agreement with experimental estimates. An alternative theoretical treatment by Hehre, based on the energies of disproportionation of the products of hydrogen-atom addition, gave values of 35–36 kcal mol^{-1}.[162] In either case, the triplet lies below the singlet in energy.

Ab initio (3-21G(*)//STO-3G) calculations by Chandrasekhar and Schleyer[163] on 1,4-disilabenzene **58**, its Dewar benzene isomer **59**, and a silylene isomer **60** showed that all three species exhibited approximately similar stabilities, the silylene **60** being 9.9 kcal mol^{-1} more stable than the planar aromatic form **58**, which was 5.9 kcal mol^{-1} more stable than the Dewar benzene form **59**.

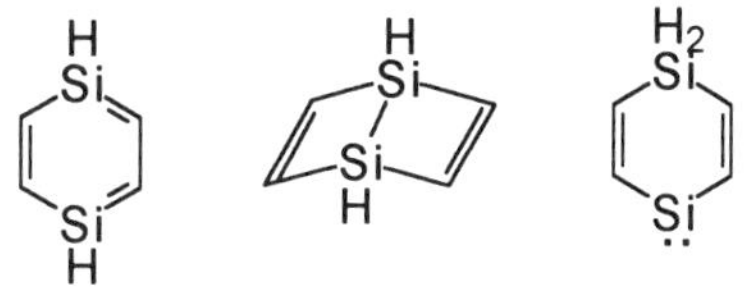

58 **59** **60**

The aromaticity in **58** is considerably lower than that in benzene itself. These findings contrast with the results for monosilabenzene and its related

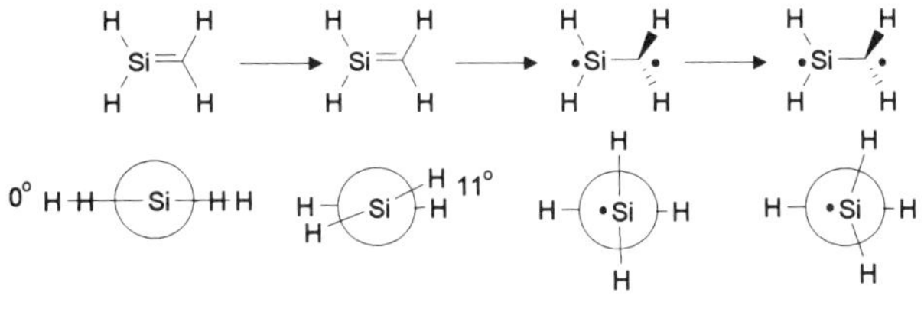

0° 11° **56** **57**

SCHEME 9

isomers, where the isomeric silylenes were about 20 kcal mol^{-1} less stable than the silaaromatic, and the Dewar benzene analog was about 38 kcal mol^{-1} less stable.[164]

A substituent study on silabenzene has been reported by Baldridge and Gordon.[165] A group of substituents (Cl, F, SH, OH, PH_2, NH_2, CH_3, SiH_3, NO_2, CN, OCH_3, and COOH) was examined in the *ipso, ortho, meta,* and *para* positions of silabenzene using the STO-2G basis set. With the exception of NH_2, lone pairs on heteroatoms bound to silicon did not enter into π-bonding interactions with silicon. The degree of aromaticity of the substituted silabenzenes was found to vary from 57 to 93% of the parent (nonsilylated) benzene. For all substituents, irrespective of their position on the ring, negative charge buildup at silicon was not predicted.

IV
CHEMICAL BEHAVIOR

A. *Background*

Prior to 1985, much had been learned about the chemistry of the silicon–carbon double bond through the study of the reactions of silenes with a wide variety of reactants. Thus it was known that all silenes studied reacted readily with alcohols (particularly methanol) by regiospecific addition across the ends of the Si═C bond in which the MeO group became attached to silicon and the alcoholic H to carbon, as in Eq. (22).

$$\mathrm{PhMeSi{=}CHCH_3} + \mathrm{MeOH} \longrightarrow \mathrm{PhMe\underset{\underset{MeO}{|}}{Si}{-}\underset{\underset{H}{|}}{C}HMe}$$

$$+ \mathrm{Me_3SiOMe} \longrightarrow \mathrm{PhMe\underset{\underset{MeO}{|}}{Si}{-}\underset{\underset{SiMe_3}{|}}{C}HMe} \qquad (22)$$

A wide variety of other polar reagents had also been shown to add across the silicon–carbon double bond, and trimethylmethoxysilane (see Eq. 22) had been found to be another rather useful reagent for trapping reactive silenes.

It was also well established that silenes could take part as the dienophile in Diels–Alder reactions. In many cases, particularly with unsymmetric dienes such as isoprene, the reactions were not clean because, in addition to formation of the [2+4] cycloadduct **61**, the possibility exists for the formation of it regioisomer **62**, products of an ene reaction **63**, and conceivably the [2+2] cycloaddition product **64**, as shown in Eq. (23). Wiberg observed from studies of the relative rates of reaction of the silene $Me_2Si{=}C(SiMe_3)_2$ with a variety of 1,3-dienes[98] that the more electron-rich the diene, the more rapid the reaction with silenes, provided that steric effects are not a major factor.

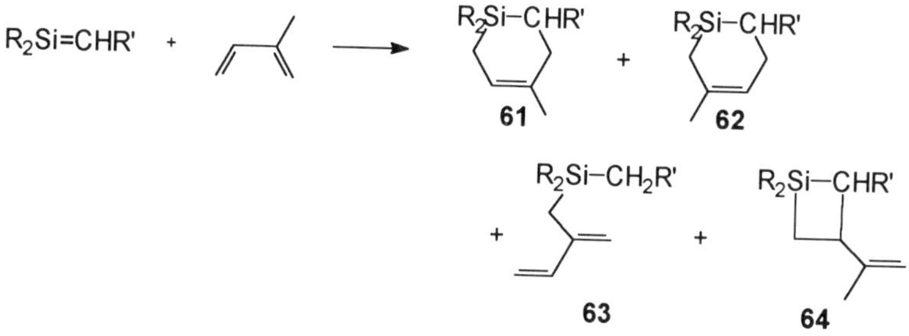

$R_2Si{=}CHR'$ + 2,3-dimethylbutadiene → **61** + **62** + **63** + **64** (23)

Reactions of silenes with carbonyl compounds had also been observed. The ene reaction yielding a product with the structure **65** was seen to predominate for carbonyl compounds that had hydrogen alpha to the carbonyl group; however, in the absence of reactive α-hydrogen, the products of [2+4] cycloaddition **66** were observed in a few cases, as with benzophenone whose aromatic ring provided two of the necessary π-electrons.[166] It was also believed that [2+2] reaction of the silene with the carbonyl group occurred in many cases, yielding initially the siloxetane **67**, which, being unstable, decomposed spontaneously to an alkene and a silanone oligomer, as shown in Eq. (24).

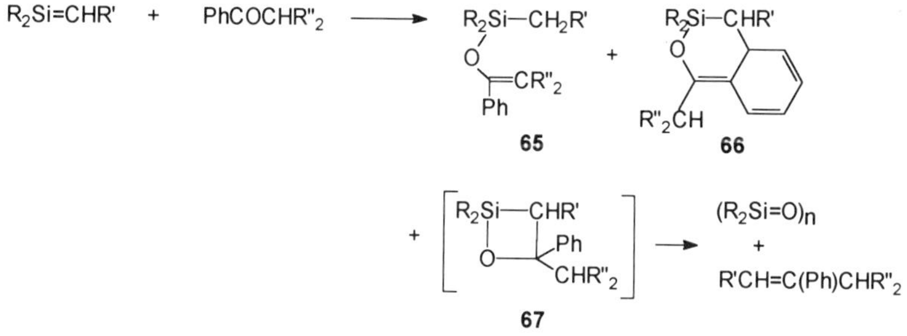

$R_2Si{=}CHR'$ + $PhCOCHR''_2$ → **65** + **66** + [**67**] → $(R_2Si{=}O)_n$ + $R'CH{=}C(Ph)CHR''_2$ (24)

Wiberg had also described a number of [2+3] cycloadditions of his silenes with 1,3-dipolarophiles such as N_2O, diazo compounds, and azides.[98]

During the last decade most of the above reactions have been studied in much further detail, utilizing a wide variety of silenes; thus a much greater understanding of the mechanisms of the reactions has been accumulated. These results will be described in some detail in the sections below.

B. *Dimerization*

Dimerization is a special case of [2+2] cycloaddition; with silenes it has been observed to occur in both a head-to-tail and in a head-to-head manner, yielding 1,3- or 1,2-disilacyclobutanes. These two cases will be discussed separately below.

1. *Calculations*

The mechanisms by which silenes dimerize, both head-to-tail and head-to-head, have been the subject of intense interest to theoreticians for as long as the dimerization process has been known. Siedl, Grev, and Schaefer[167] calculated that the barrier between the parent $H_2Si{=}CH_2$ **68** and its head-to-tail dimer **69** was 5.2 kcal mol^{-1}, with the product head-to-tail dimer **69** being 79 kcal mol^{-1} more stable than reactants. The head-to-head dimer **70** was found to be 19.8 kcal mol^{-1} less stable than **69**.

A concerted 2_s+2_s cycloaddition (*supra–supra*) is forbidden by the Woodward–Hoffman rules. However, the strong polarization of the silicon–carbon double bond, $Si^{\delta+}{—}C^{\delta-}$, results in a relaxation of the rules. A simple correlation diagram allowing for the reduction in orbital symmetry for the appropriate space group indicates that a concerted pathway for the head-to-tail dimerization becomes allowed; the concerted *supara–antara* process is also allowed, but is sterically hindered. These theoretical studies have indicated that, for the head-to-tail cycloaddition, the concerted reaction can take place. In contrast, the formation of the 1,2-disilacyclobutane was calculated to occur through a two-step biradical intermediate **71**. As a consequence of the ability of silicon to stabilize α- and β-carbon radicals, together with steric factors, the head-to-head pathway involving initial Si—Si bond formation was predicted to have a lower reaction barrier than the diradical formed by C—C coupling.

This proposal is completely at odds with the results of Bernardi *et al.*[168,169] Their calculations suggest that the $[2\pi_s+2\pi_s]$ cycloaddition cannot take place and that head-to-tail dimerization also occurs through a biradical intermediate with a barrier of 2.5 kcal mol^{-1}. This view is discussed in great detail, and the results are compared with other, related dimerizations including that of ethylene. They note that the $[2\pi_s+2\pi_s]$ cycloaddition can become the dominant process in the presence of polar solvents with polarized double-bonded species. The transition state (conical intersection) for the photoinitiated $[2_s+2_s]$ dimerization of silaethene was also described.[169] The predictions of Bernardi *et al.* have recently been challenged by Grev, who was able to calculate details of the transition states leading to concerted formation of the dimers.[169a]

2. Head-to-Tail Dimerization

In the absence of trapping agents, almost all silenes dimerize in a [2+2] manner. The few exceptions are those species having very bulky groups attached to at least one end of the Si=C double bond, where steric effects are presumed to hinder the process. As implied above, the most common pathway of most silenes is head-to-tail dimerization, which is consistent with the significant polarization of the $Si^{\delta+}=C^{\delta-}$ bond in most compounds. A strongly polarized bond appears to be a major factor favoring head-to-tail dimerization. Some typical examples of silenes forming head-to-tail dimers are shown in Scheme 10. The first examples involve silenes prepared by the addition of *t*-butyllithium to vinylchlorosilanes, where dimer-

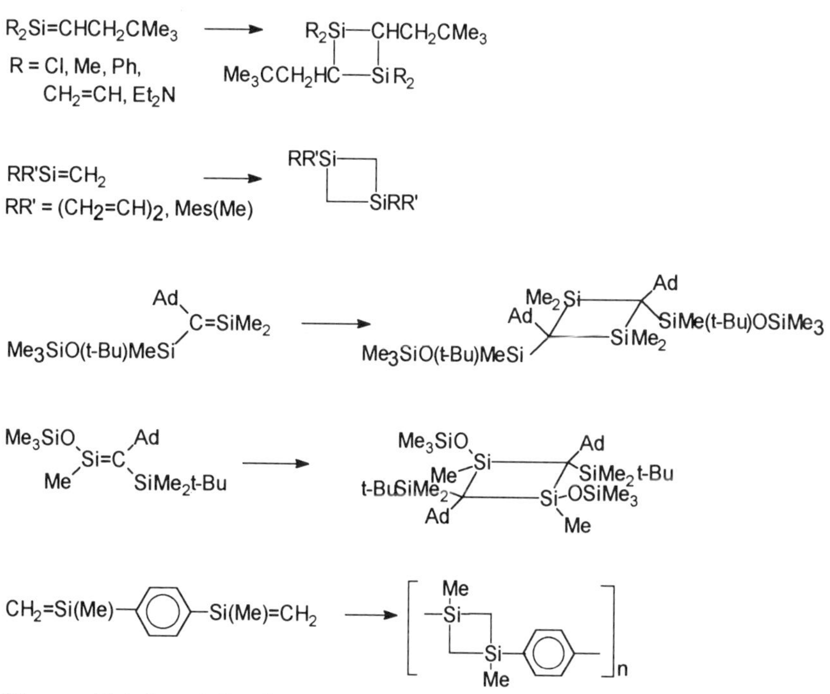

$R_2Si=CHCH_2CMe_3$ → $R_2Si-CHCH_2CMe_3$ / $Me_3CCH_2HC-SiR_2$ (four-membered ring)
R = Cl, Me, Ph, $CH_2=CH$, Et_2N

$RR'Si=CH_2$ → $RR'Si$ / $SiRR'$ (1,3-disilacyclobutane)
RR' = $(CH_2=CH)_2$, Mes(Me)

$Me_3SiO(t\text{-}Bu)MeSi(Ad)C=SiMe_2$ → Me_2Si, Ad, $SiMe(t\text{-}Bu)OSiMe_3$, Ad, $SiMe_2$, $Me_3SiO(t\text{-}Bu)MeSi$

$Me_3SiO(Me)Si=C(Ad)(SiMe_2t\text{-}Bu)$ → Me_3SiO, Si, Me, Ad, $SiMe_2t\text{-}Bu$, $t\text{-}BuSiMe_2$, Ad, $Si\text{-}OSiMe_3$, Me

$CH_2=Si(Me)-C_6H_4-Si(Me)=CH_2$ → $[-Si(Me)-CH_2-Si(Me)(-C_6H_4-)-CH_2-]_n$

Silenes which do not dimerize.

$Me_2Si=C(SiMe_3)(SiMe(t\text{-}Bu)_2)_2$

$(Me_3Si)_2Si=C(OSiMe_3)R$ R = Ad, CEt_3, Mes,

$(Me_3Si)R'Si=C(OSiMe_3)Ad$ R' = t-Bu, Mes

SCHEME 10

ization occurs spontaneously at room temperature or lower.[48,170–172] The second entry involves silenes produced by the thermolysis of silacyclobutanes at high temperature[15] or by the photolysis of di- or trisilanes,[110] whereas the third and fourth entries involve silenes formed by photochemical rearrangement of the silene (Me_3Si)-*t*-BuSi=C$(OSiMe_3)$Ad, which is itself too sterically hindered to dimerize.[87] The last entry constitutes a case in which the bis-silene was generated by thermolysis of a bis-1,4-phenylenesilacyclobutane. This compound apparently polymerized spontaneously.[94] At the bottom of the scheme are listed several silenes that do *not* undergo any form of dimerization, thereby providing a sense of the size and location of groups necessary to inhibit dimer formation.

3. *Head-to-Head Dimerization*

Head-to-head dimerization is a less common mode of dimerization. It occurs in two ways. In the first, the silene forms a 1,2-disilacyclobutane by a process believed to be initiated by silicon–silicon bond formation; this process probably involves the formation of a 1,4-diradical species whose radical centers are located on carbon. In one telling example, photolysis of the acylsilane $(Me_3Si)_3SiCOCMe_3$ **72** gave the silene $(Me_3Si)_2Si{=}C(OSiMe_3)(CMe_3)$ **73**, which is relatively stable and can be characterized by its NMR spectra: it was shown to be in a temperature-sensitive dynamic equilibrium with its head-to-head dimer **74** (Eq. 25), a solid whose crystal structure has been obtained.[93] An indication that the dimer would be relatively unstable was the finding that the length of the ring carbon–carbon bond was 1.66 Å, one of the longest carbon–carbon bonds reported.

The dimer showed no ESR signal as a solid; but when it was dissolved in pure solvent, it gave rise to a strong characterless ESR signal, which was virtually identical to that shown by the original photolysis solution. Attempts to trap the putative diradical **75** failed, but do not rule out its possible existence as an intermediate in low concentration in the system.

$(Me_3Si)_3SiCOCMe_3$ (**72**) $\xrightarrow{h\nu}$ $(Me_3Si)_2Si{=}C(OSiMe_3)(CMe_3)$ (**73**) ⇌ **74** [cyclic dimer: $(Me_3Si)_2Si$–$C(OSiMe_3)(CMe_3)$ / $(Me_3Si)_2Si$–$C(CMe_3)(OSiMe_3)$]

73 ⇌ **75** [diradical: $(Me_3Si)_2Si$–$\dot{C}(OSiMe_3)$–CMe_3 / $(Me_3Si)_2Si$–$\dot{C}(OSiMe_3)$–CMe_3] ⇌ **74**

(25)

When the Me_3C group in the above system (see Eq. 25) was replaced by the bulkier Ad group, no dimer was formed and the silene was isolated as a crystalline solid. Evidently, the increased steric hindrance inhibited dimer formation by preventing formation of the ring C—C bond.

The result of replacing one of the Me_3Si groups attached to the sp^2-hybridized silicon atom of the silene by another group depended on the bulk of the new substituent. If a methyl group or phenyl group replaced one of the Me_3Si groups, the resulting silene dimerized to a rather unstable 1,2-disilacyclobutane,[87] which readily dissociated back to silene and reacted in methanol solvent (Eq. 26).

$$(Me_3Si)RSi{=}C(OSiMe_3)Ad \rightleftharpoons \text{1,2-disilacyclobutane: } [Me_3Si(R)Si{-}C(OSiMe_3)Ad]_2 \quad R = Me, Ph$$

$$Ph_2Si{=}C(OSiMe_3)Ad \rightleftharpoons [Ph_2Si{-}C(OSiMe_3)Ad]_2 \tag{26}$$

When phenyl (Ph) groups replaced both Me_3Si groups, again a rather unstable 1,2-disilacyclobutane dimer appeared to be formed,[90] as shown by NMR data; but when *t*-butyl replaced a Me_3Si group, the silene failed to dimerize.[87] Thus, it is evident that whether or not head-to-head [2 + 2] cyclodimerization occurs depends on the bulk of the substituents on both sp^2-hybridized silicon and carbon.

The second form of head-to-head dimerization involved the formation of a linear (as distinct from a cyclic) species in which two molecules of silene form a silicon–silicon bond. If this follows the pathway suggested above in Eq. (25), the resulting 1,4-diradical must then disproportionate by hydrogen abstraction, forming a molecule saturated at one end and unsaturated at the other. Recent examples are given in Eq. (27).[86]

$$(Me_3Si)_2Si{=}C(OSiMe_3)CHRR' \rightarrow (Me_3Si)_2Si(-\dot{C}(OSiMe_3)CHRR')-Si(Me_3Si)_2-\dot{C}(OSiMe_3)CHRR' \rightarrow (Me_3Si)_2Si(-C(OSiMe_3)(H)CHRR')-Si(Me_3Si)_2-C(OSiMe_3){=}CRR'$$

$$CHRR' = Me, Et, i\text{-}Pr, CH_2Ph$$

$$\text{(alternative path via two } (Me_3Si)_2Si{=}C(OSiMe_3)CHRR' \text{ molecules with H transfer)} \tag{27}$$

In some cases, such as the isopropyl or benzyl compounds in Eq. (27), both the cyclic [2 + 2] dimer and the linear dimer were formed in appreciable amounts.[86]

Relevant to the question of the mode of dimerization, Conlin[173] has studied the kinetics of dimerization of the silene $(Me_3Si)_2Si{=}C(OSiMe_3)Me$ using laser flash photolysis in which the linear head-to-head dimer is formed. In cyclohexane at 300 K or in a 3-MP matrix at 77 K, an intense absorption was observed at 330 nm, with an extinction coefficient of at least 6500 or greater that was attributed to the silene (see also Section III.C.2). Clean second-order decay kinetics were observed on a millisecond time scale, yielding a bimolecular rate constant of $1.3 \times 10^7\ M^{-1}\ s^{-1}$, $\log A = 7 \pm 1\ s^{-1}$, and $E_{act} = 0.2 \pm 0.1$ kcal mol^{-1}. These parameters imply a strongly ordered transition state, which is probably more in accord with an ene mechanism for the reaction, as shown in the lower part of Eq. (27), than one involving a 1,4-diradical intermediate. In the same study, the rate of quenching of the silene by oxygen at 20°C was measured and found, under pseudo–first-order conditions, to be $7.3 \pm 1.5 \times 10^{-5}$ $M^{-1}\ s^{-1}$, a relatively slow process. Attempts to study the rate of reaction of the silene with 2,3-dimethylbutadiene under flash photolysis conditions failed because the reaction was so slow, even at high concentrations of the diene, so the rate constant for the reaction must be much less than $10^5\ M^{-1}\ s^{-1}$. When the acylsilane and diene were photolyzed continuously, the products obtained included the [2 + 4] cycloadduct, the ene product, and the head-to-head dimer, the latter predominating at all concentrations of diene employed.

Scheme 11 shows three other head-to-head dimerizations that were recently reported. The first two arose from reactions of $(Me_3Si)_3SiLi$ with ketones[58] in a Peterson-type reaction. Even though the former bears allylic hydrogen, the strain (Bredt rule) that would be involved in a disproportionation process is evidently too great to lead to a "linear" dimer; surprisingly, however, the head-to-head dimer was not too sterically hindered to be formed. The second example is a more typical "linear" head-to-head dimer. The third dimer resulted from treatment of $(Me_3Si)_3SiCOPh$ with MeLi.[60]

In the past decade, Ishikawa *et al.* have investigated the photochemistry of aryldisilanes.[66,73,76,84] Lately, dimers derived from these unusual silenes have been observed from the first time.[78] Thus, photolysis of 1,4-bis(pentamethyldisilyl)benzene **76** in hexane presumably gave rise to the silene **77**; but, on workup employing crystallization, two head-to-head stereoisomeric dimers **78** and **79** were obtained in about 45% yield in a 1 : 1 ratio. These were said to be formed from the silenecyclohexadienes **77** by the two alternative pathways shown in Scheme 12. Related dimers were also

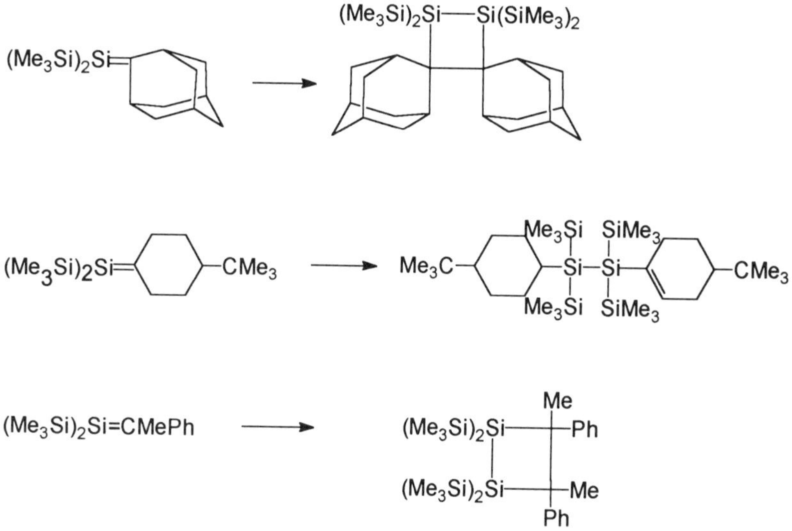

SCHEME 11

formed from the silene as a result of photolysis of 1-pentamethyldisilyl-4-trimethylsilylbenzene, and a crystal structure was obtained for one of the isomers.

Finally, Eq. (28) documents a unique situation in which a bis-silene was observed to undergo both head-to-tail and head-to-head [2 + 2] cyclodimerizations. Photolysis of the bisdiazoalkane **80** in benzene apparently

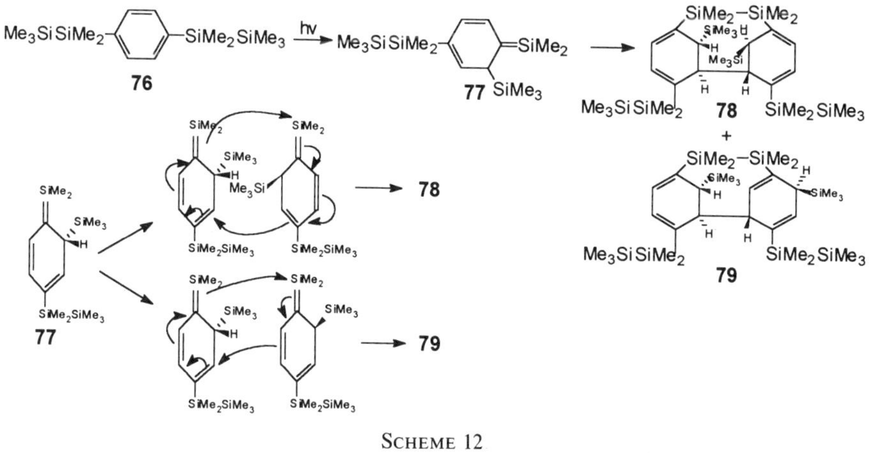

SCHEME 12

yielded the bis-silene **81**, which formed two isomeric species—the head-to-tail "crisscross" intramolecular addition compound **82** and the head-to-head "parallel" adduct **83** in low yields.[27] Thermolysis of the bisdiazo compound gave the criss cross product in 24.6% yield.

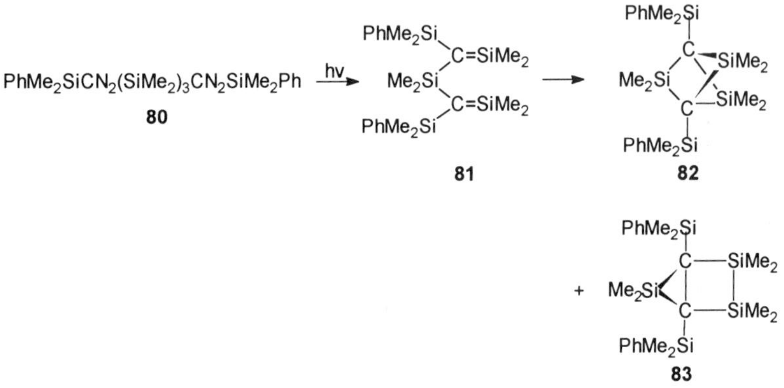

$$PhMe_2SiCN_2(SiMe_2)_3CN_2SiMe_2Ph \xrightarrow{h\nu} \mathbf{81} \longrightarrow \mathbf{82} + \mathbf{83} \qquad (28)$$

Some unusual behaviour was displayed by the benzodisilacyclobutane **84** as described by Ishikawa *et al.*[95] When thermolyzed, it appeared to form the quinodimethane bis-silene species **85** shown in Scheme 13, as confirmed by trapping reactions with *t*-butyl alcohol, alkynes, or aldehydes, all of which added in a 1,4-manner (see Scheme 13). In the absence of a trapping reagent, **85** decomposed, but not to **86** as claimed earlier.[95a]

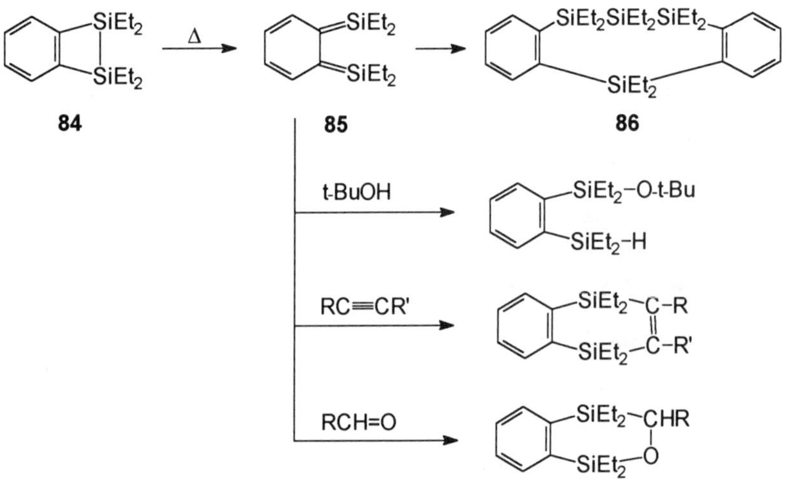

SCHEME 13

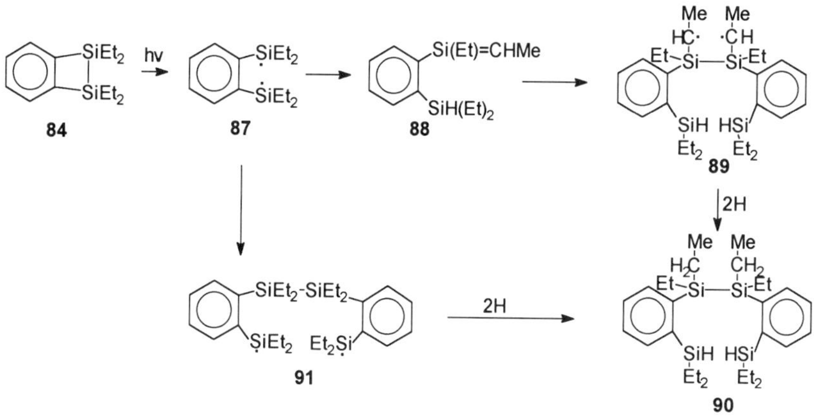

SCHEME 14

Alternatively, it was suggested that if the benzocyclobutane **84** is photolyzed, homolysis of the silicon–silicon bond leads to the diradical **87**, which then disproportionates to the silene **88**.[97] The silene was said to undergo head-to-head (silicon–silicon) coupling, giving a 1,4-diradical **89** that ultimately acquires hydrogen from an unspecified source to yield the observed final product **90**, as shown in Scheme 14. We suggest a different interpretation involving intermolecular radical coupling of **87** with itself to give thc silicon-centered diradical **91**. Failing to undergo further radical coupling leading to ring formation because of lack of proximity of the radical centers, **91** instead abstracts hydrogen from the solvent or other sources to give the observed product **90**. When the starting material was photolyzed in the presence of *t*-butyl alcohol, the adduct of the silene **88** with the alcohol was a major product, thus demonstrating that the initial diradical **87** could disproportionate to the silene in the presence of a polar solvent.

C. *Cycloaddition and Related Reactions*

Recent study of silenes has clearly established their ability to act as dienophiles in [2 + 4] Diels–Alder-type reactions involving 1,3-dienes. However, it has also been clearly demonstrated that products of the ene reaction commonly accompany the [2 + 4] cycloadducts and may on occasion be the major products. In addition, unlike those in carbon chemistry, [2 + 2] cycloadditions are often observed to occur in competition with the above processes—not only from reactions of silenes with dienes but also from reactions of silenes with alkenes, provided they do not possess allylic hydrogen (e.g., styrene): the presence of allylic hydrogen normally results in the products of the ene reaction. The silene may serve either as the enophile or the ene component. Frequently, the yields of [2 + 2] products exceed the yields of [2 + 4] products in the reactions of silenes with certain dienes, especially butadiene. Unlike the [2 + 4] and ene reactions, which are believed to occur by concerted mechanisms, the [2 + 2] reaction is generally considered to be a stepwise process involving either a 1,4-dipolar intermediate or a 1,4-diradical intermediate. Finally, some silenes readily enter into $[2\pi + 2\sigma + 2\sigma]$ processes with reactants like quadricyclane. It is evident that the relative proportions of products from the above reactions depend strongly both on electronic effects, such as the polarity of the Si=C bond and the electron density in the diene system, and on steric factors inherent in both the silene and the diene.

Two comprehensive studies of the reactions of Wiberg type and of the Jones–Auner type silenes with dienes will be described immediately below, without separating the results into the separate subsections of [2 + 4], [2 + 2], and ene reactions. The overall results of their investigations, given in one location, will allow a better appreciation of the effects that the substituents on the silene or diene have in determining which reaction takes place.

Wiberg *et al.*, in a series of papers,[30,33,174–177] have reported the results of a systematic study of the behavior of the silene $Me_2Si{=}C(SiMe_3)_2$ toward cycloaddition and related reactions, together with limited studies on other silenes. These papers contain a wealth of data about the properties and yields of the products obtained. Since the silene is not stable, it was liberated in each case by mild thermolysis at about 100°C from its [2 + 2] adduct with *N*-trimethylsilylbenzophenonimine. The results given for the various cycloadditions were generally obtained under conditions employing a large excess of diene relative to the silene precursor, as well as sufficient time and/or heat to take the reactions to completion.[174] A selection of the results is given in Table VII, but the yields should not always be taken as entirely accurate. The methods of analysis have improved over the years; moreover, in some cases, the yields may be based on NMR analyses whereas in other cases they may be yields of isolated material.

As shown in Table VII, [2 + 4] cycloaddition is the most common reaction pathway followed by $Me_2Si{=}C(SiMe_3)_2$, but it is usually accompanied by significant quantities of the product of an ene reaction. As the diene becomes more sterically hindered in its *s-cis* conformation, as in *cis*/*trans*-2,4-hexadiene, the product of an ene reaction predominates. With butadiene, where minimal steric effects are to be expected, the exclusive product of the reaction was found to be the [2 + 4] cycloaddition

TABLE VII

PRODUCTS FROM THE REACTION OF $Me_2Si{=}C(SiMe_3)_2$ WITH DIENES (% YIELDS)

Silene	Diene	[2 + 4]	Ene	Other	Reference
$Me_2Si{=}C(SiMe_3)_2$	butadiene	66X[a]			174
$Me_2Si{=}C(SiMe_3)_2$	2-methylbutadiene	72 (2 isomers)	14	15[b]	175,176
$Me_2Si{=}C(SiMe_3)_2$	2,3-dimethylbutadiene	73	25		174
$Me_2Si{=}C(SiMe_3)_2$	*trans*-piperylene	73	25.5	1.5[b]	175
$Me_2Si{=}C(SiMe_3)_2$	*trans*/*trans*-2,4-hexadiene	95	5		177
$Me_2Si{=}C(SiMe_3)_2$	*cis*/*trans*-2,4-hexadiene	20	80		177
$Me_2Si{=}C(SiMe_3)_2$	cyclopentadiene	100			174
$Me_2Si{=}C(SiMe_3)_2$	anthracene	78			174
$Me_2Si{=}C(SiMe_3)(SiMe(t\text{-}Bu)_2)$	butadiene	100			33
$Ph_2Si{=}C(SiMe_3)_2$	2,3-dimethylbutadiene	30	35	12[b]	30

[a] X = exclusive product; [b] [2 + 4]-adduct of ene product.

product.[174] When the related silene $Me_2Si{=}C(SiMe_3)(SiMe(t\text{-}Bu_2))$ was employed, the product was the [2 + 4] cycloadduct, also obtained quantitatively.[33] Reaction with the less sterically hindered *trans*-piperylene gave exclusively the [2 + 4] cyclic adduct, but the more hindered *cis*-piperylene gave only the [2 + 2] adduct.[178]

In addition to the product data given in Table VII, Wiberg established the relative rates of reaction of numerous dienes and other reactants with the silene $Me_2Si{=}C(SiMe_3)_2$. Using the rate of [2 + 4] cycloaddition with butadiene as the standard (rel. rate = 1), Wiberg conducted competitive experiments to establish the relative rates, listed in Table VIII. These data provide a valuable indication of the relative reactivities of a wide variety of reagents undergoing very different types of reaction with a standard silene.

The explanations for the relative rates of reaction have been based on three factors: (1) The rate of reaction increases as the electron density in the diene system increases; thus isoprene reacts faster than butadiene and a complex electron-rich 2-silylmethylbutadiene reacts even faster. (2) The rate of reaction increases as the steric hindrance due to the diene substituents decreases; thus *trans*-piperylene reacts more slowly than dimethylbutadiene or isoprene. (3) A decrease in the equilibrium concentration of the cisoid conformer results in a slower reaction rate; thus *cis*-piperylene or *cis*/*trans*-2,4-hexadiene react more slowly than *trans*-piperylene or *trans*/*trans*-2,4-hexadiene, respectively.[175,177]

The overall combination of these effects identifies isoprene as the linear diene that reacts most rapidly in a [2 + 4] manner with the silene $Me_2Si{=}C(SiMe_3)_2$ (involving formation of the favored 3-methyl-substituted

TABLE VIII

RELATIVE RATES OF REACTION OF $Me_2Si{=}C(SiMe_3)_2$ WITH VARIOUS REAGENTS[174]

Reagent	Type of Product	Relative Rate
i-$PrNH_2$	1,2-addition	23,000
MeOH	1,2-addition	21,000
MeCOOH	1,2-addition	17,000
$Ph_2C{=}NH$	1,2-addition	14,000
EtOH	1,2-addition	13,000
t-$BuNH_2$	1,2-addition	11,000
i-PrOH	1,2-addition	10,000
t-BuOH	1,2-addition	6,800
$Ph_2C{=}O$	[2 + 2]-cycloaddition	~3,000
n-PeOH	1,2-addition	1,700
cyclohexOH	1,2-addition	760
$PhNH_2$	1,2-addition	470
Me_2CO	ene adduct	420
PhSH	1,2-addition	280
t-BuN_3	[2 + 3], then rearrangement	230
PhOH	1,2-addition	210
isoprene	[2 + 4]-cycloaddition	4.6
2,3-dimethylbutadiene	[2 + 4]-cycloaddition	3.7
trans-1-methylbutadiene	[2 + 4]-cycloaddition	3.1
cyclopentadiene	[2 + 4]-cycloaddition	2.9
propene	ene reaction	2.9
isobutene	ene reaction	2.3
isoprene	ene reaction	1.5
butadiene	[2 + 4]-cycloaddition	1[a]
2,3-dimethylbutadiene	ene reaction	0.95
cis-2-butene	ene reaction	0.42
isoprene	[2 + 4]-cycloaddition	0.32
trans-2-butene	ene reaction	0.31
cis-piperylene	[2 + 2]-cycloaddition	~0.015
Me_3SiCl	1,2-addition	<0.015

[a] Relative rate defined as 1.

The authors are grateful to Professor Nils Wiberg, University of Munich, for allowing us to use revised and unpublished data in Table VIII.

silacyclohexene), followed by dimethylbutadiene, and then *trans*/*trans*-hexadiene, which is the slowest of the dienes in the series. Some simple sterically unhindered alkenes such as propene and isobutene react, in an ene reaction, more rapidly than butadiene (the standard compound for the series) does in a [2 + 4] manner, while more substituted and sterically hindered alkenes, e.g., *cis*- and *trans*-2-butene, react more slowly.

Jones[179] pioneered in adding *t*-butyllithium to vinylhalosilanes as a route to several silenes of the general structure $RR'Si{=}CHCH_2CMe_3$, as shown in Eq. (29).

$$R_2SiX(CH{=}CH_2) + t\text{-}BuLi \longrightarrow R_2\underset{\displaystyle X}{\underset{|}{Si}}—\underset{\displaystyle Li}{\underset{|}{C}}HCH_2CMe_3 \longrightarrow R_2Si{=}CHCH_2CMe_3 \quad (29)$$

These silenes showed typical behavior, yielding [2 + 2] plus a small amount of [2 + 4] cycloadducts with butadiene and [2 + 4] cycloadducts with 2,3-dimethylbutadiene, cyclopentadiene, and anthracene.[180] More recently, Auner, in a series of papers[35–45,51–53,171,181–185] published over the past few years, has reported the behavior of nearly two dozen silenes having the general structure $RR'Si{=}CR''CH_2CMe_3$, most of which had strongly electronegative groups, Cl, RO, or R_2N, attached to the sp^2-hybridized silicon atom. A few silenes had either alkyl or aryl groups attached to silicon, for comparative purposes. All the silenes had a neopentyl group attached to the silene sp^2-hybridized carbon atom. None of the silenes was stable or observable by spectroscopic methods. A number of silenes had substituents, such as Ph or Me_3Si, attached to the sp^2-hybridized carbon of the silene. Reactions with several dienes—e.g., butadiene, 2-methylbutadiene, 2,3-dimethylbutadiene, cyclopentadiene, pentamethylcyclopentadiene, cyclohexadiene, and norbornadiene and related compounds quadricyclane, styrene, and norbornene—were investigated, resulting in an immense amount of data. In many cases multiple products were formed which were identified only by NMR data taken from the reaction mixtures. Frequently, nonseparable mixtures of *endo*/*exo* isomers were observed so that isolation of products was very difficult and only low yields of material were obtained in many cases. Reactions with some alkynes gave rise to silacyclobutenes, but in a few cases products from an ene reaction were also observed.[186]

Table IX lists the data available for the reactions of three typical silenes, $Cl_2Si{=}CHCH_2CMe_3$, $Cl_2Si{=}C(SiMe_3)CH_2CMe_3$, and $Ph_2Si{=}CHCH_2$-CMe_3, with a variety of dienes.

The results of the investigations of the silenes $R_2Si{=}CR'CH_2CMe_3$ suggested that both the polarity of the double bond, $Si{=}C \leftrightarrow Si^{\delta+}—C^{\delta-}$, as influenced by substituents on both the sp^2-hybridized silicon and carbon atoms of the double bond, and the electron density in the diene system play important roles in determining which of the [2 + 4], [2 + 2], or ene reactions will predominate in a given reaction. Cycloadditions yielding predominantly [2 + 4] adducts were most common for silenes having hydrocarbon groups on silicon; such cycloadditions were also common when no hydrocarbon groups were present on silicon but a substituent such as Ph or Me_3Si occupied the R′ site on the sp^2-hybridized carbon atom.

TABLE IX

PRODUCTS FROM THE REACTION OF SELECTED SILENES $R_2Si{=}CR'CH_2CMe_3$ WITH DIENES (% YIELDS)

Silene	Diene	[2 + 4]	[2 + 2]	Ene	[2 + 2 + 2]	H-T Dimer	Other	Reference
$Cl_2Si{=}CHCH_2CMe_3$	1,3-cyclohexadiene	26	74					181,182
$Cl_2Si{=}CHCH_2CMe_3$	naphthalene	40						182
$Cl_2Si{=}CHCH_2CMe_3$	butadiene		80					184
$Cl_2Si{=}CHCH_2CMe_3$	2-methylbutadiene		80	2				184
$Cl_2Si{=}CHCH_2CMe_3$	2,3-dimethylbutadiene		75	2				184
$Cl_2Si{=}CHCH_2CMe_3$	*trans/trans*-1,4-diphenylbutadiene		38					184
$Cl_2Si{=}CHCH_2CMe_3$	2,5-dimethyl-2,4-hexadiene		40					184
$Cl_2Si{=}CHCH_2CMe_3$	dimethylacetylene		41	14				36
$Cl_2Si{=}CHCH_2CMe_3$	diphenylacetylene		85					36
$Cl_2Si{=}C(SiMe_3)CH_2CMe_3$	1,3-cyclohexadiene	54 X[a]						35
$Cl_2Si{=}C(SiMe_3)CH_2CMe_3$	butadiene		33					35
$Cl_2Si{=}C(SiMe_3)CH_2CMe_3$	2,3-dimethylbutadiene	10		56				35
$Cl_2Si{=}C(SiMe_3)CH_2CMe_3$	norbornadiene			trace	38			35
$Cl_2Si{=}C(SiMe_3)CH_2CMe_3$	quadricyclane				23			35
$Cl_2Si{=}C(SiMe_3)CH_2CMe_3$	norbornene			43				35
$Ph_2Si{=}CHCH_2CMe_3$	1,3-cyclohexadiene	48				10		171
$Ph_2Si{=}CHCH_2CMe_3$	2-methylbutadiene	11	3	3				171
$Ph_2Si{=}CHCH_2CMe_3$	2,3-dimethylbutadiene	20		5				171
$Ph_2Si{=}CHCH_2CMe_3$	norbornadiene			1	3	2		171
$Ph_2Si{=}CHCH_2CMe_3$	cyclopentadiene	2					8 start. mat.	171

[a] X = Exclusive product.

Reactions favoring [2 + 2] cycloaddition tended to be those that had strongly electronegative groups on the sp^2-hybridized silicon but only H and the neopentyl group on the sp^2-hybridized carbon atom. Butadiene and cyclohexadiene generally favored [2 + 2] cycloaddition with these silenes. The [2 + 2] adducts with cyclohexadiene appear to be kinetic products, since they cleanly isomerized to the Diels–Alder adducts over time.[182]

Ene reactions tended to occur as alternative reaction pathways to [2 + 4] cycloaddition, especially when sterically bulky silenes had substituents on the sp^2-hybridized carbon atom and dimethylbutadiene served as the diene component. In the ene reactions studied the silene acted as the enophile as often as it acted as the ene.

While less thoroughly investigated than the silenes of Wiberg, Jones, and Auner, the family of silenes with the general structural $(Me_3Si)_nR_{2-n}$ $Si{=}C(OSiMe_3)R'$, formed by photolysis at wavelengths greater than 360 nm of the parent acylsilane $(Me_3Si)_{n+1}R_{2-n}SiCOR'$, display interesting and unusual behavior in cycloaddition reactions with a variety of reagents.[88,187] Table X shows the results of reactions with some typical dienes, alkenes, and alkynes.

As noted previously, [2 + 4] cycloaddition was often accompanied by products of the ene reaction. With 2,3-dimethylbutadiene and isoprene, the [2 + 4] cycloadduct was the major product; but while the Wiberg results with isoprene gave the 1-sila-3-methylcyclohexene adduct as the dominant species the Brook silene gave the 1-sila-4-methylcyclohexene as the major product. Furthermore, while the Wiberg silene reaction with butadiene gave the [2 + 4] product as the only adduct formed, the Brook silene afforded two [2 + 2] products that constituted about 80% of the products together with a [2 + 4] adduct that amounted to about 20%. The proportions of products formed were shown to be only slightly sensitive to the temperature or to the solvent used, whether the silene and diene reacted in the dark or during the photolysis of the mixture of acylsilane and diene. With cyclohexadiene, the [2 + 4] adduct constituted the major product, and the [2 + 2] adduct was formed in only about 20% of the total products. In reactions with alkenes, if allylic hydrogen was present, only the products of an ene reaction were formed; in the absence of allylic hydrogen, clean [2 + 2] cyclizations occurred. The geometric isomers of $(Me_3Si)MesSi{=}C(OSiMe_3)Ad$ appeared to react both regio- and stereospecifically in a [2 + 2] manner with both phenylacetylene and trimethylsilylacetylene. Each geometric isomer reacted with butadiene to give a pair of stereoisomers, but neither reacted with 2,3-dimethylbutadiene, presumably because of steric hindrance.[88]

TABLE X

PRODUCTS FROM THE CYCLOADDITION REACTIONS OF $(Me_3Si)_nR_{2-n}Si{=}C(OSiMe_3)R'$ WITH DIENES, ALKENES, AND ALKYNES (% YIELDS)

Silene	Diene	[2 + 4]	[2 + 2]	Ene	Reference
$(Me_3Si)_2Si{=}C(OSiMe_3)Ad$	2,3-dimethylbutadiene	60		40	187
$(Me_3Si)_2Si{=}C(OSiMe_3)Ad$	butadiene	18	82 (2 stereoisomers)		187
$(Me_3Si)_2Si{=}C(OSiMe_3)Ad$	2-methylbutadiene	88 (2 regioisomers)		12	187
$(Me_3Si)_2Si{=}C(OSiMe_3)Ad$	cyclopentadiene	>95%			187
$(Me_3Si)_2Si{=}C(OSiMe_3)Ad$	cyclohexadiene	80 (2 stereoisomers)	20 (2 stereoisomers)		187
$(Me_3Si)_2Si{=}C(OSiMe_3)Ad$	1-octene			100	187
$(Me_3Si)_2Si{=}C(OSiMe_3)Ad$	styrene		100 (2 stereoisomers)		187
$(Me_3Si)_2Si{=}C(OSiMe_3)Ad$	$CH_2{=}CH$(1-naphthyl)		100 (2 stereoisomers)		187
$(Me_3Si)_2Si{=}C(OSiMe_3)Ad$	$CH_2{=}CMePh$			100	187
$(Me_3Si)_2Si{=}C(OSiMe_3)Ad$	methyleneindane			100	187
$(Me_3Si)_2Si{=}C(OSiMe_3)t$-Bu	2,3-dimethylbutadiene	60		40	187
$(Me_3Si)_2Si{=}C(OSiMe_3)t$-Bu	butadiene	20	80		187
$(Me_3Si)_2Si{=}C(OSiMe_3)t$-Bu	cyclopentadiene	>95			187
$(Me_3Si)_2Si{=}C(OSiMe_3)t$-Bu	cyclohexadiene	80 (2 stereoisomers)	20 (2 stereoisomers)		187
$(Me_3Si)_2Si{=}C(OSiMe_3)t$-Bu	$CH_2{=}CH$(1-naphthyl)		100 (2 stereoisomers)		187
$(Me_3Si)_2Si{=}C(OSiMe_3)t$-Bu	$CH_2{=}CMePh$			100	187
$(Me_3Si)_2Si{=}C(OSiMe_3)t$-Bu	methyleneindane			100	187
$(Me_3Si)_2Si{=}C(OSiMe_3)t$-Bu	methylpropyne		100[a]		88
$(Me_3Si)_2Si{=}C(OSiMe_3)Mes$	2,3-dimethylbutadiene	100			187
$(Me_3Si)_2Si{=}C(OSiMe_3)Mes$	butadiene	25 (2 stereoisomers)	75 (2 stereoisomers)		187
$(Me_3Si)_2Si{=}C(OSiMe_3)Mes$	styrene		100		187
$(Me_3Si)_2Si{=}C(OSiMe_3)Mes$	$CH_2{=}CMePh$			100	187
$(Me_3Si)_2Si{=}C(OSiMe_3)Mes$	methyleneindane			100	187
$(Me_3Si)MesSi{=}C(OSiMe_3)Ad$	butadiene		100 (4 stereoisomers)		88
$(Me_3Si)MesSi{=}C(OSiMe_3)Ad$	phenylacetylene		100		88
$(Me_3Si)MesSi{=}C(OSiMe_3)Ad$	trimethylsilylacetylene		100		88

[a] Original regiochemistry revised.

1. *[2 + 4] Cycloadditions*

It is obvious from the details above of three major families of silenes that most silenes—from the most reactive short-lived transient to the relatively stable silenes that can be isolated or observed in solution—undergo [2 + 4] cycloaddition reactions with dienes by what is considered to be a concerted mechanism. A few additional examples recently reported in the literature are listed below in Scheme 15. The first examples[58,61] illustrate that some highly sterically hindered silenes will undergo [2 + 4] cycloaddition with 2,3-dimethylbutadiene. It is interesting that in the third example the treatment of silenolates with a mixture of triethylchlorosilane and 2,3-dimethylbutadiene yielded the [2 + 4] cycloadducts of the silylated silenolate.[63] The last reaction, involving a Brook-type silene, is interesting in that when the silene $(Me_3Si)_2Si{=}C(OSiMe_3)Ph$ was generated photochemically from the acylsilane in the presence of styrene, only the [2 +

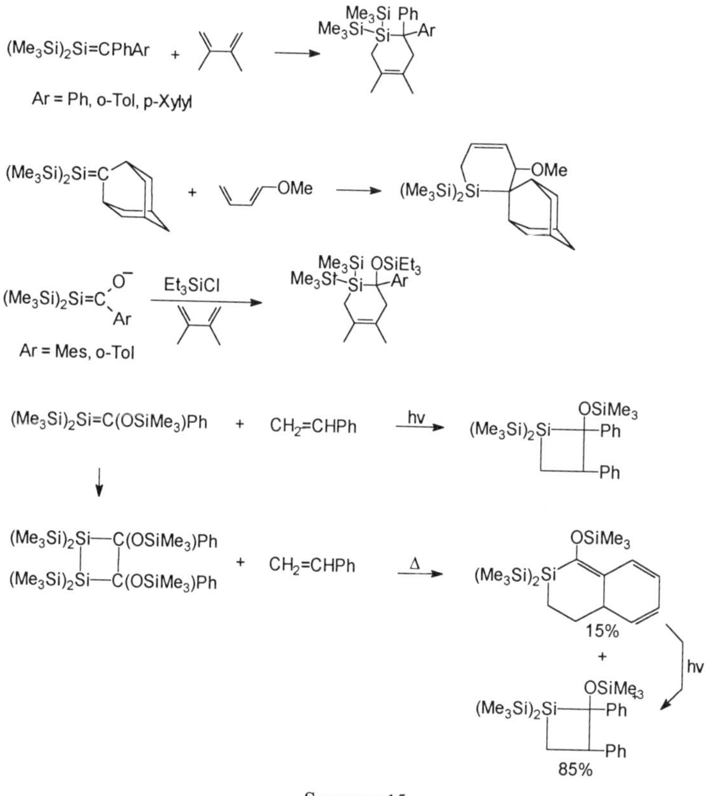

$(Me_3Si)_2Si{=}CPhAr$ +
Ar = Ph, o-Tol, p-Xylyl
Me_3Si Ph
Me_3Si–Si–Ar

$(Me_3Si)_2Si{=}C$ + OMe
$(Me_3Si)_2Si$ OMe

$(Me_3Si)_2Si{=}C(O^-)Ar$
Ar = Mes, o-Tol
Et_3SiCl
Me_3Si $OSiEt_3$
Me_3Si–Si–Ar

$(Me_3Si)_2Si{=}C(OSiMe_3)Ph$ + $CH_2{=}CHPh$ hv
$(Me_3Si)_2Si$ $OSiMe_3$ Ph Ph

$(Me_3Si)_2Si—C(OSiMe_3)Ph$
$(Me_3Si)_2Si—C(OSiMe_3)Ph$
+ $CH_2{=}CHPh$ Δ
$(Me_3Si)_2Si$ $OSiMe_3$
15%
+
$(Me_3Si)_2Si$ $OSiMe_3$ Ph Ph
85%
hv

SCHEME 15

2] cycloadduct was formed. However, when the silene was generated by mild warming of the silene head-to-head dimer in the presence of styrene, about 15% of the [2 + 4] cycloadduct was formed, involving 4π-electrons of the silene and 2π-electrons from the styrene, as well as 85% of the [2 + 2] adduct.[187] When photolyzed at 360 nm the [2 + 4] adduct rearranged to the [2 + 2] adduct. This implies that the initial (kinetic) product of the reaction of the silene with styrene is the [2 + 4] adduct, and that the thermodynamic product is the [2 + 2] adduct. As will be seen later, such mixtures of [2 + 2] and [2 + 4] adducts are common in the reactions of several silenes with benzophenone and other aromatic ketones, or with benzophenonimines, but this is the only case known to date where mixtures of adducts are formed from aromatic-substituted alkenes.

2. *[2 + 2 + 2] Cycloadditions*

There are several known examples of silenes participating in [2 + 2 + 2] cycloadditions, including reactions with norbornadiene and with quadricyclane. Examples of these reactions with norbornadiene are shown in Eq. (30), where the silene bridges the 2,6-positions.[38,51,53,171,185,188] This reaction has only been investigated using some of the Auner-type silenes. In many cases mixtures of stereoisomers were obtained, and in some cases products of an ene reaction were also observed.

$R_2Si{=}CR'CH_2CMe_3$ + (norbornadiene) ⟶ (product: R_2Si—C(R')CH_2CMe_3 bridging the 2,6-positions)

R	R'
Cl,Cl	H or Me_3Si
Ph, Ph	H
$-(CH_2)_3-$	H
$-CHMe(CH_2)_2-$	H
Me_3SiO, Me_3SiO	H

(30)

The reactions with quadricyclane, shown in Eq. (31), gave products identical to those formed by the same silene reacting in a [2 + 2] manner with norbornene. Mixtures of *exo*/*endo* isomers were frequently observed. Again, only silenes of the Auner type have been studied with this reagent,[51–53,185,188] so it is not known whether the Wiberg- or Brook-type silenes will undergo this mode of cycloaddition.

$$R_2Si{=}CR'CH_2CMe_3 \quad + \quad \text{(quadricyclane)} \longrightarrow \text{(norbornene-fused silacyclobutane: } R', CH_2CMe_3, SiR_2) \qquad (31)$$

R	R'
Me, Me	$SiCl_2Ot\text{-}Bu$
Cl,Cl	$SiMe_3$
Me_3SiO, Me_3SiO	H

3. *[2 + 2] Cycloadditions*

Many silenes react cleanly in a [2 + 2] manner with carbon–carbon multiple bonds, and several examples of such behavior have been given above. Some additional examples are listed in Scheme 16, there now being too many examples known to allow a full listing. For the cases listed,[33,65,185,189] each of which involves a polarized carbon–carbon double bond, the products were isolated and well characterized and the main products were those of a [2 + 2] reaction. Wherever regioisomerism was possible, only a single regioisomer has been observed; this includes cases in which unsymmetric alkynes were involved.

Scheme 17 lists some more speculative, but very interesting reactions in which it is uncertain whether the proposed [2 + 2] cycloaddition product, denoted in square brackets, was actually formed on the pathway to the final product. In each case, the structure of the final product seems to be

$$Me_2Si{=}C(SiMe_3)(SiMe(t\text{-}Bu)_2 \ + \ CH_2{=}CHOMe \longrightarrow Me_2Si\text{-}C(SiMe_3)(SiMe_2(t\text{-}Bu)_2)\text{-}CH(OMe)\text{-}CH_2 \text{ (ring)}$$

$$(Me_3SiO)_2Si{=}CHCH_2CMe_3 \ + \ CH_2{=}CHPh \longrightarrow (Me_3SiO)_2Si\text{-}CH(CH_2CMe_3)\text{-}CH(Ph)\text{-}CH_2 \text{ (ring)}$$

$$(Me_3Si)_2Si{=}C(OSiMe_3)R \ + \ CH_2{=}CHCN \longrightarrow (Me_3Si)_2Si\text{-}C(OSiMe_3)(R)\text{-}CH(CN)\text{-}CH_2 \text{ (ring)}$$

$$Me_2Si{=}CHCH_2SiMe_2CH{=}CH_2 \longrightarrow \text{bicyclic } Me_2Si \ / \ SiMe_2$$

SCHEME 16

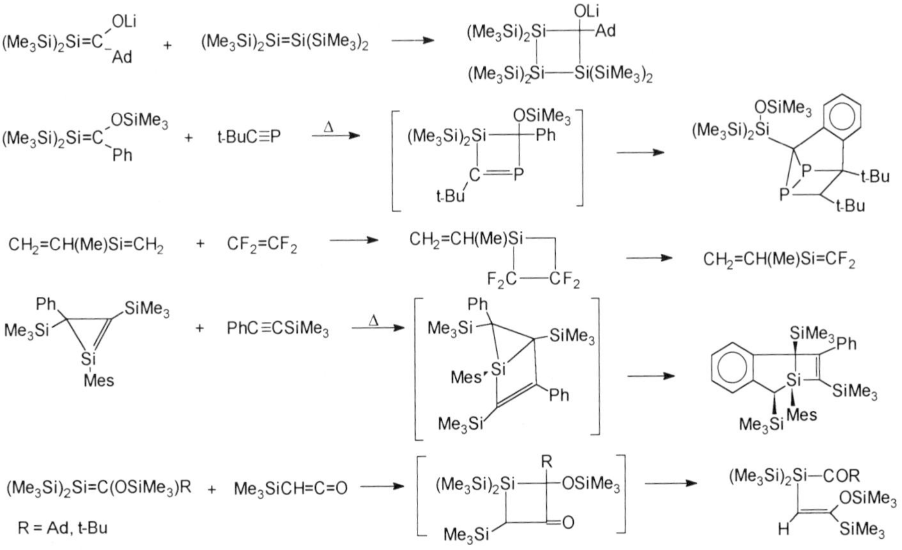

SCHEME 17

securely established. The first example[100] represents the only reported case of [2 + 2] addition of a silene with a disilene. The second, which is an unusual reaction with a phosphyne,[190] is obviously a complex process involving migration of a trimethylsiloxy group from the sp^2-hybridized carbon of the original silene to the silicon atom which was originally sp^2-hybridized. The silene bearing fluorine on carbon[18] was not directly observed in the third reaction, but was a reasonable proposal in view of the addition products observed in the reaction. The reaction described between the proposed silacyclopropene and an alkyne[191] yielded other products as well. The final reaction, the apparent [2 + 2] cycloaddition of a Brook-type silene and a ketene,[192] involves further rearrangement steps for which precedents are known.

4. *Reactions of Silenes with Carbonyl Groups*

In addition to undergoing cycloaddition reactions with alkenes and alkynes, silenes readily undergo cycloaddition reactions with heteroatom multiple bonds such as C=O and C=N, most commonly when the trapping reagent for the silene is either an aldehyde, ketone, or imine. In many of the known examples, one of the R groups attached to the sp^2-hybridized carbon atom of the trapping reagent is a aryl group, and the [2 + 2] reaction with the heteroatom C=O or C=N bond is commonly accompanied by [2 + 4] cycloaddition involving 2π-electrons of the aromatic ring.

As noted earlier, the [2 + 2] adduct of the silene $Me_2Si{=}C(SiMe_3)_2$ with *N*-trimethylsilylbenzophenonimine serves as an important source of the silene on warming to about 100°C, and the [2 + 4] adduct of the same silene with benzophenone can also serve as a "store" for the silene.[166]

a. *Calculations.* Historically, silenes were long believed to form [2 + 2] cycloadducts with ketones. For many years, however, only the products of retro-cleavage of the siloxetane ring—namely an alkene and an oligomer of a silanone $R_2Si{=}O$ (presumed to be formed from individual silanone product molecules)—were observed when the reactions were carried out. Bachrach and Streitwieser[193] investigated the cycloaddition of formaldehyde with silanone ($H_2Si{=}O$) at the SCF level (Fig. 1). The reaction to form the siloxetane was found to be exothermic by about 80 kcal mol^{-1}; however, the further reaction in which the siloxetane would decompose to ethylene and silanone was found to be endothermic to the extent of 50 kcal mol^{-1}, too large to be consistent with a unimolecular retro-[2 + 2] process. These authors suggested, before one had actually been isolated, that it should be possible to isolate a siloxetane, as was soon done experimentally.[194] Also, it was noted that decomposition of a siloxetane was too endothermic to occur by a unimolecular process, and that one or more bimolecular reactions between two siloxetanes would be a more energetically feasible way to account for the observed alkenes and silanone oligomers.

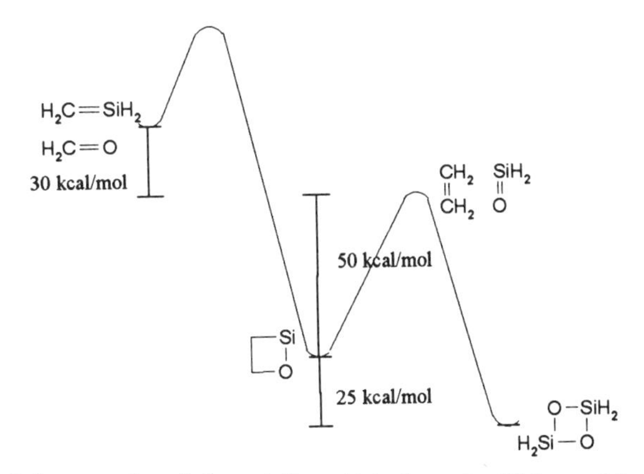

FIG. 1. Relative energies of silene + formaldehyde cycloaddition and fragmentation.

b. *Experimental Results.* More recently, Brook *et al.*[194] succeeded in preparing a number of relatively stable siloxetanes, and were able to determine the crystal structure for the compound obtained from $(Me_3Si)_2Si{=}C(OSiMe_3)Ad$ and benzophenone. The thermal decomposition of one of the stable pure siloxetanes was investigated kinetically and, consistent with Streitwieser's suggestions, was found to be a complex process greater than second-order in siloxetane.

The reactions of silenes of the family $(Me_3Si)_2Si{=}C(OSiMe_3)R$ with aromatic ketones were unpredictable. In some cases, only the [2 + 2] adduct was formed, although over time spontaneous isomerization in the dark often afforded the [2 + 4] isomer, a [4.4.0] bicyclic species as shown in Scheme 18. In other cases, the [2 + 4] adduct was the initially formed species, and it sometimes spontaneously isomerized to the [2 + 2] adduct. And, in one case, the silene $(Me_3Si)_2Si{=}C(OSiMe_3)Ph$ with benzophenone or other aromatic ketones gave a [2 + 4] cycloadduct that involved 2π-electrons from the carbonyl group of the ketone while the 4π-electrons came from the silene double bond plus the 2π-electrons from its attached phenyl group. These reactions are illustrated in Scheme 18.

In view of the evident reactivity of the Brook-type silenes toward carbonyl compounds and the fact that these silenes were prepared by the photolysis of acylsilanes, it is natural to ask why the silenes apparently did not react with their acylsilane precursors. This question has been answered recently. On the one hand, as shown in Scheme 19, the silene $Ph_2Si{=}C(OSiMe_3)Ad$ apparently did add in a [2 + 2] manner to its acylsi-

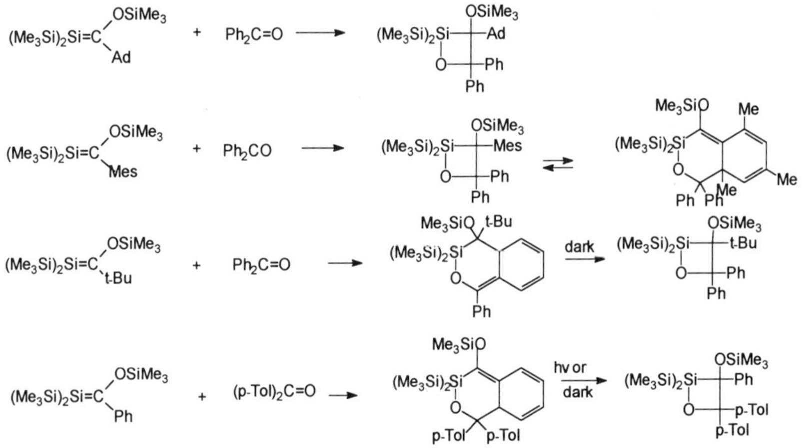

SCHEME 18

$Ph_2Si{=}C(OSiMe_3)Ad$ + $Me_3SiSiPh_2COAd$ → [Ph_2Si–O four-membered ring bearing $OSiMe_3$, Ad, $SiPh_2SiMe_3$, Ad] → $Me_3SiOSiPh_2OCAd{=}CAdSiPh_2SiMe_3$

cis and trans isomers

$(Me_3Si)_2Si{=}C(OSiMe_3)Mes$ + $Me_3SiCOPh$ → [$(Me_3Si)_2Si$–O four-membered ring bearing $OSiMe_3$, Mes, Ph, $SiMe_3$]

$(Me_3Si)_2Si{=}C(OSiMe_3)Ad$ + $Ph_3SiCOPh$ → [Me_3SiO, Ad, $(Me_3Si)_2Si$, O, $SiPh_3$]

SCHEME 19

lane precursor $Me_3SiSiPh_2COAd$,[90] but the adduct then rearranged spontaneously by a previously observed pathway[194] to give the alkene shown. Furthermore, when silenes that would not react with their own parent acylsilane were treated with less bulky acylsilanes, the expected [2 + 2] or [2 + 4] cycloadditions occurred, indicating that the failure to react with their parent acylsilanes was due to steric hindrance.[195]

The behavior of the Brook silenes with simple α,β-unsaturated carbonyl compounds was also investigated. In contrast to most cycloaddition reactions of silenes in the $(Me_3Si)_2Si{=}C(OSiMe_3)R$ family, it was observed that both possible regiochemistries were involved in the [2 + 4] cycloaddition reactions of Brook silenes with simple compounds like propenal or methyl vinyl ketone for R = Ad or *t*-Bu. With crotonaldehyde both possible cycloaddition modes, [2 + 2] and [2 + 4], occurred, and a silicon–oxygen bond was formed in each case.[196] When R = Mes, only [2 + 2] cycloaddition leading to a siloxetane occurred with cinnamaldehyde, β-phenylcinnamaldehyde, or methyl vinyl ketone; but [2 + 4] cycloaddition occurred with benzalacetophenone, again involving silicon–oxygen bond formation. Most of these results are summarized in Scheme 20.

The most logical explanation for the formation of both regioisomers during the [2 + 4] cycloadditions appears to be that two competing mechanisms are operating. The formation of compounds with a silicon–oxygen bond in the ring logically arises from attack of the nucleophilic carbonyl oxygen atom on the electrophilic sp^2-hybridized silicon atom of the silene and probably involves conventional concerted cycloaddition in a manner consistent with the anticipated polarities of the reagents. Formation of the other regioisomer in which a ring silicon–carbon bond is formed is probably best explained on the basis of a competing radical reaction, although this was not proved. It seems likely that these cycloadditions

$(Me_3Si)_2Si{=}C(OSiMe_3)R + CH_2{=}CHCH{=}O \longrightarrow$ $(Me_3Si)_2Si$–C($OSiMe_3$)(R) ring with O + $(Me_3Si)_2Si$–C($OSiMe_3$)(R) ring with O

R = Ad, t-Bu

$(Me_3Si)_2Si{=}C(OSiMe_3)R + R'CH{=}CHCH{=}O \longrightarrow$ $(Me_3Si)_2Si$–C($OSiMe_3$)(R) ring with O, –Me + $(Me_3Si)_2Si$–C($OSiMe_3$)(R), O—CH-CH=CHMe

R = Ad, t-Bu, Mes R' = Me, Ph

R = Ad, t-Bu R = Ad, t-Bu, Mes

$(Me_3Si)_2Si{=}C(OSiMe_3)R + CH_2{=}CHCOMe \longrightarrow$ $(Me_3Si)_2Si$–C($OSiMe_3$)(R) ring with O, Me + $(Me_3Si)_2Si$–C($OSiMe_3$)(R) ring with O, Me

R = Ad, t-Bu, Mes

R = Ad, t-Bu, Mes

+ $(Me_3Si)_2Si$–C($OSiMe_3$)(R), O—C(Me)-CH=CH_2

R = Mes

SCHEME 20

are also very sensitive to small changes in electronic and steric effects. The fact that adding the methyl group of crotonaldehyde to the α,β-unsaturated system results in suppression of the unexpected C—Si regioisomer suggests that a somewhat more nucleophilic oxygen favors formation of the normal isomer. When the silene bears the mesityl group on sp^2-hybridized carbon, formation of the [2 + 2]-siloxetane products becomes an important process, relative to the presence of an alkyl group such as Ad or *t*-Bu. This could be an electronic or a steric effect.

When α,β-unsaturated esters were allowed to react with the silenes, anomalous behavior was again observed in the [2 + 4] cyclizations that occurred.[189] With a series of acrylate esters only the isomer having a ring silicon–carbon bond was formed. When substituted acrylate esters were employed, the other regioisomer having a ring silicon–oxygen bond was formed, accompanied by significant amounts of products, which were nominally the result of C—H addition across the ends of the silicon–carbon bond. With fumarate and maleate esters, the same anomalous regiochemistry was observed, but the regiochemistry was reversed with cinnamate esters. Evidently, the less nucleophilic carbonyl oxygen of the esters does not drive the reactions to occur with the expected regiochemistry determined by bond polarity, except in the case of ethyl cinnamate, where the regiochemistry anticipated on the basis of bond polarities is observed.

$(Me_3Si)_2Si{=}C(OSiMe_3)R + CH_2{=}CHCOOR' \longrightarrow$ [cyclic product: $(Me_3Si)_2Si$, $OSiMe_3$, R, O, OR']

R = Ad, t-Bu R' = Me, Et, n-Bu, t-Bu

$(Me_3Si)_2Si{=}C(OSiMe_3)R + CH_3CH{=}CHCOOt\text{-}Bu \longrightarrow$ [cyclic product: $(Me_3Si)_2Si$, $OSiMe_3$, R, O, CH_3, t-BuO] + [$(Me_3Si)_2Si$, $OSiMe_3$, R, H, COOt-Bu]

R = Ad, t-Bu

$(Me_3Si)_2Si{=}C(OSiMe_3)R + EtOOCCH{=}CHCOOEt \longrightarrow$ [cyclic product: $(Me_3Si)_2Si$, $OSiMe_3$, R, EtOOC, O, OEt]

R = Ad, t-Bu

$(Me_3Si)_2Si{=}C(OSiMe_3)R + PhCH{=}CHCOOEt \longrightarrow$ [cyclic product: $(Me_3Si)_2Si$, $OSiMe_3$, R, O, Ph, EtO]

R = Ad, t-Bu

SCHEME 21

If it were not for a number of crystal structures that unambiguously confirmed the structures of the products, it would be easy to believe that some of the structural assignments were incorrect. At present no completely satisfactory explanation for these results is available. Examples of the reactions with esters are given in Scheme 21.

The reactions of $Cl_2Si{=}CHCH_2CMe_3$ with several aldehydes have been described by Auner.[197] It is believed that [2+2]-siloxetane adducts **92** were formed in the reaction but were not stable. Evidence for the formation of the siloxetanes lay in the isolation of the anticipated alkene **93** derived by a retro-[2+2] reaction with elimination of $Cl_2Si{=}O$. Interestingly, significant quantities of isomeric products **94** were observed if the reactions were carried out at $-40°C$ instead of $-78°C$, as shown in Eq. (32).

t-BuLi + $Cl_3SiCH{=}CH_2 \longrightarrow Cl_2Si{=}CHCH_2CMe_3$ + RCHO $\longrightarrow$ [Cl_2Si—$CHCH_2CMe_3$ / O—CHR] **92**

t-BuLi, RCH=O $\downarrow$ Me_3C–RCHO$^-$ **96** + $Cl_3SiCH{=}CH_2$ $\not\longrightarrow$ $RCH(CMe_3){-}O{-}SiCl_2CH{=}CH_2$ **94** $\longleftarrow$ $Cl_2\bar{Si}{-}CHCH_2CMe_3$ / $O{-}\overset{+}{C}HR$ **95** (from **92**, $-40\,^{o}C$)

92 $\xrightarrow{-78\,^{o}C}$ $(Cl_2Si{=}O)_n$ + $RCH{=}CHCH_2CMe_3$ **93**

(32)

Auner attributed their formation to dissociation of the siloxane to a zwitterionic species **95**, followed by what appears to be a truly remarkable rearrangement. An alternative explanation for the formation of **94** involves the formation of the oxyanion **96** by the addition of *t*-butyllithium to the carbonyl group of the aldehyde, which subsequently attacks the vinyltrichlorosilane to displace a chloride ion, as shown in Eq. (32). This pathway was claimed not to be operative.

Silenes derived from aromatic di- or polysilanes have been characterized in particular by the ene-type reactions they undergo when treated with isobutene or acetone. Recently, Leigh[80] observed the first reported case of one of these silenes undergoing [2+2] cycloaddition (21% yield) with acetone. The ene product, the only product previously detected from the reaction of such silenes, was formed in 41% yield, as shown in Eq. (33).

$$PhSiPh_2SiMe_3 \xrightarrow{h\nu} C_6H_5(=SiPh_2)(SiMe_3) \xrightarrow{Me_2C=O} [\text{Me}_2\text{C–O–SiPh}_2 \text{ ring}](SiMe_3) + C_6H_5(SiPh_2\text{–O–CHMe}_2)(SiMe_3) \quad (33)$$

5. *Reactions of Silenes with Imines*

Silenes undergo cyclization with imines in a manner similar to their reaction with ketones. Wiberg had observed both [2+2] and [2+4] cycloadditions of the silene $Me_2Si{=}C(SiMe_3)_2$ with *N*-trimethylsilylbenzophenonimine,[198] and [2+2] or [2+4] additions were also found when the Brook-type silenes reacted with a variety of imines that had aromatic rings attached to the sp^2-hybridized carbon atom of the C=N bond.[199] Auner found that the silene $Cl_2Si{=}CHCH_2CMe_3$ reacted with a variety of imines, forming silaazetidines by [2+2] cycloaddition.[200]

Some examples of these reactions are shown in Scheme 22.

Whether the [2+2] or [2+4] cycloaddition product is formed in the first reaction depends on the reaction conditions; however, one product is convertible to the other via the silene intermediate. In the other reactions shown, the identity of the product formed, [2+2] or [2+4], seemed to be a function of the structures of both the silene and imine. In the last case, the [2+4] product appeared to be the kinetic isomer since conversion to the [2+2] isomer slowly occurred on standing in the dark, or faster if photolyzed, even at room temperature.

6. *Ene Reactions*

As has been noted numerous times in this review, silenes of all types are very disposed to undergo ene-type reactions with a variety of reagents that have allylic hydrogen or hydrogen alpha to a carbonyl group. Depending on the structures of both the silene and the added reagent, the silene may play the role of the ene component or of the enophile. Several examples illustrating this variable behavior are shown in Scheme 23.[35,51,79]

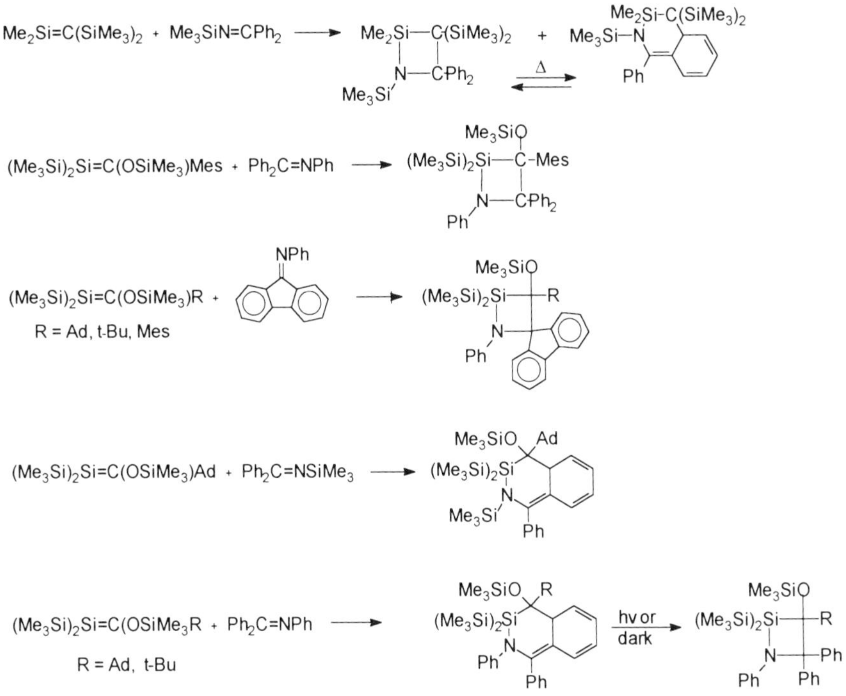

$Me_2Si{=}C(SiMe_3)_2 + Me_3SiN{=}CPh_2 \rightarrow$ (four-membered Si–C–C–N ring) $+$ (Diels–Alder type adduct); the two products interconvert on heating (Δ).

$(Me_3Si)_2Si{=}C(OSiMe_3)Mes + Ph_2C{=}NPh \rightarrow$ (four-membered Si–C–C–N ring)

$(Me_3Si)_2Si{=}C(OSiMe_3)R$ + fluorenone N-phenylimine $\rightarrow$ (spiro four-membered Si–C–C–N ring)

R = Ad, t-Bu, Mes

$(Me_3Si)_2Si{=}C(OSiMe_3)Ad + Ph_2C{=}NSiMe_3 \rightarrow$ (six-membered ring, Diels–Alder type adduct)

$(Me_3Si)_2Si{=}C(OSiMe_3R + Ph_2C{=}NPh \rightarrow$ (six-membered ring adduct) $\xrightarrow{h\nu\text{ or dark}}$ (four-membered Si–C–C–N ring)

R = Ad, t-Bu

SCHEME 22

(silene with $SiMe_2$, $Me_3SiSiMe_2$, Me_3Si and H substituents) $+ O{=}CMe_2 \rightarrow$ (benzene ring bearing $SiMe_2{-}OCHMe_2$, $SiMe_3$ and $Me_3SiSiMe_2$)

$Cl_2Si{=}C(SiMe_3)CH_2CMe_3$ + 2,3-dimethylbutadiene $\rightarrow$ $Cl_2Si{-}CH(SiMe_3)CH_2CMe_3$ (bearing a 2,3-dimethylbutadienyl-type group on Si)

$Cl_2Si{=}CPhCH_2CMe_3$ + butadiene $\rightarrow$ $Cl_2Si{-}CPh{=}CHCMe_3$ (bearing a butenyl group on Si)

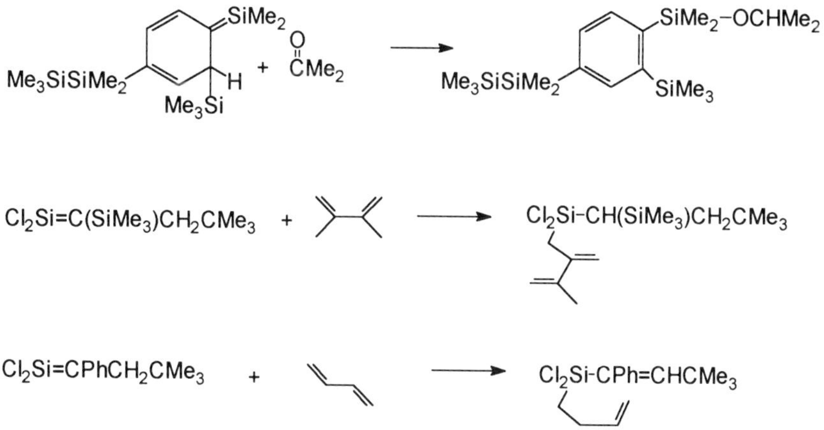

SCHEME 23

The fact that the silene acts as the ene component in the first example can be understood from the stabilization gained on restoration of the aromatic ring. More often than not, however, the silene acts as the enophile with reagents that have allylic or alpha hydrogen, as shown in the second example where the silene may assume either role. In the final case, if an ene reaction is to occur, the silene must play the role of the ene component of the reaction. In many cases, it is not obvious what determines which role is adopted, although it seems clear that a strongly polarized double bond is one of the important factors favoring the ene reaction involving silenes. Nor is it obvious in the latter case why cycloaddition was not a major reaction of the system.

7. [2+1] *Cycloadditions*

a. *Experimental Results.* Several reactions that are nominally [2+1] cycloadditions have been observed between silenes and other reagents. Brook[85] reported that the silene $(Me_3Si)_2Si{=}C(OSiMe_3)Mes$ reacted under mild conditions with hexamethylsilirane to insert the elements of dimethylsilylene across the ends of the silicon–carbon double bond, yielding the disilacyclopropane shown in Scheme 24.

Disilacyclopropanes **97** were also obtained as a significant product in the reaction of other silenes (R = Ad, *t*-Bu) with hexamethyldisilirane.[201] The major products **98** and minor products **99** observed suggest that the

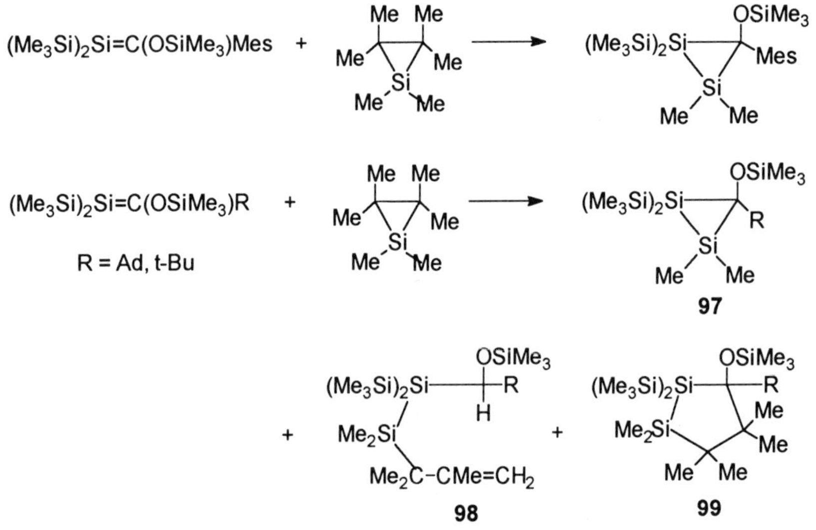

$(Me_3Si)_2Si{=}C(OSiMe_3)Mes$ + [hexamethylsilirane] → [disilacyclopropane: $(Me_3Si)_2Si$, $OSiMe_3$, Mes, $SiMe_2$]

$(Me_3Si)_2Si{=}C(OSiMe_3)R$ + [hexamethylsilirane] → **97** + **98** + **99**

R = Ad, t-Bu

SCHEME 24

reaction probably occurs by a radical pathway, where a homolytically ring-opened Si—C bond of the silirane attacks the silicon end of the silene. The resulting diradical then reacts in conventional ways, including ring closure to the disilirane with expulsion of tetramethylethylene.

Other authors have observed similar behavior of silenes with silylene sources. Thus Ishikawa[202] reported the addition of dimesitylsilylene to the silicon–carbon double bond of a silaallene, yielding the disilacyclopropene shown in Eq. (34).

$$Mes_2Si{=}C{=}C(SiMe_3)_2 + Mes_2Si: \longrightarrow Mes_2Si{-}SiMes_2 \; (\text{ring, } {=}C(SiMe_3)_2) \quad (34)$$

More recently, Pola[203] has suggested that hydroxymethylsilylene **100** is involved in cycloaddition to dimethylsilene **1**, with the resulting disilacyclopropane then rearranging to the observed cyclic disiloxane **101**.

$$Me_2Si{=}CH_2 + Me(HO)Si: \; (\mathbf{100}) \longrightarrow Me_2Si{-}Si(HO)(Me) \longrightarrow Me_2Si,\; HO^{+}{-}Si(Me),\; \bar{C}H_2 \longrightarrow Me_2Si{-}O{-}SiMe_2 \; (\mathbf{101}) \quad (35)$$

It was also observed that silenes of the family $(Me_3Si)_2Si{=}C(OSiMe_3)R$ reacted with isonitriles.[204] There was some evidence that the initial adducts are the expected silacyclopropanimines **102**, but the products isolated are silaaziridines **103** as a result of rapid rearrangements. It was suggested that the rearrangement might involve an intermediate zwitterion formed as a result of ring cleavage. In some cases with aryl substituents, the silaaziridines reacted further by inserting another isonitrile molecule into the ring to yield 1-sila-3-azacyclobutanes **104**, as shown in Eq. (36).[205]

$$(Me_3Si)_2Si{=}C(OSiMe_3)R + R'NC: \longrightarrow (Me_3Si)_2Si{-}C(OSiMe_3)R{-}C{=}NR' \; (\mathbf{102})$$

R = Ad, t-Bu, Mes R' = t-Bu, Ar

$$\mathbf{102} \longrightarrow (Me_3Si)_2Si{-}N(R'){-}C{=}C(R)OSiMe_3 \; (\mathbf{103}) \xrightarrow[\text{R or R' must be Ar}]{R'N{=}C:} (Me_3Si)_2Si{-}C({=}NR'){-}N(R'){-}C({=}C(R)OSiMe_3) \; (\mathbf{104}) \quad (36)$$

Finally, it has recently been shown[207] that some silenes can add sulfur or selenium to give silathiiranes **110**, as confirmed by a crystal structure, or silaseleniranes **111**, respectively (Eq. 37). Only sterically crowded silenes such as **112** gave stable solid adducts, and in the case of the selenium species the adducts visibly decomposed in the solid state over a few days.

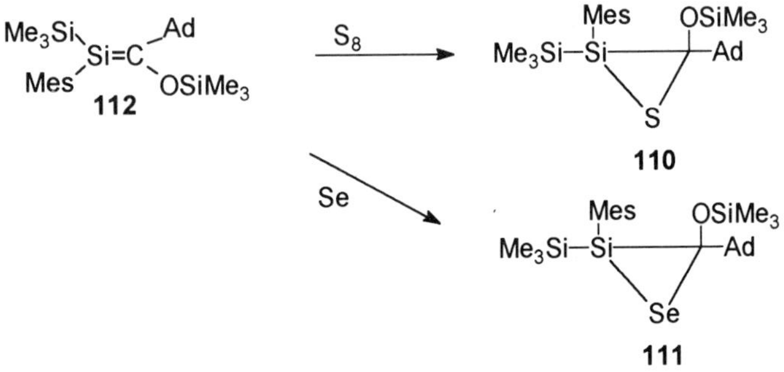

(37)

b. *Calculations*. The [2+1]-cycloaddition reaction between $H_2Si{=}CH_2$ and HN=C: has been modeled using theoretical methods[206] and compared with the experimental results described in Eq. (36). Experimentally, as mentioned above, the first formed siliranimine rearranges upon warming to the isomeric silaaziridine (Eq. 36). The calculations suggest that the reaction proceeds via the interaction of the LUMO of the silene and the HOMO of the isocyanide. The isocyanide lone pair attacks initially at the silicon, shown as species **105**, on the way to the product siliranimine **106** in a concerted, but asynchronous process.

The calculations further show that the initially formed siliranimine **106** is less stable than the isomeric silaaziridine **107** (see Eq. 38) by about 6 kcal mol^{-1}. The mechanism for the conversion suggested by the authors[206] involves a four-membered carbene **108** rather than the open but delocalized zwitterion **109** suggested by Brook.[204] Further experimental work is desirable to clarify the reaction mechanism.

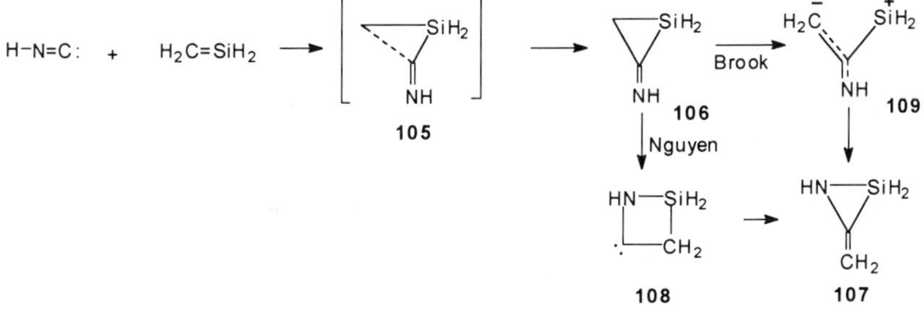

(38)

8. *[2+3] Cycloadditions*

Further details concerning cycloadditions between the silenes $Me_2Si{=}C(SiMe_3)_2$ and $Me_2Si{=}C(SiMe_3)(SiMe(t\text{-}Bu)_2)$ with a variety of azides and with N_2O have been provided by Wiberg following a preliminary publication.[98] The former silene undergoes cycloaddition with a variety of azides RN_3 to yield silatriazacyclopentenes **113**, regardless of whether the silene is generated by salt elimination[208] or thermally from its cycloadduct with *N*-trimethylsilylbenzophenonimine.[174] Some of the initially formed cycloadducts were isolable at room temperature while others spontaneously decomposed by two competing pathways, as did the stabler species on mild warming. The products of decomposition were simple silyldiazoalkanes **114** resulting from a retro-[2+3] reaction, as well as more complex silyldiazoalkanes **115** resulting from ring cleavage accompanied by a 1,3-shift of a Me_3Si group from ring carbon to ring nitrogen. An example is shown in Eq. (39).

$Me_2Si{=}C(SiMe_3)_2 + RN_3 \longrightarrow$ **113** (five-membered ring: Me_2Si—$C(SiMe_3)_2$, $N(R)$—N=N) $\longrightarrow (Me_3Si)_2C{=}N{=}N + (Me_2Si{=}NR)$ **114**

113 —(1,3-Me_3Si shift)→ **115**: $(Me_3Si)(R)N{-}SiMe_2{-}C(SiMe_3){=}N{=}N$ (39)

Identical behavior was displayed by the more complex silene $Me_2Si{=}C(SiMe_3)(SiMe(t\text{-}Bu)_2)$, and in the reactions of $Ph_2Si{=}C(SiMe_3)_2$ with $t\text{-}Bu_2MeSiN_3$.[30]

The reactions of $Me_2Si{=}C(SiMe_3)_2$ with N_2O apparently gave related [2+3] cycloadducts which were unstable and spontaneously decomposed to the silyldiazoalkane $(Me_3Si)_2C{=}N{=}N$ and oligomeric $Me_2Si{=}O$,[209] as shown in Eq. (40).

$Me_2Si{=}C(SiMe_3)_2 + N_2O \longrightarrow$ (five-membered ring: Me_2Si—$C(SiMe_3)_2$, O—N=N) $\longrightarrow (Me_3Si)_2C{=}N{=}N + (Me_2Si{=}O)_n$ (40)

D. *Other Bimolecular Reactions*

1. *Reactions with Alcohols—Including Stereochemistry*

Beginning with the earliest studies, it became well established that silenes react readily by 1,2-addition across the ends of the silicon–carbon double bond with a wide variety of reagents possessing polar bonds. Thus alcohols, carboxylic acids, amines, and thiols all react with relative ease with virtually every known silene. The reaction is believed to involve nucleophilic attack by the lone-pair electrons of the atom bearing the acidic hydrogen (deuterium), followed in a second step by transfer of hydrogen (deuterium) to the carbon atom of the original double bond (see Eq. 41).

$$R_2Si{=}CHR' + MeOD \longrightarrow \underset{\underset{Me\ddot{O}-D}{\uparrow}}{R_2Si{\cdots}CHR'} \longrightarrow \underset{Me\ddot{O}-D}{R_2Si{\cdots}CHR'} \longrightarrow \underset{MeO\quad D}{R_2Si-CHR'} \quad (41)$$

A number of results concerning the stereochemistry of addition of several reagents to the silicon–carbon double bond appeared to be inconsistent. On one hand, Jones[210] had reported that the thermolysis of the anthracene [2+4] adduct of one stereoisomer of the silene $PhMeSi{=}C(SiMe_3)CH_2CMe_3$ (see below), in the presence of excess methanol at 190°C or higher, gave a single adduct. The formation of this adduct, the result of *syn* addition across the ends of the silicon–carbon double bond, suggested that methanol addition to a silene was a stereospecific reaction. On the other hand, it had been established that mixtures of silene geometric isomers, when treated with methanol at room temperature, gave mixtures of stereoisomeric methanol adducts in proportions very different from that of the starting silenes, indicating a nonstereospecific process.[87,88] While Jones had shown that the extent of decomposition of the anthracene adducts was independent of the amount of methanol employed, this result does not necessarily define the actual circumstances of the reaction of the silene with methanol and, in our opinion, does not prove that the methanol adduct "does not arise through a bimolecular reaction of methanol" [210] with the *anthracene adduct* (emphasis added).

Sakurai *et al.* have provided what is probably the most important mechanistic finding in the area of intermolecular additions of silenes in recent years, namely a detailed proposal for the mechanism of alcohol addition to the silicon–carbon double bond.[68] A cyclic silene **116** was synthesized in the presence of various amounts of methanol and other alcohols, and varying proportions of methanol adducts **117** and **118** were obtained. It was concluded that the methanolysis involved two steps, the first being the association of the oxygen lone pairs with the sp^2-hybridized silicon atom of the silene. The second step, proton transfer, could occur in two ways. If the proton was transferred from the complexed methanol molecule (path **a**) its delivery would result in *syn* addition. However, if a second molecule of methanol participated (path **b**), it would deliver its proton from the opposite face of the silene ring, leading to overall *anti* addition to the double bond, as in Eq. (42).

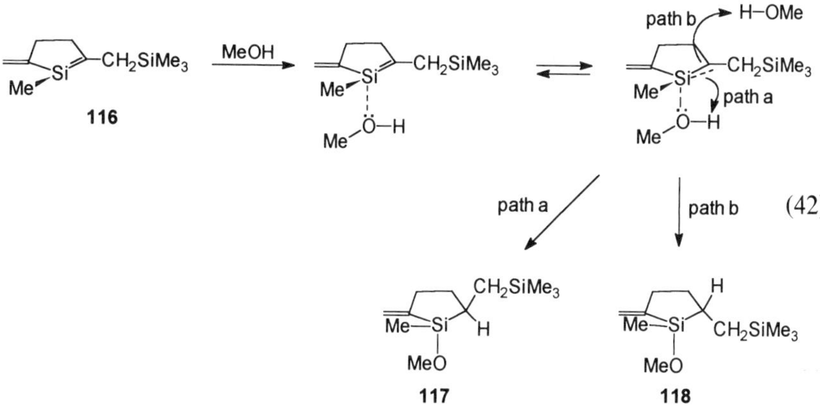

(42)

This result predicts that mixtures of *syn* and *anti* adducts would be expected from the methanolysis; the proportions of the isomers would depend on the amount of methanol present, larger amounts of methanol favoring more *anti* addition, as was shown experimentally. The evidence and arguments presented by Sakurai are very persuasive. Further studies of alcohol additions have recently been reported by Leigh.[68a]

The Jones results can be interpreted with the example depicted in Eq. (43).

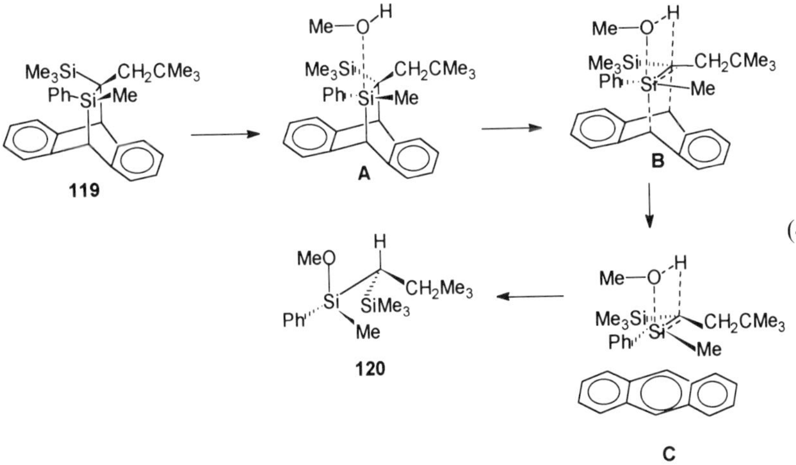

(43)

A molecule of methanol may begin to complex with the silicon atom of the adduct **119**, which is about to be sp^2-hybridized (**A**) as the silene frees itself from the anthracene, forming what is essentially a five-coordinated silicon species (**B**). Given the elevated temperatures involved and the high concentrations of methanol used (9- to 66-fold excesses), it seems probable that the overall addition should be a relatively rapid process, one in which the anthracene molecule will not move far from the newly generated silicon–carbon double bond before methanol addition is complete. Thus, one face of the silene should be effectively screened by the bulky anthracene molecule (**C**) and hence only one molecule of methanol would be involved in the addition, resulting in *syn* stereochemistry of the product **120**.

Methanol and other alcohols also add across the silicon–carbon double bond component of silaallenes. Thus, when $(Me_3Si)PhC{=}C{=}Si(SiMe_3)Mes$ was generated at 140°C in the presence of methanol, the adduct $(Me_3Si)PhC{=}CHSi(OMe)(SiMe_3)Mes$ was obtained, among other products.[72]

2. *Addition of Alkoxysilanes*

Most silenes also react readily with alkoxysilanes such as $MeO{-}SiMe_3$ and $EtO{-}SiH(OEt)_2$, except that the former reagent does not react with silenes of the family $(Me_3Si)_2Si{=}C(OSiMe_3)R$. This is probably a reflection of the fact that the double bond in these silenes is not very electrophilic owing to the electron-releasing substituents attached to the ends of the bond. Jones reported stereospecific *syn* additions by methoxytrimethylsilane to the silene $PhMeSi{=}CHCH_2CMe_3$, which was liberated at elevated temperatures from its anthracene adduct[211] or with methoxytriphenylsilane.[212] In these cases it appears improbable that more than one molecule of alkoxysilane would be involved in the reaction, in contrast to the reactions with methanol. Here again, one face of the newly liberated silene is likely to be shielded by the departing anthracene molecule. Hence *syn* addition can be expected.

Fink reported one other result of a *syn* stereospecific addition of methoxytrimethylsilane to the cyclic, highly sterically hindered silene **121**.[114] The silene was generated photochemically at 77 K in a glass that contained excess alkoxysilane, and a crystal structure of the adduct confirmed the *syn* stereochemistry of the product **122** shown in Eq. (44). When ethanol was used as trapping agent, a single diastereomer of the ethanol adduct was formed when brief irradiation to generate the silene was employed; on extended photolysis to increase the yield, both possible diastereomers were observed. It was suggested, and shown, that photoisomerization of the initially formed diastereomer occurred.

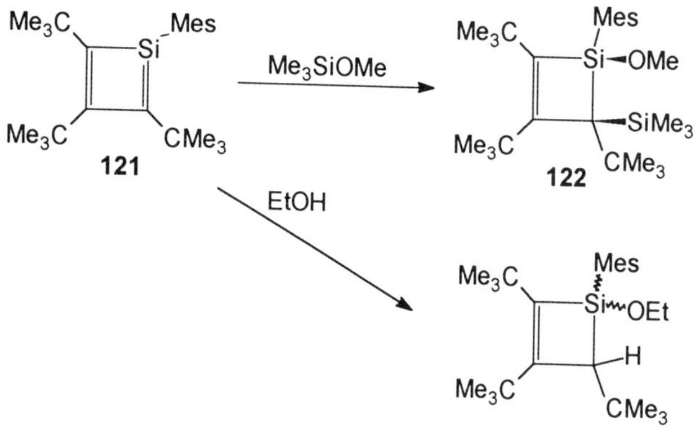

(44)

3. *Other Additions to the Silicon–Carbon Double Bond*

Wiberg has described the reactions of the silene $Me_2Si{=}C(SiMe_3)_2$ with a wide variety of reagents and has reported on their relative rates of reaction (see Table VIII).[98,174] Some silenes will add chlorogermanes and chlorostannanes[174] as well as reactive organic halides such as chloroform, carbon tetrachloride, and benzyl chloride.

More recently it has been found that lithium hydride,[39] organolithium reagents,[30,39,57,61] silyllithium reagents,[56] lithium trimethylsilanolate,[61,91] and Grignard reagents[91] add across the silicon–carbon double bond, although the latter case is complicated by further rearrangements and reactions. In each case the anionic part of the organometallic reagent adds to the silicon end of the double bond, yielding a carbanion which can be protonated or deuterated by hydrolysis, or which may rearrange prior to hydrolysis. Some examples are shown in Table XI.

TABLE XI

REACTIONS OF SILENES WITH ORGANOMETALLIC REAGENTS

Structure	Organometallic reagent	Workup	Product	Reference
$(Me_3Si)_2Si{=}CHCMe_3$	PhLi	H_3O^+	$(Me_3Si)_2PhSiCH_2CMe_3$	57
$(Me_3Si)_2Si{=}CHCMe_3$	Me_3CLi	H_3O^+	$(Me_3Si)_2(Me_3C)SiCH_2CMe_3$	57
$(Me_3Si)_2Si{=}CMe_2$	$(Me_3Si)_3SiLi$		$(Me_3Si)_3SiSi(SiMe_3)_2Me_2C^-Li^+$	56
$RR''Si{=}C(SiMe_3)R'$[a]	$R''MgBr$	H_3O^+	$RR''_2Si{-}CHR'(SiMe_3)$	91
$(Me_3Si)_2Si{=}CPhAr$[b]	ArLi	H_3O^+	$(Me_3Si)_2ArSi{-}CHPhAr$	61
$(Me_3Si)_2Si{=}CPhAr$[b]	Me_3SiOLi	H_3O^+	$(Me_3Si)_2(Me_3SiO)Si{-}CHPhAr$	61

[a] R = Me_3Si, Ph; R′ = Ad, *t*-Bu, Ph; R″ = Me, CD_3, Et, $PhCH_2$, Ph, Me_3SiO.
[b] Ar = Ph, *o*-Tol, *p*-Xylyl.

4. *Reactions with Oxygen*

Silenes are very sensitive toward oxygen, and "stable" silenes exposed to air instantaneously smoke and decompose. Under more controlled conditions, the stable silene $(Me_3Si)_2Si{=}C(OSiMe_3)Ad$ was observed to give $AdCOOSiMe_3$ and a cyclic trimer of $(Me_3Si)_2Si{=}O$.[5] When the silenes $(Me_3Si)MesSi{=}C(OSiMe_3)Ad$ were exposed to dry air, the ultimate product on workup was the silyl ester $(Me_3Si)(Me_3SiO)MesSiOCOAd$.[88] The mechanisms of these reactions have not been studied.

Sander[26] generated the silene $Me_2Si{=}CHMe$ in an argon matrix doped with oxygen. Based on infrared data, the structures of two products were assigned, as shown in Eq. (45).

$$Me_2Si{=}CHMe + O_2 \longrightarrow Me_2\overset{\overset{\large OOH}{|}}{Si}{-}CH{=}CH_2 + Me_2\overset{\overset{\large OH}{|}}{Si}{-}O{-}CH{=}CH_2 \quad (45)$$

Further studies of the reactions of other silenes with oxygen in argon matrices have recently been reported.[212a]

E. *Silene Rearrangement Reactions*

Ever since their discovery in 1967, there has been interest in the kinds of rearrangements that silenes might undergo and curiosity about the behavior of the silicon–carbon double bond as compared to that of the carbon–carbon double bond.

Recently, a number of new and more exotic silenes have been prepared. These silenes have been observed to undergo interesting rearrangements: some under the conditions of silene formation, which may be either thermal or photochemical; and some in which stable, isolable silenes are subjected to either thermal or photochemical treatment. The diversity of the rearrangements observed and the dearth of mechanistic details make it difficult to categorize some of the observed rearrangements. For simplicity, they will be classified below as (1) silene–silylene rearrangements; (2) rearrangements where, overall, a C—H or X—Y bond nominally adds across the ends of a silicon–carbon double bond; (3) 1,2-shifts of an R group from one end of a silicon–carbon double bond to the other end; and (4) 1,3-shifts of an R group involving the silicon–carbon double bond.

1. *Silene–Silylene Rearrangements*

a. *Calculations.* Much attention has been given to the details of the interconversion of simple silenes, such as methylsilene with its isomer dimethylsilylene or silene $H_2Si{=}CH_2$ and methylsilylene. Earlier studies, which have been well summarized by Raabe and Michl[6] and later by Apeloig,[153] clearly show that the former reaction is essentially a thermoneutral process and that, under appropriate conditions, significant proportions of both species can coexist, as trapping reactions reveal. In Gordon's[9] and Grev's[11] reviews, the difference in energy between the two species is agreed to be about 4 kcal mol^{-1}, with the silene being the more stable of the two, having a forward energetic barrier of about 40 kcal mol^{-1}. Goddard's recent paper suggested a larger energy difference between the two isomers,[159] but this has been contested at equally high levels of theory by Schaefer *et al.*[160] The latter paper offers a summary of the energies of the two isomers calculated by a wide variety (16 !) of theoretical means. The energy difference of 4 kcal mol^{-1} seems to be the value on which there is the greatest consensus.

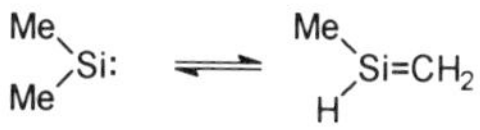

$$\mathrm{Me_2Si:} \rightleftharpoons \mathrm{Me(H)Si{=}CH_2}$$

b. *Experimental Results.* Conlin and Kwak undertook trapping experiments to evaluate the facility for the methylsilene–dimethylsilylene rearrangement.[148] The thermolysis of 1-methylsilacyclobutane **123** in the presence of butadiene under pseudo–first-order conditions led to the formation of several products. The decomposition had the activation parameters $A = 10^{14.9}\ \mathrm{s^{-1}}$, $E_{act} = 59.1$ kcal mol^{-1}. The products of primary interest are the silacyclohexene (**124**), derived from methylsilene, and the 1,1-dimethyl-2-silacyclopentene (**125**) and -3-silacyclopentene (**126**), derived from dimethylsilylene. By using the ratio of these products obtained at several temperatures, the Arrhenius parameters of $A = 9.6\ \mathrm{s^{-1}}$, $E_{act} =$ 30.4 kcal mol^{-1} can be determined for the silene-to-silylene rearrangement.

$$\mathbf{123} \xrightarrow{\Delta} \mathrm{CH_2{=}SiHMe} \longrightarrow \mathrm{Me_2Si:}$$

124 **125** **126**

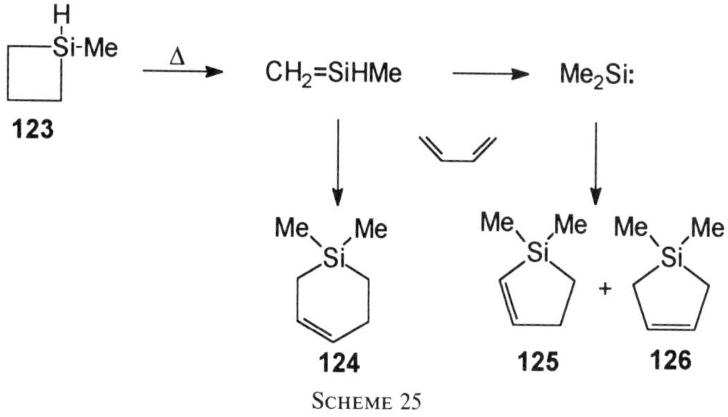

Scheme 25

2. *Intramolecular Rearrangements Involving C—H or X—Y Additions to the Si═C Double Bond*

In a number of rearrangements of silenes the Si═C bnd has been observed to react with a C—H bond of a methyl group that is usually attached at the *ortho* position of an adjacent mesityl group. Formally, these can be regarded as $2\pi + 2\sigma$ reactions, although other descriptions may be possible. For example, a 1,5-H shift followed by an electrocyclic rearrangement of a Si═C with a C═C would effect the same results. Little is known about the mechanisms involved. Several examples of these types of reaction are described below, some being effected photochemically and some thermally.

Brook *et al.*[85,87] observed such reactions during the formation of silenes by photolysis. Using radiation with $\lambda > 360$ nm, they photolyzed acylsilanes such as **127**, which bears a mesityl group attached to the carbonyl carbon. On prolonged photolysis of the initially formed silene **128**, the C—H bond of the *ortho* methyl group of the mesityl group added to the silicon–carbon double bond to form the benzocyclobutane **129**. Alternatively a 1,5-H shift would lead to the species **130**, which would also yield the benzocyclobutane on electrocyclic rearrangement.

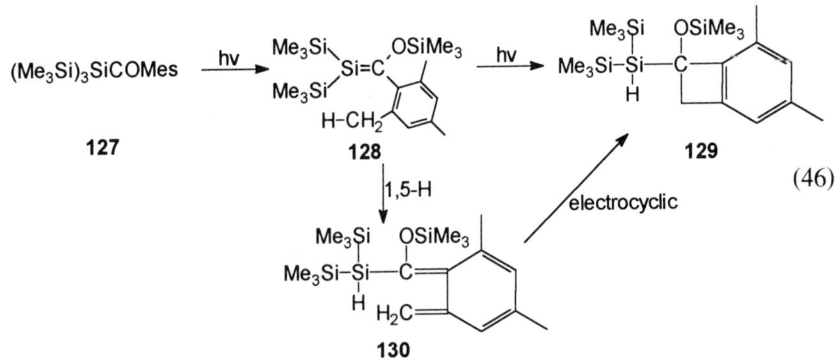

$(Me_3Si)_3SiCOMes$ (**127**) —hν→ $(Me_3Si)_2Si{=}C(OSiMe_3)$(mesityl, H-CH₂) (**128**) —hν→ **129**; **128** —1,5-H→ **130** —electrocyclic→ **129** (46)

Klingebiel[32] observed a related insertion with the silene **131** (Eq. 47), which gave the product **132** under the thermal conditions of its formation. Surprisingly, the regiochemistry of the C—H addition to the Si═C bond was opposite to the case in Eq. (46).

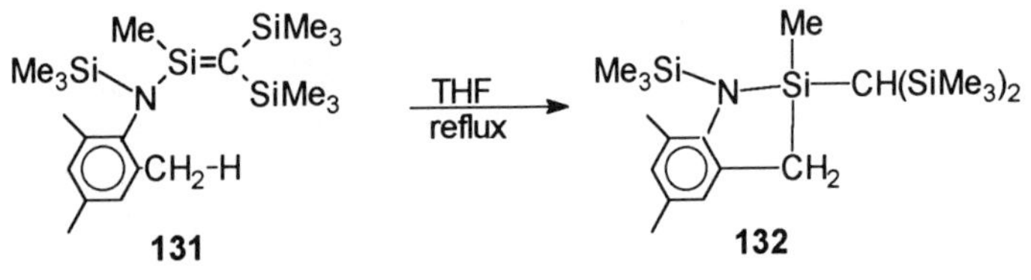

131 —THF, reflux→ **132** (47)

The remarkably stable silaallene **133** described recently by West[213,214] showed related behavior. Photolysis of **133** is reported to result in the C—H bond addition, across the Si=C bond, of a methyl group on the aromatic ring attached to the carbon end of the silaallene, resulting in the polycyclic compound **134**. Alternatively, treatment of the silaallene with acid yields compound **135** by the nominal addition that occurs when one of the C—H bonds of an *ortho* *t*-Bu group of the supermesityl group attached to silicon adds to the ends of the Si=C bond with the opposite regiochemistry.

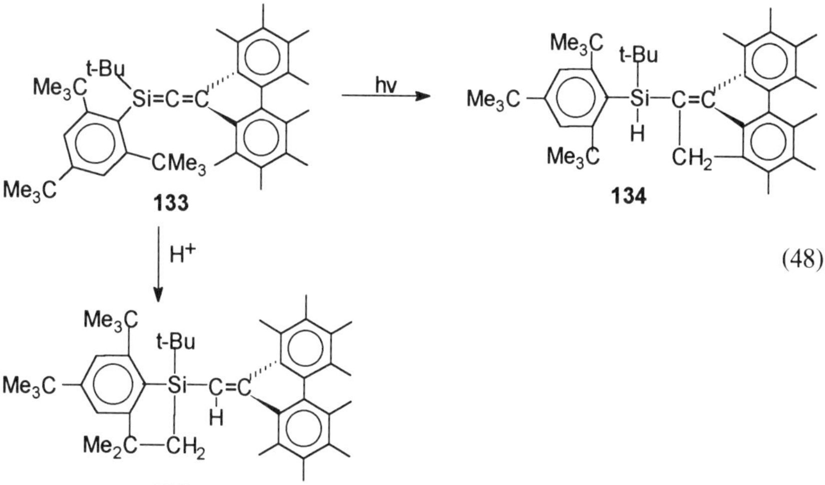

(48)

Another case of silene rearrangement under mild conditions, shown in Eq. (49), involves the 1-methyl-1-*N*,*N*,*N*′-tris(trimethylsilyl)ethylenediamino-2-neopentylsilene **136**. At about room temperature **136** reacts in a nominal $2\pi + 2\sigma$ manner, involving N—Si bond formation and migration of a trimethylsilyl group to the original sp^2-hybridized carbon atom of the silene, to give the cyclic compound **137**.[45] This was a major product in most reactions that involved attempted trapping of the silene **136** with various dienes. The rearrangement probably involves initial coordination by nitrogen of the silene silicon atom.

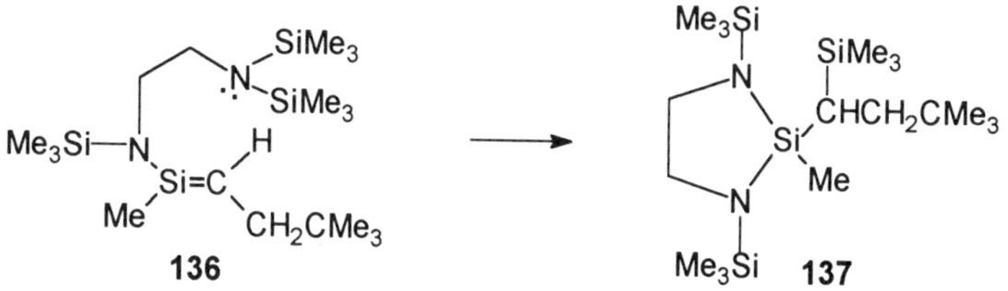

(49)

A number of rearrangements of transient silenes have been observed in the course of their preparation by low-pressure gas-phase thermolysis. Gusel'nikov[16] generated a variety of silenes containing aryl groups directly or indirectly attached to the sp^2-hybridized silicon atom of silenes by low-pressure gas-phase cyclodecomposition of monosilacyclobutanes (Eq. 50). The *o*-tolylmethylsilene **138** gave rise to a significant yield of the dimethylbenzosilacyclobutane **139**, said to be the result of a 1,5-H shift, while the related phenyl species **140** gave the monomethylbenzosilacyclobutane **141** from a 1,3-H shift. With the benzyl analog **142**, 2-methyl-2-silaindane **143** was said to be formed due to a 1,4-H shift. These processes presumably involve diradical intermediates which subsequently couple, although the first two reactions could be attributed to electrocyclic rearrangement reactions followed or preceded by the appropriate 1,*n*-H shift.

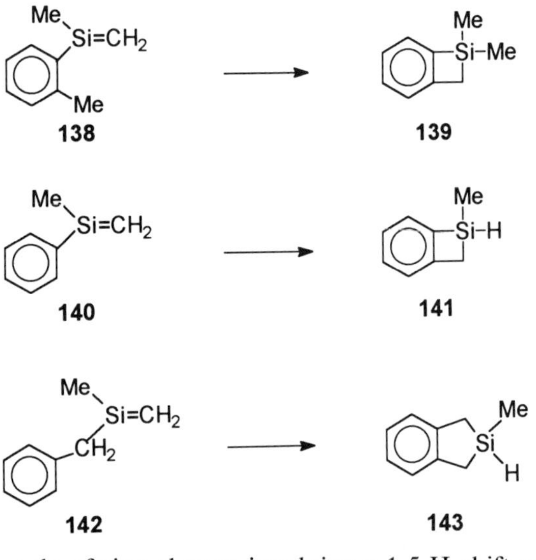

Me Si=CH2 Me → Me Si–Me

138 **139**

Me Si=CH2 → Me Si–H (50)

140 **141**

Me Si=CH2 CH2 → Me Si H

142 **143**

Another example of ring closure involving a 1,5-H shift appears to be that provided by Jung,[119] who reported that the heteroatom-substituted silenes **144** rearranged to give 1,3-disilacyclobutanes **145** via a diradical intermediate (Eq. 51). When R = Cl the yield was 30%, and with R = MeO the yield of the disilacyclobutane was 44%.

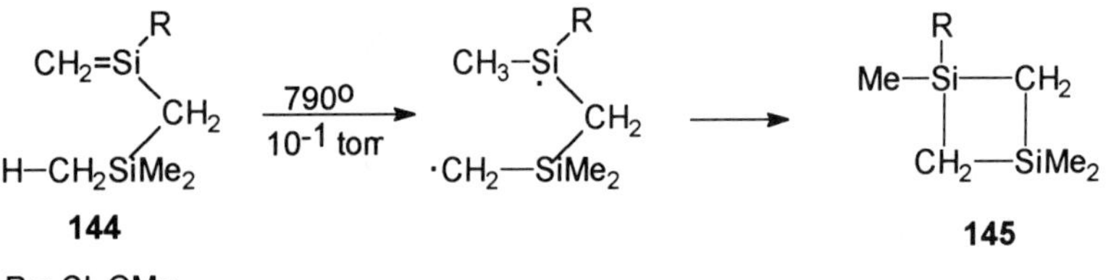

R CH2=Si CH2 H–CH2SiMe2 —790°, 10^-1 torr→ R CH3–Si· CH2 ·CH2—SiMe2 → R Me—Si—CH2 CH2—SiMe2 (51)

144 **145**

R = Cl, OMe

3. *1,2-Shifts of Groups or Atoms from End to End of the Silicon–Carbon Double Bond*

A number of examples have been found recently in which silenes of the family $(Me_3Si)_2Si{=}C(OSiMe_3)R$ undergo further photochemical rearrangement during the course of their formation by the photolysis of polysilylacylsilanes. When mesityl groups are attached to the silicon atom of the Si═C bond as in compound **146** (Eq. 52), it is clear from the structures of the products **148** that the siloxy group originally on carbon has undergone 1,2-migration to silicon. If this is a stepwise process, as proposed,[90] the silylcarbene **147** that would be formed could then insert into the spatially accessible C—H bond of the *ortho* methyl of the mesityl group, thus forming the observed products **148**, whose structures were confirmed by a crystal structure determination.[88,90] Although attempts to intercept the proposed carbene intermediate **147** were unsuccessful, the proposed mechanism is not vitiated given the known efficiency of intramolecular processes in general.

$Me_3SiSi(Mes)RCOAd$ —hν→ **146** —hν→ **147** → **148**

R = Me_3Si, Mes

(52)

When there is no adjacent, reactive C—H bond into which the carbene could insert, insertion into a silicon–silicon bond may occur, resulting in a 1,2-trimethylsilyl migration from silicon to carbon. This has been proposed on the basis of the known structures of several compounds.[87] A sequence of further steps, each of which has been separately observed, is required to explain the structures of the final products.

As an example (Eq. 53), the *t*-butyl-substituted silene **149**, on photolysis, may undergo sequential steps of 1,2-trimethylsiloxy and 1,2-trimethylsilyl migrations, involving the carbene intermediate **150**.[89,90] It has also been suggested that these two steps may occur simultaneously as a dyotropic process,[87] yielding the new silene **151**. This silene evidently rearranges with a 1,3-methyl shift from sp^3-hybridized silicon to sp^2-hybridized silicon more rapidly than it dimerizes, thus leading to the isomeric silene **152**, the monomer responsible for one of the two observed head-to-tail silene dimers, **153**. Such 1,3-methyl shifts have been observed by Eaborn[215] and by Wiberg[216] as facile processes. The other monomeric silene, **154**, responsible for the formation of the other observed head-to-tail silene dimers, **155**, could be formed from **152** by a 1,3-*t*-butyl silicon-to-silicon migration or by a 1,3-trimethylsiloxy silicon-to-silicon migration followed by a 1,3-methyl silicon-to-silicon migration. It has not been possible to establish the full details of these complex rearrangements, but the crystal structures of the head-to-tail dimers **153** and **155** clearly establish the structures of the precursor monomers.

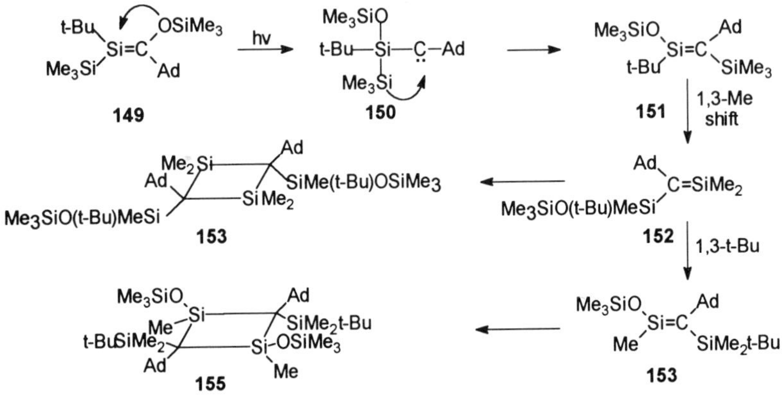

(53)

Analogous behavior was followed by the phenyl-substituted silene **156**. The initially formed silene **157** underwent 1,3-methyl migration to give the silene **158**, which then dimerized in a head-to-tail manner to yield three different stereoisomeric dimers **159**, two of which were characterized by crystal structures. Again, the exchange of trimethylsilyl and trimethylsiloxy groups at the ends of the Si=C bond occurred, followed by 1,3-methyl silicon-to-silicon rearrangements. The steps are summarized in Eq. (54).

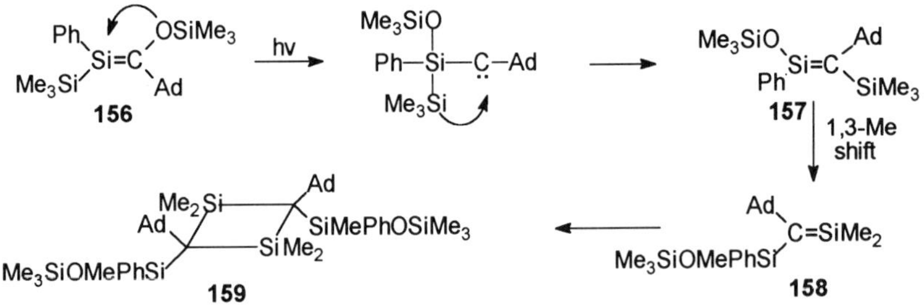

(54)

Ishikawa[71,191] has described the thermal rearrangement at 280°C of the silacyclopropene **160** to the silaallene **161**. It was suggested that a 1,2-trimethylsilyl rearrangement from silicon to carbon could lead to the silylene **162**; on insertion into the C—H bond of one of the trimethylsilyl methyl groups **162** would give **163**, which was isolated in 49% yield.

Alternatively, a radical process involving the diradical **164** was proposed to explain the formation of the 28% yield of **165** isolated.

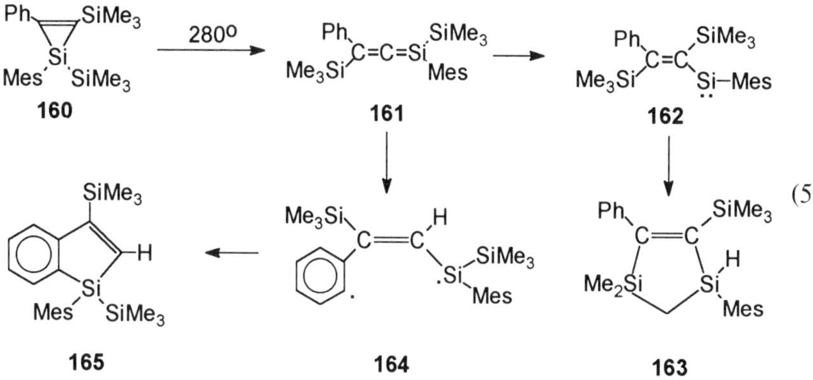

(55)

Grobe[15] has described the pyrolysis of 1-methyl-1-vinyl- and 1,1-divinyl-1-silacyclobutanes **166** which led to the formation of methylvinylsilene and divinylsilene, respectively. Under the experimental conditions used, it was suggested that the silenes rearrange to *exo*-methylene-1-silacyclopropanes **167** which extrude methylsilylene or vinylsilylene, respectively. In support of this proposal, when the reactions were carried out in the presence of 2,3-dimethylbutadiene, the anticipated silylenes were trapped as their respective 1-silacyclopent-3-enes **168**.

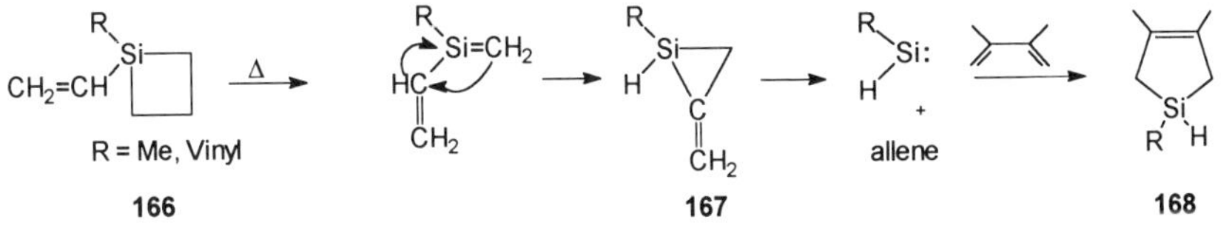

(56)

4. *1,3-Shifts of Groups or Atoms*

As mentioned and illustrated above in Eqs. (53) and (54), a 1,3-shift, particularly of a methyl group, from sp^3- to sp^2-hybridized silicon is a common and facile process that occurs with silenes. Thus it was clearly established by Wiberg,[98,216] using NMR techniques, that the short-lived silene $t\text{-}Bu_2Si{=}C(SiMe_3)_2$ rapidly rearranged to the stabler species $Me_2Si{=}C(SiMe_3)(SiMe(t\text{-}Bu)_2)$ and that the CD_3 groups in $(CD_3)_2Si{=}C(SiMe_3)_2$ were completely scrambled into all locations when the silene precursor adduct was heated to 120°C. Other examples are also well known.

Eaborn *et al.*[215] observed skeletal rearrangements in the course of the thermal elimination of Me_3SiF from compound **169**. The anticipated silene $Ph_2Si{=}C(SiMe_3)_2$ **170** obviously underwent a 1,3-methyl silicon-to-silicon migration leading to silene **171** prior to undergoing electrophilic attack by silicon of the canonical structure **172** on one of the phenyl groups. This resulted in the cyclic species **173**, one of several related compounds detected in the reaction mixture (Scheme 26). This was one of the earliest reported 1,3-methyl migrations from sp^3-hybridized silicon to sp^2-hybridized silicon.

Similar rearrangements are also observed during the photolysis (and often thermolysis) of silyldiazoalkanes such as **174**. When photolyzed in benzene compounds, **174** initially forms carbenes **175**, which then isomerize to acylsilenes **176**, as in Scheme 27. In the cases reported to date, the silene formed normally undergoes spontaneous electrocyclic rearrangement with the acyl carbonyl group (in the absence of traps) or reaction with trapping reagents. It is reasonably assumed that these subsequent rearrangements are nonphotochemical processes. Thus, Ando[25] and Maas *et al.*[21–23] have described the formation of acylsilenes **176** of the general structure $Me_2Si{=}C(R)COR'$ which, in the absence of trapping agents, spontaneously cyclize to the four-membered siloxatene ring species **177** when R′ is bulky (Ad or *t*-Bu)—the steric demands of the bulky groups evidently favoring cyclization to the four-membered ring. However, when R′ is smaller (*i*-Pr or Me), the eight-membered cyclic system **178** is formed instead (this latter species may be formed directly from the silene or by dimerization of the siloxatene **177**).[21,22,25,217] Accompanying the siloxatenes were small amounts of bis-silylketenes **179**, the products of a 1,3-shift of the alkyl group R′ from carbon to silicon.[22] A similar migration of an alkoxyl group from carbon to silicon to give an alkoxysilylketene had

$Ph_2FSi{-}C(SiMe_3)_3$ **169** $\xrightarrow{\Delta}$ $Ph_2Si{=}C(SiMe_3)_2$ **170** $\xrightarrow{\text{1,3-Me shift}}$ $Ph_2MeSi{-}C(SiMe_3){=}SiMe_2$ **171** ⟷ **172** (PhMeSi(Ph)–$\bar{C}$(SiMe$_3$)–$\overset{+}{Si}Me_2$) → **173** (PhMeSi, SiMe$_2$, SiMe$_3$)

SCHEME 26

$$Me_3SiSiMe_2CN_2COR \xrightarrow{h\nu} Me_3SiSiMe_2\ddot{C}COR \longrightarrow Me_2Si{=}C(SiMe_3)\text{—}C(R){=}O$$

174 **175** **176**

176 → $Me_2Si\text{—}CSiMe_3$ / $O\text{—}CR$ (four-membered ring); R = Ad, t-Bu **177** → $(Me_2Si{=}O)_n$ + $Me_3SiC{\equiv}CR$

176 → **178** (eight-membered ring: $Me_3Si(R)C{=}C\text{—}O\text{—}SiMe_2\text{—}C(SiMe_3){=}C(R)\text{—}O\text{—}SiMe_2$); R = Me, i-Pr

176 → $(Me_3Si)(Me_2RSi)C{=}C{=}O$ **179**

SCHEME 27

been reported earlier by Ando.[25] The siloxatenes are generally not very stable, decomposing at 120°C or lower temperatures to silylalkynes and oligomers of dimethylsilanone.

Different rearrangements were observed in other cases. Thus, Maas[22] reported that when photolyzed in benzene the polysilyldiazoketone **180** gave the isomeric ketene **181**, the product of a Wolff rearrangement (a 1,2 carbon-to-carbon rearrangement) of the initially formed carbene **182** (Eq. 57). The isomeric bis-silylketene **183** was not observed, but the siloxatene **184** was also a product of the reaction.

$$(Me_3Si)_3SiSiMe_2CN_2COAd \xrightarrow[\text{benzene}]{h\nu} (Me_3Si)_3SiSiMe_2\ddot{C}COAd \longrightarrow (Me_3Si)_3SiSiMe_2(Ad)C{=}C{=}O$$

180 **181** **182**

181 ↛ $(Me_3Si)_3Si(AdMe_2Si)C{=}C{=}O$ **183**

182 + $(Me_3Si)_3SiC{=}C(Ad)\text{—}O\text{—}SiMe_2$ (ring) **184**

(57)

Wolff rearrangements were also observed when most of the same acylsilyldiazoalkanes were photolyzed in acetone instead of benzene.[21] The ketenes **185** resulting from a 1,3-methyl migration of the silene were detected in addition to the expected ene product **186** derived from the reaction of the silene with acetone (or other enolizable ketones) (Eq. 58). When R′ = Ad, only the cyclic siloxatene **187** was formed under the same conditions, indicating that intramolecular ring closure of the silene to **187** is much faster than the intermolecular ene reaction with solvent that leads to **186**. The influence of the solvent on these rearrangements is not understood.

$$Me_3SiSiMe_2CN_2COR \xrightarrow[\text{acetone, R = Me}]{h\nu} Me_2Si{=}C(SiMe_3)(COR) \xrightarrow[\text{1,3-Me}]{\text{R = Me}} (Me_3Si)_2C{=}C{=}O\ (\mathbf{185}) + Me_2Si(OC(Me){=}CH_2){-}CH(SiMe_3)(COMe)\ (\mathbf{186})$$

$$Me_2Si{=}C(SiMe_3)(COR) \xrightarrow{\text{R = Ad}} \underset{\mathbf{187}}{Me_2Si{-}C(SiMe_3){=}C(Ad){-}O\ (\text{ring})} \tag{58}$$

Another example of an intramolecular photochemical rearrangement occurs when 1,1,2-trimethylsilene is irradiated at 250 nm, isomerizing to vinyldimethylsilane as the result of a 1,3-H shift.[26]

Auner[52,53] has described the attempted formation of 1,1-dichloro-1-silaethenes **188**, which bear alkoxymethylsilyl groups on the sp^2-hybridized carbon atom. These were found to rearrange spontaneously at about 0°C, as shown in Eq. (59), by a 1,3-alkoxy silicon-to-silicon migration, yielding the isomeric 1,1-dimethylsilenes **189** which were trapped by a variety of reagents. Mixed 1,1-chloro-*t*-butoxy and 1,1-di-*t*-butoxy analogs of silene **188** rearranged similarly. While 1,3-rearrangements are observed fairly commonly in organosilicon chemistry, these appear to be the first reported cases of alkoxy migrations from silicon to silicon, and are notable for the mild conditions under which the rearrangements occur.

$$((t\text{-BuO})Me_2Si)(Cl_3Si)C{=}CH_2 \xrightarrow[\text{-78°, -LiCl}]{t\text{-BuLi}} \underset{\mathbf{188}}{Cl_2Si{=}C(CH_2CMe_3)(SiMe_2(Ot\text{-Bu}))} \longrightarrow \underset{\mathbf{189}}{Me_2Si{=}C(CH_2CMe_3)(SiCl_2(Ot\text{-Bu}))} \tag{59}$$

5. *Other Rearrangements*

Jones[218] has described an unusual photochemically initiated rearrangement of a silene–anthracene adduct to a silene which is part of an eight-membered ring (Eq. 60). Photolysis of the adduct **190** was believed to form the silaallylic diradical **191**, whose canonical form **192** affords the novel silene **193**. The silene was appropriately trapped, and thus characterized. This is the only known case of a silicon–carbon double bond in an eight-membered ring.

(60)

F. *Rates of Reaction Studies*

In the course of many studies concerning silenes, a number of measurements of the rates of various reactions have been obtained. Some of these have been reported above in other locations. In many cases, activation parameters have also been obtained. Some of the studies are summarized briefly below, but the original papers should be consulted for the full details.

Conlin studied the photochemical ring opening of 1,1-dimethyl-2-phenylcyclobut-3-ene to yield a siladiene and its recyclization (Eq. 61). The E_{act} for cyclization was 9.4 kcal mol^{-1}.[113]

$$Me_2Si=CPh-CH=CH_2 \quad (61)$$

Related studies on the thermal conversion of 1,1-dimethyl-1-silacyclobut-2-ene to 1,1-dimethyl-1-silabutadiene and its reaction with ethylene and *cis*- and *trans*-2-butene were also described.[219]

Wiberg has studied the kinetics of several systems involving the silene $Me_2Si{=}C(SiMe_3)_2$. The kinetics for the complex system of the silene with *N*-trimethylsilylbenzophenonimine, namely [2+4] adduct <=> silene + imine <=> [2+2] adduct as shown in Eq. (62), were measured[174,198] as were the data for the corresponding system with benzophenone, *viz*. [2+4] adduct <=> silene + benzophenone <=> [2+2] adduct.[220]

$Me_2Si\text{-}C(SiMe_3)_2$, $Me_3Si\text{-}N$, Ph $\rightleftharpoons^{\Delta}$ $Me_2Si{=}C(SiMe_3)_2$ + $Me_3SiN{=}CPh_2$ $\rightleftharpoons$ $Me_2Si\text{—}C(SiMe_3)_2$ / $Me_3Si\text{-}N\text{—}CPh_2$

$Me_2Si\text{-}C(SiMe_3)_2$, O, Ph $\rightleftharpoons^{\Delta}$ $Me_2Si{=}C(SiMe_3)_2$ + $O{=}CPh_2$ $\rightleftharpoons$ $Me_2Si\text{—}C(SiMe_3)_2$ / $O\text{—}CPh_2$

(62)

The relative rates of reaction of the silene $Me_2Si{=}C(SiMe_3)_2$ with a series of amines, alcohols, phenols, thiophenols, dienes, and alkenes were obtained[174] and are reported in Table VIII Section IV.C.

A study of the dimerization of $(Me_3Si)_2Si{=}C(OSiMe_3)Me$ was carried out by Conlin.[173] The reaction followed clean second-order kinetics and had an $E_{act} = 0.2 \pm 0.1$ kcal mol^{-1} and log $A = 7 \pm 1$ s^{-1}. The study did not entirely resolve the question of whether the dimerization involves an ene mechanism or a biradical intermediate.

G. *Silaaromatics*

Aromatic rings containing sp^2-hybridized silicon atoms also constitute members of the silene family. Interest in these compounds largely focuses on the possible delocalization of electron density in the rings, the extent to which they display *aromatic* character, and their relative stabilities. Raabe and Michl[6] have reported much data, which will not be repeated here; other computational studies are reported in Section III.D. A few interesting additions to our knowledge of these systems have been reported in recent years.

Routes for the synthesis of several of silaaromatics from silyldiazoalkanes were shown in Eqs. (15) and (16) in Section II.B.13.

In a nice review of the preparative methods for silaaromatics, Maier[221] suggested that the flash vacuum pyrolytic techniques that had been so useful for the monosilyl compound could be utilized to prepare 1,4-disilabenzene. Thermolysis of 1,4-disila-2,5-cyclohexadiene led to formation of 1,4-disilabenzene, which had a UV absorption maximum at 405 nm.

West[222] developed a synthesis using the thermal or photochemical retro-Diels–Alder reaction of its anthracene adduct **194** to obtain hexamethyl-1,4-disilabenzene **195**. As summarized in Scheme 28, compound **195** added a single molecule of methanol in a 1,4-manner, yielding **196**, and reacted in a [2+4] manner with several alkynes to give 1,4-disilabarrelenes such as **197**. Disilabenzene **195** also added an oxygen atom in a 1,4-manner,

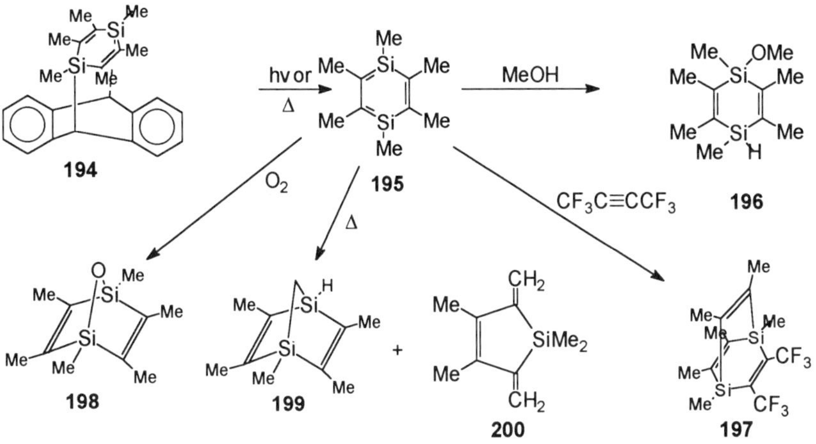

SCHEME 28

giving **198**. Finally, **195** thermally rearranged in two ways, yielding on the one hand a 1,4-methylene-bridged species **199** whose methylene bridge was derived from the methyl group originally on silicon, and on the other hand a silacyclopentene derivative **200** by a very complex process.

V
CONCLUSIONS

In the less than three decades since silenes were first described by Gusel'nikov and Flowers, an impressive amount of knowledge concerning silenes and the behavior of the silicon–carbon double bond has been discovered and reported. Several hundred papers have been published dealing with silenes in one context or another, and it is clear that the status of silenes has changed from that of rare oddity to a not uncommon occurrence. Much has been learned about their reactions, although much remains to be learned about the finer details of the mechanisms of some of their reactions.

It may well be that activity in the area of silene chemistry has peaked. Certainly, the number of papers currently published is lower than it was a few years ago. This may be due, at least in part, to the fact that many of the easier and more obvious experiments have been attempted. However, only the discovery of important applications is required for the field to be revitalized. In particular, the successful synthesis of useful polymers from silenes could lead to important materials for semiconductor and nonlinear optical applications. For recent studies of poly(silaethylene), see Interrente *et al.*[223]

It is recognized that, of the hundreds of reports involving silenes published even in the last ten years, not every one has been mentioned in this review. The authors selected those publications that appeared to be the most significant or interesting. If any publications have been misinterpreted in the review the authors regret the error: the responsibility lies fully with the first-named author.

References

(1) Gusel'nikov, L. E.; Flowers, M. C. *J. Chem. Soc., Chem. Commun.* **1967,** 864.
(2) Gusel'nikov, L. E.; Flowers, M. C. *J. Chem. Soc. B* **1968,** 419, 1396.
(3) Brook, A. G. *Abstracts of the XIVth Organosilicon Symposium,* Duke University, Durham, NC, March *1981.*
(4) Brook, A. G.; Abdesaken, F.; Gutekunst, B.; Gutekunst, G.; Kallury, R. K. M. R. *J. Chem. Soc., Chem. Commun.* **1981,** 191.
(5) Brook, A. G.; Nyburg, S. C.; Abdesaken, F.; Gutekunst, B.; Gutekunst, G.; Kallury, R. K. M. R.; Poon, Y. C.; Chang, Y.-M.; Wong-Ng, W. *J. Am. Chem. Soc.* **1982,** *104*, 5667.
(6) Raabe, G.; Michl, J. *Chem. Rev.* **1985,** *85*, 419.
(7) Brook, A. G.; Baines, K. M. *Adv. Organomet. Chem.* **1986,** *25*, 1.
(8) Cowley, A. H.; Norman, N. C. *Prog. Inorg. Chem.* **1986,** *1*, 34.
(9) Gordon, M. S. In *Molecular Structure and Energetics;* VCH: Deerfield Beach, Florida, 1986; Chapter 4, p. 101.
(10) Shklover, V. E.; Struchkov, Y. T.; Voronkov, M. G. *Main Group Met. Chem.* **1988,** *11*, 109.
(11) Grev, R. S. *Adv. Organomet. Chem.* **1991,** *33*, 125.
(12) Lickiss, P. D. *Chem. Soc. Rev.* **1992,** *21*, 271.
(13) *The Chemistry of Organic Silicon Compounds;* Patai, S.; Rappoport, Z., Eds.; Wiley: New York, 1989.
(14) Yeh, M.-H.; Linder, L.; Hoffmann, D. K.; Barton, T. J. *J. Am. Chem. Soc.* **1986,** *108*, 7849.
(15) Grobe, J.; Ziemer, H. *Z. Naturforsch., B: Chem. Sci.* **1993,** *48B*, 1193.
(16) Gusel'nikov, L. E.; Volkova, V. V.; Volnina, E. A.; Buravtseva, E. N. *Abstract A-31 from XXVII Organosilicon Conference,* Troy, NY, March, *1994.*
(17) Jung, I. N.; Pae, D. H.; Yoo, B. R.; Lee, M. E.; Jones, P. R. *Organometallics* **1989,** *8*, 2017.
(18) Pola, J.; Volnina, E. A.; Gusel'nikov, L. E. *J. Organomet. Chem.* **1990,** *391*, 275.
(19) Sekiguchi, A.; Ando, W. *Organometallics* **1987,** *6*, 1857.
(20) Sekiguchi, A.; Ando, W. *Chem. Lett.* **1986,** 2025.
(21) Maas, G.; Schneider, K.; Ando, W. *J. Chem. Soc., Chem. Commun.* **1988,** 72.
(22) Schneider, K.: Daucher, B.; Fronda, A.; Maas, G. *Chem. Ber.* **1990,** *123*, 589.
(23) Maas, G.; Alt, M.; Schneider, K.; Fronda, A. *Chem. Ber.* **1991,** *124*, 1295.
(24) Sekiguchi, A.; Ando, W.; Honda, K. *Tetrahedron Lett.* **1985,** *26*, 2337.
(25) Sekiguchi, A.; Sato, T.; Ando, W. *Organometallics* **1987,** *6*, 2337.
(26) Sander, W.; Trommer, M. *Chem. Ber.* **1992,** *125*, 2813.

(27) Ando, W.; Kurishima, K.; Sugiyama, M. *J. Am. Chem. Soc.* **1991,** *113*, 7790.
(28) Wiberg, N.; Preiner, G.; Schieda, O. *Chem. Ber.* **1981,** *114*, 2087.
(29) Wiberg, N.; Wagner, G.; Müller, G. *Angew. Chem., Int. Ed. Engl.* **1985,** *24*, 229.
(30) Wiberg, N.; Link, M.; Fischer, G. *Chem. Ber.* **1989,** *122*, 409.
(31) Wiberg, N.; Joo, K. S.; Polborn, K. *Chem. Ber.* **1993,** *126*, 67.
(32) Klingebiel, U.; Pohlmann, S.; Skoda, L. *J. Organomet. Chem.* **1985,** *291*, 277.
(33) Wiberg, N.; Wagner, G. *Chem. Ber.* **1986,** *119*, 1467.
(34) Wiberg, N.; Wagner, G.; Müller, G.; Riede, J. *J. Organomet. Chem.* **1984,** *271*, 381.
(35) Ziche, W.; Auner, N.; Behm, J. *Organometallics* **1992,** *11*, 2494.
(36) Auner, N.; Heikenwalder, C.-R.; Wagner, C. *Organometallics* **1993,** *12*, 4135.
(37) Auner, N. *J. Organomet. Chem.* **1989,** *377*, 175.
(38) Auner, N. *J. Organomet. Chem.* **1987,** *336*, 83.
(39) Auner, N. *Z. Anorg. Allg. Chem.* **1988,** *558*, 87.
(40) Auner, N.; Weingartner, A. W.; Herdtweck, E. *Z. Naturforsch., B: Chem. Sci.* **1993,** *48B*, 318.
(41) Auner, N. *J. Organomet. Chem.* **1987,** *336*, 59.
(42) Auner, N.; Penzenstadler, E. *Z. Naturforsch., B: Chem. Sci.* **1992,** *47B*, 795.
(43) Auner, N.; Penzenstadler, E. *Z. Naturforsch., B: Chem. Sci.* **1992,** *47B*, 217.
(44) Auner, N.; Penzenstadler, E. *Z. Naturforsch., B: Chem. Sci.* **1992,** *47B*, 805.
(45) Auner, N.; Penzenstadler, E.; Herdtweck, E. *Z. Naturforsch., B: Chem. Sci.* **1992,** *47B*, 1377.
(46) Ohshita, J.; Naka, A.; Ishikawa, M. *Organometallics* **1992,** *11*, 602.
(47) Auner, N.; Grobe, J.; Schaefer, T.; Krebs, B.; Dartman, M. *J. Organomet. Chem.* **1989,** *363*, 7.
(48) Auner, N.; Gleixner, R. *J. Organomet. Chem.* **1990,** *393*, 33.
(49) Jones, P. R.; Lim, T. F. O. *J. Am. Chem. Soc.* **1977,** *99*, 2013.
(50) Auner, N. *Z. Anorg. Allg. Chem.* **1988,** *558*, 55.
(51) Auner, N.; Wagner, C.; Ziche, W. *Z. Naturforsch., B: Chem. Sci.* **1994,** *49B*, 831.
(52) Ziche, W.; Auner, N.; Behm, J. *Organometallics* **1992,** *11*, 3805.
(53) Ziche, W.; Auner, N.; Kiprof, P. *J. Am. Chem. Soc.* **1992,** *114*, 4910.
(54) Oehme, H.; Wustrack, R. *Z. Anorg. Allg. Chem.* **1987,** *552*, 215.
(55) Wustrack, R.; Oehme, H. J. *J. Organomet. Chem.* **1988,** *352*, 95.
(56) Oehme, H.; Wustrack, R.; Heine, A.; Sheldrick, G. M.; Stalke, D. *J. Organomet. Chem.* **1993,** *452*, 33.
(57) Krempner, C.; Oehme, H. *J. Organomet. Chem.* **1994,** *464*, C7.
(58) Bravo-Zhivotovskii, D.; Braude, V.; Stanger, A.; Kapon, M.; Apeloig, Y. *Organometallics* **1992,** *11*, 2326.
(59) Bravo-Zhivotovskii, D.; Zharov, I.; Apeloig, Y. *Abstracts Xth International Symposium on Organosilicon Chemistry,* Poznan, Poland, *1993*; p. 295.
(60) Ohshita, J.; Masaoka, Y.; Ishikawa, M. *Organometallics* **1991,** *10*, 3775.
(61) Ohshita, J.; Masaoka, Y.; Ishikawa, M.; Takeuchi, T. *Organometallics* **1993,** *12*, 876.
(62) Biltveva, I. S.; Bravo-Zhivitovskii, D. A.; Kalikhman, J. D.; Vitkovskii, V. Y.; Shevchenko, S. G.; Vyazankin, N. S.; Voronkov, M. G. *J. Organomet. Chem.* **1989,** *368*, 163.
(63) Ohshita, J.; Masaoka, Y.; Masaoka, S.; Ishikawa, M.; Akimoto, T.; Yano, T.; Yamabe, T. *J. Organomet. Chem.* **1994,** *473*, 15.
(64) Krempner, C.; Oehme, H. *Abstracts Xth International Symposium on Organosilicon Chemistry,* Poznan, Poland, *1993*; p. 234.
(65) Conlin, R. T.; Bobbitt, K. L. *Organometallics* **1987,** *6*, 1406.
(66) Ishikawa, M.; Nishimura, Y.; Sakamoto, H. *Organometallics* **1991,** *10*, 2701.

(67) Takaki, K.; Sakamoto, H.; Nishimura, Y.; Sugihara, Y.; Ishikawa, M. *Organometallics* **1991,** *10*, 888.
(68) Kira, M.; Maruyama, T.; Sakurai, H. *J. Am. Chem. Soc.* **1991,** *113*, 3986.
(68a) Leigh, W. H.; Sluggett, G. W. *J. Am. Chem. Soc.* **1994,** *116*, 10468.
(69) Ishikawa, M.; Sugisawa, H.; Matsuzawa, S.; Hirotsu, K.; Higushi, T. *Organometallics* **1986,** *5*, 182.
(70) Ishikawa, M.; Matsuzawa, S.; Sugisawa, H.; Yano, F.; Kamitori, S.; Higuchi, T. *J. Am. Chem. Soc.* **1985,** *107*, 7706.
(71) Ishikawa, M.; Yuzuriha, Y.; Horio, T.; Kunai, A. *J. Organomet. Chem.* **1991,** *402*, C20.
(72) Oshita, J.; Isomura, Y.; Ishikawa, M. *Organometallics* **1989,** *8*, 2050.
(73) Ishikawa, M.; Sakamoto, H. *J. Organomet. Chem.* **1991,** *414*, 1.
(74) Sluggett, G. W.; Leigh, W. J. *Organometallics* **1992,** *11*, 3731.
(75) Sluggett, G. W.; Leigh, W. J. *J. Am. Chem. Soc.* **1992,** *114*, 1195.
(76) Ohshita, J.; Ohsaki, H.; Ishikawa, M. *Organometallics* **1991,** *10*, 2695.
(77) Ohshita, J.; Ohsaki, H.; Ishikawa, M.; Tochibana, A.; Kurosaki, Y.; Yamabe, T.; Minato, A. *Organometallics* **1991,** *10*, 880.
(78) Ishikawa, M.; Kikuchi, M.; Kunai, A.; Takeuchi, T.; Tsukihara, T.; Kido, M. *Organometallics* **1993,** *12*, 3474.
(79) Ishikawa, M.; Sakamoto, H.; Kanetani, F. *Organometallics* **1989,** *8*, 2767.
(80) Leigh, W. J.; Sluggett, G. W. *Organometallics* **1993,** *13*, 269.
(81) Leigh, W. J.; Sluggett, G. W. *Abstract P-34 of XXVII Organosilicon Conference,* Troy, NY, March 18, *1994*; also idem *Organometallics* **1994,** *13*, 1005.
(82) Sakurai, H. In *Silicon Chemistry*; Corey, E. R.; Corey, J. Y.; Gaspar, P. P., Eds.; Ellis Horwood: Chichester, UK, 1987; p. 163.
(83) Gaspar, P. P.; Holten, D.; Konieczny, S.; Corey, J. Y. *Acc. Chem. Res.* **1987,** *20*, 329.
(84) Ohshita, J.; Ohsaki, H.; Ishikawa, M.; Tachibani, A.; Kurosaki, Y.; Yamabe, T.; Tsukihara, T.; Takahashi, K.; Kiso, Y. *Organometallics* **1991,** *10*, 2685.
(85) Brook, A. G.; Wessely, H.-J. *Organometallics* **1985,** *4*, 1487.
(86) Baines, K. M.; Brook, A. G. *Organometallics* **1987,** *6*, 692.
(87) Baines, K. M.; Brook, A. G.; Ford, R. R. ; Lickiss, P. D.; Saxena, A. K.; Chatterton, W. J.; Sawyer, J. F.; Behnam, B. A. *Organometallics* **1989,** *8*, 693.
(88) Brook, A. G.; Baumegger, A.; Lough, A. J. *Organometallics* **1992,** 11, 3088.
(89) Brook, A. G.; Safa, K. D.; Lickiss, P. D.; Baines, K. M. *J. Am. Chem. Soc.* **1985,** *107*, 4338.
(90) Brook, A. G.; Baumegger, A.; Lough, A. J. *Organometallics* **1992,** *11*, 310.
(91) Brook, A. G.; Chiu, P.; McClenaghan, J.; Lough, A. J. *Organometallics* **1991,** *10*, 3292.
(92) Brook, A. G.; Harris, J. W.; Lennon, J.; El Sheikh, M. *J. Am. Chem. Soc.* **1979,** *101*, 83.
(93) Brook, A. G.; Nyburg, S. C.; Reynolds, W. F.; Poon, Y. C.; Chang, Y.-M.; Lee, J.-S.; Picard, J.-P. *J. Am. Chem. Soc.* **1979,** *101*, 6750.
(94) Volkova, V. V.; Volnina, E. A.; Buravtseva, E. N.; Gusel'nikov, L. E. *Abstracts Xth International Symposium on Organosilicon Chemistry,* Poznan, Poland, *1993*; p. 170.
(95) Ishikawa, M.; Sakamoto, H.; Tabuchi, T. *Organometallics* **1991,** *10*, 3173.
(95a) Naka, A.; Hayashi, M.; Okazaki, S.; Ishikawa, M. *Organometallics* **1994,** *13*, 4994.
(96) Ishikawa, M. *Abstracts Xth International Symposium on Organosilicon Chemistry,* Poznan, Poland, *1993*; p. 35.
(97) Sakamoto, H.; Ishikawa, M. *Organometallics* **1992,** *11*, 2580.
(98) Wiberg, N. *J. Organomet. Chem.* **1984,** *273*, 141.

(98a) Wiberg, N.; Wagner, G. *Angew. Chem., Int. Ed. Engl.* **1983,** *22*, 1005.
(99) Wiberg, N.; Preiner, G.; Schieda, O. *Chem. Ber.* **1981,** *114*, 3518.
(99a) Delpon-Lacraze, G.; Couret, C. *J. Organomet. Chem.* **1994,** *480*, C14.
(100) Bravo-Zhivotovskii, D.; Apeloig, Y.; Ovchinnikov, Y.; Igonin, V.; Struchkov, Y. T. *J. Organomet. Chem.* **1993,** *446*, 123.
(101) Sakurai, H.; Kamiyama, Y.; Nakadaira, Y. *J. Am. Chem. Soc.* **1976,** *98*, 1976.
(102) Ishikawa, M.; Kovar, D.; Fuchikami, T.; Nishimura, K.; Kumada, M. *J. Am. Chem. Soc.* **1981,** *103*, 2324.
(103) Ishikawa, M.; Fuchikami, T.; Kumada, M. *J. Organomet. Chem.* **1979,** *173*, 117.
(104) Nyburg, S. C.; Brook, A. G.; Abdesaken, F.; Wong-Ng, W. *Acta Crystallogr., Sect. C: Cryst. Struct. Commun.* **1985,** *C41*, 1632.
(105) Brook, A. G.; Abdesaken, F.; Gutekunst, G.; Plavac, N. *Organometallics* **1982,** *1*, 994.
(106) Brook, A. G.; Lassacher, P. *Organometallics* **1995,** in press.
(107) Sekiguchi, A.; Tanikawa, H.; Ando, W. *Organometallics* **1985,** *4*, 584.
(108) Märkl, G.; Schlosser, W. *Angew. Chem., Int. Ed. Engl.* **1988,** *27*, 963.
(109) Märkl, G.; Schlosser, W.; Sheldrick, W. S. *Tetrahedron Lett.* **1988,** *29*, 467.
(110) Braddock-Wilking, J.; Chiang, M. Y.; Gaspar, P. P. *Organometallics* **1993,** *12*, 197.
(111) Leigh, W. J.; Bradaric, C. J.; Sluggett, G. W. *J. Am. Chem. Soc.* **1993,** *115*, 5332.
(112) Fajgar, R.; Bastl, Z.; Pola, J. *Abstracts Xth International Symposium on Organosilicon Chemistry,* Poznan, Poland, *1993*; p. 160.
(113) Conlin, R. T.; Zhang, S.; Namavari, M.; Bobbitt, K. L. *Organometallics* **1989,** *8*, 571.
(114) Fink, M. J.; Puranik, D. B.; Pointer Johnson, M. *J. Am. Chem. Soc.* **1988,** *110*, 1315.
(115) Khabashesku, V. N.; Balaji, V.; Bogano, S. E.; Nefedov, O. M.; Michl, J. *J. Am. Chem. Soc.* **1994,** *116*, 320.
(116) Lei, D.; Chen, Y. S.; Boo, B. H.; Frueh, J.; Svoboda, D. L.; Gaspar, P. P. *Organometallics* **1992,** *11*, 559.
(117) Barton, T. J.; Groh, B. L. *J. Am. Chem. Soc.* **1985,** *107*, 7221.
(118) Barton, T. J.; Groh, B. L. *Organometallics* **1985,** *4*, 575.
(119) Hong, J.-H.; Han, J. S.; Lee, G.-H.; Jung, I. N. *J. Organomet. Chem.* **1992,** *437*, 265.
(120) Tillman, N.; Barton, T. J. *Main Group Met. Chem.* **1987,** *10*, 301.
(121) Pannell, K. H. *J. Organomet. Chem.* **1984,** *273*, 141.
(122) Lewis, C.; Wrighton, M. S. *J. Am. Chem. Soc.* **1983,** *105*, 7768.
(123) Randolph, C. L.; Wrighton, M. S. *Organometallics* **1987,** *6*, 365.
(124) Procopio, L. J.; Mayer, B.; Ploesl, K.; Berry, D. H. *Polym. Prepr., Am. Chem. Soc., Div. Polym. Chem.* **1992,** *33*, 1241.
(125) Djurovich, P. I.; Carroll, P. J.; Berry, D. H. *Organometallics* **1994,** *13*, 2551.
(126) Berry, D. H.; Procipio, L. J. *J. Am. Chem. Soc.* **1989,** *111*, 4099.
(127) Sharma, S.; Pannell, K. *Organometallics* **1993,** *12*, 3979.
(128) Campion, B. K.; Heyn, R. H.; Tilley, T. D. *J. Am. Chem. Soc.* **1988,** *110*, 7558.
(129) Campion, B. K.; Heyn, R. H.; Tilley, T. D. *J. Am. Chem. Soc.* **1990,** *112*, 4079.
(130) Koloski, T. S.; Carroll, P. J.; Berry, D. H. *J. Am. Chem. Soc.* **1990,** *112*, 6405.
(130a) Freeman, W. P.; Tilley, T. D.; Rheingold, A. L. *J. Am. Chem. Soc.* **1994,** *116*, 8428.
(131) Ishikawa, M.; Matsuzawa, S.; Higuchi, T.; Kamitori, S.; Hirotsu, K. *Organometallics* **1985,** *4*, 2040.
(132) Ishikawa, M.; Ohshita, J.; Ito, Y. *Organometallics* **1986,** *5*, 1518.
(133) Ishikawa, M.; Ohshita, J.; Ito, Y.; Iyoda, J. *J. Am. Chem. Soc.* **1986,** *108*, 7417.
(133a) Kunai, A.; Yuzuriha, Y.; Naka, A.; Ishikawa, M. *J. Organomet. Chem.* **1993,** *455*,77.
(133b) Ishikawa, M.; Nomura, Y.; Tozaki, E.; Kunai, A.; Ohshita, J. *J. Organomet. Chem.* **1990,** *399*, 205.

(134) Ishikawa, M.; Ono, T.; Sahaki, Y.; Minato, A.; Okinoshima, H. *J. Organomet. Chem.* **1989,** *363*, C1.
(135) Ishikawa, M.; Nishimura, Y.; Sakamoto, H.; Ono, T.; Ohshita, J. *Organometallics* **1992,** *11*, 483.
(136) Ishikawa, M.; Naka, A.; Ohshita, J. *Organometallics* **1992,** *11*, 3004.
(137) Ohshita, J.; Haseke, H.; Masaoka, Y.; Ishikawa, M. *Organometallics* **1994,** *13*, 1064.
(138) Trost, B. M. *Pure Appl. Chem.* **1988,** *60*, 1615.
(139) Trost, B. M. *Angew. Chem.* **1986,** *98*, 1.
(140) Ando, W.; Yamamoto, T.; Saso, H.; Kabe, Y. *J. Am. Chem. Soc.* **1991,** *113*, 2791.
(140a) Kabe, Y.; Yamamoto, T.; Ando, W. *Organometallics* **1994,** *13*, 4606.
(141) Wiberg, N.; Wagner, G.; Reber, G.; Riede, J.; Mueller, G. *Organometallics* **1987,** *6*, 35.
(142) Auner, N.; Davidson, I. M. T.; Ijadi-Maghsoodi, S. *Organomettalics* **1985,** *4*, 2210.
(143) Auner, N.; Davidson, I. M. T.; Ijadi-Maghsoodi, S.; Lawrence, F. T. *Organometallics* **1986,** *5*, 431.
(144) Conlin, R. T.; Namavari, M.; Chickos, J. S.; Walsh, R. *Organometallics* **1989,** *8*, 168.
(145) Basu, S.; Davidson, I. M. T.; Laupert, R.; Potzinger, P. *Ber. Bunsenqes. Phys. Chem.* **1979,** *83*, 1282.
(146) Gusel'nikov, L. E. *8th International Symposium on Organosilicon Chemistry,* St. Louis, MO, June, 1987.
(147) Gusel'nikov, L. E.; Nametkin, N. S.; Dogopolov, N. N. *J. Organomet. Chem.* **1979,** *79*, 59.
(148) Conlin, R. T.; Kwak, Y.-W. *J. Am. Chem. Soc.* **1986,** *108*, 834.
(149) Apeloig, Y.; Karni, M. *J. Am. Chem. Soc.* **1984,** *106*, 6676.
(149a) Wiberg, N.; Wagner, G.; Riede, J.; Müller, G. *Organometallics* **1987,** *6*, 32.
(149b) Wiberg, N.; Wagner, G. *Chem. Ber.* **1986,** *119*, 1467.
(150) Wiberg, N.; Köpf, H. *J. Organomet. Chem.* **1986,** *315*, 9.
(151) Jones, P. R.; Lee, M. E. *Silicon, Germanium, Tin Lead Compd.* **1986,** *9*, 11.
(151a) Jones, P. R.; Lee, M. E. *J. Organomet. Chem.* **1984,** *271*, 299.
(152) West, R.; Fink, M. J.; Michl, J. *Science* **1981,** *214*, 1343.
(153) Apeloig, Y. In *The Chemistry of Organosilicon Compounds;* Patai, S.; Rappoport, Z., Eds.; Wiley: New York, 1989; Chapter 2, p. 57.
(154) Raabe, G.; Vancik, H.; West, R.; Michl, J. *J. Am. Chem. Soc.* **1986,** *108*, 671.
(155) Gutowsky, H. S.; Chen, J.; Hajduk, P. J.; Keen, J. P.; Emilsson, T. *J. Am. Chem. Soc.* **1989,** *111*, 1901.
(156) Gutowsky, H. S.; Chen, J.; Hajduk, P. J.; Keen, J. D.; Chuang, C.; Emilsson, T. *J. Am. Chem. Soc.* **1991,** *113*, 4747.
(157) Mahaffy, P. G.; Gutowsky, R.; Montgomery, L. K. *J. Am. Chem. Soc.* **1980,** *102*, 2854.
(158) Schaefer, H. F., III. *Acc. Chem. Res.* **1982,** *15*, 283.
(159) Shin, S. K.; Inkura, K. K.; Beauchamp, T. L.; Goddard, W. A. *J. Am. Chem. Soc.* **1988,** *110*, 24.
(160) Grev, R. S.; Scuseria, G. E.; Scheiner, A. C.; Schaefer, H. F., III; Gordon, M. S. *J. Am. Chem. Soc.* **1988,** *110*, 7337.
(161) Schmidt, M. W.; Gordon, M. S.; Dupuis, M. *J. Am. Chem. Soc.* **1985,** *107*, 2585.
(162) Dobbs, K. D.; Hehre, W. J. *Organometallics* **1986,** *5*, 2057.
(163) Chandrasekhar, J.; Schleyer, P. von R. *J. Organomet. Chem.* **1985,** *289*, 51.
(164) Chandrasekhar, J.; Schleyer, P. von R.; Baumgärtner, R. O. W.; Reetz, M. T. *J. Org. Chem.* **1983,** *48*, 3453.
(165) Baldridge, K. K.; Gordon, M. S. *Organometallics* **1988,** *7*, 144.

(166) Wiberg, N.; Preiner, G.; Schieda, O. *Chem. Ber.* **1981,** *114*, 3518.
(167) Seidl, E. T.; Grev, R. S.; Schaefer, H. F., III. *J. Am. Chem. Soc.* **1992,** *114*, 3643.
(168) Bernardi, F.; Bottoni, A.; Olivucci, M.; Robb, M. A.; Venturini, A. *J. Am. Chem. Soc.* **1993,** *115*, 3322.
(169) Bernardi, F.; Bottoni, A.; Olivucci, M.; Venturini, A.; Robb, M. *J. Chem. Soc., Faraday Trans.* **1994,** *90*, 1617.
(169a) Grev, R. *Abstracts of the XXVIII Organosilicon Conference,* Gainesville, FL, March 31–April 1, *1995*; p. A29.
(170) Auner, N. *J. Organomet. Chem.* **1988,** *353*, 275.
(171) Auner, N.; Ziche, W.; Herdtweck, E. *J. Organomet. Chem.* **1992,** *426*, 1.
(172) Jones, P. R.; Cheng, A. H.-B.; Albanesi, T. E. *Organometallics* **1984,** *3*, 78.
(173) Zhang, S.; Conlin, R. T.; McGarry, P. F.; Sciano, J. C. *Organometallics* **1992,** *11*, 2317.
(174) Wiberg, N.; Preiner, G.; Wagner, G.; Köpf, H. *Z. Naturforsch, B: Chem. Sci.* **1987,** *42B*, 1062.
(175) Wiberg, N.; Wagner, S.; Fischer, G. *Chem. Ber.* **1991,** *124*, 1981.
(176) Wiberg, N.; Schurz, K.; Fischer, G. *Chem. Ber.* **1986,** *119*, 3498.
(177) Wiberg, N.; Fischer, G.; Wagner, S. *Chem. Ber.* **1991,** *124*, 769.
(178) Wiberg, N.; Fischer, G.; Schurz, K. *Chem. Ber.* **1987,** *120*, 1605.
(179) Jones, P. R.; Lin, T. F. O. *J. Am. Chem. Soc.* **1977,** *99*, 2013, 8447.
(180) Jones, P. R.; Lim, T. F. O.; Pierce, R. A. *J. Am. Chem. Soc.* **1980,** *102*, 4970.
(181) Auner, N.; Seidenschwarz, C.; Herdtweck, E.; Sewald, N. *Angew. Chem., Int. Ed. Engl.* **1991,** *30*, 444.
(182) Auner, N.; Seidenschwarz, C.; Sewald, N. *Organometallics* **1992,** *11*, 1137.
(183) Auner, N.; Wolff, A. *Chem. Ber.* **1993,** *126*, 575.
(184) Sewald, N.; Ziche, W.; Wolff, A.; Auner, N. *Organometallics* **1993,** *12*, 4123.
(185) Auner, N.; Herkenwälder, C.-R.; Ziche, W. *Chem. Ber.* **1993,** *126*, 2177.
(186) Auner, N.; Seidenschwarz, C.; Herdtweck, E. *Angew. Chem., Int. Ed. Engl.* **1991,** *30*, 1151.
(187) Brook, A. G.; Vorspohl, K.; Ford, R. R.; Hesse, M.; Chatterton, W. J. *Organometallics* **1987,** *6*, 2128.
(188) Ziche, W.; Auner, N.; Behm, J. *Organometallics,* **1992,** *11*, 2494.
(189) Brook, A. G.; Hu, S. S.; Saxena, A. K.; Lough, A. J. *Organometallics* **1991,** *10*, 2758.
(190) Breit, B.; Boese, R.; Regitz, M. *J. Organomet. Chem.* **1994,** *464*, 41.
(191) Ishikawa, M.; Horio, T.; Yuzuriha, Y.; Kunai, A.; Tsukihara, T.; Naitou, H. *Organometallics* **1992,** *11*, 597.
(192) Brook, A. G.; Baumegger, A. *J. Organomet. Chem.* **1993,** *446*, C9.
(193) Bachrach, S.; Streitwieser, A., Jr. *J. Am. Chem. Soc.* **1985,** *107*, 1186.
(194) Brook, A. G.; Chatterton, W. J.; Sawyer, J. F.; Hughes, D. W.; Vorspohl, K. *Organometallics* **1987,** *6*, 1246.
(195) Brook, A. G.; Kumarathasan, R.; Chatterton, W. J. *Organometallics* **1993,** *12*, 4085.
(196) Brook, A. G.; Hu, S. S.; Chatterton, W. J.; Lough, A. J. *Organometallics* **1991,** *10*, 2752.
(197) Auner, N.; Seidenschwarz, C. *Z. Naturforsch., B: Chem. Sci.* **1990,** *45B*, 909.
(198) Wiberg, N.; Preiner, G.; Wagner, G.; Köpf, H.; Fischer, G. *Z. Naturforsch., B: Chem. Sci.* **1987,** *42B*, 1055.
(199) Brook, A. G.; Chatterton, W. J.; Kumarathasan, R. *Organometallics* **1993,** *12*, 3666.
(200) Auner, N.; Weingartner, A. W.; Bertrand, G. *Chem. Ber.* **1993,** *126*, 581.
(201) Brook, A. G.; Chatterton, W. J. unpublished studies.
(202) Ishikawa, M.; Matsuzawa, S. *J. Chem. Soc., Chem. Commun.* **1985,** 588.

(203) Pola, J.; Cukanova, D.; Ponec, R.; Bürger, H.; Beckers, H.; Stanzyk, W.; Diaz, L.; Siguenza, C.; Gonzales-Diaz, P. F. *Abstracts Xth International Symposium on Organosilicon Chemistry,* Poznan, Poland, *1993*; p. 148.

(204) Brook, A. G.; Kong, Y. K.; Saxena, A. K.; Sawyer, J. F. *Organometallics* **1988,** *7*, 2245.

(205) Brook, A. G.; Saxena, A. K.; Sawyer, J. F. *Organometallics* **1989,** *8*, 850.

(206) Nguyen, M. T.; Vansweevelt, H.; De Neef, A.; Vanquickenborne, L. G. *J. Org. Chem.* **1994,** *59*, 8015.

(207) Brook, A. G.; Kumarathasan, R.; Lough, A. J. *Organometallics* **1994,** *13*, 424.

(208) Wiberg, N.; Karampatses, P.; Kim, C.-K. *Chem. Ber.* **1987,** *120*, 1203.

(209) Wiberg, N.; Preiner, G.; Schurz, K. *Chem. Ber.* **1988,** *121*, 1407.

(210) Jones, P. R.; Bates, T. F. *J. Am. Chem. Soc.* **1987,** *109*, 913.

(211) Jones, P. R.; Lee, M. E. *J. Am. Chem. Soc.* **1983,** *105*, 6725.

(212) Jones, P. R.; Bates, T. F.; Cowley, A. T.; Arif, A. M. *J. Am. Chem. Soc.* **1986,** *108*, 3122.

(212a) Trommer, M.; Sander, W.; Patyk, A. *J. Am. Chem. Soc.* **1993,** *115*, 11775.

(213) Miracle, G.; West, R.; Ball, J. L.; Powell, D. R. *Abstracts Xth International Symposium on Organosilicon Chemistry,* Poznan, Poland, *1993*; p. 33.

(214) Miracle, G. E.; Ball, J. L.; Powell, D. R.; West, R. *J. Am. Chem. Soc.* **1993,** *115*, 11598.

(215) Eaborn, C.; Happer, D. A. R.; Hitchcock, P. B.; Hopper, S. P.; Safa, K. D.; Washburne, S. S.; Walton, D. A. R. *J. Organomet. Chem.* **1980,** *186*, 309.

(216) Wiberg, N.; Köpf, H. *Chem. Ber.* **1987,** *120*, 653.

(217) Sekiguchi, A.; Ando, W. *J. Am. Chem. Soc.* **1984,** *106*, 1486.

(218) Jung, I. N.; Yoo, B. R.; Lee, M. E.; Jones, P. R. *Organometallics* **1991,** *10*, 2529.

(219) Conlin, R. T.; Namavari, M. *J. Am. Chem. Soc.* **1988,** *110*, 3689.

(220) Wiberg, N.; Preiner, G.; Schurz, K.; Fischer, G. *Z. Naturforsch, B: Chem. Sci.* **1988,** *43B*, 1468.

(221) Maier, G. *Pure Appl. Chem.* **1986,** *58*, 95.

(222) Welsh, K. M.; Rich, J. D.; West, R.; Michl, J. *J. Organomet. Chem.* **1989,** *325*, 105.

(223) Interrente, L. V.; Wu, H.-J.; Apple, T.; Shen, Q.; Ziemann, B.; Narsavage, D. M.; Smith, K. *J. Am. Chem. Soc.* **1994,** *116*, 12085.

Iminosilanes and Related Compounds—Synthesis and Reactions

INA HEMME and UWE KLINGEBIEL

Institute of Inorganic Chemistry, University of Goettingen
D-37077 Goettingen, Germany

I
INTRODUCTION

Compounds of multiple bond systems involving heavier main group elements were long considered to be unstable and synthetically inaccessible. In particular, the so-called double bond rule, which forbade the formation of ($p\pi$–$p\pi$) multiple bonds between silicon and other elements, hindered the development of the chemistry of low-coordinate silicon compounds containing Si═X (X = C, N, Si, P) double bonds for some years.

After decades of unsuccessful attempted syntheses, Gusel'nikov and Flowers in 1967 reported the first compelling evidence for the existence of silenes, compounds containing a double bond between silicon and carbon. This initiated a renewed interest in the synthesis and behavior of stable silenes,[a] disilenes,[b] iminosilanes,[c] phosphasilenes[d] and their heavier homologs.

The first silaethenes and disilenes were synthesized at the beginning of the 1980s, and the first iminosilanes were reported in the mid-1980s. The difference in the electronegativity between Si (1.8) and N (3.0) gives the iminosilanes an ylidic nature, which makes them susceptible to oligomerization and dimerization.

$$>Si{=}N{-} \longleftrightarrow >\overset{\oplus}{Si}{-}\overset{\ominus}{\underline{\overline{N}}}{-} \longrightarrow \frac{1}{n}\left(>Si{-}N{-}\right)_n$$

However, monomers can be stabilized by the presence of bulky substituents. In this review (see also reviews[c] by Walter and Klingebiel and by Raabe and Michl), we examine the methods of preparing iminosilanes, together with their structural features and some of their chemistry, that have been reported since 1985.

II
PREPARATION

A. *Detection of Transient Iminosilanes*

The photolysis of azidosilanes often leads to products that can be considered iminosilane intermediates. These intermediates can be observed in solution at low temperature, isolated in an inert matrix, or detected as chemical trapping products. Irradiation of trimesitylazidosilane (Mes_3SiN_3) at 77 K in a 3-methylpentane (3-MP) glass yields the yellow $Mes_2Si{=}NMes$ and a CH-insertion product (Eq. 1). In the UV/visible-spectrum three new bands at 296 nm, 444 nm, and 257 nm arise. When EtOH or BuOH is used as the trapping agent, the bands at 296 nm and 444 nm disappear to give the expected trapping products, but the band at 257 nm from the CH-insertion product remains.[1,2]

$$Mes_3SiN_3 \xrightarrow[3\text{-MP},\ 77\ K]{h\nu,\ 254\ nm} Mes_2Si{=}NMes + [\text{CH-insertion product: } Si(Mes_2){-}N(H)\text{ ring}] \xrightarrow{+ROH} Mes_2Si(OR){-}N(H)Mes$$

R = Et, *t*-Bu

Mes = 2,4,6-trimethylphenyl (1)

Under the same reaction conditions, other hindered azidosilanes such as $Mes_2Si(N_3)SiPh_2$-*t*-Bu or $R_2Si(N_3)SiMes_2$-*t*-Bu (R = Me, *i*-Pr) form only the iminosilanes and/or the alcohol addition products.[1]

Upon irradiation of (dimesityl)(trimethylsilyl)silylazide at 254 nm in a cyclohexane/*t*-butanol or a cyclohexane/ethanol solution, migration of the trimethylsilyl group to the nitrene center is observed (Eq. 2). The products are equivalent to the addition products mentioned above.

$$\mathrm{Me_3Si{-}Si(Mes)_2N_3} \xrightarrow[-N_2]{h\nu,\ 254\ nm} \mathrm{Me_3Si{-}Si(Mes)_2{-}\overline{N}|} \xrightarrow{\sim SiMe_3}$$

$$\mathrm{Mes_2Si{=}NSiMe_3} \xrightarrow{+ROH} \mathrm{Mes_2Si(OR){-}N(H)SiMe_3} \quad (2)$$

R = Et, *t*-Bu

Photolysis (254 nm) of this azide in an isopentane/methylcyclohexane matrix at 77 K produced an orange color, with the appearance of two bands at 272 nm and 474 nm in the UV/visible-spectrum. Irradiation in an argon matrix at 20 K gave the same results.[3]

Photolysis of silyl azide in an argon matrix at 12 K with a low-pressure mercury lamp (254 nm) forms the aminosilylene, which can be converted into the silaisonitrile and molecular hydrogen by irradiation with wavelengths >300 nm. Using 254-nm light again, this reaction is reversible. (This silaisonitrile was isolated in 1966 by Ogilvie and Cradock.[4,5]

Phenylisosilacyanide is produced either by irradiation of triazidophenylsilane in matrix isolation or by pyrolysis followed by trapping in noble gas matrix (Eq. 3). The reaction with *t*-butanol leads to the expected product[6]:

$$\mathrm{PhSi(N_3)_3} \xrightarrow[-4N_2]{h\nu \text{ or } \Delta} \mathrm{PhN=Si|} \xrightarrow{t\text{-BuOH}} \mathrm{PhNHSiH(O\text{-}\mathit{t}\text{-}Bu)_2} \quad (3)$$

The intermediacy of benzenesilanitrile, the isomeric species with a —Si≡N| triple bond, is strongly expected from the starting material, but has not been proved by trapping. *Ab initio* calculations at the fully optimized MP2/6-31G* level including zero-point corrections show that the SiN double bond product is 55 kcal mol^{-1} more stable than the triple bond product.[6]

The first two unhindered iminosilanes were identified in two different ways[7]:

1. Retro-ene reaction of an *N*-allylsilanamine under flash vacuum thermolysis (FVT) conditions. The iminosilane has been identified not only by reaction with the trapping agent *t*-butyl alcohol, but also directly by coupling the oven with a high-resolution mass spectrometer.[7]

$$\mathrm{CH_2{=}CHCH_2N(Ph)(HSiMe_2)} \xrightarrow[(-C_3H_6)]{\text{FVT}(900°C/10^{-4}\ hPa)\ \text{retro-ene}} \mathrm{Me_2Si{=}NPh} \xrightarrow{+t\text{-BuOH}} \mathrm{Me_2Si(OCMe_3){-}N(H)Ph} \quad (4)$$

2. Dehydrochlorination of a 1-chlorosilanamine by vacuum gas solid reaction (VGSR). The presence of the iminosilane was confirmed by direct analysis of the gaseous flow.[7]

$$\underset{\displaystyle \text{Cl}\qquad\quad \text{H}}{\text{Me}_2\text{Si}\text{—}\text{NCHMe}_2} \xrightarrow[(\text{Ph}_2\text{MeSi})_2\text{NK}]{\text{VGSR},\ 10^{-4}\ \text{hPa}} \text{Me}_2\text{Si}{=}\text{NCHMe}_2 \qquad (5)$$

Using potassium bis(trimethylsilyl)amide as a solid base, *N*′-isopropyl-1,1-dimethyl-*N*,*N*′-bis(trimethylsilyl)silanediimine and 2,4-diisopropyl-1,1,3,3-tetramethylcyclodisilazane were formed, accompanied by hexamethyldisilazane.

The formation of the cyclodisilazane ring system can be explained by dimerization of the unstable iminosilane in the cold trap.[7]

B. *Calculations*

Ab initio calculations on the $H_2Si{=}N{-}SiH_3$ molecule (6-31G*, with *d*-functions on both N and Si) give a SiNSi angle of 175.6°. The SiN single bond length is 168.8 pm and the double bond length is 154.9 pm. The nearly linear arrangement of SiNSi is more strongly dependent on electronic effects than on steric effects. The total strength of the SiN double bond is 143.3 kcal mol^{-1}, only 44.6 kcal mol^{-1} higher than that of a SiN single bond. The molecule is calculated to have a C_s symmetry.[8]

$$H_2Si \overset{154.9\ \text{pm}}{=} N \overset{168.8\ \text{pm}}{—} SiH_3 \qquad \angle HSiH = 110.1^\circ,\ \angle SiNSi = 175.6^\circ$$

Ab initio calculations on $H_2Si{=}NH$ (HF/6-31G*) show a planar molecular geometry with a SiN double bond length of 157.6 pm and an Si=N—H angle of 125.2°.

$$H_2Si \underset{157.6\ \text{pm}}{=} N{-}H \qquad \angle SiNH = 125.2^\circ$$

The inversion barrier for syn/anti isomerization of $H_2Si{=}NH$ is only 5.6 kcal mol^{-1}, whereas the internal rotation energy is 37.9 kcal mol^{-1} (SOCI level of calculation). The rotation barrier can be equated to the π-bond strength. The inversion transition state has an even shorter SiN bond length of 153.2 pm. The symmetry is C_{2V}.[9,10]

$$\begin{array}{c} H \\ \quad \diagdown \; 153.2\ pm \\ \quad Si{=}N{-}H \\ \quad \diagup \; 125.1^\circ \\ H \end{array}$$

Schleyer *et al.* calculated zero-point energies of 17.8 kcal mol^{-1} (data at 6-31G* basis, scaled by 0.89) for the $H_2Si{=}NH$ molecule. The point group is C_s, the double bond length is 157.3 pm, and the bond order (NLMO/NPA) is 1.006, as defined by Reed and Schleyer. In terms of theoretical π-bond energies (E_π), the SiN double bond is weaker than two of the corresponding single linkages.[11] Calculations also show that iminosilanes with a carbon substituent at the nitrogen atom have a smaller valence angle at the nitrogen atom and an even longer silicon-nitrogen double bond.[12]

C. *Stable Iminosilanes*

In 1985/86 the first iminosilanes that are kinetically stable at room temperature were synthesized. Two groups demonstrated two different processes of preparation.

Wiberg *et al.* prepared an *N*-silyl substituted iminosilane by means of an N_2/NaCl and -LiCl elimination. None of the intermediate products was isolated. The iminosilane was formed in the reaction of azido-di-*t*-butylchlorosilane and tri-*t*-butylsilylsodium in dibutyl ether at −78°C.[13–16]

$$(t\text{-Bu})_2SiClN_3 + NaSi(t\text{-Bu})_3 \xrightarrow[-N_2]{-NaCl} (t\text{-Bu})_2Si{=}N{-}Si(t\text{-Bu})_3 \tag{6}$$

Preparation routes:

$$(t\text{-Bu})_2SiHCl + NaN_3\,(THF,\ reflux) \rightarrow (t\text{-Bu})_2SiHN_3 + NaCl$$
$$(t\text{-Bu})_2SiHN_3 + Cl_2\,(CH_2Cl_2,\ -20^\circ C) \rightarrow (t\text{-Bu})_2SiClN_3 + HCl$$
$$(t\text{-Bu})_3SiBr + 2\,Na\,(Bu_2O,\ 80^\circ C) \rightarrow (t\text{-Bu})_3SiNa + NaBr$$

The iminosilane product has a slightly yellow color and undergoes a melting decomposition at 85°C. The silicon atom bonded to three *t*-butyl groups shows an NMR signal at $\delta\ ^{29}Si = -7.72$ ppm, whereas the unsaturated silicon atom has a large downfield shift with a signal at $\delta\ ^{29}Si = 78.29$ ppm.[15] The structure of this iminosilane has been determined by X-ray crystallography.[15,16] In the solid state the iminosilane is monomeric and the Si=N—Si skeleton is nearly linear (Si—N—Si 177.8°). The SiN bond lengths are drastically different. The SiN single bond is about 169.5 pm, and the double bond is 156.8 pm.[15,16]

$C_3(C2)$, $C_3(C1)$ > Si(1)=N—Si(2)—C(5)C_3 (with C(4)C_3, C(3)C_3 on Si(2))

Bond lengths (pm)	
Si(1)—N	156.8
Si(1)—C(1)	189.9
Si(1)—C(2)	189.4
Si(2)—N	169.5
Si(2)—C(3)	194.9
Si(2)—C(4)	194.1
Si(2)—C(5)	194.3

Bond angles (deg.)	
Si(1)—N—Si(2)	177.8
N—Si(1)—C(1)	119.9
N—Si(1)—C(2)	122.8

The second preparation for a free iminosilane was found by the Klingebiel group. It is based on a fluorine/chlorine exchange of lithiated fluorosilylamines with subsequent thermal LiCl elimination.[17]

$(Me_2HC)_2SiF_2$ + HLiN–$C_6H_2(CMe_3)_3$ $\xrightarrow{-LiF}$ $(Me_2HC)_2Si(F)$—N(H)–$C_6H_2(CMe_3)_3$

$\xrightarrow[-MeH]{+MeLi}$ $(Me_2HC)_2Si$=N–$C_6H_2(CMe_3)_3$ (Si–F–$Li(THF)_3$) $\xrightarrow[-Me_3SiF,\ -THF]{+Me_3SiCl}$ $(Me_2HC)_2Si(Cl)$—N($Li(THF)_2$)–$C_6H_2(CMe_3)_3$

$\xrightarrow[-LiCl]{>80°C/0.01\ mbar}$ $(Me_2HC)_2Si$=N–$C_6H_2(CMe_3)_3$ (7)

The lithium intermediates in this synthesis were identifed by their X-ray structures.[17,18] The crystal structure of $(Me_2HC)_2SiF(THF)_3NC_6H_2(CMe_3)_3$ shows that the lithium has completely migrated to the stronger Lewis base fluorine, resulting in a short LiF bond (182.2 pm) and a long SiF bond (168.6 pm). The SiN double bond legnth is 161.9 pm, and the Si—N—C angle of 172° are typical for an imine.[14–16] Because of the existing iminosilane structure in this lithium derivative, the compound reacts like an unsaturated molecule. Therefore, it must be considered a LiF adduct of an iminosilane. Further adducts will be mentioned in the following section.

After a fluorine/chlorine exchange, a lithiated aminochlorosilane is obtained. Lithium is bound to nitrogen, the strongest Lewis base in this compound. The SiN bond is enlarged (164.2 pm) and the SiNC angle of 138.7° shows the character of the amine.[17,18]

```
          Li(THF)3        CC3
             \              \
              F              C — C
               \            /      \
   C2C(3) — Si == N — C(1)          C — CC3
    ↑          |           \       /
    └────────→ C2C(2)        C — C
                            /
                          CC3
```

Bond lengths (pm)	
Si=N	161.9
Si—F	168.6
Si—C(3)	188.9
N—C(1)	141.1
F—Li	182.2

Bond angles (deg.)	
N—Si—F	113.8
N—Si—C(3)	115.4
N—Si—C(2)	117.9
Si—F—Li	162.0
C(3)—Si—C(2)	109.3

```
   C2C(2)  C2C(3)   CC3
        \  /          \
         Si            C — C
        /  \          /      \
      Cl    N — C(1)          C — CC3
            |  ,'  \         /
       (THF)2Li      C — C
                    /
                  CC3
```

Bond lengths (pm)	
Si—N	164.2
Si—Cl	206.5
Si—C(3)	190.0
Si—C(2)	192.8
N—Li	199.3
N—C(1)	139.1
Li—C(1)	246.6

Bond angles (deg.)	
N—Si—Cl	112.2
N—Si—C(3)	118.7
N—Si—C(2)	113.0
Si—N—C(1)	138.7
Li—N—C(1)	91.8
Si—N—Li	129.4

When the $(Me_2HC)_2SiClNLi(THF)_2C_6H_2(CMe_3)_3$ is warmed to >80°C (vacuum, 0.01 mbar), LiCl is eliminated and the iminosilane sublimes, forming orange crystals, that melt into a red liquid. The downfield shift of $\delta\ ^{29}Si = 60.3$ ppm is characteristic for unsaturated silicon–nitrogen compounds.[19]

Two additional kinetically stable iminosilanes, $(Me_3C)_2Si{=}N\text{-}2,4,6\text{-}(CMe_3)_3C_6H_2$ and $(Me_3C)_2Si{=}N{-}Si(CMe_3)_2Ph$, have been synthesized by a fluorine/chlorine exchange.[20,21] Comparing the ^{29}Si NMR signals of the known free iminosilanes, it seems probable that the unsaturated silicon atom has greater Lewis acid character in the *N*-silyl-substituted compounds than in the *N*-aryl-substituted compounds.

$R_2Si{=}N{-}2,4,6\text{-}(CMe_3)_3C_6H_2$

$\delta^{29}Si = 60.3$ ppm ($R = CHMe_2$)[19]
$= 63.1$ ppm ($R = CMe_3$)[20]

$(Me_3C)_2Si{=}N{-}Si(CMe_3)_2{-}R$

$\delta^{29}Si = 78.3$ ppm ($R = CMe_3$)[14–16]
$= 80.4$ ppm ($R = Ph$)[21]

The third preparation of a stable iminosilane is the photochemical reaction of di-*t*-butylsilylene with tri-*t*-butylsilylazide.[22]

$$(Me_3C)_2Si: + N_3Si(CMe_3)_3 \xrightarrow[-N_2]{h\nu} (Me_3C)_2Si{=}N{-}Si(CMe_3)_3 \tag{8}$$

The product is the iminosilane, first isolated by Wiberg *et al.* in 1985/1986.[13–16]

A fourth preparation for a stable iminosilane as the THF adduct is the reaction of a silylene with trityl azide.[23]

$$(HC{=}CH)(N\text{-}CMe_3)_2SiCl_2 \xrightarrow[65^\circ C]{2.2\ K,\ THF} (HC{=}CH)(N\text{-}CMe_3)_2Si| \xrightarrow[-N_2]{+Ph_3CN_3} (HC{=}CH)(N\text{-}CMe_3)_2Si(THF){=}N{-}CPh_3 \tag{9}$$

D. *Iminosilane Adducts*

Many iminosilane adducts with very different donors have been isolated. What are the characteristics of these adducts?

The unsaturated silicon atoms form five bonds because of the donor. They have a tetrahedral environment and are four-coordinate. The adducts can be hydrolyzed very easily and are kinetically stable under normal conditions. One of the adducts, the THF adduct of $Me_2Si{=}N{-}Si(CMe_3)_3$, has been isolated by Wiberg *et al.*, as shown in Eq. (10).[15]

$$THF \cdot Me_2Si{=}C(SiMe_3)(SiMe(CMe_3)_2) \xrightarrow{+t\text{-}Bu_3SiN_3} THF \cdot Me_2Si{=}N{-}Si(CMe_3)_3 + ((CMe_3)_2MeSi)(Me_3Si)CN_2 \quad (10)$$

There are two crystallographically independent molecules of $Me_2Si(THF)NSi(CMe_3)_3$.[15]

```
             (CH2)4
               O
                \
     C(1)—Si(1)—N    CC3
           /       \ /
       C(2)        Si(2)—CC3
                    |
                   CC3
```

Bond lengths (pm)		Bond angles (deg.)	
Si(1)—N	158.8/157.4	Si(1)—N—Si(2)	161.5/161
Si(2)—N	165.9/166.7	N(1)—Si(1)—O	106.4/106.3
Si(1)—C(1)	188/182	N(1)—Si(1)—C(1)	119.9/120.4
Si(1)—C(2)	184/183	N(1)—Si(1)—C(2)	121.5/122.5
Si(1)—O	188.8/186.6		

Beyond these characteristics the SiN double bond is slightly elongated and the SiN single bond shortened. This is an accord with a more zwitterionic description (positive charge at O, negative charge at N), equivalent to a reduction in bond order for the SiN double bond, while the buildup of the negative charge at N would allow for better delocalization of the lone pair. The Si—N—Si angle is about 161°, showing that the environment of the unsaturated silicon atom has become more pyramidal.

The THF adduct of the fully *t*-butyl-coordinated iminosilane shows a remarkable upfield shift of the unsaturated silicon atom ($\delta\ ^{29}Si = 1.08$ ppm) compared to the free iminosilane described in the previous section.[14–16,24]

$$(Me_3C)_2Si{=}N{-}Si(CMe_3)_2 \xrightarrow{+THF} \underset{\substack{\uparrow \\ THF}}{(Me_3C)_2Si}{=}N{-}Si(CMe_3)_3 \quad (11)$$

The crystal structure of another THF adduct, $(Me_3C)_2Si(THF){=}N{-}SiMe(CMe_3)_2$, has been determined in the Klingebiel group.[18,25]

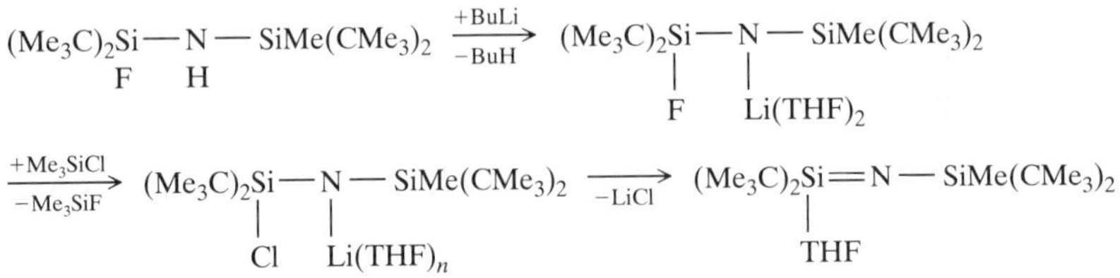

$$(Me_3C)_2Si(F)—N(H)—SiMe(CMe_3)_2 \xrightarrow[-BuH]{+BuLi} (Me_3C)_2Si(F)—N(Li(THF)_2)—SiMe(CMe_3)_2$$

$$\xrightarrow[-Me_3SiF]{+Me_3SiCl} (Me_3C)_2Si(Cl)—N(Li(THF)_n)—SiMe(CMe_3)_2 \xrightarrow{-LiCl} (Me_3C)_2Si(THF)=N—SiMe(CMe_3)_2 \quad (12)$$

C—C
C C (THF ring, O)
$C_3(C2)$ — Si(1)—N—Si(2) — $C_3C(3)$, C(5), $C_3C(4)$
$C_3C(1)$

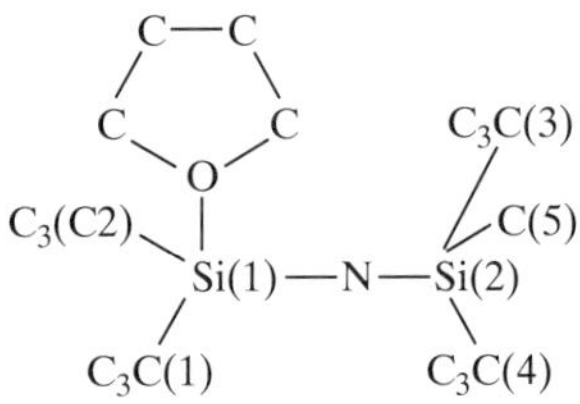

Bond lengths (pm)	
Si(1)—N	159.6
Si(1)—C(1)	190.7
Si(1)—C(2)	191.1
Si(1)—O	190.2
Si(2)—N	166.1
Si(2)—C(3)	192.7
Si(2)—C(4)	192.1
Si(2)—C(5)	189.7

Bond angles (deg.)	
Si(1)—N—Si(2)	174.3
N—Si(1)—O	104.2
N—Si(1)—C(1)	117.3
N—Si(1)—C(2)	115.1
C(1)—Si(1)—C(2)	115.3

This compound can be distilled in vacuum and recrystallized from benzene. This THF adduct of an iminosilane has, contrary to the previously described $Me_2Si(THF)=N—Si(CMe_3)_2$, a nearly linearly coordinated nitrogen atom (174.3°), with two significantly different SiN bond lengths. Although the donor coordination results in a slight elongation of the double bond and a shortening of the single bond, the SiN bond lengths of 166.1 pm and 159.6 pm reflect the difference between a SiN single and double bond. The unsaturated Si(1) atom has a distorted tetrahedral geometry; the two *t*-butyl groups and the nitrogen atom tend to form a trigonal-planar environment around Si(1) (bond angle sum 347.7°).

In the ^{13}C-NMR spectrum of the iminosilane, the C_2O signal of the THF molecule appears at 73.78 ppm—a downfield shift of nearly 6 ppm compared with that for free THF. This reflects the strong Lewis acid character of three-coordinated silicon in the iminosilane.

Several other THF adducts have been isolated by Wiberg *et al.* using the following method of synthesis[14,24]:

$$\underset{\displaystyle \text{Cl}}{R_2Si}\!-\!\underset{\displaystyle \text{LiD}_n}{NSiR'_3} \;\underset{+(n-1)\text{D}}{\overset{\text{Solvent}}{\rightleftharpoons}}\; \underset{\displaystyle \uparrow \atop \displaystyle \text{D}}{R_2Si}{=}NSiR'_3 + LiCl \tag{13}$$

D = Et_2O, THF, NEt_3, NMe_2Et
R′ = R = Me, *t*-Bu

The iminosilane $Me_2Si{=}NSi(CMe_3)_3$, which is unstable under normal conditions with regard to dimerization, forms metastable adducts D · $Me_2Si{=}NSi(CMe_3)_3$ that decompose thermally to give $Me_2Si{=}NSi(CMe_3)_3$ and D. The adducts can thus serve as sources of $Me_2Si{=}NSi(CMe_3)_3$. Lewis basicity of D, relative to $Me_2Si{=}NSi(CMe_3)_3$, increases in the order Et_2O < THF < NEt_3, < Cl^- < NMe_2Et < F^-. Resistance of D · $Me_2Si{=}NSi(CMe_3)_3$ to decomposition into the dimer of $Me_2Si{=}NSi(CMe_3)_3$ and D increases in the same order. Adducts D · $Me_2Si{=}NSi(CMe_3)_3$ also decompose by the action of excess donor. Et_2O · $Me_2Si{=}NSi(CMe_3)_3$ decomposes in Et_2O into ethylene and Me_2SiOEt—NH-$Si(CMe_3)_3$, $EtMe_2N$ · $Me_2Si{=}NSi(CMe_3)_3$ decomposes in $EtMe_2N$ under Stevens migration into $EtMeNCH_2SiMe_2$—$NHSi(CMe_3)_3$. Reaction of adducts D · $Me_2Si{=}NSi(CMe_3)_3$ with water, alcohols, and amines; with organic enes (propene, isobutene, dimethylbutadiene, cyclopentadiene); with silylazides; or with benzophenone gives, respectively, the OH and NH bond insertion products, ene reaction products, [2 + 2] cycloadducts, or a dimer of the iminosilane.[24]

When the steric size of the substituents increases, no adducts with bulky (Et_3N) or less basic (Et_2O) donors are formed with $(Me_3C)_2Si{=}NSi(CMe_3)_3$.[14] The structure of the benzophenone adduct is quite remarkable.[16] Instead of the expected cycloaddition product, the carbonyl oxygen atom is coordinated to the undersaturated silicon atom.

The benzophenone adduct Ph_2CO · $(CMe_3)_2Si{=}NSi(CMe_3)_3$ and the tetrahydrofuran adducts[14,25] contain donor molecules coordinated exclusively to the unsaturated silicon atoms; these adopt a distorted tetrahedral coordination geometry upon coordination, with characteristically elongated SiN double bonds. The SiN single bonds are shortened. The solid-state structure of the benzophenone adduct strongly suggests that [2 + 2] cycloadditions of Si=N double bonds with aldehydes or ketones proceed in a nonconcerted manner via prior formation of donor Si bonds.[16] It seems as if the formation of the [2 + 2] cycloadduct of a ketone with an iminosilane stops at the adduct stage for steric reasons.

```
                 C(Ph)2
                /
               O                 C(3)
                \               /
      C(1) — Si(1) — N — Si(2)
                /          | \ C(4)
             C(2)         C(5)
```

Bond lengths (pm)	
Si(1)—N	160.1
Si(1)—O	192.7
Si(1)—C(1)	193.0
Si(1)—C(2)	192.3
Si(2)—N	167.8
Si(2)—C(3)	195.0

Bond angles (deg.)	
Si(1)—N—Si(2)	169.3
N—Si(1)—O	106.9
N—Si(1)—C(1)	120.4
N—Si(1)—C(2)	117.5
O—Si(1)—C(1)	91.3

One LiF adduct of an iminosilane, —$(Me_2HC)_2(THF)_3LiFSi{=}N{-}C_6H_2(CMe_3)_3$—, has already been described as a product of the iminosilane synthesis.[17–19] Other LiF adducts have been prepared by the Klingebiel group. They are not only substituted by aryl groups,[17–19,26] but also by silyl groups.[21] For example, the LiF adduct of the silyl-substituted iminosilane has been prepared as shown in Eq. (14).

$$(Me_3C)_2SiF_2 + LiNHSiR(CMe_3)_2 \rightarrow (Me_3C)_2SiF{-}NHSiR(CMe_3)_2 \xrightarrow[-BuH]{+BuLi} (CMe_3)_2SiFLiNSiR(CMe_3)_2 \quad (14)$$

R = Me, Ph

The lithium derivative of the Ph-substituted aminofluorosilane crystallizes from THF as a monomeric LiF adduct of an iminosilane, $(THF)_3LiF \cdot (CMe_3)_2Si{=}NSiPh(CMe_3)_3$.[21]

```
                     C(1)C3          Ph
                       |              |
(THF)3Li — F — Si(1) = N — Si(2) — C(3)C3
                       |              |
                     C(2)C3         C(4)C3
```

Bond lengths (pm)	
Si(1)—N	160.8
Si(2)—N	165.2
Si(1)—F	169.2
F—Li	184.5
O—Li	192.9

Bond angles (deg.)	
Si(1)—N—Si(2)	176.3
N—Si(1)—F	111.8
Si(1)—F—Li	167.5

The SiNSi skeleton of this molecule is nearly linear (176.1°), and the SiN bonds are of two different lengths. This is the most linear LiF adduct to date. In the presence of TMEDA a salt-like amide is formed, $F(CMe_3)_2Si{=}N{-}SiPh(CMe_3)_2^{\ominus}\ Li^{\oplus}(TMEDA)_2$.[21]

The lithium derivative of the aminosilane $(Me_3C)_2SiFNHSiMe(CMe_3)_2$ forms a dimeric LiF adduct of the iminosilane $[(Me_3C)_2SiFLiNSiMe(CMe_3)_2]_2$ associated via Li—F contacts.

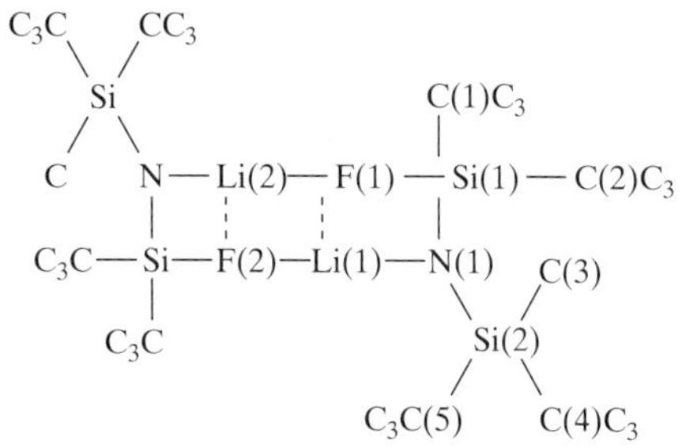

```
C3C    CC3
   \  /
    Si                        C(1)C3
   /  \                         |
  C    N — Li(2)— F(1) — Si(1) — C(2)C3
       |     :       :      |
C3C — Si — F(2) — Li(1) — N(1)    C(3)
       |                      \   /
      C3C                      Si(2)
                              /   \
                          C3C(5)   C(4)C3
```

Bond lengths (pm)	
Si(1)—N(1)	164.7
Si(1)—F(1)	169.9
N(1)—Si(2)	169.4
N(1)—Li(1)	190.2
Li(1)—F(2)	177.1

Bond angles (deg.)	
Si(1)—N(1)—Si(2)	148.9
F(1)—Si(1)—N(1)	103.6
Si(1)—F(1)—Li(2)	170.9
N(1)—Li(1)—F(2)	154.3
$\Sigma < C_2Si(1)N$	349.5

E. *Reactions of Iminosilanes*

1. *Kinetically Stable Iminosilanes*

In 1988 Wiberg *et al.* presented reactions of a free, stable iminosilane.[14,27] Since that time the number of products has rapidly increased.[28] Although the double bond is sterically shielded, the reactivity of iminosilanes is immense.

a. In reactions with active proton compounds R′—H, insertion into the SiN bond is observed [see (aa)–(ad) and (ba)]. These compounds include not only water or alcohols, but also hydrogen halides and even benzene (Eq. 15).

(aa) The reaction with water leads to the silanol in the first step. This product reacts again with the iminosilane to give the sterically overloaded disiloxane.[21] The Si—O—Si unit in these sterically overloaded siloxanes is linear (∢ Si—O—Si = 180°).

(ab)–(ad) The iminosilane reacts with other active proton compounds at the R′—H bond to give insertion products.

$$\begin{array}{c}
(\mathrm{ab})\ +\mathrm{R'OH}: \ >\mathrm{Si}(\mathrm{OR'})-\mathrm{N(H)}-\mathrm{Si(CMe_3)_2R} \quad \mathrm{R'} = \mathrm{Me, Et} \\
(\mathrm{ac})\ +\mathrm{HCl}: \ >\mathrm{Si(Cl)}-\mathrm{N(H)}-\mathrm{Si(CMe_3)_2R} \quad \mathrm{R'} = \mathrm{CMe_3, Ph} \\
>\mathrm{Si{=}N}-\mathrm{Si(CMe_3)_2R} \\
(\mathrm{aa})\ +\mathrm{H_2O}: \ >\mathrm{Si(OH)}-\mathrm{N(H)}-\mathrm{Si(CMe_3)_2R} \\
(\mathrm{ad})\ +\mathrm{C_6H_6}: \ >\mathrm{Si(C_6H_5)}-\mathrm{N(H)}-\mathrm{Si(CMe_3)_2R} \\
>\mathrm{Si(OH)}-\mathrm{N(H)}-\mathrm{Si(CMe_3)_2R} \xrightarrow{>\mathrm{Si{=}N-Si(CMe_3)_2R}} [\mathrm{R(CMe_3)_2SiN(H)Si}<]_2\mathrm{O}
\end{array} \tag{15}$$

b. Compounds such as isobutene, propene, and acetone that have hydrogens in α positions prefer to react with the SiN double bond in an ene reaction (Eq. 16).

$$\begin{array}{c}
(\mathrm{ba})\ \mathrm{Me_2C{=}O}: \ >\mathrm{Si(OC(Me){=}CH_2)}-\mathrm{N(H)}-\mathrm{SiMe_2R} \\
(\mathrm{bb})\ +\ \text{diene}: \ >\mathrm{Si}-\mathrm{N(H)}-\mathrm{Si(CMe_3)_2R} \\
>\mathrm{Si{=}N}-\mathrm{Si(CMe_3)_2R} \\
(\mathrm{bc})\ \mathrm{CH_2{=}C(Me)X}: \ >\mathrm{Si(CH_2C({=}CH_2)X)}-\mathrm{N(H)}-\mathrm{Si(CMe_3)_2R} \\
\mathrm{R} = \mathrm{CMe_3, Ph} \\
\mathrm{X} = \mathrm{H, Me, CH_2Si(CMe_3)_2NSi(CMe_3)_3}
\end{array} \tag{16}$$

Even the 2,3-dimethyl-1,3-butadiene reacts as an enophile, not as a diene[14,28,29]:

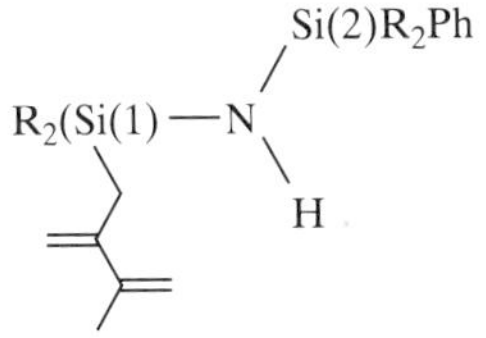

$Si(2)R_2Ph$

$R_2(Si(1)$—N

H

R = CMe_3

Si(1)—N	174.9 pm
Si(2)—N	174.7 pm
Si(1)—N—Si(2)	147.1°

In the case of isobutene, the ene product reacts with the iminosilane once again to give the bulkier twofold-substituted compound.[14]

c. If the reaction compounds are not too bulky, [2 + 2]-cycloaddition reactions of the iminosilane take place (Eq. 17).

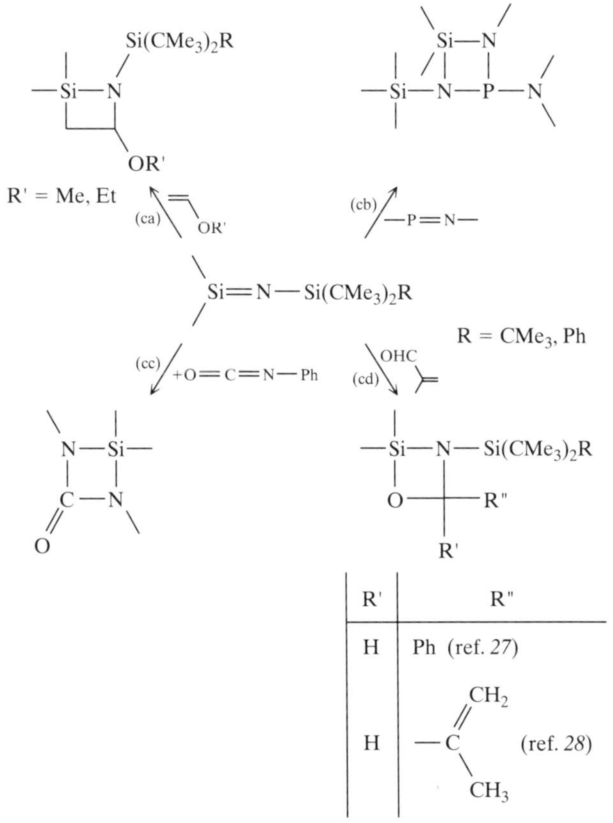

$Si(CMe_3)_2R$

—Si—N

OR'

Si—N

—Si—N—P—N

R' = Me, Et

(ca) OR'

(cb) —P=N—

$Si{=}N{-}Si(CMe_3)_2R$

R = CMe_3, Ph

(cc) +O=C=N—Ph

(cd) OHC

N—Si—

C—N

O

$-Si-N-Si(CMe_3)_2R$

O—R"

R'

R'	R"
H	Ph (ref. *27*)
H	—C(=CH_2)CH_3 (ref. *28*)

(17)

(ca) Using vinyl ethers, the expected [2 + 2] cycloadducts are obtained.[14,28] For example,

```
   Me3C               Si(2)Ph(CMe3)2
    |                /
Me3C—Si(1)—N
    |       |
   H2C(1) — C(2)—OC2H5
            |
            H
```

Bond lengths (pm)	
Si(1)—N	178.0
Si(2)—N	175.2
Si(1)—C(1)	187.9
C(2)—N	147.3

Bond angles (deg.)	
Si(1)—N—Si(2)	144.6
C(1)—Si(1)—N(1)	79.5

The angle C(1)—Si(1)—N is found to be very small in this four-membered SiNCC ring system.

(cb) In the reaction of an iminosilane with an iminophosphane, a 1,3-diaza-2-sila-4-phosphacyclobutane is formed in a head-to-tail [2 + 2] dimerization.[29] For example,

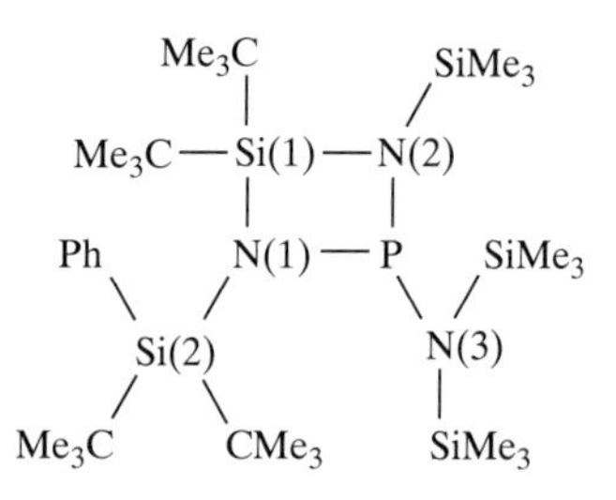

Bond lengths (pm)	
Si(1)—N(1)	179.4
Si(2)—N(1)	177.7
Si(1)—N(2)	175.5
P—N(1)	179.0
P—N(2)	174.8
P—N(3)	172.4

Bond angles (deg.)	
Si(1)—N(1)—Si(2)	140.2
N(1)—Si(1)—N(2)	85.5
N(1)—P—N(2)	85.8
N(2)—P—N(3)	108.8
N(1)—P—N(3)	116.2

Because of the bulkiness of the substituents, the SiN and PN ring bonds are remarkably elongated. At ambient temperature, this compound shows hindered rotation about the P—N$(SiMe_3)_2$ bond.

(cc) With phenylisocyanate, the usual (SiNCO) four-membered-ring system is not formed, but rather the (SiNCN) ring system,[29] for example,

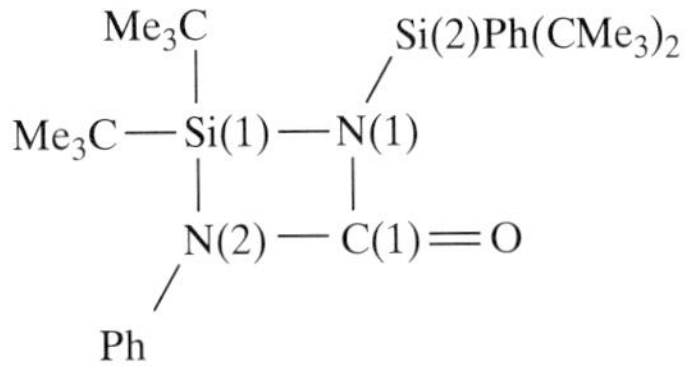

Me_3C
$Me_3C—Si(1)—N(1)—Si(2)Ph(CMe_3)_2$
$N(2)—C(1)=O$
Ph

Bond lengths (pm)	
Si(1)—N(1)	177.9
Si(1)—N(2)	176.9
N(1)—Si(2)	177.6
C(1)—N(2)	139.4
N(1)—C(1)	140.9
C(1)—O	120.9

Bond angles (deg.)	
N(1)—Si(1)—N(2)	76.24
Si(1)—N(1)—Si(2)	148.17
N(1)—C(1)—N(2)	102.5
C(1)—N(2)—Si(1)	91.2

(cd) Oxaazasilacyclobutanes (SiNCO ring) are formed in the reaction of the iminosilane with methacrolein or benzaldehyde.[27,29]

d. The [2 + 3]-cycloaddition products of iminosilanes are shown in Eq. (18).

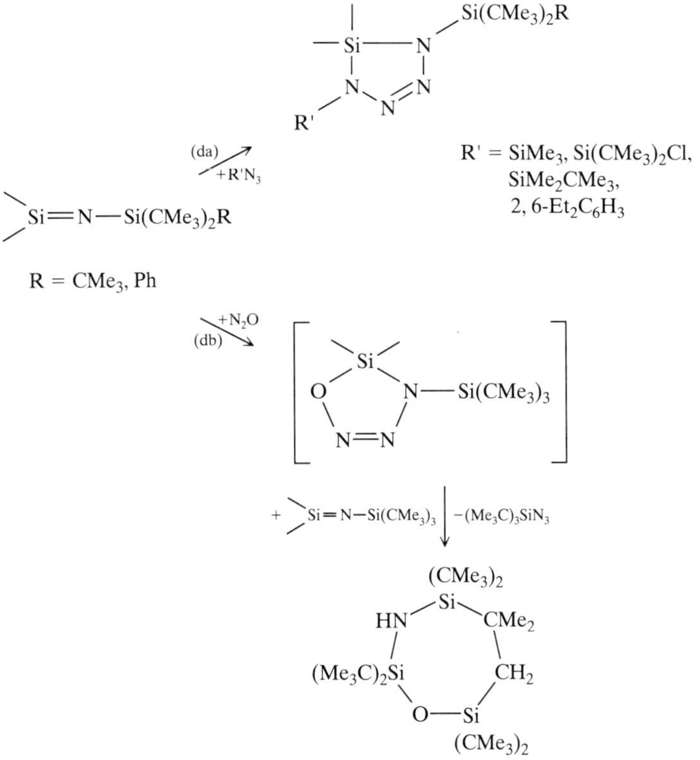

$>Si=N—Si(CMe_3)_2R$

R = CMe_3, Ph

(da) $+R'N_3$

$R' = SiMe_3$, $Si(CMe_3)_2Cl$, $SiMe_2CMe_3$, 2, 6-$Et_2C_6H_3$

(db) $+N_2O$

$N—Si(CMe_3)_3$

$+ >Si=N—Si(CMe_3)_3$ $-(Me_3C)_3SiN_3$

$(CMe_3)_2$ Si, HN, CMe_2, $(Me_3C)_2Si$, CH_2, O—Si, $(CMe_3)_2$

(18)

(da) The iminosilanes react with azides (RN_3, R = alkyl, silyl, aryl) to give the [2 + 3]-cycloaddition products.[21,28–30] For example, Grosskopf *et al.*[21b]:

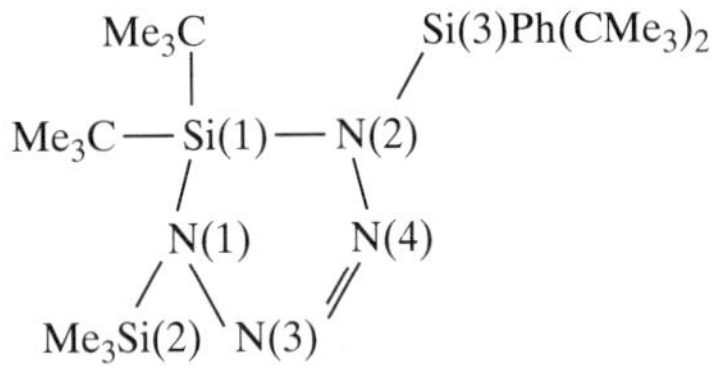

$$\begin{array}{c} \text{Me}_3\text{C} \quad\quad\quad\quad \text{Si(3)Ph(CMe}_3)_2 \\ | \quad\quad\quad\quad / \\ \text{Me}_3\text{C}-\text{Si(1)}-\text{N(2)} \\ / \quad\quad \backslash \\ \text{N(1)} \quad\quad \text{N(4)} \\ / \ \backslash \quad\quad // \\ \text{Me}_3\text{Si(2)} \quad \text{N(3)} \end{array}$$

Bond lengths (pm)	
Si(1)—N(1)	176.2
Si(1)—N(2)	178.4
Si(2)—N(1)	177.3
Si(3)—N(2)	179.1
N(1)—N(3)	141.2
N(3)—N(4)	125.1
N(2)—N(4)	142.3

Bond angles (deg.)	
N(1)—Si(1)—N(2)	88.54
Si(1)—N(1)—Si(2)	143.61
Si(1)—N(2)—Si(3)	147.66

(db) Analogously, the [2 + 3]-cycloaddition product with N_2O is formed, but it is not stable and reacts with one more iminosilane to give a seven-membered-ring system.[14,30]

2. *Reactions of Kinetically Unstable Iminosilanes*

The reactions of kinetically unstable iminosilanes have already been discussed by Raabe and Michl in 1985.[31] Basically, they react in the same way that stable iminosilanes do. Further reactions are described as follows.

a. The reaction of photochemically generated dimesitylsilylene with tri-*t*-butylsilylazide gives the iminosilane, which isomerizes spontaneously to the silylcyclobutane derivative.[22]

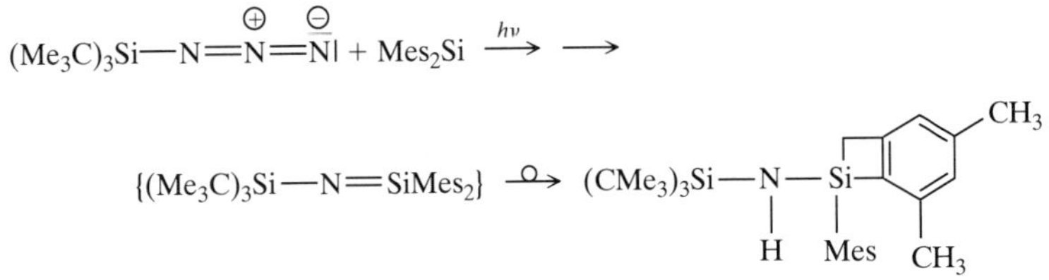

$$(\text{Me}_3\text{C})_3\text{Si}-\text{N}{=}\overset{\oplus}{\text{N}}{=}\overset{\ominus}{\text{N}}| + \text{Mes}_2\text{Si} \xrightarrow{h\nu} \longrightarrow$$

$$\{(\text{Me}_3\text{C})_3\text{Si}-\text{N}{=}\text{SiMes}_2\} \longrightarrow (\text{CMe}_3)_3\text{Si}-\underset{\text{H}}{\text{N}}-\underset{\text{Mes}}{\text{Si}}\text{(benzocyclobutene; CH}_3\text{, CH}_3) \quad (19)$$

b. Many intermediate iminosilanes dimerize in a head-to-tail [2 + 2] cycloaddition.[18,32–35]

$$\begin{array}{c} >Si{=}N{-} \\ + \\ {-}N{=}Si< \end{array} \longrightarrow \begin{array}{c} {-}Si{-}N< \\ | \quad\ | \\ >N{-}Si{-} \end{array} \qquad (20)$$

c. Besides the previously mentioned active proton compounds, amines also participate in insertion reactions.[33]

$$Me_2Si{=}N{-}Si(CMe_3)_3 \xrightarrow{+H_2NR} Me_2Si(NHR){-}N(H){-}Si(CMe_3)_3 \qquad (21)$$

$R = CHMe_2, CMe_3, Ph$

d. It is also possible to isolate insertion products of iminosilanes into polar single bonds without participation of hydrogen.[33] For example,

$$Me_2Si{=}N{-}R + a{-}R' \longrightarrow Me_2Si(a){-}NR(R') \qquad (22)$$

R	a—R′	R	a—R′
$SiMe_3$	$Cl{-}SiMe_3$	$SiMe_2CMe_3$	$N_3{-}SiMe_3$
	$Cl{-}SiClMe_2$		$N_3{-}SiMe_2CMe_3$
	$Cl{-}SiCl_2Me$	$Si(CMe_3)_3$	$EtO{-}SiH(OEt)_2$
	$Cl{-}SiCl_3$	*p*-Tol	$Cl{-}SiMe_3$
	$Cl{-}GeMe_3$		$N_3{-}SiMe_3$
	$Cl{-}SnMe_3$		
	$N_3{-}SiMe_3$		
	$MeO{-}SiMe_3$		
	$Me_2N{-}SiMe_3$		

F. *Reactions of the Donor Adducts of Iminosilanes*

1. With water, alcohols, amines and other compounds having a polar single bond, donor adducts react in the same way as free iminosilanes.[18,24,25,33] For example[24]:

$$Me_2Si(D){=}NSi(CMe_3)_3 \xrightarrow[-D]{+RH} Me_2Si(R){-}N(H){-}Si(CMe_3)_3$$

$R = OH, OMe, OEt, O\text{-}iPr_2, O\text{-}t\text{-}Bu, OPh, NH\text{-}iPr, NH\text{-}t\text{-}Bu, NHPh \qquad (23)$

2. Ene reaction products are obtained with compounds that have hydrogen atoms in the α position.[18,24,33] For example,[24]

$Me_2Si{=}NSi(CMe_3)_3$ (with D on Si) $\xrightarrow{-D}$ + cyclopentadiene → $-Si(Cp)-N(H)-$; + isobutene → $-Si(CH_2C(R){=}CH_2)-N(H)-$

R = H, Me, CH_2CMe (24)

3. Head-to-tail [2 + 2]-cycloaddition products[24] are formed with methyl vinyl ether[14] or benzophenone.[14] For example,[14,18,24]

$Si{=}N-$ (with D on Si) $\xrightarrow{-D}$ ×2 → cyclo-$(Si-N)_2$; + methyl vinyl ether → $-Si-N-$ / $-C-C(H)(OMe)$ four-membered ring; + $OCPh_2$ → $-Si-N-$ / $O-CPh_2$ four-membered ring (25)

4. When the iminosilane adducts react with silylazides, the [2 + 3] cycloadduct is obtained.[14,18,24,33] For example,[33]

$$Me_2Si{=}N{-}R \xrightarrow{+R'N_3} Me_2Si(NR)N{=}N{-}N(R')$$ (cyclic: Me_2Si–N(R)–N=N–N(R')–)

R = silyl, germyl; R' = CMe_3, silyl (26)

In some special cases the adducts react in a different way in the presence of THF.

5. With N_2O, the iminosilanes $Me_2Si{=}NSi(CMe_3)_3$ and $(Me_3C)_2Si{=}NSi(CMe_3)_3$ form unstable [2 + 3] cycloadducts; with the participation of the iminosilanes, these cycloadducts react to yield products composed of one molecule of the silanone and one molecule of the iminosilane.[14,30] For example,[30]

$$R_2Si{=}N{-}R' \xrightarrow[THF]{N_2O} \left[\text{cyclo-}\{R_2Si{-}O{-}N{=}N{-}N(R')\} \right] \xrightarrow{-R'N_3} \left[R_2Si{=}O \right]$$

$$\xrightarrow{+\ R_2Si{=}N{-}R'} \text{cyclo-}\{R_2Si{-}O{-}SiR_2{-}NH{-}SiR_2{-}CMe_2{-}CH_2\} \qquad R = CMe_3 \qquad (27)$$

$(Me_3C)_2Si{=}O$, generated in benzene from $(Me_3C)_2Si(THF){=}N{-}Si(CMe_3)_3$ and N_2O in the presence of equimolar amounts of THF, forms an adduct $(Me_3C)_2Si(THF){=}O$, which is trapped by excess of the iminosilane.

$$(Me_3C)_2Si(\leftarrow THF){=}NSi(CMe_3)_3 \xrightarrow[THF]{+N_2O,\ -N_3Si(CMe_3)_3} (Me_3C)_2Si(\leftarrow THF){=}O \xrightarrow{+\ {>}Si{=}N{-}}$$

$$(Me_3C)_2Si(\leftarrow THF){-}O \longrightarrow {>}Si{=}N{-} \longrightarrow (Me_3C)_2Si(\leftarrow THF){-}O{-}Si(CMe_3)_2{-}N(H){-}Si(CMe_3)_3 \qquad (28)$$

6. In the reaction of ethyldimethylamine with the iminosilane THF adduct the donor THF again is involved in the formation of the product[14]: $Me_2N{-}(CH_2)_4{-}OSi(CMe_3)_2{-}NH{-}Si(CMe_3)_3$ is formed.

$$(Me_3C)_2Si(\leftarrow THF)=NSi(CMe_3)_3 \xrightarrow[-CH_2=CH_2]{NMe_2Et} Me_2N-(CH_2)_4-OSi(CMe_3)_2-NH-Si(CMe_3)_3 \quad (29)$$

The amine opens the THF ring system because of a nucleophilic attack at the α-C atom of the coordinated THF. Owing to the movement of a proton from the ethyl to the amide group with simultaneous ethene elimination, a betaine structure appears.

$$EtMe_2\overset{\oplus}{N}-(CH_2)_4-O-Si(CMe_3)_2-\overset{\ominus}{N}-Si(CMe_3)_3$$

G. *Reactions of LiF Adducts of Iminosilanes*

In general, almost all lithiated aminofluorosilanes—the LiF adducts of iminosilanes—react like iminosilanes.[18] Only exceptional reactions are mentioned here.[18]

1. The lithium derivative of di-*t*-butylfluorosilyl-2,6-diisopropylphenylamine reacts with 2-methyl-2-propenal in two competing ways. In a [2 + 4] cycloaddition, an oxa-3-aza-2-sila-5-cyclohexene is formed; and in a [2 + 2] cycloaddition, 2-methyl-2-propenyl-*N*-(2,6-diisopropylphenyl)-imine is generated via an (SiNCO)-ring intermediate.[18,36]

$$(THF)_3LiF-SiR_2=N-R' \xrightarrow[-LiF]{+O=CH-C(Me)=CH_2} \begin{cases} \text{6-ring: } R_2Si-N(R')-CH_2-C(Me)=CH-O \text{ (ring)} \\ \text{4-ring: } R_2Si-N(R)-CH(C(Me)=CH_2)-O \text{ (ring)} \end{cases} \quad (30)$$

$$\text{4-ring } R_2Si-N(R')-CH(C(Me)=CH_2)-O \xrightarrow[-1/3\,(R_2SiO)_3]{\Delta} R'N=CH-C(Me)=CH_2$$

$R = CMe_3$, $R' = 2,6\text{-}C_6H_3(CHMe_2)_2$ (31)

2. Four-membered rings are formed in the reaction with benzophenone and phenylaldehyde. The oxaazasilacyclobutane ($((Me_3C)_2Si—NCMe_3—CHC_6H_5—O)$) was purified by distillation and crystallized from *n*-hexane. The thermal cleavage of the (SiNCO) rings leads to carboimines and cyclosiloxanes.[18,36]

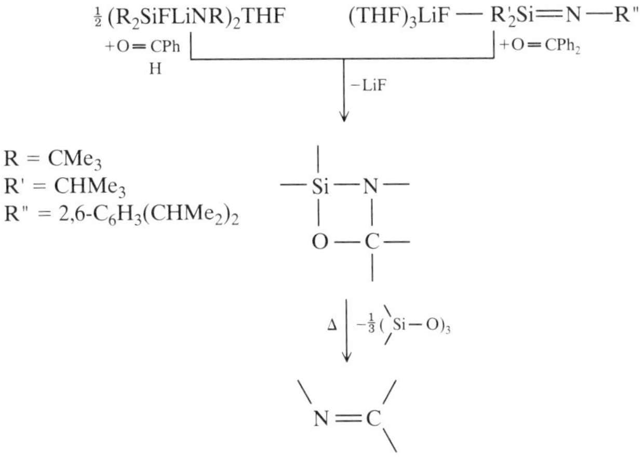

$\frac{1}{2}(R_2SiFLiNR)_2THF$ + O=CPh(H); $(THF)_3LiF—R'_2Si=N—R''$ + $O=CPh_2$; −LiF → four-membered Si—N—C—O ring; Δ, $-\frac{1}{3}(\text{>Si—O})_3$ → >N=C< (32)

R = CMe_3
R' = $CHMe_3$
R'' = 2,6-$C_6H_3(CHMe_2)_2$

The four-membered ring is not planar. The angle between the planes C(1)—N—O and Si—N—O amounts 17°.

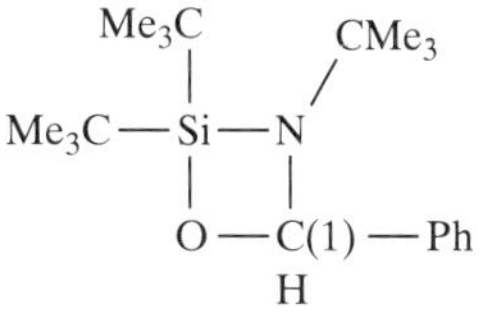

$(Me_3C)_2Si$—N(CMe_3)—C(1)H(Ph)—O (four-membered ring)

Si—O	168.6 pm
N—C(1)	147.0 pm
Si—N	172.3 pm
C(1)—O	145.6 pm
O—Si—N	80°

3. Lithiated aminofluorosilanes are often very stable and can be purified by distillation. Attempts to eliminate LiF lead to dimerization or rearrangement of the polar iminosilane formed. The limits of dimerization have been reached in the dimeric $(Me_2HC)_2Si{=}N\text{-}2,4,6\text{-}C_6H_2(CMe_3)_3$.[18–20]

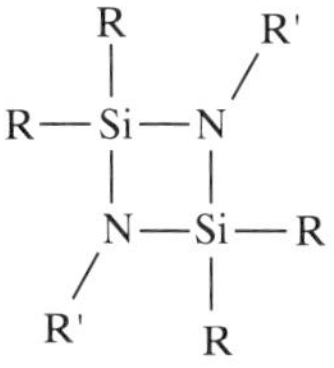

$(R_2Si—NR')_2$ (four-membered Si—N—Si—N ring)

R = $CHMe_2$
R' = 2,4,6-$C_6H_2(CMe_3)_3$

The dimer is an asymmetric stretched four-membered ring. The isopropyl groups are twisted.

Because of the bulkiness of the substituents, bis(silyl)aminoorganyliminosilanes cyclize with 1,3-migration of a silyl group from one nitrogen atom to the other. The silyl-group migration can be explained by the propensity of silyl groups to be bonded to the more negatively charged atom.

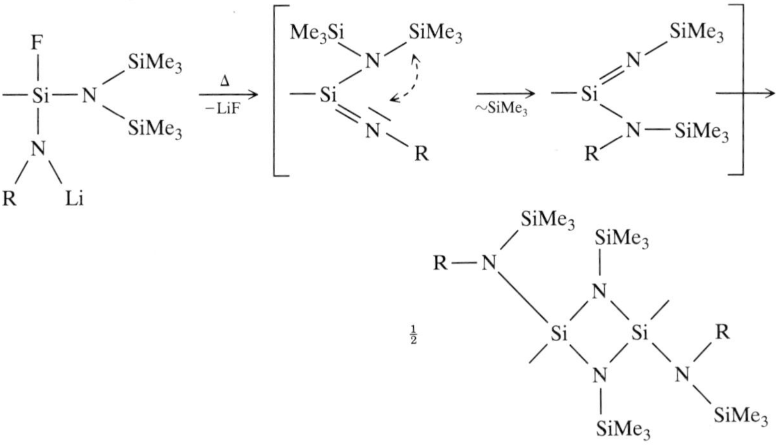

(33)

Thermally enforced LiF elimination of LiF adducts of organyl(silyl)aminoorganyliminosilanes leads to another surprising ring-closure reaction. For example,[18,20]

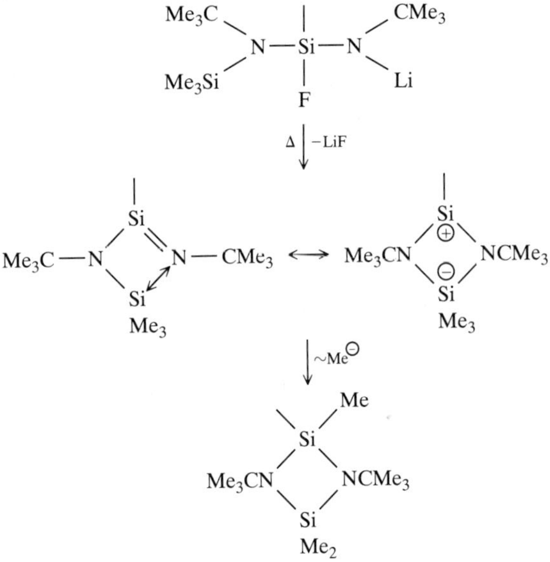

(34)

The formation of the $(SiN)_2$ ring can be explained by a nucleophilic 1,3-rearrangement of a methanide ion at silicon. The cyclic ylide is formed because the two nitrogen atoms are identically substituted and a silyl group migration would be without energy profit in such a molecule.

Cross-dimerization products are formed by thermal LiF elimination from two LiF adducts.[36]

$$R_2Si(F)-N(Li)R' + R''_2Si(F)-N(Li)R''' \xrightarrow{-2\,LiF} R_2Si\langle N(R')N(R''')\rangle SiR''_2 \quad (35)$$

H. *Silanediimines—Silicon Analogs of Carbodiimides*

In recent years considerable research has been done on the photochemistry and thermal chemistry of main group element azides,[37] especially those of silicon. Monoazidosilanes have long been employed in the photochemical generation of iminosilanes.[38,39] These products arise from the thermal or photochemical loss of molecular nitrogen accompanied by a 1,2-shift of a substituent from silicon to nitrogen.

In 1987 the first silanediimine, the *N,N'*-bis(trimethylsilyl)silanediimine, was isolated[40] by photolysis of matrix-isolated 2,2-diazidohexamethyltrisilane. It has been identified by UV spectroscopy and trapping experiments.

$$(Me_3Si)_2Si(N_3)_2 \xrightarrow[77\,K]{h\nu\,(254\,nm)} [(Me_3Si)Si(N_3){=}NSiMe_3]$$

$$[(Me_3Si)Si(N_3){=}NSiMe_3] \xrightarrow{Me_3SiOMe} Me_3SiSi(OMe)(N_3)N(SiMe_3)_2$$

$$[(Me_3Si)Si(N_3){=}NSiMe_3] \xrightarrow{h\nu\,270\,nm} Me_3Si{-}N{=}Si{=}N{-}SiMe_3 \xrightarrow{+Me_3SiOMe} [(Me_3Si)_2N]_2Si(OMe)_2 \quad (36)$$

Irradiation of 2,2-diazidohexamethyltrisilane in 3-MP at 77 K ($\lambda = 254$ nm) shows two new bands in the UV at 274 nm and 324 nm. Longer wavelength irradiation ($\lambda > 270$ nm) leads to an increase of the 324-nm band, while the intensity of the 274-nm band decreases. The azidosilane-imine ($\lambda = 274$ nm) is a photochemical precursor of the silanediimine (λ = 324 nm). Both products have been identified by trapping experiments with Me_3SiOMe and analysis of the photolysate by gas chromatography (GC).[40]

The photolysis of matrix-isolated di-*t*-butyldiazidosilane in 3-MP at 77 K or in argon matrix at 10 K yields several products, the *N*,*N*′-di-*t*-butylsilane diimine,[41] among others. This pale yellow compound is characterized by UV absorption bands at 240 nm and 385 nm.

Chemical trapping experiments demonstrate that this silicon analog of a carbodiimide is formed in 10% yield. Several trapping experiments have been carried out.

$$(Me_3C)N{=}Si{=}N(CMe_3) \xrightarrow{+2\,MeOH} (MeO)_2Si(\underset{H}{N}CMe_3)_2$$

$$(Me_3C)N{=}Si{=}N(CMe_3) \xrightarrow{+2\,t\text{-}BuOH} (Me_3CO)_2Si(\underset{H}{N}CMe_3)_2$$

$$(Me_3C)N{=}Si{=}N(CMe_3) \xrightarrow{+2\,Me_3SiOMe} (MeO)_2Si(\underset{SiMe_3}{N}CMe_3)_2 \qquad (37)$$

I. *Molecular Geometry of Silanediimine—Theoretical Studies*

The geometry of dimethylsilanediimine was optimized using MNDO.[41] The NSiN bond angle is 172°, and the CNSi bond angle is 153°. The molecule has a planar, essentially W-shaped, geometry and a computed carbon-to-carbon dihedral angle of 3°. The SiN double-bond length is calculated to be 151 pm, which is shorter than the length of 157 pm found in a stable iminosilane.[15] An *ab initio* 3-21 G SCF calculation for the HN=Si=NH yields results similar to those for the described compound, but the optimized geometry is planar and trans-bent.[41,42]

The HN=Si=NH geometry is calculated to be trans-planar with a HNSi bond angle of 157.5° and a Si=N bond length of 151.6 pm. Optimization leads to the complete linear geometry with a SiN bond lengths of 150.5 pm.

The linear silanediimide geometry (SCF/6-31 G (d)) is only 0.2 kcal mol^{-1} higher in energy than the trans-planar geometry.

The silanediimide geometry has also been reoptimized with MP2/6-31 G (d) wavefunctions in C_2 symmetry. Here, it has a nonlinear N=Si=N angle of 156.9° and an HNSi angle about 125.8° with noncoplanar hydrogens. The SiN double-bond length increases to 159.7 pm. Reoptimization in C_2 symmetry leads to the linearity of the NSiN bond again. This structure is 2.1 kcal mol^{-1} higher in energy than the nonlinear siladiimide.

In conclusion, whether or not this species crystallizes in an allene-like or a trans-planar structure may depend on the crystal packing.[43]

RHF/6-31 G (d) structures for silanediimide:

H—N=Si=N—H (98.4 pm; 151.6 pm; 157.5°) C_{2h} (min)

H—N=Si=N—H (98.1 pm; 150.5 pm) D_h (dsp)

MP2/6-31 G (d) structure for silanediimide:

H–N–Si–N–H (125.8°; 156.9°; 101.5 pm)

dihedral HNSiN = ± 141.5°

J. *Silaamidide Salts—Another Form of SiN Double Bonds*

Although silicon analogs of amidides $R_2N—SiR=NR$ are still unknown, the isolation and characterization of silaamidide ions as salts have been described.[44,45]

The preparation of the lithium-*N*,*N'*-bis(2,4,6-tri-*t*-butylphenyl)phenylsilaamidide is presented as an example for the preparation of the other silaamidides.[44,45]

$(ArNH)_2SiClPh$ —(2 *t*-BuLi/Et_2O, −LiCl)→ [Et_2O·Li⊕(ArN)₂Si⊖Ph]

+[15]-crown-5 → $[(ArN)_2SiPh]^{\ominus}$ $[Li[15]crown-5]^{\oplus}$

160°C/vacuum, $-Et_2O$ ⇌ $+Et_2O$: Li⊕(ArN)₂Si⊖Ph (38)

Even with a weak potassium base ($KN(SiMe_3)_2$), the potassium silaamidide was isolated.[44]

All the silaamidides are moisture-sensitive, but they are thermally stable. Some of them lose ether when they are heated above 100°C under vacuum to give the unsolvated salts.

The lithium *N,N'*-bis(2,4,6-tri-*t*-butylphenyl)-*t*-butylsilaamidide (THF) (12-crown-4) etherate has been isolated as single crystals and studied by X-ray crystallography.

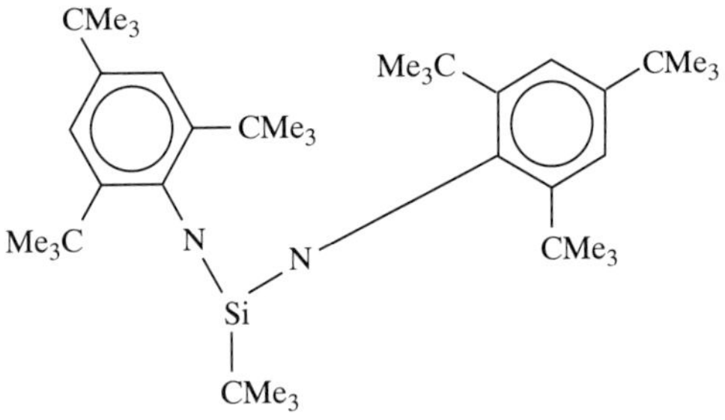

CMe_3 Me_3C CMe_3 CMe_3 Me_3C N N CMe_3 Si CMe_3

Li(THF)(12-crown-4)

The structure of this silaamidide salt consists of well-separated, noninteracting lithium cations $[Li(12\text{-crown-4})THF]^{\oplus}$ and *N,N'*-bis(2,4,6-tri-*t*-butylphenyl)-*t*-butylsilaamidide anions. In the anion, the silicon is tricoordinated to a carbon atom of the *t*-butyl group and to two nitrogen atoms.

The silicon–nitrogen bond distances in this molecule are 159.4 pm and 162.6 pm. These values are intermediate between those for a SiN double bond (156.8 pm)[15] and those for a SiN single bond (174.8 pm).[46]

1. *Addition Reactions*

a. With alcohols or amines, the organyl-substituted lithium silaamidides react by protonation at one nitrogen and addition across the double bond.[44]

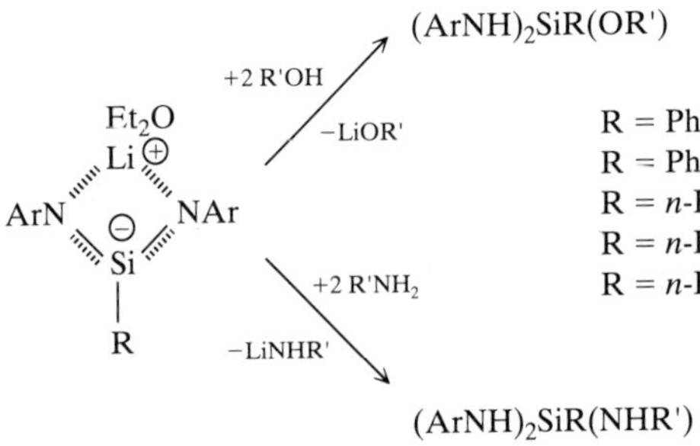

Et_2O $Li^{\oplus}$ ArN $\ominus$ NAr Si R

$\xrightarrow[-LiOR']{+2\,R'OH}$ $(ArNH)_2SiR(OR')$

R = Ph, R' = H
R = Ph, R' = *t*-Bu
R = *n*-Bu, R' = Me
R = *n*-Bu, R' = *i*Pr
R = *n*-Bu, R' = *t*-Bu

$\xrightarrow[-LiNHR']{+2\,R'NH_2}$ $(ArNH)_2SiR(NHR')$

Ar = 2,4,6-tri-*t*-BuPh

R = *n*-Bu, R' = *n*-Bu
R = *n*-Bu, R' = Ph (39)

b. Lithium and potassium chlorosilaamide react with two equivalents of alcohol or amine to give the double addition products.[44]

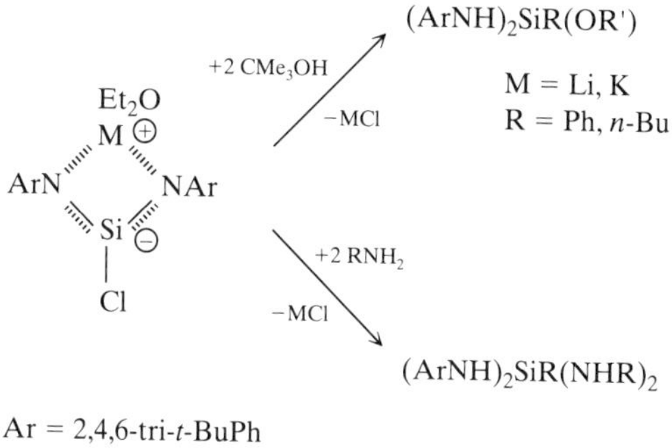

$$\text{ArN}(\text{M}^{\oplus}(\text{Et}_2\text{O}))\text{NAr}\text{Si}^{\ominus}\text{Cl} \xrightarrow[-\text{MCl}]{+2\,\text{CMe}_3\text{OH}} (\text{ArNH})_2\text{SiR}(\text{OR}')$$

$$\text{ArN}(\text{M}^{\oplus}(\text{Et}_2\text{O}))\text{NAr}\text{Si}^{\ominus}\text{Cl} \xrightarrow[-\text{MCl}]{+2\,\text{RNH}_2} (\text{ArNH})_2\text{SiR}(\text{NHR})_2$$

M = Li, K
R = Ph, *n*-Bu
Ar = 2,4,6-tri-*t*-BuPh (40)

c. Silaamidides react with HBr (as the HBr · PPh_3 salt) to give bromodiaminosilanes (Eq. 41).

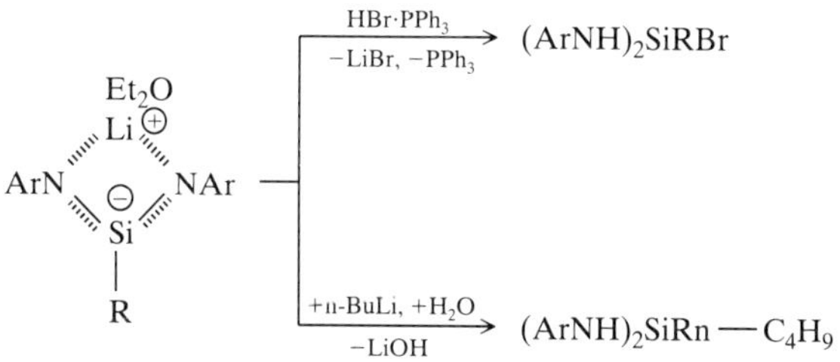

$$\text{ArN}(\text{Li}^{\oplus}(\text{Et}_2\text{O}))\text{NAr}\text{Si}^{\ominus}\text{R} \xrightarrow[-\text{LiBr},\,-\text{PPh}_3]{\text{HBr}\cdot\text{PPh}_3} (\text{ArNH})_2\text{SiRBr}$$

$$\text{ArN}(\text{Li}^{\oplus}(\text{Et}_2\text{O}))\text{NAr}\text{Si}^{\ominus}\text{R} \xrightarrow[-\text{LiOH}]{+n\text{-BuLi},\,+\text{H}_2\text{O}} (\text{ArNH})_2\text{SiR}n\text{—C}_4\text{H}_9 \qquad (41)$$

d. Treatment with *n*-BuLi results in addition across the Si=N double bond, which then yields, after protonation with water, aminosilanes (Eq. 41).[44]

e. Treatment of an organyl-substituted lithium silaamidide ($(ArN)_2Si(n\text{-}Bu)Li(Et_2O)$) with benzaldehyde gives the [2+2]-cycloaddition product. After protonation with water and recrystallization, a stable 1-oxa-3-aza-2-silacyclobutane is obtained.[44] Similar cycloadditions have already been reported.[27,36]

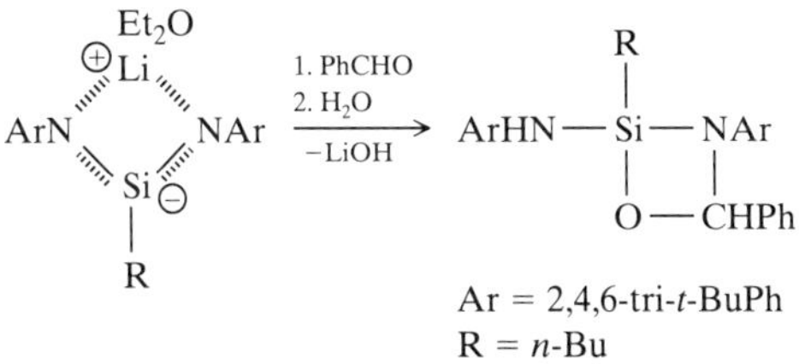

$$\text{ArN}(\text{Li}^{\oplus}(\text{Et}_2\text{O}))\text{NAr}\text{Si}^{\ominus}\text{R} \xrightarrow[-\text{LiOH}]{1.\ \text{PhCHO};\ 2.\ \text{H}_2\text{O}} \text{ArHN—Si(R)—NAr—CHPh—O (ring)}$$

Ar = 2,4,6-tri-*t*-BuPh
R = *n*-Bu (42)

f. In 1994, the first dimeric silaamidide was described.[47] It is formed in a [2+2] cycloaddition only if the substituents are not bulky enough to stabilize the monomeric compound.

$$2\,(RNH)_2SiFPh \xrightarrow[-4\,t\text{-}BuH]{+4\,t\text{-}BuLi} 2\,[(RNLi)_2SiFPh] \xrightarrow[-2\,LiF]{\Delta} 2\,[\text{PhSi}(\mu\text{-NR})_2\text{Li}] \xrightarrow[{[2+2]\ \text{cycloaddition}}]{+4\,THF} [(\text{PhSi–NR})_2\text{LiNR}_2]^{\ominus}\,[Li(THF)_4]^{\oplus} \qquad (43)$$

$R = SiMe_2CMe_3$

Bis(t-butyldimethylsilylamino)fluorophenylsilane reacts with t-BuLi in a 1 : 2 molar ratio to give a dilithium derivative, $PhSiF(NLiR)_2$. LiF elimination in the presence of THF leads to the formation of a silaamidide, which is isolated as a four-membered cyclosilazane anion, $(PhSi\text{—}NR)_2LiNR_2^{\ominus}$, and $Li(THF)_4^{\oplus}$ cation. The endocyclic Si—N bonds are much longer (174.9, 176.7 pm) than the exocyclic ones (165.0, 165.6 pm). The $(Si\text{—}N)_2$ ring is not planar and the sum of angles at the nitrogen atoms differs from 360° (356.4°, 352.4°).

K. *Silaisonitrile and Silanitrile*

1. *Silaisonitrile*

In 1966, the first silaisonitrile HN=Si was described.[4,48] It was produced by photolysis of H_3SiN_3 in a 4 K argon matrix (Eq. 44).

$$H_3Si\text{—}N_3 \xrightarrow[-N_2,\,-H_2]{h\nu} H\text{—}N{=}Si \qquad (44)$$

According to *ab initio* calculations[49] the hydrogen isosilanitrile HN=Si: contains a divalent silicon atom and an :Si=N double bond and is far more stable than the isomeric hydrogen silanitrile HSi≡N|. Similar *ab initio* results have been obtained for the relative stabilities of RN=Si: and R—Si≡N|, where R = CH_3 and C_6H_5.[50] The emission spectrum of the fundamental vibration–rotation ν_1 band (NH) of NHSi has been observed near 2.7 μm by high-resolution Fourier-transform spectroscopy from a mixture of N_2 + SiH_4 excited in a radiofrequency discharge. This is the first spectroscopic observation of this molecule in the gas phase.[51]

The total energy, the structure, and the ionization pattern of the more stable phenyl derivative have been calculated.[52,53] The controlled explosion of triazide(phenyl)silane in a flash-pyrolysis apparatus under approximately unimolecular conditions ($p \approx 10^{-3}$ Pa, $c \approx 10^{-3}$ mM) can be carried out without danger.[52,53]

$$\text{Ph—Si}(N_3)_3 \xrightarrow[-4\,N_2]{1100\ \text{K}} \text{Ph—N}\equiv\text{Si} \qquad (45)$$

MNDO: C_6H_5—N≡Si (136, 152 pm)

Real-time photoelectron-spectroscopic analysis of the band intensities revealed that the optimum reaction conditions required a temperature of 1100 K for the complete elimination of four molecules of N_2. The formation of the valence isomer Ph—Si≡N, which is predicted to be 400 kJ mol^{-1} less stable, can be excluded.[53]

The reaction of matrix-isolated triazidophenylsilane with *t*-BuOH led to the formation of *N*-(di-*t*-butoxysilyl)aniline and amino(di-*t*-butoxy)phenylsilane, the trapping products of the nitrile and isonitrile.[6]

$$PhSi(N_3)_3 \xrightarrow[-N_2]{h\nu \text{ or } \Delta} PhSi(NH_2)(O\text{-}t\text{-}Bu)_2,\ PhNHSiH(O\text{-}t\text{-}Bu)_2 \qquad (46)$$

2. *Silanitrile*

A compound with a formal SiN triple bond has been spectroscopically identified. The matrix irradiation of the silylazide H_3SiN_3 with a 193-nm ArF-excimer laser gives the silanitrile H—Si≡N.

$$H_3SiN_3 \xrightarrow[-N_2]{h\nu\ (193\ \text{nm})} HSi\equiv N\cdot H_2 + H\text{—}Si\equiv N \qquad (47)$$

$$H\text{—}Si\equiv N \xrightarrow[-H]{h\nu\ (193\ \text{nm})} {}^{\cdot}Si\equiv N \qquad (48)$$

The spectroscopic data are as follows: UV: $\lambda_{max} = 238, 258, 266, 350$ nm. IR: $\nu = 2151, 2149$ ($\nu_{Si—H}$); 1163, 1161 ($\nu_{Si\equiv N}$) cm^{-1}.

Irradiation at the wavelength of 193 nm with a ArF-excimer laser for a long time leads to the loss of the hydrogen atom. The well known SiN radical appears.[55]

III

CONCLUSION

The synthesis of iminosilanes has opened up a new field of silicon–nitrogen chemistry. These compounds have an interesting chemistry. As shown, many addition and cycloaddition reactions are possible. Many new and surprising results are likely in the future.

Reviews

For reviews on the chemistry of silenes see the following:

(a) Brook, A. G.; Baines, K. M. *Adv. Organomet. Chem.* **1986,** *25*, 1. Raabe, G.; Michl, J. *Chem. Rev.* **1985,** *85*, 419. Wiberg, N. *J. Organomet. Chem.* **1984,** *273*, 141.

(b) West, R. *Angew. Chem.* **1987,** *99*, 1231. *Angew. Chem., Int. Ed. Engl.* **1987,** *26*, 1201. Tsumuraya, T.; Batcheller, S. A.; Masamune, S. *Angew. Chem.* **1991,** *103*, 916. *Angew. Chem., Int. Ed. Engl.* **1991,** *30*, 902.

(c) Walter, S.; Klingebiel, U. *Coord. Chem. Rev.* **1994,** *130*, 481. Raabe, G.; Michl, J. *Chem. Rev.* **1985,** *85*, 419.

(d) Driess, M. *Chem. Unserer Zeit* **1993,** *27*, 141.

References

(1) Zigler, S. S.; Johnson, L. M.; West, R. *J. Organomet. Chem.* **1988,** *341*, 187.

(2) Zigler, S. S.; West, R.; Michl, J. *Chem. Lett.* **1986,** 1025.

(3) Sekiguchi, A.; Ando, W.; Honda, K. *Chem. Lett.* **1986,** 1029.

(4) Ogilvie, J. F.; Cradock, S. *J. Chem. Soc., Chem. Commun.* **1966,** 364.

(5) Maier, G.; Glatthaar, J.; Reisenauer, H. P. *Chem. Ber.* **1987,** *120*, 1213.

(6) Radziszewski, J. G.; Littmann, D.; Balaji, V.; Fabry, L.; Gross, G.; Michl, J. *Organometallics* **1993,** *12*, 4816.

(7) Denis, J.-M.; Guénot, P.; Letulle, M.; Pellerin, B.; Ripoll, J.-L. *Chem. Ber.* **1992,** *125*, 1397.

(8) Schleyer, P. von R.; Stout, P. D. *J. Chem. Soc., Chem. Commun.* **1986,** 1373.

(9) Truong, T. N.; Gordon, M. S. *J. Am. Chem. Soc.* **1986,** *108*, 1775.

(10) Schmidt, M. W.; Truong, P. N.; Gordon, M. S. *J. Am. Chem. Soc.* **1987,** *109*, 5217.

(11) Schleyer, P. von R.; Kost, D. *J. Am. Chem. Soc.* **1988,** *110*, 2105.

(12) Reed, A. E.; Schade, C.; Schleyer, P. von R.; Kamath, P. V.; Chandrasekhar, J. *J. Chem. Soc., Chem. Commun.* **1988,** 67.

(13) Wiberg, N.; Schurz, K.; Fischer, G. *Angew. Chem.* **1985,** *97*, 1058. *Angew. Chem., Int. Ed. Engl.* **1985,** *24*, 1053.

(14) Wiberg, N.; Schurz, K. *Chem. Ber.* **1988,** *121*, 581.
(15) Wiberg, N.; Schurz, K.; Reber, G.; Müller, G. *J. Chem. Soc., Chem. Commun.* **1986,** 591.
(16) Reber, G.; Riede, J.; Wiberg, N.; Schurz, K.; Müller, G. *Z. Naturforsch., B: Chem. Sci.* **1989,** *44B*, 786.
(17) Boese, R.; Klingebiel, U. *J. Organomet. Chem.* **1986,** *315*, C17–C21.
(18) Walter, S.; Klingebiel, U. *Coord. Chem. Rev.* **1994,** *130*, 481.
(19) Hesse, M.; Klingebiel, U. *Angew. Chem.* **1986,** *98*, 638. *Angew. Chem., Int. Ed. Engl.* **1986,** *25*, 649.
(20) Stalke, D.; Keweloh, N.; Klingebiel, U.; Noltemeyer, M.; Sheldrick, G. M. *Z. Naturforsch., B: Chem. Sci.* **1987,** *42B, 1237.*
(21a) Grosskopf, D.; Marcus, L.; Klingebiel, U.; Noltemeyer, M. *Phosphorus, Sulfur, Silicon* **1994,** *97*, 113.
(21b) Grosskopf, D.; Klingebiel, U.; Belgardt, T.; Noltemeyer, M. *Phosphorus, Sulfur, Silicon* **1994,** *91*, 241.
(22) Weidenbruch, M.; Brand-Roth, B.; Pohl, S.; Saak, W. *J. Organomet. Chem.* **1989,** *379*, 217.
(23) Denk, M.; West, R.; Hayashi, R. K. *7th International Symposium on Inorganic Ring Systems,* Banff, Canada, *1994*. Denk, M.; Lennon, R.; Hayashi, R.; West, R.; Belyakov, A. V.; Verne, H. P.; Haaland, A.; Wagner, M.; Metzler, N. *J. Am. Chem. Soc.* **1994,** *116*, 2694. Denk, M.; Hayashi, R.; West, R. *J. Chem. Soc., Chem. Commun.* **1994,** 33.
(24) Wiberg, N.; Schurz, K. *J. Organomet. Chem.* **1988,** *341*, 145.
(25) Walter, S.; Klingebiel, U.; Schmidt-Bäse, D. *J. Organomet. Chem.* **1991,** *412*, 319.
(26) Stalke, D.; Pieper, U.; Vollbrecht, S.; Klingebiel, U. *Z. Naturforsch., B: Chem. Sci.* **1990,** *45B*, 1513.
(27) Wiberg, N.; Schurz, K.; Müller, G.; Riede, J. *Angew. Chem.* **1988,** *100*, 979. *Angew. Chem., Int. Ed. Engl.* **1988,** *27*, 935.
(28) Niesmann, J.; Grosskopf, D.; Klingebiel, U., II. *Münchner Silicontage, 1994.*
(29) Niesmann, J.; Klingebiel, U.; Rudolph, S.; Herbst-Irmer, R.; Noltemeyer, M. *J. Organomet. Chem.,* in press.
(30) Wiberg, N.; Preiner, G.; Schurz, K. *Chem. Ber.* **1988,** *121*, 1407.
(31) Raabe, G.; Michl, J. *Chem. Rev.* **1985,** *85*, 419.
(32) A. G. Brook, Ed. *Heteroatom Chemistry:* VCH: Weinheim, 1989; p. 105.
(33) Wiberg, N.; Preiner, G.; Karampatses, P.; Kim, Ch.-K.; Schurz, K. *Chem. Ber.* **1987,** *120*, 1357.
(34) Wiberg, N.; Karampatses, P.; Kim, Ch.-K. *Chem. Ber.* **1987,** *120*, 1213.
(35) Wiberg, N.; Karampatses, P.; Kim, Ch.-K. *Chem. Ber.* **1987,** *120*, 1203.
(36) Vollbrecht, S.; Klingebiel, U.; Schmidt-Bäse, D. *Z. Naturforsch., B: Chem. Sci.* **1991,** *46B*, 709. Klingebiel, U. *Chem. Ber.* **1978,** *111*, 2735.
(37) Bertrand, G.; Majoral, J.-P.; Baceiredo, A. *Acc. Chem. Res.* **1986,** *19*, 17.
(38) Zigler, S. S.; West, R.; Michl, J. *J. Chem. Lett.* **1986,** 1025.
(39) Sekiguchi, A.; Ando, W.; Honda, J. *Chem. Lett.* **1986,** 1029.
(40) Zigler, S. S.; Welsh, K. M.; West, R. *J. Am. Chem. Soc.* **1987,** *109*, 187.
(41) Welsh, K. M.; Michl, J.; West, R. *J. Am. Chem. Soc.* **1988,** *110*, 6689.
(42) Thompson, C.; Glidewell, C. *J. Comput. Chem.* **1983,** 4, 1.
(43) Gordon, M. S.; Schmidt, M. W.; Koseki, S. *Inorg. Chem.* **1989,** *28*, 2161.
(44) Underiner, G. E.; Tan, R. P.; Powell, D. R.; West, R. *J. Am. Chem. Soc.* **1991,** *113*, 8437.
(45) Underiner, G. E.; West, R. *Angew. Chem.* **1990,** *102*, 579. *Angew Chem., Int. Ed. Engl.* **1990,** *29*, 529.

(46) Allen, F. H.; Kennard, O.; Watson, D. G.; Brammer, L.; Orpen, A. G.; Taylor, R. J. *J. Chem. Soc., Perkin Trans. 2* **1987,** S1–S19.
(47) Hemme, I.; Schäfer, M.; Herbst-Irmer, R.; Klingebiel, U. *J. Organomet. Chem.* **1995,** *493*, 223.
(48) Maier, G.; Glatthaar, J.; Reisenauer, H. P. *Chem. Ber.* **1989,** *122*, 2403.
(49) Goldberg, N.; Iraqi, M.; Hrusak, H.; Schwarz, H. *Int. J. Mass Spectrom. Ion Processes* **1993,** *125*, 267.
(50) El-Shall, M. S. *Chem. Phys. Lett.* **1989,** *159*, 21.
(51) Elhanine, H.; Farrenq, R.; Guelachvili, G. *J. Chem. Phys.* **1991,** *94*, 2529. Elhanine, M.; Hanoune, B.; Guelachvili, G. *J. Chem. Phys.* **1993,** *99*, 4970.
(52) Bock, H.; Dammel, R. *Angew. Chem.* **1985,** *97*, 128. *Angew. Chem., Int. Ed. Engl.* **1985,** *24*, 111.
(53) Bock, H. *Angew. Chem.* **1989,** *101*, 1659. *Angew. Chem., Int. Ed. Engl.* **1989,** *28*, 1627.
(54) Maier, G.; Glatthaar, J. *Angew. Chem.* **1994,** *106*, 486. *Angew. Chem., Int. Ed. Engl.* **1994,** *33*, 473.
(55) Mullikan, R. S. *Phys. Rev.* **1925,** *26*, 319.

Silicon–Phosphorus and Silicon–Arsenic Multiple Bonds

MATTHIAS DRIESS

Anorganisch-chemisches Institut der Universität Heidelberg
Im Neuenheimer Feld 270, D-69120 Heidelberg, Germany

I

INTRODUCTION

Since the discovery of metastable alkylidenephosphanes (P=C, phosphaalkenes), i.e., compounds bearing phosphorus–carbon π-bonds, in the basic work of Dimroth and Hoffmann,[1a] Märkl,[1b] and Becker,[1c] it has been well established that phosphorus forms multiple bonds to most of the main group elements of the second and third row. Compounds having P=X and P≡X bonds and low-coordinate phosphorus are efficiently stabilized only by sterically demanding organic groups attached to the low-coordinate centers or by special electronic effects (π-conjugation). It has been clearly demonstrated, however, that phosphorus (like carbon) forms strong σ- and π-bonds relative to other members of its row. This carbon-like behavior of phosphorus was recognized in the heuristic terms of a diagonal relationship long ago.[2] Most of the resemblance observed in phosphorus and carbon chemistry may be explained by their similar spectroscopic electronegativity[3] (carbon 2.544, phosphorus 2.253). Thus, the relatively high electronegativity of phosphorus seems to result in its capacity to form stronger σ- and π-bonds compared with its electropositive neighbor silicon (1.916).

Despite the position of silicon in the periodic table, silicon chemistry differs tremendously from that of carbon. This was realized early in the development of organosilicon chemistry, owing to many failures in preparing stable, olefin-like, low-valent silicon compounds.[4] L. C. Allen has suggested that the spectroscopic electronegativity of atoms is a very useful third dimension of the periodic table, as it offers a chemically more reasonable order of the main-group elements. For example, silicon is more closely related to boron than to carbon.[3] Silicon represents the most electropositive member of the metalloid series (B, Si, Ge, As, Te); that is, it is intrinsically less likely to form multiple bonds to itself and other elements than is a "real" nonmetal. Moreover, quantum-chemical calculations have shown that silicon should form the weakest multiple bonds (X=X, X=Y) possible in the element series C, N, O, Si, P, S.[5] During the last ten years, however, significant developments in synthetic methods have provided many new stable compounds containing low-coordinate, heavier main-group elements originally thought to be nonexistent.[4,6] Perhaps the most spectacular breakthrough is represented by the synthesis of stable silylidenesilanes (disilenes), which have been known since the work of West *et al.* in 1981.[7] Their metastabilization was achieved by steric protection of the low-coordinate silicon atoms, i.e., by employing sterically demanding organic groups.

Thus far, it has been shown that alkylidenesilanes (Si=C, silenes),[8] iminosilanes (Si=N, silanimines),[9] silylidenephosphanes (Si=P, phosphasilenes),[10a,b,c] silylidenesulfanes (Si=S, thiasilanones),[11] and silylidenearsanes (Si=As, arsasilenes)[12] exist in the form of stable derivatives. Of special interest for comparisons were the Si=X bonds in which X represents a nitrogen atom. The synthesis of iminosilanes was first reported in the mid-1980s.[9] Owing to the high difference in the spectroscopic electronegativity of silicon (1.916) and nitrogen (3.066), the intrinsically more stable Si=N π-system is extremely polarized relative to a Si=Si bond and therefore undergoes rapid dimerization or polymerization. The high Lewis acidity of the low-coordinate silicon center, as in the case of alkylidenesilanes (Si=C), favors the formation of adducts with donor molecules[8] and therefore hinders the synthesis of "free" iminosilanes, which proceeds via salt elimination from suitable alkali metal–amido(halogensilanes). However, it has been shown that lithiumamido(fluorosilanes) may be regarded as masked Si=N derivatives, reacting like iminosilanes.[13]

The Si=P bond is significantly less polar than the Si=N bond, thereby providing greater kinetic stability and, at the same time, lower Lewis acidity on silicon. However, the isolation of moderately metastable phosphasilenes (Si=P) remained elusive for almost ten years after they were first observed in solution. In the case of Si=P bonds, it appeared that, steric protection on silicon was more important than on phosphorus. In 1991 a smooth method for the synthesis of phosphasilenes by thermal abstraction of LiF from lithium(fluorosilyl)phosphanylides was reported.[10b,14] Recently, it has been demonstrated that the analogous synthetic route is also suitable for the preparation of remarkably stable arsasilenes.[12,14]

The task of this review is to summarize theoretical and experimental aspects of phospha- and arsasilene chemistry.

II

THEORETICAL ASPECTS

A. *Molecular and Electronic Structures of Parent Compounds (Si=P, Si=As)*

Calculations (SCF/3-21G*, MP3/6-31G*) on Si=P multiple bonds were first reported by Gordon *et al.* in 1985.[15] The nominal double bond in closed-shell, planar $H_2Si{=}PH$ was found to be 13 kcal mol^{-1} more stable than that in its isomer silanediylphosphane $HSi(PH_2)$, with a barrier for the 1,2-hydrogen shift of 40 kcal mol^{-1}. The calculated Si=P bond length of 2.044 Å predicted the Si—P distance to be 9.4% shorter than the predicted single-bond value for $H_3Si{-}PH_2$ (2.258 Å), and the phosphorus to be quasi-nonhybridized (bond angle at phosphorus 93.7°). Recently, more sophisticated *ab initio* calculations at the MCSCF level of theory within the CASSCF formalism have been performed by Driess and Janoschek, employing MP2/DZ + POL, as well as HF/6-31G* basis sets.[16] This work includes results on electronically excited states, calculated by means of the CI(S)/6-31G* method. Of particular interest are the results of changes in net atomic charges, orbital energies, electronic transitions, and rotational barriers (π-bond strength) in going from $H_2Si{=}PH$ to the silylated system $H_2Si{=}P(SiH_3)$. Results of geometric optimization and Mulliken net atomic charges are given in Fig. 1.

As expected, the Si=P bond in $H_2Si{=}P(SiH_3)$ is significantly shorter than in $H_2Si{=}PH$ due to strengthening by hyperconjugation. Both compounds have singlet ground states without trans-bending. The relatively small bond angle at phosphorus in $H_2Si{=}PH$ (93.7°) and $H_2Si{=}P(SiH_3)$ (100°) may indicate that phosphasilenes can be regarded as adducts formed by silanediyl and phosphanediyl fragments.[16,17] Interestingly, the silyl

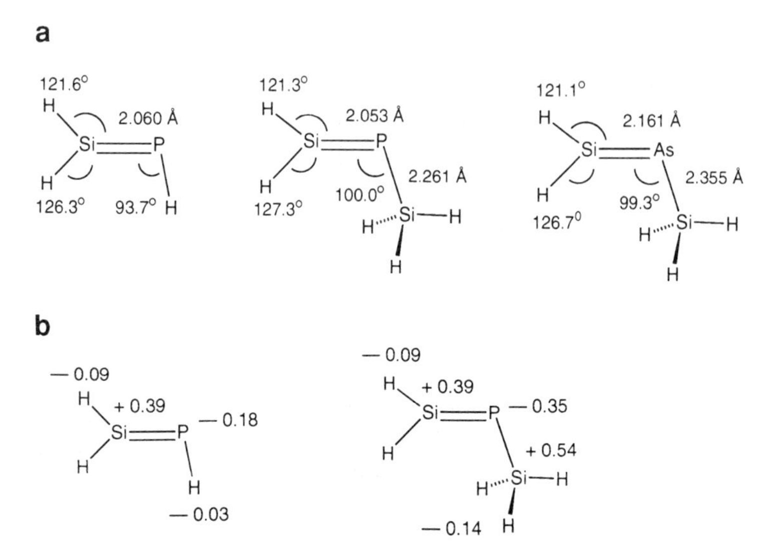

FIG. 1. (a) *Ab initio* calculated structures (HF/6-31G*) of $H_2Si{=}PH$, $H_2Si{=}P(SiH_3)$, and $H_2Si{=}As(SiH_3)$; (b) Mulliken net-atomic charges of $H_2Si{=}PH$ and $H_2Si{=}P(SiH_3)$. Data from Driess *et al.*[16]

group in $H_2Si{=}P(SiH_3)$ increases the negative charge on phosphorus by 94% relative to $H_2Si{=}PH$, although the charge at the low-coordinate silicon center remains almost unchanged (Fig. 1). Analogous calculations on the arsasilene $H_2Si{=}As(SiH_3)$ revealed molecular and electronic features similar to those found for its phosphorus homolog (Fig. 1). Differences of bond dissociation energies ΔH between Si=X and Si—X (X = P, As) clearly indicate that the σ-bond in Si=X is much weaker than in the corresponding saturated Si—X compounds.

The rotation barriers are substantially different for $H_2Si{=}PH$ and $H_2Si{=}P(SiH_3)$. The SiH_3 group significantly influences the transition structure. This structure is already reached after distortion of the Si=P bond to about 65° and requires 27 kcal mol^{-1}; further rotation to 90° leads to a minimum at 18 kcal mol^{-1}. This effect is caused by a favorable Si—H interaction between the low-coordinate silicon and a hydrogen atom of the silyl group during the rotational process. In contrast, the height of the Si=P barrier in $H_2Si{=}PH$ was found to be 34 kcal mol^{-1}, which is in agreement with values previously reported.[5,15,16] For $H_2Si{=}AsH$, the height of the rotation barrier is 30 kcal mol^{-1}. Furthermore, the calculations indicated that cis/trans isomerization could not proceed via phosphorus inversion, although this is observed for iminosilanes. The inversion barriers in $H_2Si{=}PH$ and $H_2Si{=}P(SiH_3)$ were calculated to be 52 kcal mol^{-1}.[16]

B. *Calculations of ^{31}P and ^{29}Si Chemical Shifts of Parent Phosphasilenes*

NMR chemical shifts of ^{31}P and ^{29}Si nuclei of the parent compounds $H_2Si{=}PH$ and $H_2Si{=}P(SiH_3)$ have been predicted by applying the IGLO method (individual gauge for localized orbitals).[16,18] Remarkably, the trend of different negative net charges on phosphorus, with nearly unchanged values on the low-coordinate silicon, in $H_2Si{=}PH$ and $H_2Si{=}P(SiH_3)$ is clearly reflected by the distinctly different ^{31}P chemical shifts ($H_2Si{=}PH$ $\delta = 85$; $H_2Si{=}P(SiH_3)$ $\delta = 48$) but little changed ^{29}Si chemical shifts ($\delta =$ 188 vs. 175).

III

TRANSIENT SILYLIDENEPHOSPHANES (PHOSPHASILENES)

The first evidence for a phosphasilene was reported in 1979.[19] Thermolysis of 1,2-phosphasiletane **1** at 100°C gave the transient silylidenephosphane $Me_2Si{=}PPh$ **2**, which undergoes head-to-tail [2 + 2]-cyclodimerization, as well as an insertion reaction into the Si—P bond in **1**, leading to **3** and **4** (Scheme 1).

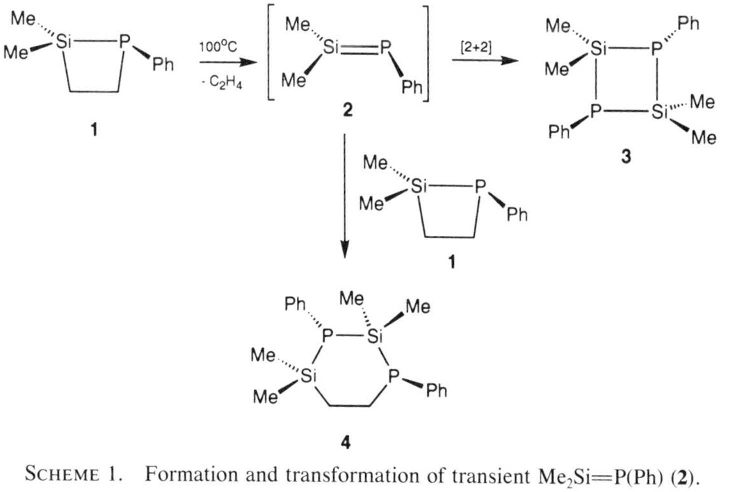

SCHEME 1. Formation and transformation of transient $Me_2Si{=}P(Ph)$ (**2**).

Other indications for the generation of phosphasilenes involved compounds **5**[20] and **6**.[21] Both derivatives were generated by thermally induced elimination of LiF/THF from corresponding lithium(fluorosilyl)phosphanylides. Whereas **5** was directly observed by means of ^{31}P-NMR spectroscopy, surprisingly, the sterically more hindered phosphasilene **6** was not detected due to its lower thermostability. Therefore, evidence for the existence of **6** was achieved solely by trapping experiments. Due to the more crowded substitution in **5** compared with **2**, the dimerization process to give **7** is relatively slow (Eq. 1) but is complete within a few hours.

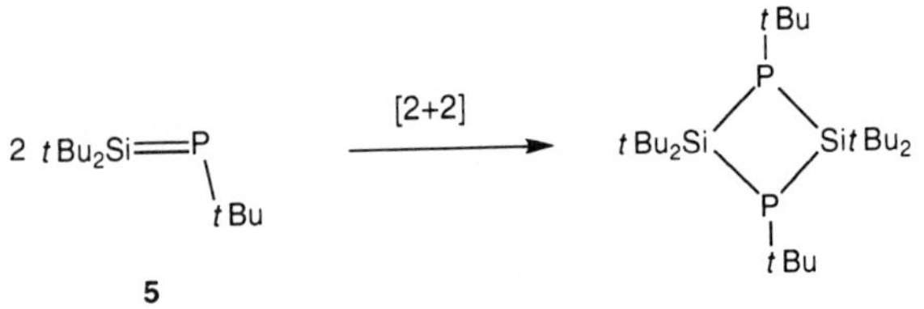

$$2\ t\mathrm{Bu_2Si{=}P}(t\mathrm{Bu})\ (\mathbf{5}) \xrightarrow{[2+2]} [t\mathrm{Bu_2Si}\text{—}\mathrm{P}(t\mathrm{Bu})]_2\ (\mathbf{7}) \qquad (1)$$

The existence of **6** was proved by trapping reactions, using the corresponding lithium(fluorosilyl)phosphanylide as precursor. Thus, thermolysis of the latter at 80–120°C in the presence of $t\mathrm{Bu_3Si{-}N_3}$, butadiene, isobutene, and lithium-tri-*tert*-butylsilylphosphanide led to the products **8–11** (Scheme 2).

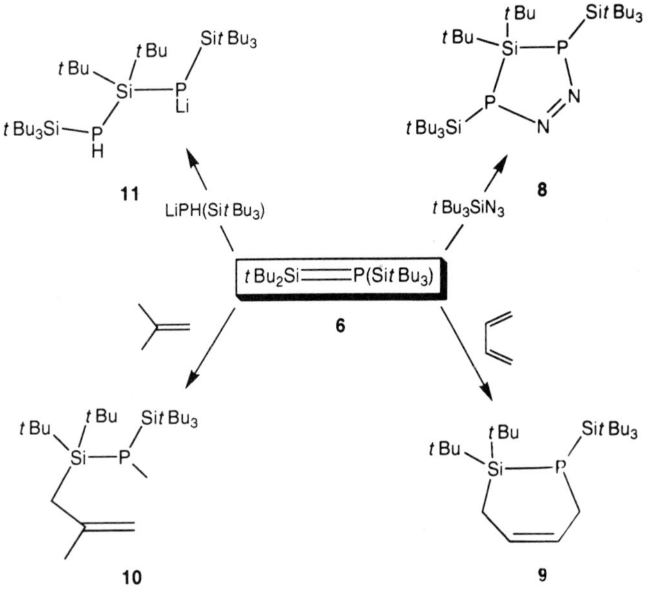

tBu$_2$Si=P(SitBu$_3$) **6**; reagents: LiPH(SitBu$_3$) → **11**; tBu$_3$SiN$_3$ → **8**; isobutene → **10**; butadiene → **9**

SCHEME 2. Trapping reactions of the transient phosphasilene $t\mathrm{Bu_2Si{=}P(Si}t\mathrm{Bu_3)}$ (**6**), leading to **8**, **9**, **10**, and **11**.

The formation of the hetero-Diels–Alder product **3** is remarkable, since the Si=N bond in iminosilanes does not yield analogous adducts.

IV

METASTABLE SILYLIDENEPHOSPHANES

A. *Synthesis of P-Organyl-Substituted Derivatives*

The first moderately stable phosphasilene derivatives **12** were synthesized by Bickelhaupt *et al.* in 1984 (Eq. 2).[10a] They possess limited stability (up to 60°C), and were characterized by NMR spectroscopy (see Section IV.C) and chemical reactions (see Section IV.E).

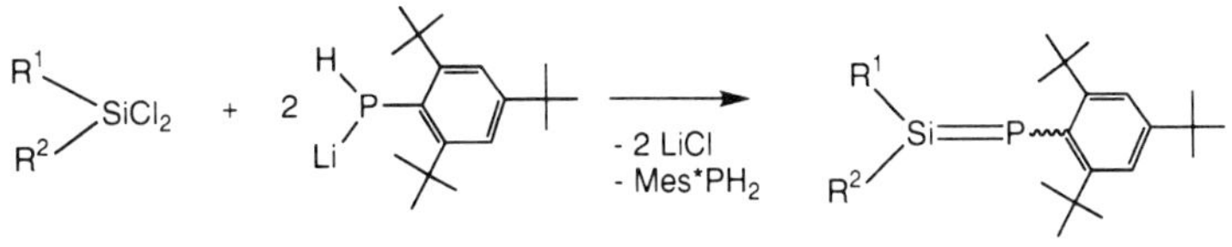

$R^1R^2SiCl_2$ + 2 Mes*P(H)Li → (− 2 LiCl, − Mes*PH_2) $R^1R^2Si{=}P{-}Mes^*$ **12** (2)

An alternative approach was achieved by the one-pot reaction of dichlorosilane with RPH_2 in the presence of two equivalents of *n*BuLi. However, neither synthetic method yields crystalline phosphasilenes. The first synthesis of the crystalline phosphasilene **13** was achieved by a two-step reaction of *t*$BuSiCl_3$ with 4 equivalents of RPHLi (R = 2,4,6-*t*$Bu_3C_6H_2$) and Ph_2PCl by Niecke *et al.* in 1993 (Scheme 3).[10c] This reaction proceeds

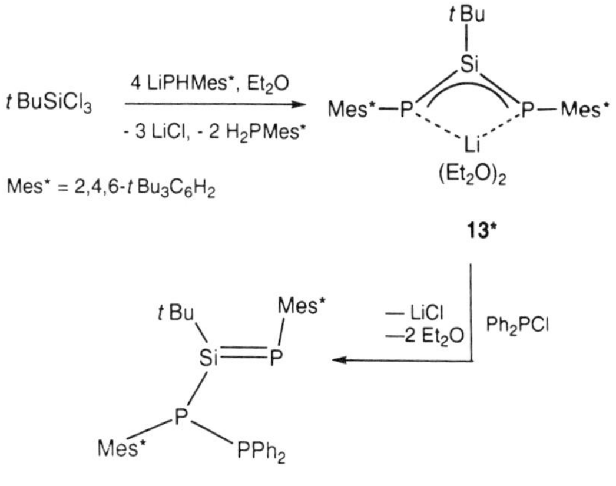

t$BuSiCl_3$ —(4 LiPHMes*, Et_2O; − 3 LiCl, − 2 H_2PMes^*)→ **13*** —(Ph_2PCl; — LiCl, —2 Et_2O)→ **13**

Mes* = 2,4,6-*t*$Bu_3C_6H_2$

SCHEME 3. Synthesis of the phosphasilene **13**, proceeding via the 1,3-diphospha-2-silaallylide derivative **13***.

via the remarkable diphosphasilaallyl salt **13***, which represents one of the rare examples of structurally characterized, conjugated sila-π-systems.[22]

Another slightly modified route to the phosphasilene **12** ($R^1 = R^2 =$ Mes) resulted from the reaction of Mes_2SiCl_2 with lithium(trimethylsilyl)-arylphosphanide (aryl = $tBu_3C_6H_2$), which was reported by Markovskij *et al.*[23] However, **12** could not be separated from the diphosphene Mes*P═PMes* by-product. A rather unusual type of phosphasilene, compound **14**, was prepared by Corriu *et al.*[24] Owing to the intramolecular N—Si donor–acceptor bond and the unusually large shielding of the ^{31}P and ^{29}Si nuclei in the NMR spectra, its true electronic structure seems better represented by the polar resonance form **14a**.

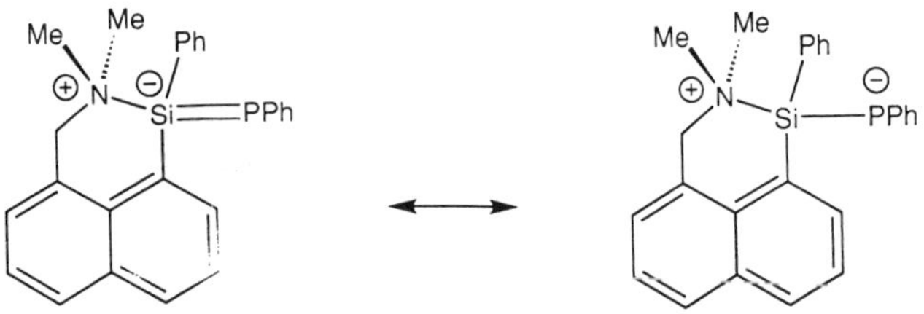

14 ⟷ **14a**

B. *Synthesis of* P-*Silyl/Germyl-Substituted Derivatives*

In order to overcome the difficulties of thermal instability, the phosphasilene derivatives **15**, which bear a silyl or germyl group attached to phosphorus, were synthesized. Indeed, they proved to have stronger Si═P bonds (stable up to 100°C), thus allowing studies of their structures and reactivity.[10b,14] Phosphasilenes **15a–15i** were synthesized from the corresponding lithium(fluorosilyl)phosphanides **16a–16i** by the thermally induced elimination of LiF (see Scheme 4).[10b] It has been shown that excellent steric protection of the highly reactive Si═P bond in **15** is provided by the 2,4,6-triisopropylphenyl (Is = isityl) substituent attached to the low-coordinate silicon center. The appropriate precursors **16a–16i** were synthesized in a multiple-step procedure, starting from **17** (Scheme 4).[10b,14]

Compounds **17** were formed by the reaction of $Is(R)SiF_2$ (R = Is, *t*Bu) with two equivalents of $[LiPH_2(DME)]$ (DME = 1,2-dimethoxyethane). They do not show any tendency to eliminate LiF in THF solutions, surely because of the strong Si—F bond. Similar stability was observed for the nitrogen analogs.[13]

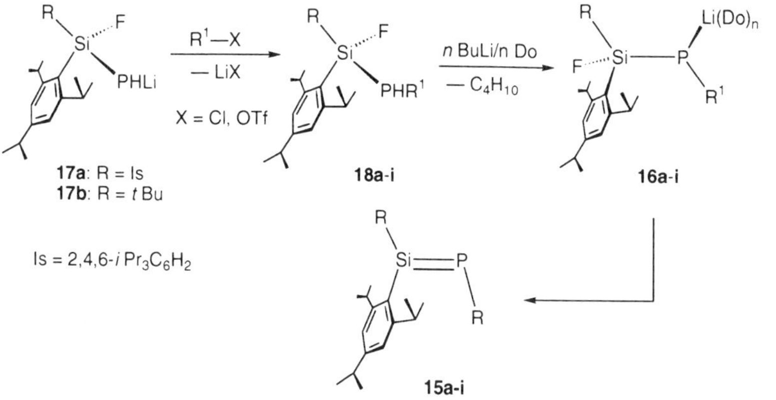

Is = 2,4,6-iPr$_3$C$_6$H$_2$

15,16,18	a	b	c	d	e	f	g	h	i
R	Is	Is	Is	Is	Is	Is	Is	Is	tBu
R^1	SiiPr$_3$	SiMe$_3$	SitBuMe$_2$	SiPh$_2$Me	SiPh$_3$	Si(Naph)$_3$	SitBu$_2$H	GetBu$_2$H	SiiPr$_3$

SCHEME 4. Synthesis of the stable *P*-silylated phosphasilenes **15a–i**, proceeding via **16–18**.

On silylation/germylation of **17**, compounds **18** were formed, which were subsequently lithiated on phosphorus to yield **16**. It appears that the success of the next step from **16** to **15** depends on different factors, but the following three parameters may be regarded as critical: (1) The reaction temperature for the elimination of LiF should not exceed 80°C; (2) the concentration of **16** dissolved in hexane or toluene should be about <0.1 mol L^{-1}; (3) the steric bulk and electronic influence of the substituent attached to phosphorus should be optimized. In the first place, the nature of the substituent at phosphorus determines the structures of the precursors **16**. This has been demonstrated by a study of a structure–reactivity

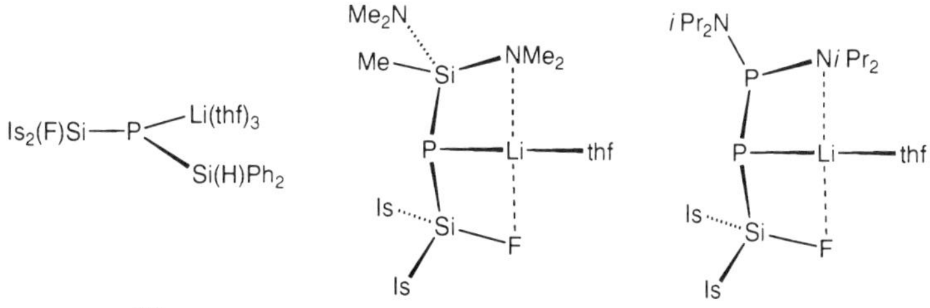

16j **16k** **16l**

relationship for several compounds **16**, including the derivatives **16j–16l**.[25] The latter investigation revealed that lithium phosphanides **16** are monomeric both in the solid state and in solution. In solution, the number (n) of the donor (Do) solvent molecules (THF) attached to the lithium center ($n = 1$–3) is strongly dependent on the nature of the substituent at phosphorus. It was further observed that phenyl-substituted silyl groups attached to phosphorus in **16** cause a $P(n)/\pi^*$-hyperconjugation that seems to stabilize the trigonal-planar geometry around phosphorus and results in remarkably short Si—P distances in **16j** (2.16, 2.18 Å) (see Fig. 2). The results of X-ray structure determinations of the related phosphanides **16a** and **16f** are depicted in Figs. 3 and 4.

In contast to the planar coordination geometry on phosphorus in **16j**, the geometry of the P atoms in **16a** and **16f**[14] is pyramidal and the sum of the phosphorus bond angles is 326.0 and 350.4°, respectively. For **16l**[25] the sum of the bond angles at phosphorus is 238.4°. Evidently, the sterically demanding tri(1-naphthyl)silyl group at phosphorus in **16f** prevents an effective hyperconjugative interaction, which seems to be crucial for favoring the trigonal-planar arrangement around phosphorus.

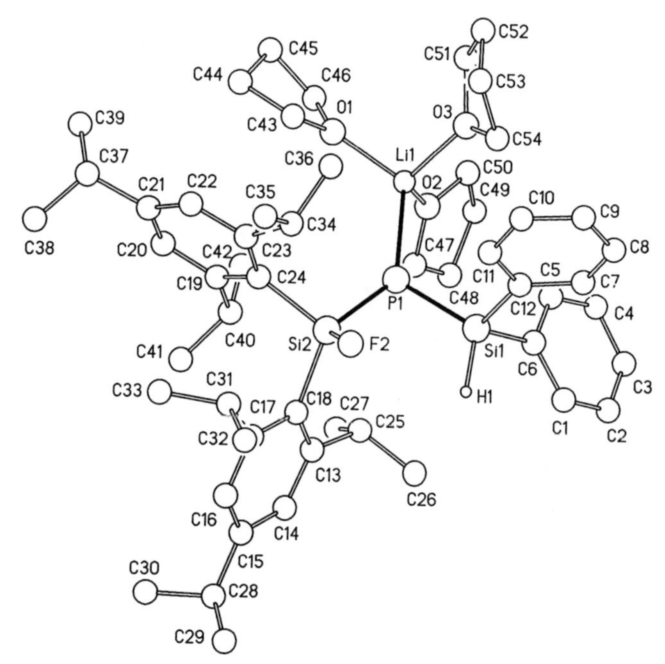

FIG. 2. Solid-state structure of **16j**. [Reproduced with permission of Driess *et al.*[25]]

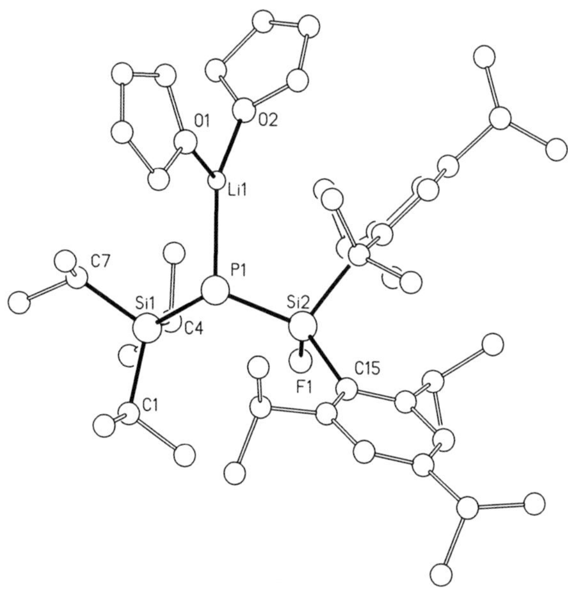

FIG. 3. X-Ray crystal structure of **16a**. Bond distances: Si—F 1.618(3), Si1—P1 2.180(3), Si2—P1 2.218(3), Li—P1 2.39(2) Å.

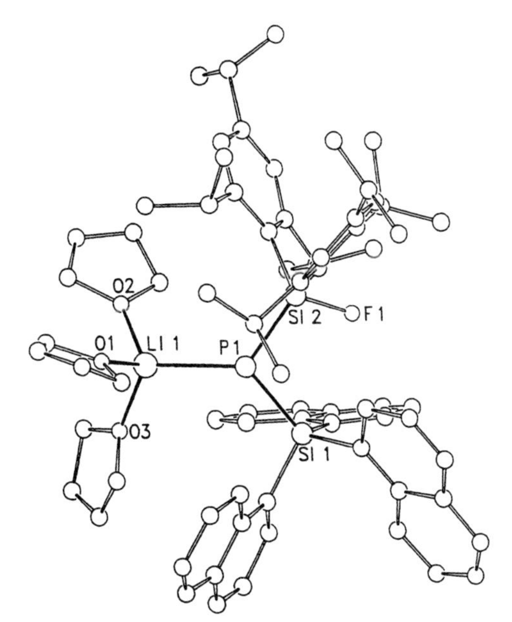

FIG. 4. X-Ray crystal structure of **16f**. [Reproduced with permission of Driess *et al.*[14]] Bond distances: P1—Si1 2.205(4), P1—Si2 2.195(4), P1—Li1 2.65(2), Si2—F1 1.623(6).

Surprising structural facts were also observed in the X-ray crystal structure determination of the di-*tert*-butylgermyl-substituted lithium phosphanide **16h** (see Fig. 5).[14] In the latter compound, the lithium center that is chelated by one molecule of DME is simultaneously attached to the phosphorus and fluorine atoms, as observed in **16k, l**. The result is a large pyramidalization on phosphorus (sum of bond angles 268.4°). The Li—F distance is 1.983(6) Å, significantly shorter than that in **16l** (2.060(8) Å). The phosphasilenes **15a–i** were formed by heating solutions (60–80°C) of the corresponding derivatives **16**. With the exception of **15b** (R = $SiMe_3$) and **15e** (R = $SiPh_3$), these derivatives have been isolated as yellow or orange-red oils and, in the case of **15i**, as yellow crystals.[26]

C. *Spectroscopic Properties and Cyclovoltammetric Measurements*

The derivatives **15a–i** exhibit characteristic ^{31}P NMR spectroscopic data (see Table I), which are distinctly different from those observed for the *P*-organo-substituted derivatives **12** and **13** ($\delta(^{31}P)$ = 65.8–136.0).[27] However, the similar ^{29}Si chemical shifts and ^{1}J(Si, P) coupling constants observed for **12**, **13**, and **15** clearly show the identical electronic nature of the low-coordinate silicon centers in these derivatives.

Especially noteworthy is the relative large shielding for the ^{31}P nucleus in the ^{31}P NMR spectra of **15**, which is quite unusual for two-coordinate phosphorus. Evidently, the latter is caused by the strong σ-donor ability of the silyl and germyl groups, which is also reflected in the calcu-

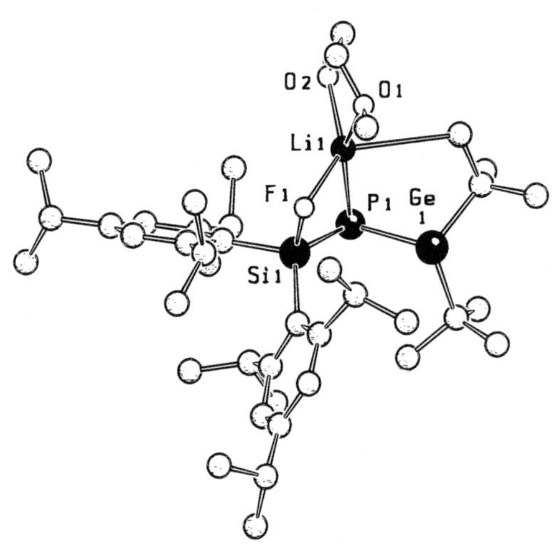

FIG. 5. Solid-state structure of **16h**. [Reproduced with permission of Driess *et al.*[14]]

lated ^{31}P chemical shifts for the parent compounds $H_2Si{=}PH$ ($\delta = 85$) and $H_2Si{=}P(SiH_3)$ ($\delta = 48$) (see Section II.A). In the series of phosphasilenes containing two isityl groups, the greatest ^{31}P shielding is found for **15g** ($\delta = -7.8$). An even more strongly shielded ^{31}P-nucleus was observed for **15i** ($\delta = -29.9$), where one isityl group was replaced by a *tert*-butyl group.[26] This unusual finding for **15g** and **15i** indicated that both steric and electronic effects may have a considerable influence on the ^{31}P chemical shifts, although electronic effects appear to be more important.

The ^{29}Si-NMR spectra of **15a–h** (Table I) exhibit doublet signals for the low-coordinate Si atoms at very low field ($\delta = 167.8$–181.7).[10b,14] The unusually large shielding of the phosphorus nucleus in **15i** evidently corresponds to the large deshielding of the three-coordinate silicon atom at $\delta = 213.2$, the lowest value known to date in the Si=P series.[26] The $^1J(Si, P)$ coupling constants of **15a–h** (149–160 Hz) are diagnostic for Si—P π-bonds, whereas much smaller values were obtained for saturated silylphosphanes.[28] In contrast, the exceptionally large $^1J(Si,P)$ value of 203 Hz of **13** indicates the special electronic nature of this derivative. This huge value may be interpreted as the consequence of a strong interaction of

TABLE I

^{31}P AND ^{29}Si NMR DATA FOR **15a–i**

	$\delta(^{31}P)$	$\delta(^{29}Si)$	$^1J(Si{=}P)$ (Hz)	$^1J(Si{-}P)$ (Hz)
15a	11.1[a]	167.8 (d, Si=P)	160	—
		20.8 (d, Si—P)	—	78
15b	28.1[a]	172.4 (d, Si=P)	152	—
		21.4 (d, Si—P)	—	75
15c	17.7[a]	178.4 (d, Si=P)	149	—
		14.2 (d, Si—P)	—	72
15d	6.8[b]	180.6 (d, Si=P)	154	—
		−3.7 (d, Si=P)	—	70
15e	3.5[b]	181.7 (Si=P)	155	—
		−5.0 (d, Si—P)	—	71
15f	14.4[b]	not observed	n.o.	—
15g	−7.8[b]	179.6 (d, Si=P)	153	—
		17.1 (d, Si—P)	—	76
15h	4.0[b]	178.9 (d, Si=P)	154	—
15i	−29.9[c]	213.2 (d, Si=P)	161	—
		21.0 (d, Si=P)	—	75

[a] Bickelhaupt *et al.*[10a]
[b] Driess *et al.*[14]
[c] Driess *et al.*[26]

the Si=P bond with the lone pair at the λ^3,σ^3-coordinate phosphorus attached to silicon. Such an interaction is described in terms of a second-order Jahn–Teller distortion.[29] In line with this, the molecular structure of **13** (see Section IV.D) shows unusual features.

The resonance signals of the ^{29}Si nuclei of the SiR_3 groups lie in the expected range for saturated silicon compounds, although the ^1J(Si, P) coupling constants are found to be significantly larger (70–78 Hz) due to the low coordination of phosphorus. Furthermore, all ^{31}P- and ^{29}Si-chemical shifts appear to be significantly temperature-dependent.

Compounds **15a**, **15c**, and **15d** were also investigated by UV/visible spectroscopy.[14] The λ_{max} values obtained are as follows: 343 (**15a**, log $\varepsilon = 2.0$), 338 (**15c**, log $\varepsilon = 2.7$), and 331 nm (**15d**, log $\varepsilon = 2.2$). These values probably correspond to n–π^* transitions. This assignment is consistent with the calculated values for electron transitions of $H_2Si=P(SiH_3)$[14,16]: 271 nm for the n–π^* transition and 266 nm for the π–π^* transition.[16] Hypsochromic shifts of π–π^* transitions in comparison to n–π^* transitions were also found for phosphaalkenes (P=C).[30]

Cyclovoltammetric measurements were performed for the derivatives **15a** and **15d**. Interestingly, a reversible one-electron reduction was observed (−80°C, DME, nBu$_4$NPF$_6$) at $E = -1.81$ (**15a**) and −1.74 V (**15d**).[14] In contrast, the analogous arsasilenes (Si=As) undergo only irreversible reduction. Compounds **15a,d** show a further one-electron reduction, which is irreversible. This remarkably different reduction behavior of Si=P and Si=As bonds is probably caused by the small, but significant difference in the electronegativities of phosphorus and arsenic.

D. *X-Ray Structure Determinations*

Compound **13** was the first crystalline phosphasilene derivative whose molecular structure was elucidated by X-ray diffraction (Fig. 6a,b).[10c] This compound exists in the (E)-isomeric form of a 1,3,4-triphospha-2-silabut-1-ene. The most interesting features of this structure, however, are the rather long Si=P distance (2.094(3) Å) and the significant pyramidalization of the low-coordinate silicon atom (sum of silicon bond angles 356.7°). Theoretical investigations predicted shorter lengths (2.04–2.06 Å) and trigonal-planar geometry around the low-coordinate silicon for the parent compounds $H_2Si=PH$ and $H_2Si=P(SiH_3)$ (see Section II.A).[16] It was proposed that this distortion is due to steric crowding and/or a n–π^* interaction between the P2 atom and the Si=P bond. However, this structural distortion may be better explained in terms of a second-order Jahn–Teller

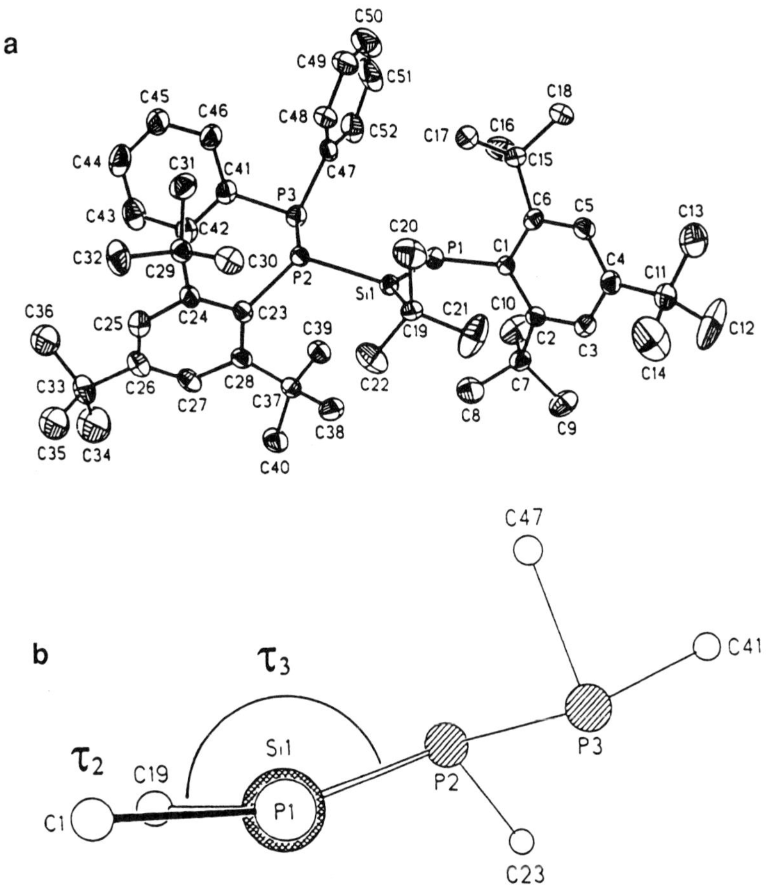

FIG. 6. (a) X-Ray crystal structure of **13**. Bond distances: Si1—P1 2.094(3), Si1—P2 2.254(3), P2—P3 2.251(2) Å; bond angle: C1—P1—Si1 104.2(2)°. (b) View of **13** along the Si=P vector; $\tau 2$ torsion angle C1—P1—Si1—C19, 3.4(3)°; $\tau 3$ C1—P1—Si1—P2, 161.0(2)°. [Reprinted with permission of Niecke *et al.*[10c] Copyright (1993) American Chemical Society.]

distortion, involving mixing of the HOMO (π Si=P) and LUMO (σ^* Si—P) of **13**, which are separated only by a small energy gap.[29]

As discussed in Section II.A, theoretical studies predicted that phosphasilenes with silyl substituents attached to phosphorus should have planar, trigonally coordinated silicon, with the Si—P π-bonds strengthened by the hyperconjugative influence of the silyl group.[16] Recently, this was proved by a single-crystal X-ray structure determination of the derivative

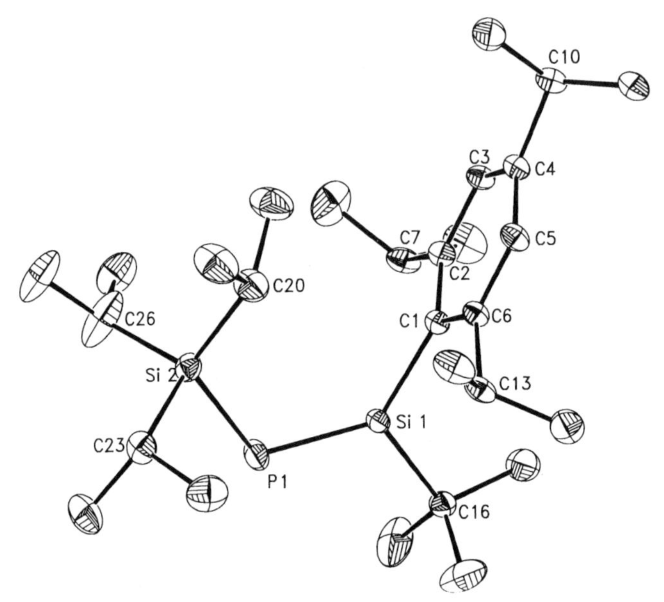

FIG. 7. Solid-state structure of **15i**. Bond distances: Si1—P1 2.062(1), Si2—P1 2.255(1) Å; bond angle at phosphorus: 112.79(2)°. [Reproduced with permission of Driess *et al.*[26]]

15i (Fig. 7).[26] Indeed, the Si═P distance in **15i** (2.062(1) Å) is significantly shorter than that in **13**, and the low-coordinate silicon is planar coordinated. The Si2—P1 single bond length (2.255(1) Å) is in the range for "normal" silylphosphanes. Both distances are in perfect agreement with values obtained from *ab initio* calculations of high theoretical level. Consistent with the results of ^{31}P-CP-MAS solid-state and solution NMR spectroscopy of **15i**, the X-ray analysis proved that the (*Z*)-isomer was formed exclusively. Particularly remarkable is the angle at the two-coordinate phosphorus (112.79(2)°), which is considerably larger than the value predicted for $H_2Si{=}P(SiH_3)$ (100°) and that observed in **13** (104.2°). This seems surprising in view of the Si═P bond length in **15i**, which is not significantly elongated by steric hindrance. Also of considerable interest are the Si2—P1—Si1—C1 (−13.7°) and Si2—P1—Si1—C16 torsion angles (169.5°) (Fig. 8) and the distinct P1—Si1—C1 (131.36(9)°) and P1—Si1—C16 bond angles (114.59(7)°). The latter surely differ for steric reasons. The geometric situation in **15i** compared to that in **13** clearly indicates that the Si═P distance and the coordination geometry at silicon are more strongly influenced by electronic than steric effects.

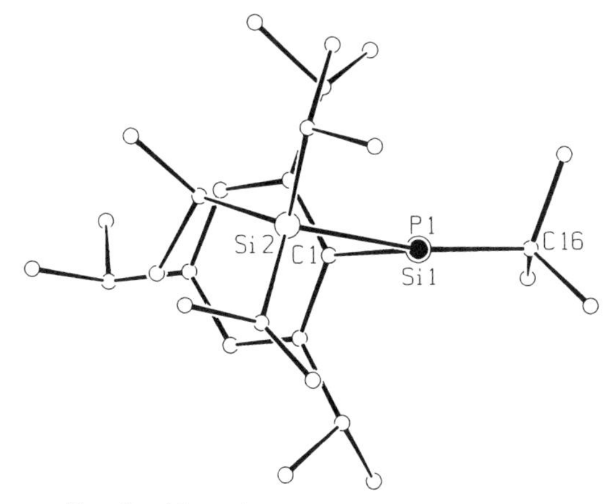

FIG. 8. View of **15i** along the Si═P vector.[26]

E. *Reactivity*

Only a few chemical reactions of the metastable phosphasilenes Mes*P═SiR^{1}R^2 (**12**) have been reported. The remarkably high thermal stability (up to 100°C) of the phosphasilenes Is(R^1)Si═PR^2 (R^1 = Is, *t*Bu; R^2 = SiR$_3$, GeR$_3$) (**15a–i**, except for **15b** and **15e**) has made possible extensive studies on the reactivity of the Si═P bond.[31]

1. *Thermolyses*

The thermal decomposition of Mes*P═Si(*t* Bu)Is (**12a**) liberates isobutene to form the bicyclic compound **19**[10a] (Eq. 3).

$$\mathbf{12a} \xrightarrow[-\ C_4H_8]{\Delta} \mathbf{19} \qquad \text{Is} = 2,4,6\text{-}i\text{Pr}_3\text{C}_6\text{H}_2 \tag{3}$$

The analogous decomposition process was observed for related germylidenephosphanes.[32] Although the thermolysis of **12a** was carried out in deuterated solvents, no deuterium incorporation was observed. Further work is needed in order to explore the reaction mechanism. In contrast to **12a**, the thermolysis behavior of phosphasilenes **15** is quite different.

For instance, on heating **15a**, the bicyclic compound **20** was formed (Eq. 4).

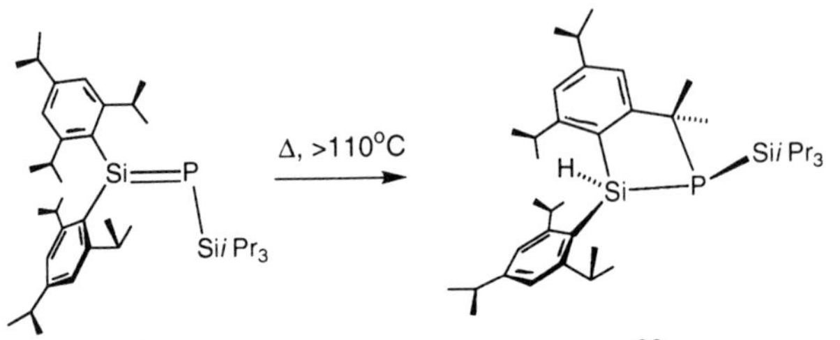

(4)

This reaction proceeds via an intramolecular hydrogen shift, that is, insertion of the Si—P π-bond into the C—H bond of the ortho isopropyl group of one isityl substituent.

2. *Protolysis Reactions*

As expected, the high reactivity of the polar Si=P bond favors protolysis reactions with HX (X = OH, OMe, Cl), in which the phosphorus is protonated and the silicon center bonds to X (Eq. 5).[10a]

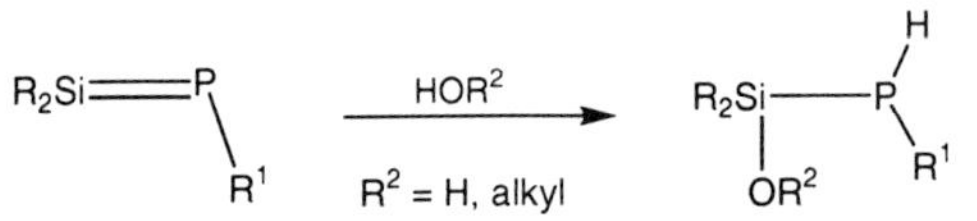

(5)

3. *Reactivity toward White Phosphorus, Elemental Sulfur, and Tellurium*

Besides disilenes, the metastable phosphasilenes **15** (and the analogous arsasilenes) are the only known compounds with $(p–p)$ π-bonds between main-group elements that react with P_4 under relatively mild conditions (40°C).[10b,33] Thus **15a** reacts with P_4 in a 2 : 1 molar ratio to build up the SiP_3 butterfly-like compound **22** (Eq. 6).

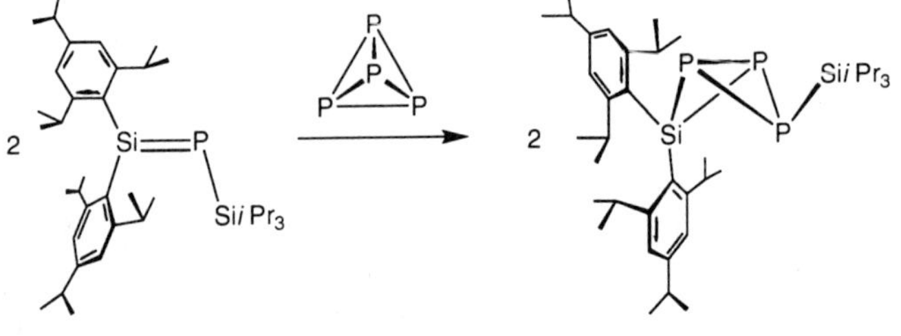

(6)

The ^{31}P- and ^{29}Si-NMR specta of **22** confirmed that the exo configuration is preferred at 25°C.[14] At higher temperatures (above 38°C), inversion of the configuration of the peripheral phosphorus atom was observed in the ^{31}P-NMR spectrum; that is, **22** rearranges into the endo isomer **22*** and vice versa (Scheme 5).

For this process an inversional barrier of 23 kcal mol^{-1} has been estimated. Following results from MO calculations for P_2Si_2 bicyclo[1.1.0]butanes,[34] we assume that the SiP_3 system also has a very high barrier to ring inversion (see Scheme 5). Therefore, the temperature dependence of the ^{1}H-, ^{31}P-, and ^{29}Si-NMR spectra of **22** seems to be caused by dynamic effects of the isityl groups at silicon (hindered rotation of aryl rings), as well as phosphorus inversion of the peripheral P atom.

The oxidation of the phosphasilenes **12a** and **15a** with elemental sulfur and tellurium led the corresponding chalcogen heterocycles **23** (Eqs. 7, 8).[10a,14]

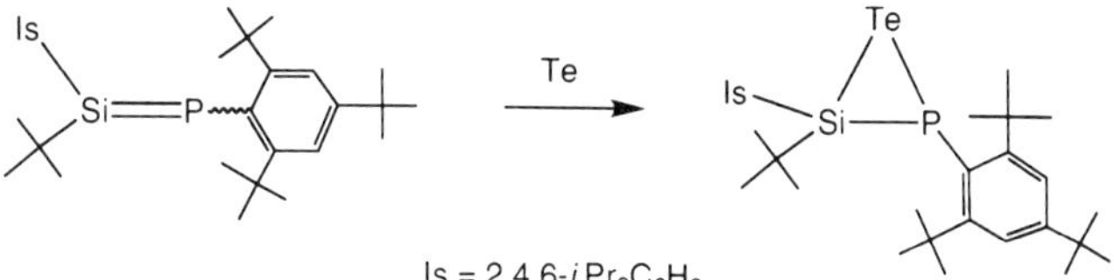

12a $\xrightarrow{Te}$ **23a** (7)

Is = 2,4,6-i-$Pr_3C_6H_2$

15a $\xrightarrow{S_8,\ Te}$ **23b,c** (8)

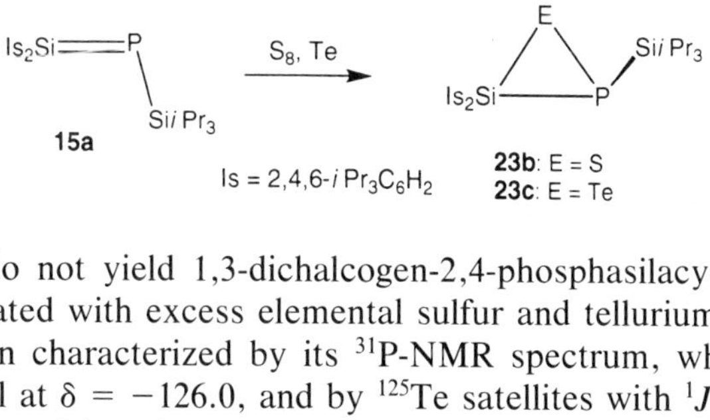

Is = 2,4,6-i-$Pr_3C_6H_2$

23b: E = S
23c: E = Te

The latter do not yield 1,3-dichalcogen-2,4-phosphasilacyclobutanes if they are treated with excess elemental sulfur and tellurium. Compound **23a** has been characterized by its ^{31}P-NMR spectrum, which shows a singlet signal at $\delta = -126.0$, and by ^{125}Te satellites with $^{1}J(P,Te) = 222$ Hz. The composition of **23b,c** was proved by mass spectrometry and their

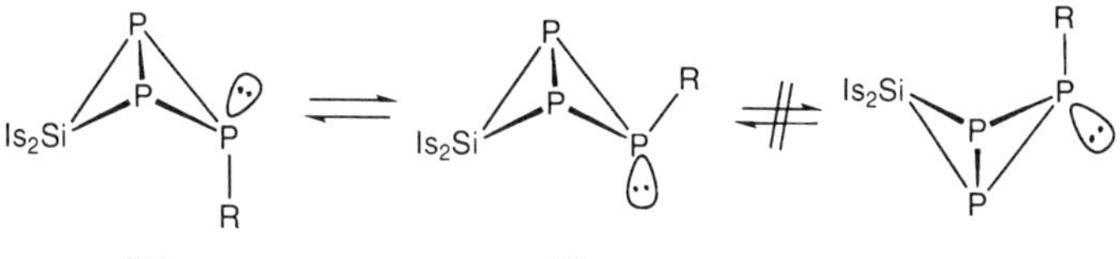

22* ⇌ **22** ⇸ **22***

SCHEME 5. Isomerization of **22** by phosphorus inversion.

constitution was deduced from ^{31}P-, ^{29}Si-, and ^{125}Te-NMR spectroscopic data.[14] Interestingly, the shielding of the ^{31}P nucleus of **23b** ($\delta = -267.6$) and the value of the ^{1}J(P,Te) coupling constant (488 Hz) is much larger than that observed for **23a**.[10a]

4. *Reactions with Alkynes and Cyclopentadiene*

The behavior of the Si—P π-bond toward a C≡C triple bond was examined in the case of **15a** by employing differently substituted alkynes.[14] It appeared that **15a** does not react with dialkyl, diaryl-, or disilyl-substituted alkynes at 110°C; even cyclooctyne, usually a very reactive alkyne, does not react. However, when **15a** was stirred with phenylacetylene at 80°C in toluene, the C—H insertion product **24** was isolated as colorless crystals (Eq. 9).[14] Its molecular structure has been elucidated by single-crystal X-ray diffraction (Fig. 9).

$$Is_2Si{=}P{-}Si\,iPr_3 \ (\mathbf{15a}) \xrightarrow{HC\equiv CPh} Is_2Si(C{\equiv}CPh){-}P(H){-}Si\,iPr_3 \ (\mathbf{24}) \ \overset{AIBN}{\not\longrightarrow} \ \mathbf{24^*} \quad (9)$$

(**24***: four-membered ring Is$_2$Si—P(SiiPr$_3$)—C(H)=C(Ph)—)

Consistent with the bond polarity in the Si=P bond (Si$^+$—P$^-$), the acetylene moiety appears to add to the Si atom while the phosphorus is protonated. The sterically demanding SiiPr$_3$ group at phosphorus causes an unusually large Si1—P1—Si2 bond angle (116.0(2)°). It is interesting that the reaction of tetramesityldisilene (Si=Si) with phenylacetylene yielded exclusively the [2 + 2]-cycloaddition product 1,1,2,2-tetramesityl-3-phenyldisilacyclobut-3-ene.[35] Experiments to convert **24** into the corresponding cyclic silaphosphaheterocyclobut-3-ene **24*** (Eq. 9), by using azodiisobutyronitrile (AIBN) as radical initiator for hydrophosphination, were unsuccessful.

The ability of the Si=P bond to serve as a dienophile for a [2+4]-cycloaddition (hetero-Diels–Alder reaction) has been verified by the reaction of **15a** with cyclopentadiene (Eq. 10).[14] The components react at 60°C in benzene to give a quantitative yield of **25**, which was isolated as a colorless solid. Interestingly, a Brønsted acid (C—H)/base (Si=P) reaction, as in the case of **15a** and phenylacetylene, was not observed. In the mass spectrum of **25**, the molecular peak but also the "free" phosphasilene **15a**$^+$ have been detected. Similar behavior was observed for related ad-

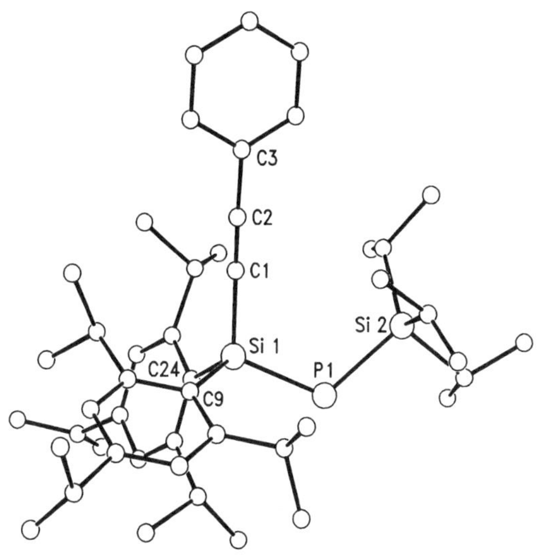

FIG. 9. X-Ray crystal structure of **24**. [Reproduced with permission of Driess et al.[14]]

ducts (like those from Ge=P bonds onto heterobutadienes[32]) in the mass spectrometer.

$$\underset{\textbf{15a}}{Is_2Si{=}P{-}Si\,iPr_3} \xrightarrow{C_5H_6} \textbf{25} \qquad (10)$$

(**25**: bicyclic adduct bearing $SiIs_2$ and $P{-}Si\,iPr_3$ units.)

5. *Reactions with Benzophenone, 1,2-Diphenyl-1,2-diketone, and 3,5-Di*-tert-*butyl-1,2*-o-*quinone*

The Si=P bond in **15a** readily reacts with benzophenone at −80°C to form the corresponding 4-oxa-1-phospha-2-silacyclobutane (**26**).[14] In contrast, no reaction of the Si=P bond occurs with tBu$_2$C=O. The molecular structure of **26** was determined by X-ray analysis (Fig. 10), which revealed that the compound is isotypic to the arsenic homolog.[12] The four-membered SiOPC framework is puckered (16.9° along the SiC axis) and trapezoidally distorted due to the different bond lengths in this skeleton. A remarkably long P—C distance (1.945(6) Å is observed, which is surely

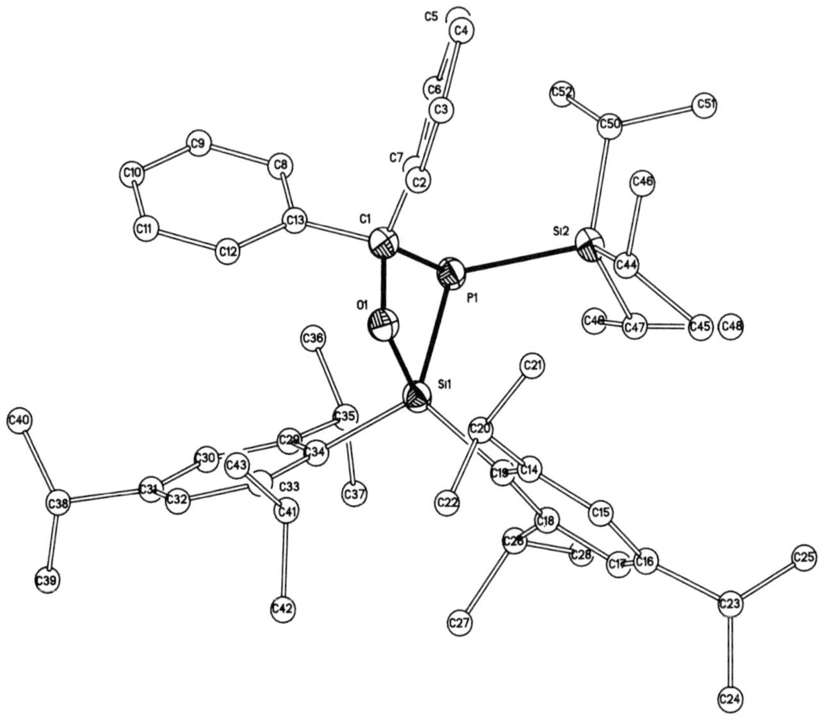

FIG. 10. Solid-state structure of **26**. [Reproduced with permission of Driess *et al.*[14]]

caused by steric hindrance. Usually, endocyclic P—C bonds in 1,3-diphosphacyclobutanes are in the range of 1.87–1.91 Å.[36] The sum of the bond angles at phosphorus in **26** is 290.8°; that is, the P atom is pyramidal.

It appeared that if heated at 160°C in a sila-Wittig-type reaction, **26** actually decomposes into the transient silanone $Is_2Si{=}O$, which immediately dimerizes to **27** and the phosphaalkenes **28** and **28*** (Scheme 6).[14]

Compounds **27** (1,3,2,4-disiladioxetanes) are well known from the work of West *et al.*[37] Due to drastic reaction conditions, it is understandable that **28** ($\delta = 293.7$) partially isomerizes into **28*** ($\delta = 281.9$). However, such an isomerization process was previously unknown for phosphaalkenes.

Furthermore, the Si=P bond possesses a remarkable reactivity toward diphenyldiketone. When **15a** was allowed to react with PhC(O)—C(O)Ph at 25°C, surprisingly, the [2+4]-cycloadduct **29** ($\delta(^{31}P) = 116.3$) was formed only in 10% yield (Scheme 7).[14] As the major product of the latter reaction, the [2+2]-cycloadduct **30** ($\delta(^{31}P) = -32.9$) was isolated in diastereomerically pure form as colorless plates.

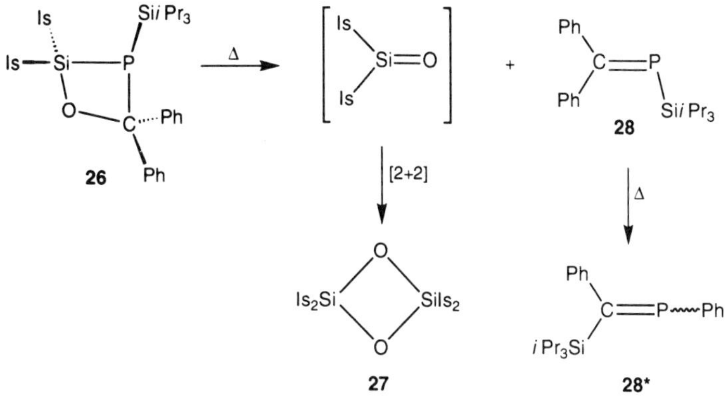

SCHEME 6. Transformation of **26** into the 1,3,2,4-dioxadisiletane **27** and the diastereomeric phosphaalkenes **28** and **28***, respectively.

On heating of the reaction mixture at 110°C, compound **29** was completely rearranged into **30** (^{31}P-NMR); that is, **30** is the thermodynamic product. It has been postulated that this preference is due to ring strain (C=C bond in the six-membered C_2O_2SiP ring). Interestingly, the formation of the four-membered CPSiO framework, as in **30**, could be suppressed if **15a** was reacted with 4,6-di-*tert*-butyl-*o*-quinone (Scheme 7).[14] The resulting benzo-condensed heterocycle **31** shows a ^{31}P-chemical shift

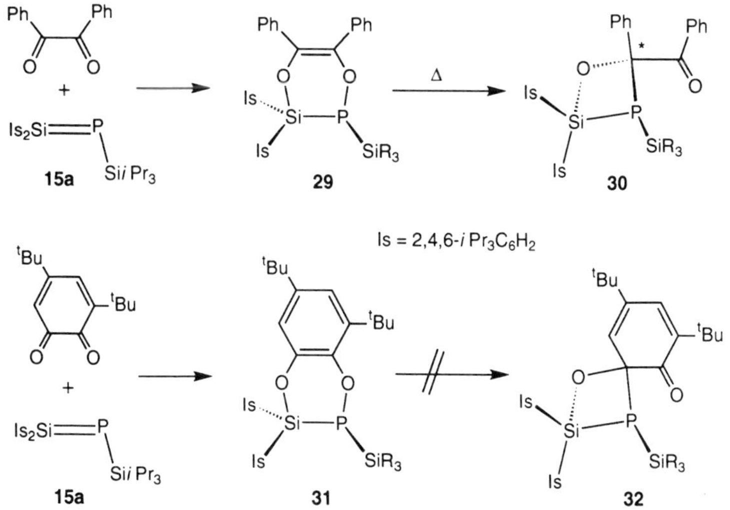

Is = 2,4,6-*i* $Pr_3C_6H_2$

SCHEME 7. Synthesis of **29**, **30**, and **31**.

of 114.3, which is almost identical to the value (116.3) observed for **29**. An isomerization of **31** (ring contraction) into **32** would be thermodynamically very unfavorable due to loss of the aromatic π-system.

6. *Cycloaddition Reactions with Mesityl Azide, Diphenyldiazomethane, Isocyanides, and Benzonitrile*

Compound **15a** reacts with mesitylazide readily at $-80°C$ in toluene, producing a deep red-purple solution. The ^{31}P-NMR spectrum of this solution revealed that the adducts **33** and **34** were produced (Eq. 11).[14]

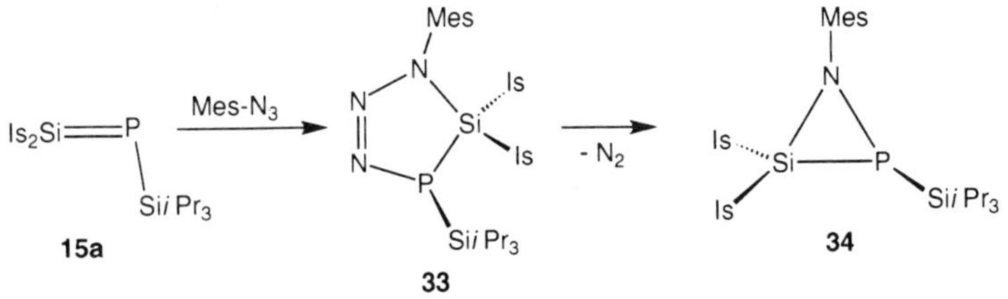

(11)

Compound **33** was completely transformed into **34** on heating of the product mixture at 40°C.

Surprisingly, in contrast to the reaction of the Si=P bond with mesityl azide, the reaction of **15a** with diphenyldiazomethane resulted in the formation of the [2+1]-cycloadduct **35** (Scheme 8).[38] The bonding situation in **35** (Fig. 11) may be described in terms of a π-complex, by employing the Dewar–Chatt–Duncanson model, in which the Si=P bond acts as π donor and acceptor at the same time (Scheme 9). The corresponding [2+3]-cycloaddition product **36** was generated only on thermal activation of **35**.

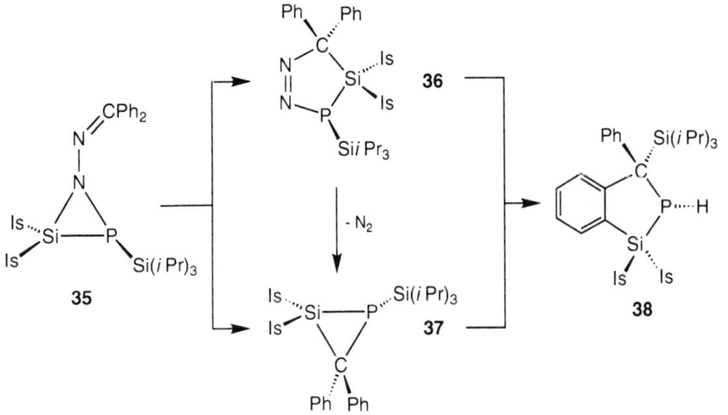

SCHEME 8. Formation of **35** and its rearrangement reactions into **36**, **37**, and **38**.

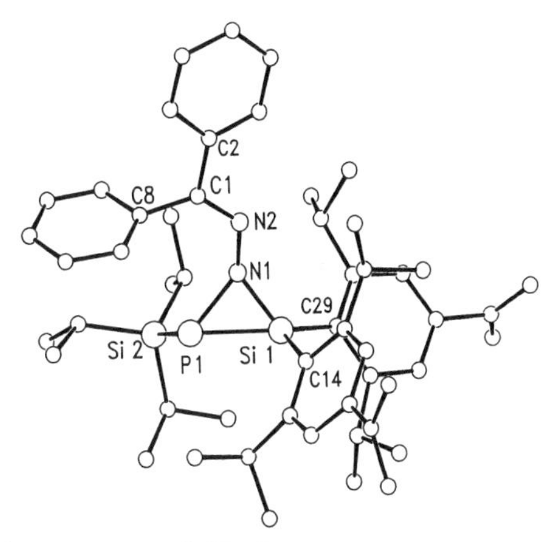

FIG. 11. Molecular structure of **35**. [Reproduced with permission of Driess and Pritzkow.[38]]

Further heating of **36** at 110°C liberated N_2 and gave the heterocyclopropane **37**, which finally gave the unusual constitutional isomer **38** in quantitative yield (Scheme 8).[38] The structure of **38** was also established by X-ray diffraction analysis (Fig. 12).

The Si=P bond in **15a** also reacted with mesityl isocyanide.[39] In this process, two equivalents of isocyanide were consumed, even if **15a** was taken in an equivalent molar ratio, and the unusual heterocycle **39**, containing an exocyclic imino group as well as a phosphaalkenylidene group, was formed (Scheme 10). Analogous results were obtained by the reaction of **15a** with 1,6-diisocyanohexane, from which the macroheterocycle **40** has been isolated (Fig. 13).[40] The P=C distance (1.698(4) Å) in **40** indicates that the π-bond is slightly stabilized by conjugation with the exocyclic C=N bond. The latter results indicated that the reactions of phosphasilenes are amazingly similar to those of silenes (Si=C) (note the similar

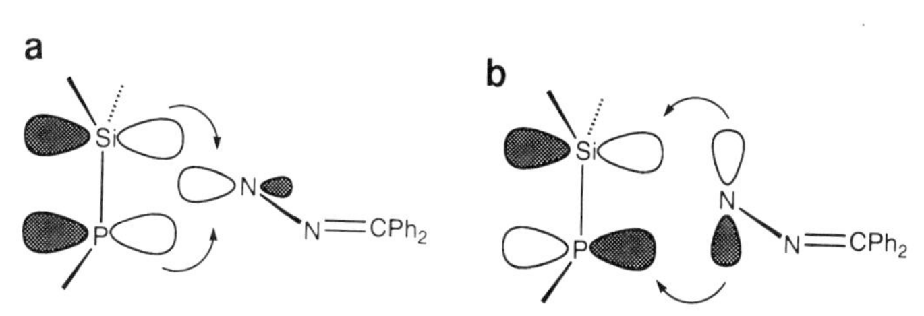

SCHEME 9. Description of **35** as π-complex (Dewar–Chatt–Duncanson model): (a) Si=P bond as π-donor; (b) Si=P bond as π-acceptor.

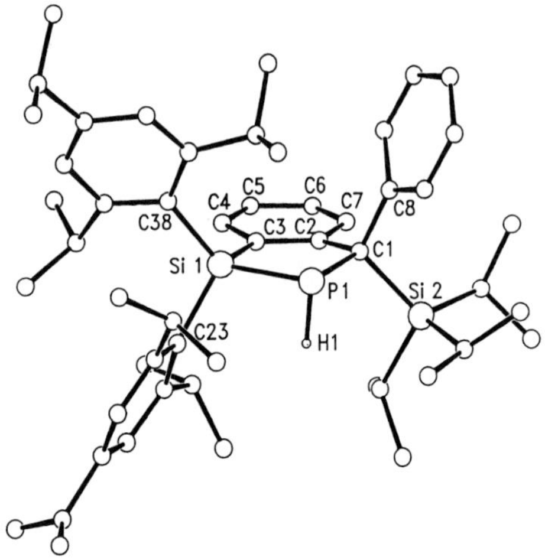

FIG. 12. X-Ray crystal structure of **38**. [Reproduced with permission of Driess and Pritzkow.[38]]

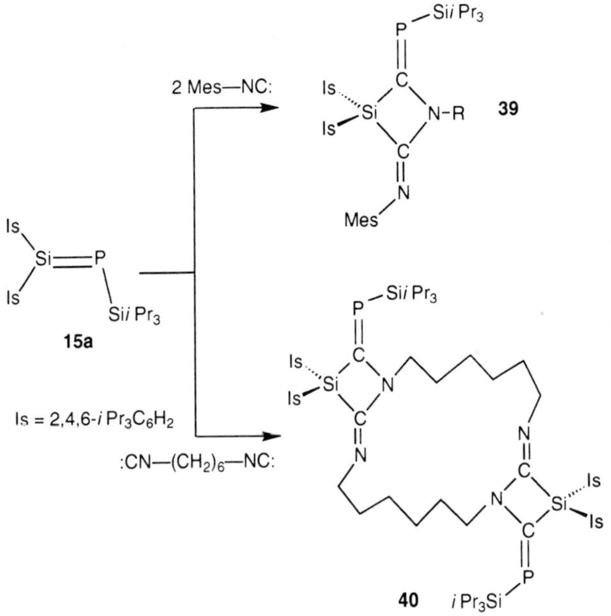

SCHEME 10. Formation of the phosphaalkenes **39** and **40** by reaction of the phosphasilene (**15a**) with mesitylisocyanide and 1,6-diisocyanohexane, respectively.

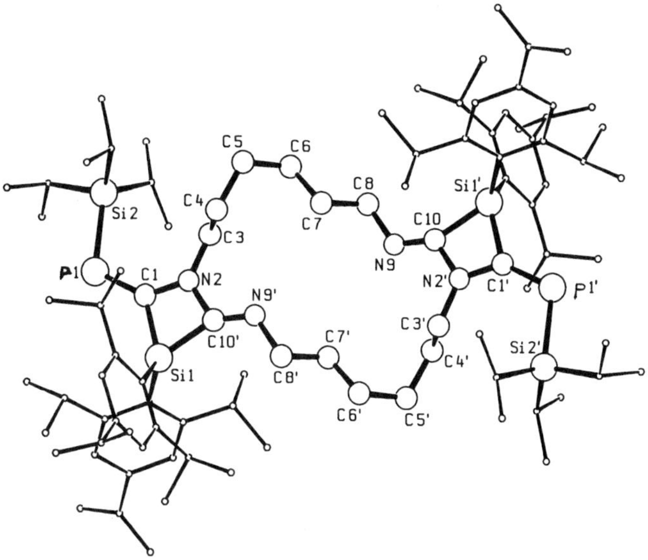

FIG. 13. Solid-state structure of **40**. [Reproduced with permission of Driess and Pritzkow.[40]]

polarities of the Si=E bonds!).[41] Interestingly, during the analogous reactions of silenes with isocyanides, the intermediates were isolated. The Si=Si bond in disilenes, however, reacts with two equivalents of isocyanide in a stepwise process to yield first the silacyclopropanimines[42] and then the 1,3-disilacyclobutane-2,4-diimines.[43]

Furthermore, the Si=P bond in **15a** reacts with organocyanides in a [2+2]-cycloaddition. Thus, the reaction of **15a** with benzonitrile gave the phosphaazasilacyclobut-3-ene **41** (Eq. 12).[14]

$$\mathrm{Is_2Si{=}P{-}Si\mathit{i}Pr_3}\ (\mathbf{15a}) \xrightarrow{\mathrm{PhCN}} \mathbf{41} \qquad (12)$$

Analogous adducts, generated by the reaction of cyanides with disilenes, are also known.[44]

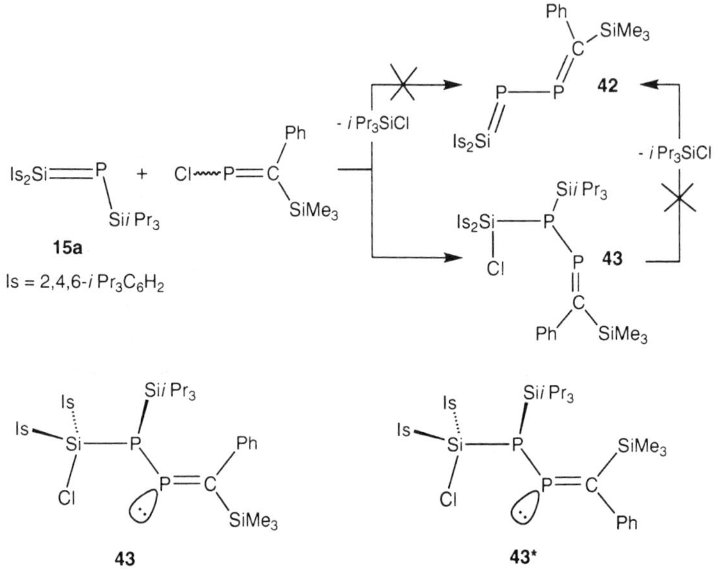

SCHEME 11. Formation of **43** and **43*** by reaction of **15a** with Cl—P=C($SiMe_3$)Ph.

7. *Reaction of the Si=P Bond with Cl—P=C(SiMe$_3$)Ph*

The aim of the reaction of **15a** with Cl—P=C($SiMe_3$)Ph was to synthesize the 2,3,1-diphosphasilabuta-1,3-diene derivative **42** (Scheme 11). However, it turned out that the Cl—P bond of the phosphaalkene exclusively adds to the Si=P bond, forming the addition product **43**, which probably has the (*E*)-isomeric structure.[45] The corresponding (*Z*)-isomer **43*** was not formed. On heating of **43**, even for 8 hours at 110°C, no elimination of iPr$_3$SiCl was observed. Instead, (*E*)/(*Z*)-isomerization simply occurred (**43** and **43*** in the molar ratio of 1 : 0.8).

V

METASTABLE SILYLIDENEARSANES (ARSASILENES)

A. *Synthesis and Spectroscopic Properties*

The synthesis of compounds **44** containing Si=As bonds was achieved by a route analogous to that used for the phosphasilenes **15** (Eq. 13).[12,14,26] The thermolysis of colorless solutions of **45a–d** in hexane at 60°C yielded the desired products **44a–d** as orange-red oils or orange solids. The starting materials **45a–d** were generated as in the phosphasilene syntheses of **15**.[10b,14]

(13)

45a-d Is = 2,4,6-i-$Pr_3C_6H_2$ **44a-d**

44, 45	a	b	c	d
R	Is	Is	Is	*t*Bu
R^1	Si*i*Pr_3	$SiPh_2Me$	$SiCy_2Me$	Si*i*Pr_3

The arsasilenes **44a–d** can be heated to 100°C for several days without decomposition. Their structures were determined by ^{29}Si-NMR and UV/visible spectroscopy.[12,14,26] For the low-coordinate Si atom, as in the analogous phosphasilenes, singlet signals were observed at very low field: $\delta(^{29}Si$, 296 K) = 179.1 (**44a**), 183.6 (**44b**), 187.0 (**44c**). The greatest deshielding for low-coordinate silicon in arsasilenes known to date was detected for **44d** (δ = 228.8).[26] As observed for phosphasilenes, the ^{29}Si-chemical shifts of **44a–d** are strongly temperature-dependent. Remarkably, the Lewis donor ability of the solvent does not significantly influence the ^{29}Si-chemical shifts; the same is true for Si=P compounds. This is in contrast to the behavior of silanimines (Si=N), which form donor solvent adducts.[9] No (*E*)/(*Z*)-isomerization has yet been observed for either phospha- or arsasilenes. We assume that this process is hampered by steric hindrance in these derivatives.

The similar electronic nature of Si=As and Si=P bonds was also reflected in their UV/visible spectra. Through the expected red shifts of the n–π^* and π–π^* transitions on going from Si=P to Si=As, both transitions for **44a–c** have been detected.[14] The λ_{max} values for the n–π^* transition were observed at 350 (log ε = 2.3 (**44a**)), 331 (log ε = 2.2 (**44b**)), and 346 nm (log ε = 2.2 (**44c**)), whereas the π–π^* transitions were detected at 291 (log ε = 1.8 (**44a**)), 289 (log ε = 2.1 (**44b**)), and 292 nm (log ε = 1.8 (**44c**)).

B. *Molecular Structure of a Derivative*

The X-ray structure determination of the first crystalline arsasilene **44d** (Fig. 14) revealed a Si=As distance of 2.164 Å and a Si—As single bond

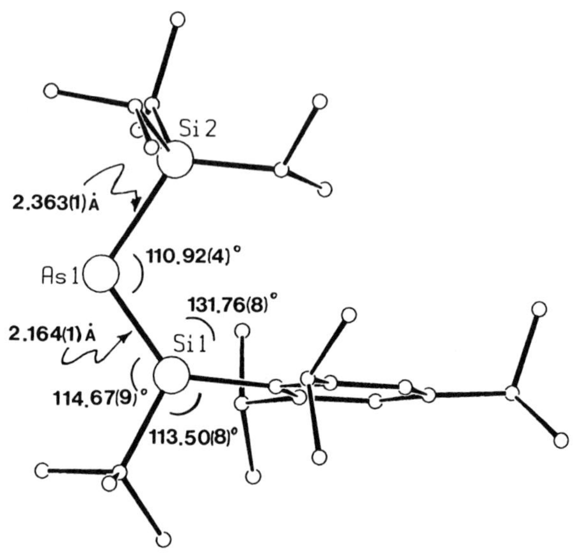

FIG. 14. X-Ray crystal structure of the arsasilene **44d**; data from Driess *et al.*[26]

distance of 2.363 Å.[26] These values are in perfect agreement with the respective calculated bond lengths in $H_2Si{=}As(SiH_3)$ (see Section II.A).

The low-coordinate silicon has a trigonal-planar configuration as in the case of the phosphorus analog **15i**. The bond angle at arsenic (110.92(4)°) is slightly smaller than the corresponding value at phosphorus in **15i** (112.79(4)°), probably because the arsenic atom is larger than phosphorus. However, the bond angles at silicon in **44d** remain unchanged from the values observed in **15i**.

C. *Reactivity*

1. *Thermolysis of the Si=As Bond*

On heating, both σ- and π-bond cleavage of Si—As occurred; the products from this process have not yet been identified. Interestingly, no insertion product of the Si—As π-bond into a C—H bond of one ortho isopropyl group, as obtained in the case of phosphasilenes (compound **20**; Eq. 4), was formed. However, when the thermolysis was carried out in the presence of triethylsilane, compounds **46** and **47** were generated, thereby clearly demonstrating the formation of silanediyl and arsanediyl as transient intermediates during this process (Scheme 12). This finding is quite interesting since it demonstrates the intrinsically lower stability of Si=As

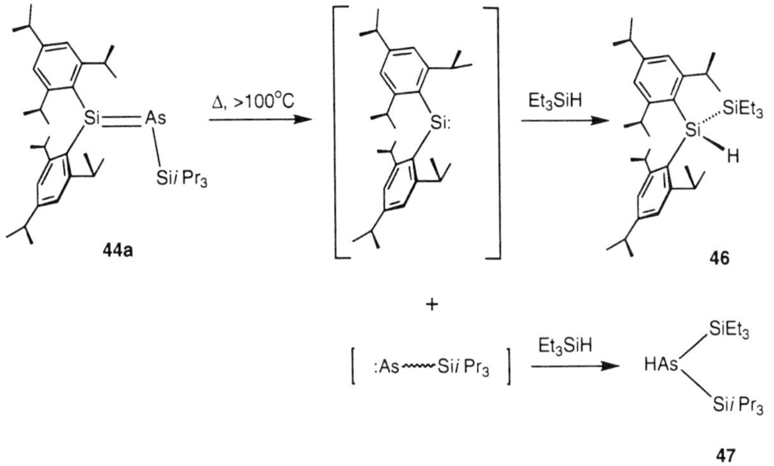

SCHEME 12. Thermolysis of the arsasilene (**44a**) in the presence of Et_3SiH, leading to **46** and **47**.

bonds relative to Si=P bonds; moreover, it reflects the fact that "double bonds" (X=Y) between heavier main-group elements are often less stable than the corresponding single bonds (X—Y).

2. *Oxidation with White Phosphorus*

The transformation of **44a** with P_4 in a 2 : 1 molar ratio readily occurred at 40°C, forming a complicated product mixture. On heating this mixture for 2 days at 100°C, the butterfly-like compounds **48** and **49** were generated (Eq. 14).[14]

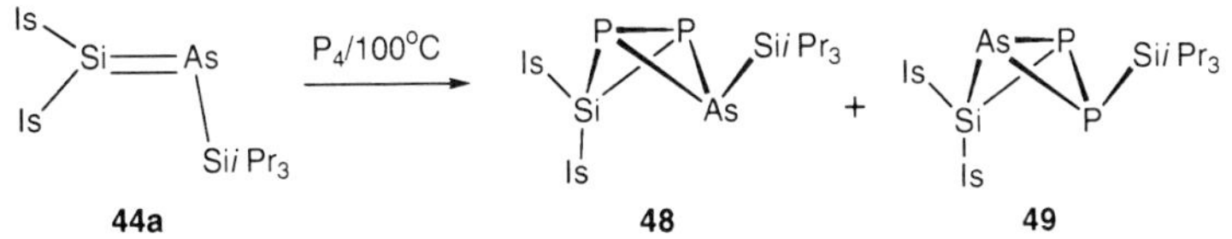

(14)

The molar ratio of these compounds was approximately 3 : 2. They could not be transformed into each other, and separation of the isomers has not yet been successful. Evidently, the initial product mixture, containing phosphorus-rich polycyclic compounds, was degraded into the bicyclic **48** and **49**. Similar product mixtures, for which structures have been established, were also observed by the analogous transformation of (*E*)-1,2-di-*tert*-butyl-1,2-dimesityldisilene and tetramesityldisilene with white phosphorus and yellow arsenic (As_4), respectively.[46]

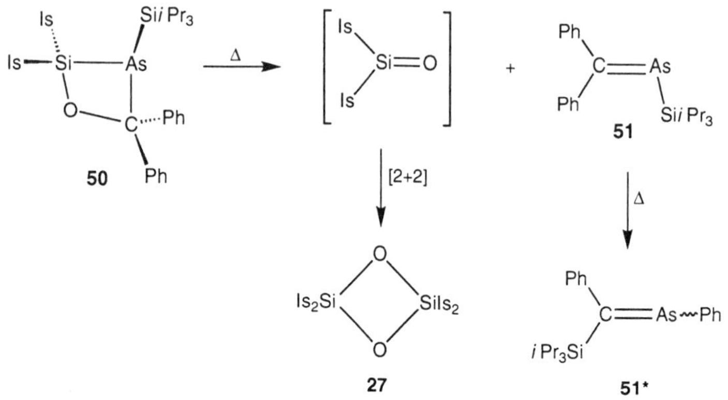

SCHEME 13. Transformation of **50** into the 1,3,2,4-dioxadisiletane **27** and the diastereomeric arsaalkenes **51** and **51***, respectively.

3. *Cycloaddition Reaction with Benzophenone and Thermolysis of the Product*

Compound **44a** reacts with $Ph_2C{=}O$ to yield the expected [2+2]-cycloaddition product **50**.[12] Its structure has been established by X-ray diffraction analysis, showing that, as a result of the crowded substitution, the As—C distance (2.057(7) Å) exceeds the normal value observed for As—C single bonds (1.95 Å). The molecular structure is isotypic with that observed for the phosphorus analog **26**. Under drastic conditions compound **50**, analogous to its phosphorus homologue **26** (see Scheme 6), decomposes to give 2,4,1,3-dioxadisiletane **27** and the isomeric arsaalkenes **51** and **51*** (Scheme 13).[12]

4. *Reactions with Diphenyldiazomethane, Isocyanides, Benzonitrile, and* tert-*Butylphosphaacetylene*

During the addition of Ph_2CN_2 to a solutions of **44a** in toluene at −78°C, the intense purple color of the diazomethane disappeared, yielding the yellow, crystalline [2+1]-cycloadduct **52** (Eq. 15).[14] Its molecular structure was elucidated by X-ray diffraction analysis (Fig. 15).

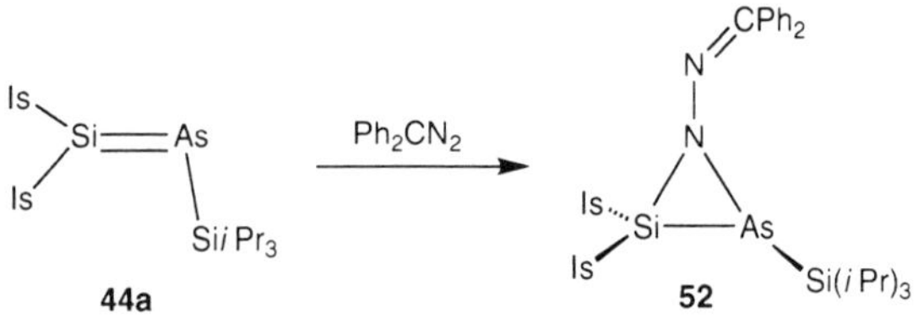

$$\mathbf{44a} \xrightarrow{Ph_2CN_2} \mathbf{52} \qquad (15)$$

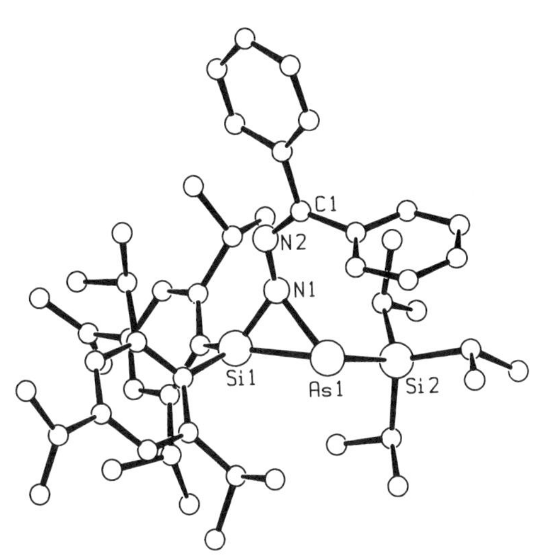

FIG. 15. Solid-state structure of the adduct **52**. [Reproduced with permission of Driess *et al.*[14]]

Compound **52** is isotypic with the analogous phosphorus derivative **35**. The Si1—As1 bond length in **52** is slightly shorter (2.353(5) Å) than the Si—As distance in the [2+2]-cycloadduct derived from **44a** and benzophenone (2.38 Å).[12]

The Si=As bond reacts with organic isocyanides in the same way that the Si=P bond does. Thus, with mesityl isocyanide and 1,6-diisocyanohexane **44a** forms the unusual arsaalkene derivatives **53** and **54**, according to Scheme 14. Their molecular structures were determined by X-ray crystallography.[39,40]

The reactions of **44a,c** with benzonitrile and *tert*-butylphosphaacetylene show that the Si=As bond easily forms [2+2]-cycloaddition products with main-group-element triple bonds.[14] However, as in the case of Si=P bonds, the Si=As bond does not react with the C≡C triple bond in alkynes.

The structures of compounds **55a,c** and **56a,c** were established by means of NMR spectroscopy and mass spectrometry. Due to the different polarity of the C≡N and C≡P triple bonds, the silicon ring atom in **55a,c** is bound to the nitrogen atom, and in **56a,c** to the carbon atom of the C=P moiety. The molecular structure of **55a** was further determined by single-crystal X-ray diffraction analysis (Fig. 16).[14] The four-membered SiNAsC framework is slightly puckered (folding angle N—Si—C/Si—C—As 7°), and

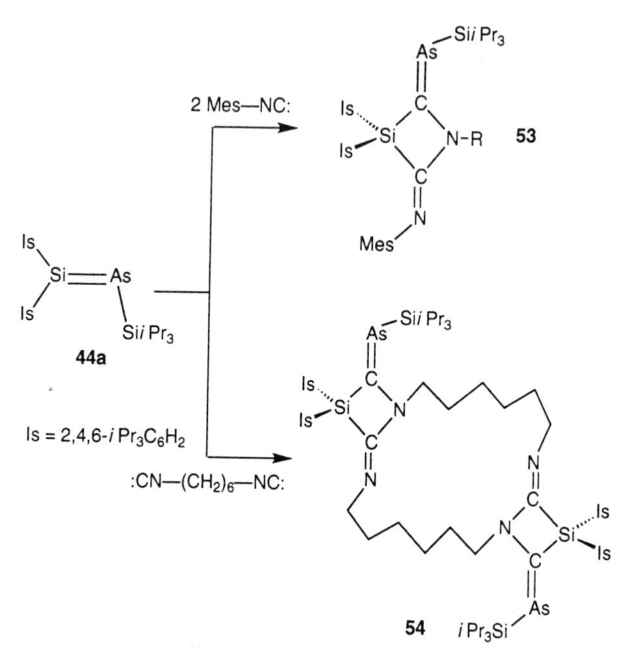

SCHEME 14. Formation of the arsaalkenes **53** and **54** by reaction of the arsasilene (**44a**) with mesitylisocyanide and 1,6-diisocyanohexane, respectively.

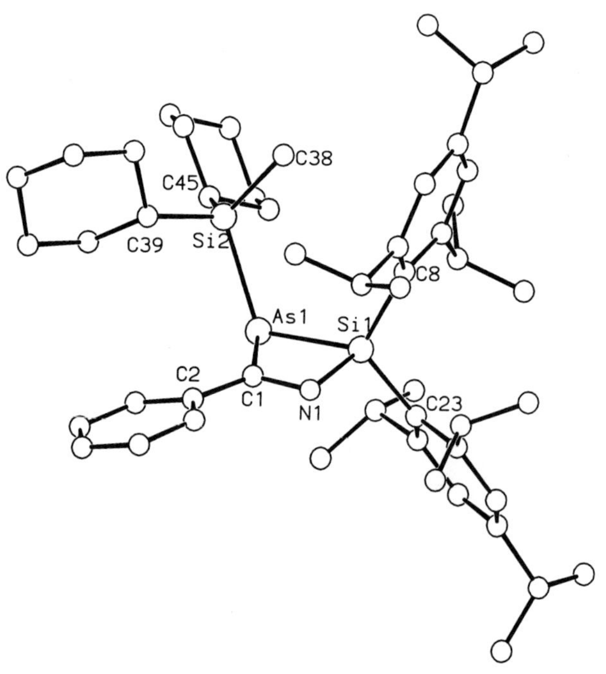

FIG. 16. X-Ray crystal structure of **55a**. [Reproduced with permission of Driess *et al.*[14]]

the As—Si distances (2.402(2) and 2.397(2) Å) are in the expected range of single bonds.

$tBuCP$ ← **44a**: R^1 = Si*i*Pr$_3$; **44c**: R^1 = $SiCy_2Me$ → PhCN

56a, c ← **44a, c** → **55a, c** (16)

Acknowledgments

This work was generously supported by the Deutsche Forschungsgemeinschaft (SFB 247) and the Fonds der Chemischen Industrie. The author is grateful to his co-workers, U. Winkler and S. Rell, and sincerely thanks the crystallographers, Dr. H. Pritzkow and Dr. L. Zsolnai for their attentive engagement in this work. Furthermore, the author is indebted to Profs. W. Siebert and G. Huttner for support and many helpful discussions.

References

(1a) Dimroth, K.; Hoffmann, P. *Angew. Chem., Int. Ed. Engl.* **1964,** *3*, 384.

(1b) Märkl, G. *Angew. Chem., Int. Ed. Engl.* **1966,** *5*, 846.

(1c) Becker, G. *Z. Anorg. Allg. Chem.* **1976,** *423*, 242. Review: Appel, R. In *Multiple Bonds and Low Coordination in Phosphorus Chemistry;* Thieme: Stuttgart, 1990; p. 157 and references herein.

(2) Hollemann, A. F.; Wiberg, E. *Lehrbuch für Anorganische Chemie;* de Gruyter: Berlin, 1985; pp. 99–100. Grimm, H. G. *Z. Elektrochem. Angew. Phys. Chem.* **1925,** *31*, 474.

(3) Allen, L. C. *J. Am. Chem. Soc.* **1989,** *111*, 9003.

(4) Corey, J. Y. In *The Chemistry of Organic Silicon Compounds;* Patai, S.; Rappoport, Z., Eds.; Wiley: New York, 1991; Part 1, p. 1. Janoschek, R. *Chem. Unserer Zeit* **1988,** *21*, 128. Driess, M. *Chem. Unserer Zeit* **1993,** *27*, 141.

(5) Kutzelnigg, W. *Angew. Chem., Int. Ed. Engl.* **1984,** *23*, 272. Schleyer, P. von R.; Kost, D. *J. Am. Chem. Soc.* **1988,** *110*, 2105.

(6) Guselnikov, L. E.; Nametkin, N. S. *Chem. Rev.* **1979,** *79*, 529.

(7) Reviews: West, R. *Angew. Chem., Int. Ed. Engl.* **1987,** *26*, 1201. Tsumaraya, T.; Batcheller, S. A.; Masamune, S. *Angew. Chem.* **1991,** *103*, 916. Weidenbruch, M. *Coord. Chem. Rev.* **1994,** *130*, 275.

(8) Brook, A. G.; Abdesaken, F.; Gutekunst, B.; Gutekunst, G.; Kallury, R. K. *J. Chem. Soc., Chem. Commun.* **1981,** 191. Brook, A. G.; Nyburg, S. C.; Abdesaken, F.; Gutekunst, B.; Gutekunst, G.; Kallury, R. K.; Pon, Y. C.; Chang, Y.-M.; Wong-Ng, W. *J. Am. Chem. Soc.* **1982,** *104*, 5667. Wiberg, N.; Wagner, G. *Angew. Chem., Int. Ed. Engl.* **1983,** *22*, 1005. Wiberg, N.; Wagner, G.; Müller, G. *Angew. Chem., Int. Ed. Engl.* **1985,** *24*, 229. Brook, A. G.; Baines, K. M. *Adv. Organomet. Chem.* **1986,** *25*, 1.

(9) Wiberg, N.; Schurz, K.; Fischer, G. *Angew. Chem., Int. Ed. Engl.* **1985,** *24*, 1053. Wiberg, N.; Schurz, K.; Reber, G.: Müller, G. *J. Chem. Soc., Chem. Commun.* **1986,** 591. Hesse, M.; Klingebiel, U. *Angew. Chem., Int. Ed. Engl.* **1986,** *25*, 649. Review: Armitage, D. A. In *The Silicon-Heteroatom Bond;* Patai, S.; Rappoport, Z., Eds.; Wiley: New York, 1991; pp. 431, 478.

(10a) Smit, C. N.; Look, F. M.; Bickelhaupt, F. *Tetrahedron Lett.* **1984,** *25*, 3011. Smit, C. N.; Bickelhaupt, F. *Organometallics* **1987,** *6*, 1156. van den Winkel, Y.; Bastiaans, H. M.; Bickelhaupt, F. *J. Organomet. Chem.* **1991,** *405*, 183.
(10b) Driess, M. *Angew. Chem., Int. Ed. Engl.* **1991,** *30*, 1022.
(10c) Bender, H. R. G.; Niecke, E.; Nieger, M. *J. Am. Chem. Soc.* **1993,** *115*, 3314.
(11) Tokitoh, N.; Suzuki, H.; Okazaki, R. *Xth International Symposium on Organosilicon Chemistry,* Poznan, Poland, *1993*, Abstracts of Lecturers, O-62, p. 110.
(12) Driess, M.; Pritzkow, H. *Angew. Chem., Int. Ed. Engl.* **1992,** *31*, 316.
(13) Walter, S.; Klingebiel, U. *Coord. Chem. Rev.* **1994,** *130*, 481; see also ref. *9*.
(14) Driess, M.; Pritzkow, H.; Rell, S.; Winkler, U. *Organometallics,* submitted for publication.
(15) Dykema, K. J.; Truong, T. N.; Gordon, M. *J. Am. Chem. Soc.* **1985,** *107*, 4535. Lee, L. G.; Boggs, J. E.; Cowley, A. H. *J. Chem. Soc., Chem. Commun.* **1985,** 773.
(16) Driess, M.; Janoschek, R. *J. Mol. Struct.* (*THEOCHEM*) **1994,** *313*, 129.
(17) Trinquier, G.; Malrieu, J. P. In *The Chemistry of Double-bonded Functional Groups;* Patai, S.; Ed.; Wiley: New York, 1989, p. 1.
(18) Kutzelnigg, W.; Fleischer, U.; Schindler, M. *NMR: Basic Princ. Prog.* **1991,** *23*.
(19) Couret, C.; Escudié, J.; Satgé, J.; Andriamizaka, J. D.; Saint-Roch, B. *J. Organomet. Chem.* **1979,** *182*, 9.
(20) Klingebiel, U.; Boese, R.; Bläser, D.; Andrianarison, M. *Z. Naturforsch., B: Chem. Sci.* **1989,** *44B*, 265.
(21) Wiberg, N.; Schuster, H. *Chem. Ber.* **1991,** *124*, 93.
(22) Niecke, E.; Klein, E.; Nieger, M. *Angew. Chem., Int. Ed. Engl.* **1989,** *28*, 751.
(23) Romanenko, V. D.; Ruban, A. V.; Drapailo, A. B.; Markovskij, L. N. *J. Gen. Chem. USSR* (*Engl. Transl.*) **1985,** *55*, 2486.
(24) Corriu, R.; Lanneau, G.; Priou, C. *Angew. Chem., Int. Ed. Engl.* **1991,** *30*, 1130.
(25) Driess, M.; Winkler, U.; Imhoff, W.; Zsolnai, L.; Huttner, G. *Chem. Ber.* **1994,** *127*, 1031.
(26) Driess, M.; Rell, S.; Pritzkow, H. *J. Chem. Soc., Chem. Commun.,* **1995,** 253.
(27) Bickelhaupt, F. In *Multiple Bonds and Low Coordination in Phosphorus Chemistry;* Thieme: Stuttgart, 1990; p. 288; see also ref. *10a*.
(28) Williams, E. A. In *The Chemistry of Organic Silicon Compounds;* Patai, S.; Rappoport; Z., Eds.; Wiley: New York, 1989; p. 511. Verkade, J. G.; Mosbo, J. A. In *Phosphorus-31 NMR Spectroscopy in Stereochemical Analysis;* Verkade, J. G.; Quin, L. D., Eds.; VCH Publishers: Weinheim, 1987, p. 446.
(29) Albright, T. A.; Burdett, J. K.; Whangbo, M.-H. In *Molecular Orbitals and Geometrical Perturbation;* Wiley: New York, 1985, p. 95.
(30) Gudat, D., Niecke, E.; Sachs, W.; Rademacher, P. *Z. Anorg. Allg. Chem.* **1987,** *545*, 7.
(31) Driess, M.; Pritzkow, H. *Phosphorus, Sulfur, and Silicon* **1993,** *76*, 57.
(32) Review: Escudié, J.; Couret, C.; Ranaivonjatovo, H.; Satgé, J. *Coord. Chem. Rev.* **1994,** *130*, 427.
(33) Driess, M.; Fanta, A. D.; Powell, D.; West, R. *Angew. Chem., Int. Ed. Engl.* **1989,** *28*, 1038.
(34) Driess, M.; Janoschek, R.; Pritzkow, H. *Angew. Chem., Int. Ed. Engl.* **1992,** *31*, 460.
(35) DeYoung, D. J.; Fink, M. J.; Michl, J.; West, R. *Main Group Metal Chem.* **1987,** *1*, 19.
(36) Appel, R.; Lambach, B. *Tetrahedron Lett.* **1980,** *21*, 2495; Becker, G.; Riffel, H.; Uhl, W.; Wessely, H. J. *Z. Anorg. Allg. Chem.* **1986,** *534*, 31.
(37) Review: West, R. In *The Chemistry of Inorganic Ring Systems;* Steudel, R., Ed.; Elsevier: Amsterdam, 1992, p. 35.
(38) Driess, M.; Pritzkow, H. *Angew. Chem., Int. Ed. Engl.* **1992,** *31*, 751.

(39) Driess, M.; Pritzkow, H.; Sander, M. *Angew. Chem., Int. Ed. Engl.* **1993,** *32*, 283.
(40) Driess, M.; Pritzkow, H. *J. Chem. Soc. Chem. Commun.* **1993,** 1585.
(41) Brook, A. G.; Kong, Y. K.; Saxena, A. K.; Sawyer, J. F. *Organometallics* **1988,** *7*, 2245; Brook, A. G.; Kong, Y. K.; Saxena, A. K. *Organometallics* **1989,** *8*, 850.
(42) Yokelson, H. B.; Millevolte, A. J.; Haller, K. J.; West, R. *J. Chem. Soc. Chem. Commun.* **1987,** 1605.
(43) Review: Weidenbruch, M. In *The Chemistry of Inorganic Ring Systems;* Steudel, R., Ed.; Elsevier: Amsterdam, 1992, p. 51.
(44) Weidenbruch, M.; Flintjer, B.; Pohl, S.: Saak, W. *Angew. Chem., Int. Ed. Engl.* **1989,** *28*, 95.
(45) Driess, M., unpublished results.
(46) Fanta, A. D.; Tan, R. P.; Comerlato, N. M.; Driess, M.; Powell, D. R.; West, R. *Inorg. Chim. Acta* **1992,** *198–200*, 733; Tan, R. P.; Comerlato, N. M.; Powell, D. R.; West, R. *Angew. Chem., Int. Ed. Engl.* **1992,** *31*, 1217.

Chemistry of Stable Disilenes

RENJI OKAZAKI

Department of Chemistry
Graduate School of Science
The University of Tokyo
Hongo, Tokyo 113, Japan

ROBERT WEST

Department of Chemistry
University of Wisconsin
Madison, Wisconsin

I

INTRODUCTION

The classic problem of the silicon–silicon double bond dates back at least to the early part of this century, when F. S. Kipping and his students attempted unsuccessfully to synthesize disilenes. Evidence for the probable transient existence of disilenes began to appear in the 1970s, but it was the isolation of the stable disilene **1** in 1981[1] that opened up modern disilene chemistry. The early history of this discovery has been recounted in a review[2a]; several other reviews covering Si═Si double bonds have been published.[2–5]

Since 1981, a great deal of progress has been made in disilene chemistry. Known stable disilenes now number more than two dozen, and structural information is available for about half that number. Important findings have developed, and continue to emerge concerning the chemical bonding in disilenes. Nowhere have developments been more exciting than in the chemical reactions of disilenes, which have led to a panoply of compounds with hitherto-unknown structures. This review is intended to describe advances in the chemistry of stable (and marginally stable) disilenes, especially those made after a review by one of us in 1987.[3a]

II

SYNTHESIS

The method employed for the first synthesis of stable disilene **1**,[1] the photolysis of linear trisilanes, has been the most widely used synthetic method for disilenes. However, photolysis of cyclic trisilanes and dehalogenation of dihalosilanes and 1,2-dihalodisilanes are also good routes in some cases (Table I).

A. *Photolysis of Linear Trisilanes (Method A)*

The disilene synthesis by the photolysis of linear trisilanes proceeds via initial formation of a silylene followed by its dimerization (Eq. 1). Disilene **1** has now become a common organometallic reagent. A detailed synthetic procedure employing photolysis of the corresponding linear trisilane is described in *Inorganic Syntheses* (Eq. 2).[6]

$$2\,R_2Si(SiMe_3)_2 \xrightarrow{h\nu} 2\,R_2Si: \longrightarrow R_2Si{=}SiR_2 \qquad (1)$$

$$2\ R_2Si(SiMe_3)_2 \xrightarrow{h\nu} R_2Si{=}SiR_2 + Me_3SiSiMe_3 \quad (2)$$

1: R = Mes
2: R = Tip

Disilene **2** having a bulkier substituent, 2,4,6-triisopropylphenyl, was obtained similarly.[7] When trisilanes bearing two different substituents on the central silicon atoms are used, a mixture of *E*- and *Z*-isomers of disilenes is obtained (Eq. 3).[8–10] Attempts to synthesize disilenes with other substituents using trisilane $Tip(R)Si(SiMe_3)_2$ (R = H, F, Cl, or 1-pyrroyl) were unsuccessful.[10]

$$2\ RR'Si(SiMe_3)_2 \xrightarrow[-\ Me_3SiSiMe_3]{h\nu} RR'Si{=}SiRR' \quad (3)$$

3: R = Mes, R' = *t*-Bu
4: R = Mes, R' = 1-Ad
5: R = Mes, R' = Xyl
6: R = Mes, R' = Dmt
7: R = Mes, R' = $N(SiMe_3)_2$
8: R = Tip, R' = Mes
9: R = Tip, R' = $SiMe_3$
10: R = Tip, R' = *t*-Bu

Cophotolysis of $A_2Si(SiMe_3)_2$ and $B_2Si(SiMe_3)_2$ led to the formation of a disilene having a new substitution pattern, $A_2Si{=}SiB_2$, in addition to $A_2Si{=}SiA_2$ and $B_2Si{=}SiB_2$ (Eq. 4). The resulting "mixed" disilenes were studied spectroscopically, but the isolation of a pure specimen of $A_2Si{=}SiB_2$ was impossible (for the synthesis of this type of disilenes, see Section II.D).[11,12] $A_2Si{=}SiAB$-type disilenes were also produced, according to the reactions of Eq. (5).

$$Mes_2Si(SiMe_3)_2 + R_2Si(SiMe_3)_2 \xrightarrow{h\nu}$$

$$Mes_2Si{=}SiMes_2 + R_2Si{=}SiR_2 + Mes_2Si{=}SiR_2 \quad (4)$$

1

11: R = Xyl
12: R = Dmt
2: R = Tip

13: R = Xyl
14: R = Dmt
15: R = Tip

$$R_2Si(SiMe_3)_2 + R'R''Si(SiMe_3)_2 \xrightarrow{h\nu}$$

$$R_2Si{=}SiR_2 + R_2Si{=}SiR'R'' + R'R''Si{=}SiR'R'' \quad (5)$$

16: R = R' = Mes, R'' = Xyl
17: R = R' = Xyl, R'' = Dmt
18: R = Xyl, R' = R'' = Dmt

TABLE I

STABLE DISILENES[a]

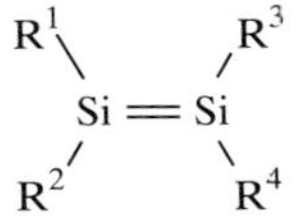

$$R^1R^2Si{=}SiR^3R^4$$

Compound	R^1	R^2	R^3	R^4	Method	λ_{max} (nm)	$\delta(^{29}Si)$	Reference
1	Mes	Mes	Mes	Mes	A, C	343, 420	63.68	41
2	Tip	Tip	Tip	Tip	A, C, D	266, 347, 432	53.4, 53.41	12, 20
(*E*)-**3**	*t*-Bu	Mes	Mes	*t*-Bu	A, B	338, 400 (sh)	90.3	41
(*Z*)-**3**	*t*-Bu	Mes	*t*-Bu	Mes	A, B[b]		94.7	8
(*E*)-**4**	Ad	Mes	Mes	Ad	A	338, 394 (sh)	87.1	35
(*Z*)-**4**	Ad	Mes	Ad	Mes	A		92.6	35
(*E*)-**5**	Mes	Xyl	Xyl	Mes	A		63.86	11
(*Z*)-**5**	Mes	Xyl	Mes	Xyl	A		63.87	11
(*E*)-**6**	Mes	Dmt	Dmt	Mes	A		64.15, 64.37[b]	11
(*Z*)-**6**	Mes	Dmt	Mes	Dmt	A			11
(*E*)-**7**	$N(SiMe_3)_2$	Mes	Mes	$N(SiMe_3)_2$	A[b]	362, 468	49.4	8
(*Z*)-**7**	$N(SiMe_3)_2$	Mes	$N(SiMe_3)_2$	Mes	A	351, 483	61.9	8
(*E*)-**8**	Mes	Tip	Tip	Mes	A	431[c]	58.19	12
(*Z*)-**8**	Mes	Tip	Mes	Tip	A[b]		59.44	12
(*E*)-**9**	$SiMe_3$	Tip	Tip	$SiMe_3$	A	294 (sh), 394	97.75	10
(*Z*)-**9**	$SiMe_3$	Tip	$SiMe_3$	Tip	A	304 (sh), 398	97.68	10
(*E*)-**10**	*t*-Bu	Tip	Tip	*t*-Bu	A	337, 410 (sh)	87.39	10
(*Z*)-**10**	*t*-Bu	Tip	*t*-Bu	Tip	A	372, 452 (sh)	96.93	10
11	Xyl	Xyl	Xyl	Xyl	A, B	277, 340, 422	64.06	14, 30

12	Dmt	Dmt	Dmt	Dmt	A		65.15	11
13	Mes	Mes	Xyl	Xyl	A		62.56, 65.19	11, 37
14	Mes	Mes	Dmt	Dmt	A		64.11, 64.22	11
15	Mes	Mes	Tip	Tip	A	400	57.86, 59.40	12, 26
16	Mes	Mes	Xyl	Mes	A		63.14, 64.42	11
17	Xyl	Xyl	Xyl	Dmt	A		64.63, 63.30	11
18	Xyl	Xyl	Dmt	Dmt	A		63.24, 65.82	11
19	Dep	Dep	Dep	Dep	B	354, 426		15a
(*E*)-**20**	Dep	Mes	Mes	Dep	B	345, 426[c]		23b
(*Z*)-**20**	Dep	Mes	Dep	Mes	B			23b
21	*t*-Bu	*t*-Bu	*t*-Bu	*t*-Bu	B, E	305, 433		27
22	*t*-$BuMe_2Si$	*t*-$BuMe_2Si$	*t*-$BuMe_2Si$	*t*-$BuMe_2Si$	B	290, 361, 424	142.1	21
23	*i*-Pr_2MeSi	*i*-Pr_2MeSi	*i*-Pr_2MeSi	*i*-Pr_2MeSi	C	293, 357, 412	144.5	21
24	*i*-Pr_3Si	*i*-Pr_3Si	*i*-Pr_3Si	*i*-Pr_3Si	C	296, 370, 425, 480	154.5	21
25	Dis	Dis	Dis	Dis	C	357, 393		23a
(*E*)-**26**	Dip	Mes	Mes	Dip	C			23b
(*Z*)-**26**	Dip	Mes	Dip	Mes	C	336, 430[d]		23b
(*E*)-**27**	Tbt	Mes	Mes	Tbt	C	368, 425, 460 (sh)	66.49	22, 24
(*Z*)-**27**	Tbt	Mes	Tbt	Mes	C	378, 403	56.16, 56.74, 57.12, 58.12[e]	22, 24

[a] The following abreviations are used for the substituents. Ad = 1-adamantyl; Dep = 2,6-diethylphenyl; Dip = 2,6-diisopropylphenyl; Dis = bis(trimethylsilyl)methyl; Dmt = 4-*t*-butyl-2,6-dimethylphenyl; Mes = 2,4,6-trimethylphenyl; Tbt = 2,4,6-tris[bis(trimethylsilyl)methyl]phenyl; Tip = 2,4,6-triisopropylphenyl; Xyl = 2,6-dimethylphenyl.

[b] By UV irradiation of the *E*-isomers.

[c] Mixture of *E*- and *Z*-isomers.

[d] Presumed *Z*.

[e] Presumed mixture of two conformers.

In the photolysis of trisilanes $RR'Si(SiMe_3)_2$ (R = R′ = Mes; R = Mes, R′ = Tip; R = R′ = Tip), the rate of disilene formation was predicted to increase in this order because the Si—Si—Si bond angles decrease in the same order, as revealed by X-ray structural analyses, thus facilitating the formation of $Me_3SiSiMe_3$. Although the predicted trend in the rates of photolysis was found, the differences were small.[12]

The dimerization process of silylene Mes_2Si to **1** was studied using matrices with varying viscosity. The dimerization rate was found to be dependent on the viscosity. Interestingly, dimerization took place even in the matrix at 77 K without annealing when a soft matrix (e.g., isopentane/3-methylpentane = 9/1) was used.[13]

B. *Photolysis of Cyclic Trisilanes (Method B)*

The first synthesis of a disilene by photolysis of the corresponding cyclic trisilane was reported by Masamune *et al.* in 1982.[14] This method has been adapted for the synthesis of a variety of stable and marginally stable disilenes (Eq. 6).[15–20] More recently, Kira *et al.* synthesized the first example of a stable tetrakis(trialkylsilyl)disilene **22** by this method.[21]

$$2\,(RR'Si)_3 \xrightarrow{h\nu} 3\,RR'Si{=}SiRR' \qquad (6)$$

19: R = R′ = Dep
20: R = Mes, R′ = Dep
21: R = R′ = *t*-Bu
22: R = R′ = *t*-BuMe$_2$Si

C. *Dehalogenation of Dihalosilanes (Method C)*

Dehalogenation of dihalosilanes was thought to produce only cyclic polysilanes, $(R_2Si)_n$ with $n \geq 4$, until Masamune *et al.* found that the use of appropriately large groups such as 2,6-dimethylphenyl and judicious selection of reaction conditions led to formation of a cyclotrisilane derivative.[14] Although under some reaction conditions disilene **1** was formed directly from the corresponding dichloride, e.g., Mes_2SiCl_2, this method was found to give rather impure disilene.[2a] However, with even larger substituents on the silicon atoms, this method proved to be a good route to stable disilenes (Eq. 7).[20–24]

D. *Dehalogenation of 1,2-Dihalodisilanes (Method D)*

When a 1,2-dihalodisilane is available as a starting material, its reductive dehalogenation can also be a synthetic approach to a disilene (Eq. 8).[25,26]

$$2\ RR'SiX_2 \xrightarrow[\text{or Na}]{LiC_{10}H_8} RR'Si{=}SiRR' \qquad (7)$$

2: R = R' = Tip
23: R = R' = i-Pr_2MeSi
24: R = R' = i-Pr_3Si
25: R = R' = $CH(SiMe_3)_2$
26: R = Dip, R' = Mes
27: R = Tbt, R' = Mes

This method proved useful for the synthesis of a disilene having different substituents on each silicon atom, although it was successful only for a tetraaryldisilene.[15] Reduction of the chlorine to hydrogen gave 1,2-dihydrodisilane when R was *i*-Pr or *t*-Bu.[26]

$$Tip_2Si(Cl){-}SiR_2(Cl) \xrightarrow[\text{or Li}]{LiC_{10}H_8} Tip_2Si{=}SiR_2 \qquad (8)$$

3: R = Tip
15: R = Mes

E. *Photolytic [4+2] Cycloreversion (Method E)*

Photolytic [4+2] cycloreversion of a disilabicyclo[2.2.2]octadiene precursor is considered to be a general method for the synthesis of disilenes of varying stability. Several examples of this method have been reported (Eq. 9).[27–31] First used in reactions producing an unstable disilene, $Me_2Si{=}SiMe_2$,[27] this method is also successful for synthesizing the marginally stable t-$Bu_2Si{=}Si(t$-$Bu)_2$.[28] However, it has not yet been applied to the synthesis of disilenes that are fully stable at room temperature, probably because the appropriate precursors are inaccessible.

t-Bu t-Bu Si t-Bu Si t-Bu

$$\xrightarrow{h\nu} t\text{-}Bu_2Si{=}Si(t\text{-}Bu)_2 \qquad (9)$$

21

F. *From Silylene Complexes*

When silylenes are generated photochemically in hydrocarbon matrices in the presence of electron-pair donors, they may form Lewis acid–base complexes that act as intermediates in the silylene dimerization to disilenes.[32,33] In a typical example, $Mes_2Si(SiMe_3)_2$ was photolyzed in 3-methylpentane (3-MP) matrix containing 5% of 2-methyltetrahydrofuran. At 77 K, dimesitylsilylene (λ_{max} 577 nm) was formed. When the matrix was

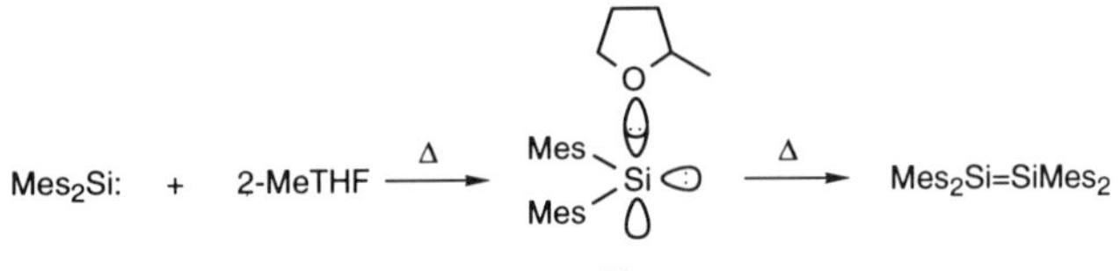

$$Mes_2Si: + \text{2-MeTHF} \xrightarrow{\Delta} \underset{\mathbf{28}}{Mes_2Si \leftarrow O(\text{2-MeTHF})} \xrightarrow{\Delta} Mes_2Si{=}SiMes_2$$

SCHEME 1

annealed, the silylene absorption band was replaced by one at 326 nm, due to the 2-MeTHF–silylene complex **28**. Further warming melted the matrix and destroyed the complex, producing the disilene (Scheme 1).

Similar experiments were done with a variety of bases, n-Bu_3N, Et_3N, Et_2O, t-Bu_2S, CO,[34] n-Bu_3P, Et_3P, and i-PrOH, all of which formed base–silylene complexes.[32] In most cases melting of the matrix led to the disilene (**1** or **3**); however, i-PrOH reacted with the silylene to give the usual addition compound (see Section VIII.B) and the phosphine–silylene complexes decomposed to give unknown products.

G. *A Possible Bridged Disilene*

Jutzi *et al.* reported that dodecamethylsilicocene (**29**) reacted with tetrafluoroboric acid to give cyclotetrasilane **31** via a dimerization of disilene **30**.[35] The intermediate disilene was not stable, but could be detected by NMR spectroscopy at −70°C (Scheme 2). However, Maxka and Apeloig later pointed out[36] that the ^{29}Si NMR data were more consistent with a bridged structure, **32**. Calculations showed that the model bridged structures, $HSi(\mu\text{-}F)_2SiH$ and $FSi(\mu\text{-}H)_2SiF$, are minima on the $Si_2H_2F_2$ potential energy surface and are only about 5 kcal mol^{-1} higher in energy than disilene HFSi=SiHF. If this compound actually has structure **32,** it represents the first example of a bridged (though unstable) disilene. This provides a good example of the utility of theoretical calculations in identifying exotic compounds.

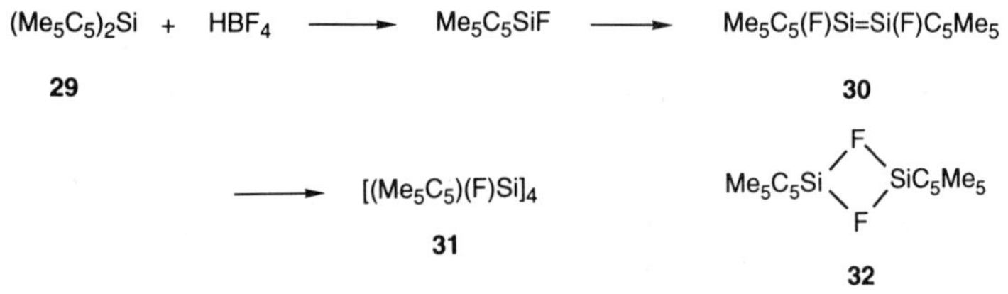

$$\underset{\mathbf{29}}{(Me_5C_5)_2Si} + HBF_4 \longrightarrow Me_5C_5SiF \longrightarrow \underset{\mathbf{30}}{Me_5C_5(F)Si{=}Si(F)C_5Me_5}$$

$$\longrightarrow \underset{\mathbf{31}}{[(Me_5C_5)(F)Si]_4} \qquad \underset{\mathbf{32}}{Me_5C_5Si(\mu\text{-}F)_2SiC_5Me_5}$$

SCHEME 2

III

THEORY

Because the theory of multiply bonded silicon has been reviewed elsewhere,[37] only those aspects relevant to stable disilenes will be discussed here. Disilene and the related hydrides of other group 14 elements, $H_2M{=}MH_2$, have attracted much attention from theoreticians, in part because of the unusual trans-pyramidal geometry around the M atoms, predicted from MO calculations and observed in many $R_2M{=}MR_2$ compounds.[4] Recently, in an important paper dealing with $H_2Si{=}SiH_2$ and related compounds, Jacobsen and Ziegler concluded from density-function theory that the intraatomic Pauli repulsion between inner electron shells and valence electrons weakens the π-bonding and promotes trans-bending.[38] This paper also gives extensive references to earlier theoretical literature.

Although most theoretical work has been done on $H_2Si{=}SiH_2$, one paper reported MO calculations for the tetramethyl compound $Me_2Si{=}SiMe_2$ as well as Si_2F_4 and Si_2Li_4.[39] Like disilene itself, tetramethyldisilene is predicted to have a very flat potential energy surface for distortion either by mutual trans-bending of the Me_2Si groups or by twisting of the Si—Si bond. Apeloig and Karni have published a communication on substitution effects on the Si=Si bond.[40] The strength of the π-bond and the tendency to distortion are closely related to the singlet–triplet excitation energies E_{S-T} of the corresponding silylenes. As E_{S-T} becomes larger, the π-bond is weakened and distortion increases. Thus both electronegative and π-donating substituents (F, OR, NR_2) should weaken Si—Si π-bonding. This prediction accords with the fact that disilenes with RO or halogen substituents are unknown and the only *N*-substituted disilene (**7**) contains $(Me_3Si)_2N$ groups, which are poor π-donors.

IV

PHYSICAL PROPERTIES

A. *Stability toward Oxygen and Heat*

In general, an increase in steric demand of the substituents on silicon atoms appears to lead to the formation of more stable disilenes. For example, **3**, which has two *t*-butyl groups, is more stable than **1** while **4**, which has two 1-adamantyl groups, is more stable than **3**.[9] Disilene **4** survives unchanged up to its melting temperature of 270–274°C and for several minutes in the molten state at 280°C, whereas decomposition temperatures are 175°C for **1** and 225°C for **3**.[9]

Disilenes generally undergo rapid reaction with dioxygen in solution. In the solid phase, the rate of oxygen oxidation may depend greatly on the state of subdivision; here, we will compare reports on the oxidation rates of finely divided samples. The half-lives of disilenes **1** and **2** undergoing oxidation are just a few minutes. The corresponding half-lives for **3** and **4** are a few hours and ~2 days, respectively.[9] Disilenes **22–24** bearing four trialkylsilyl groups are highly air- and moisture-sensitive in solution, but they survived for a day in the solid state exposed to air.[21] Disilene **27** bearing two very bulky Tbt groups is surprisingly stable, indicating the high steric protecting ability of the Tbt group. Neither (*E*)- nor (*Z*)-**27** undergoes significant decomposition during chromatography without protection from the air, and the half-life of the decomposition of a finely powdered sample of (*Z*)-**27** in the air is 40 days.[22,24] Despite its high stability toward oxidation, **27** is thermally rather unstable, undergoing dissociation into the corresponding silylenes (see Section VII.B).

V
SPECTROSCOPY

A. *Electronic Spectroscopy and Thermochromism*

The stable disilenes are pale yellow to orange-red in the solid state and have electronic absorption maxima in solution between 390 and 480 nm (Table I). The longest wavelength absorptions have been assigned to the π–π^* transition.[2a]

In general, tetraalkyldisilenes have absorption maxima at approximately 400 nm, whereas those for tetraaryldisilenes are red-shifted by 20–30 nm, probably due to a weak conjugation by the aryl groups. The absorption maxima for dialkyldiaryldisilenes are similar to those for tetraalkyldisilenes, suggesting the prohibition of the aryl conjugation by the bulky alkyl groups, as evidenced by the nearly orthogonal arrangement of the aryl groups with respect to the disilene framework.[9,41] The bathochromic shift with aryl substitution noted above for stable disilenes is mirrored in the absorption spectra of unstable disilenes isolated in a 3-MP matrix.[31] The λ_{max} values (nm) recorded were as follows: $Me_2Si{=}SiMe_2$, 344; PhMe Si=$SiMe_2$, 386; (*E*)- and (*Z*)-PhMeSi=SiMePh, 417 and 423.

An almost 60-nm red shift on going from *t*-butyl-substituted disilene

(*E*)-**10** to trimethylsilyl-substituted disilene (*E*)-**9** is most likely due to the electronic effect of the Me_3Si substituents. Very large bathochromic shifts are also observed for the *Z*- and *E*-isomers of the $(Me_3Si)_2N$-substituted disilene **7**. A possible cause is mixing of the nitrogen lone pair with the Si=Si π-HOMO, shifting the latter up in energy.

Each of the three tetrakis(trialkylsilyl)disilenes **22–24** has an absorption band near 420 nm; but compound **24**, which bears the largest substituents, also exhibits a rather strong absorption at exceptionally low energy, λ_{max} 480 nm.[21] The likely explanation of this anomaly is that a strongly twisted conformation of **24** may exist in solution. Among the tetraalkyldisilenes, the unusually long wavelength absorption at 433 nm for **21** may also result from a twisted conformation.[4]

Thermochromism was already noted for the first stable disilene **1**.[2a] In dialkyl- or bis(trimethylsilyl)diaryldisilenes, the size of the aryl groups affects the thermochromism: Tip-substituted ones (**9**, **10**) are thermochromic,[10] while Mes-substituted ones (**3**, **4**) are not.[9] Interestingly, tetrakis(trialkylsilyl)disilanes **22** and **23** are highly thermochromic: hexane solutions of **22** and **23** are light yellow below 0°C but dark red above 50°C.[21] These observations were interpreted in terms of a thermal equilibrium between the bent and twisted conformations. Compound **24**, having an absorption band at 480 nm, is dark red in solution even at room temperature.

The emission spectrum has been reported only for **1**.[41] A plot of the absorption, emission, and excitation spectra for **1** is shown in Fig. 1.

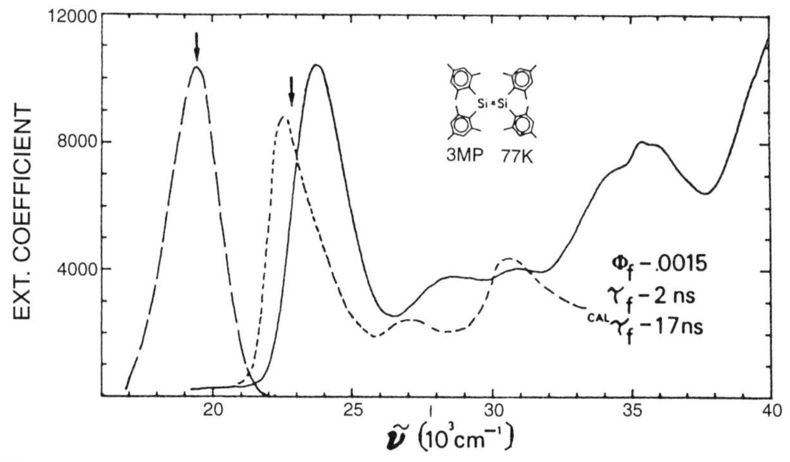

FIG. 1. Absorption (solid line), emission (broken line) and excitation (dashed line) spectra for **1** in 3-methylpentane at 77 K.

The emission and excitation spectra have identical polarizations of 0.5, indicating that a single electronic transition is responsible for both. A striking feature of the spectra is that the excitation spectrum, although having a contour similar to the absorption band, is shifted to longer wavelength. In addition, the fluorescence lifetime of 2 ns is much shorter than the theoretical lifetime of 17 ns calculated from the intensity of the absorption band. These anomalies can be explained by the presence of more than one conformer in the sample. Conformers that do not fluoresce are mainly responsible for the absorption band, while the emission spectrum arises from a small amount of a strongly fluorescent conformer that absorbs at longer wavelength.

B. *NMR Spectroscopy*

Disilenes exhibit the relatively low-field (δ = 49–155) ^{29}Si chemical shifts characteristic of low-coordinate silicon compounds (Table I); thus ^{29}Si NMR spectra are very important in their characterization. This deshielding is similar to that observed in the ^{13}C chemical shifts of doubly bonded carbons relative to those of their saturated counterparts.

The unusually low-field ^{29}Si NMR resonances (δ 142–155) observed for tetrakis(trialkylsilyl)disilenes **22–24** are interesting,[21] but whether these effects result from the trialkylsilyl substituents is open to question in view of their similarity to the chemical shifts between diarylbis(trimethylsilyl)-disilene **9** and dialkyldiaryldisilene **10** ((*Z*)-**9** (δ 97.68) vs. (*Z*)-**10** (δ 96.93)).

Quite early it was shown that the ^{29}Si resonance in **1** has a large chemical-shift anisotropy (CSA) compared with that of tetracoordinate silicon compounds.[42] Similar differences are found for the ^{13}C resonance in carbon compounds; that is, the CSA is small for tetracoordinate carbons but large for doubly bonded carbons. Recently, several other disilenes have been studied by solid-state NMR[43]; the values for their chemical shift tensors are shown in Table II. Large spreads (σ_{11}–σ_{33}) of the CSA, from 175 to 203 ppm, are found for all of the disilenes. The CSA spread for the nonconjugated disilene (*E*)-**3** is slightly smaller than that for the tetraaryl-disilenes, owing to lower shielding of the most shielded tensor component, σ_{33}.

Several calculations of the chemical shielding tensor have been carried out by different methods for simple model disilenes. For $H_2Si{=}SiH_2$, calculated values of 300,[44] 291,[45] and 195 ppm[46] have been obtained for $\Delta\sigma$, the CSA spread. A value of 255 ppm has been calculated for $Me_2Si{=}SiMe_2$,[46] and a value of 120 ppm was obtained for $(CH_2{=}CH)_2Si{=}Si$

TABLE II

PRINCIPAL VALUES OF ^{29}Si CHEMICAL SHIFT TENSORS FOR DISILENES (ppm)

Disilene	σ_{11}	σ_{22}	σ_{33}	$\Delta\sigma$
1	183	28	−20	203
1 · toluene	165	40	−25	190
2	155	30	−31	186
(*E*)-**3**	178	77	3	175
$Mes_2SiHSiHMes_2$	−31	−56	−73	42

$(CH{=}CH_2)_2$, which was taken as a model for tetraaryldisilenes.[47] While these values are all higher than those for tetraccordinate silicon, they are scattered and not very close to the experimental values for stable disilenes, as shown in Table II.

The large downfield CSA for disilenes, and indeed the large isotropic chemical shift, is caused mainly by the great deshielding of one component of the tensor, σ_{11}. Tossell and Lazzaretti propose that this deshielding results from a low-energy electronic transition between a σ-bonding orbital in the molecular plane and the Si═Si π^*-orbital.[45]

The coupling constant $^1J_{Si=Si}$ is also informative concerning the nature of the Si═Si bond. The $^1J_{Si=Si}$ observed for **13** is 155–158 Hz and that for **15** is 160 Hz, as compared with about 85 Hz for typical aryldisilanes.[48] This indicates much greater *s* character in the σ_{Si-Si} bond in disilenes than in disilanes, consistent with approximate sp^2 hybridization of the Si atoms in the disilenes. Theoretical calculations on the parent disilene predict a $^1J_{Si=Si}$ value of 183.6 Hz.[49]

C. *Raman Spectroscopy*

Vibrational spectroscopy should provide useful information on the nature of disilenes, but only two studies have been done so far on the vibrational spectra using resonance Raman spectroscopy.[50] The Raman lines appearing at 539 cm^{-1} for **1** and at 525 cm^{-1} for **3** were assigned to the Si═Si vibrations.[3a] These frequencies are about 30% higher than those for Si—Si single-bond stretching vibrations, consistent with increased Si—Si bond strength in the disilenes. In a later study, a Raman line at much higher frequency, 685 cm^{-1}, was assigned to the Si═Si stretching mode in **2**.[20] The discrepancy in these values remains to be resolved.

VI
MOLECULAR STRUCTURES

X-ray crystal structures have been determined for eleven disilenes, some of whose structural parameters are summarized in Table III. Some important structural characteristics follow. (1) The Si=Si bond distances

TABLE III

SELECTED STRUCTURAL PARAMETERS FOR DISILENES

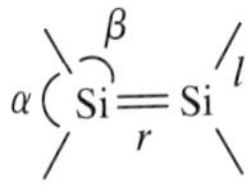

Disilene	r (Å)	l (Å)	Bond angles (deg.) α	β	γ	δ	Reference
1	2.143	1.879	116.8	120.7	3	13	51
1 · C_7H_8	2.160(1)	1.880(2)		113.9(1)	6.5(1)	18	41
		1.871(2)		126.8(1)			
2	2.144	1.885	117.5	120.8	3	0	20
		1.894		121.6			
(*E*)-**3**	2.143(1)	1.884(2)	113.2(1)	123.86(8)	0	0	41
		1.904(3)		122.77(8)			
(*E*)-**4**	2.138(2)	1.903(5)	115.4(2)	123.4(1)	0	2.8	9
		1.890(4)		121.2(2)			
(*E*)-**9**	2.152(3)	1.898(5)			0	0	10
		2.334(2)					
(*E*)-**10**	2.157(2)	1.880(3)			0	0	10
		1.918(4)					
19	2.140(3)	1.886(4)	117.6(2)	117.6(2)	0	10	15a
		1.878(4)		124.8(2)			
22	2.202(1)	2.377(2)	112.5(0)	112.9(0)	8.9(1)	0.1(0)	21
		2.383(2)		124.5(0)			
23	2.228(2)	2.374(2)	115.3(1)	120.3(1)	0	5.4(0)	21
		2.366(4)		124.1(1)			
24	2.251(1)	2.400(2)	114.9(0)	126.3(0)	0	10.2(0)	21
		2.411(2)		117.5(0)			
(*Z*)-**27**	2.195(4)	1.899(9)	113.5(4)	136.3(3)	14(1)	9.8(4)	24
		1.91(1)	114.5(4)	134.6(3)		7.6(4)	
		1.90(1)		109.5(3)			
		1.90(1)		109.7(3)			
(*E*)-**27** · $C_{10}H_8$	2.228(3)	1.895(8)	108.8(3)	132.2(2)	8.7(8)	14.6(3)	22, 24
		1.922(7)	109.2(3)	130.0(2)		9.4(3)	
		1.927(7)		120.1(2)			
		1.894(7)		115.8(2)			

for carbon-substituted disilenes range from 2.138 to 2.228 Å, with the average being 2.16 Å. Silicon-substituted disilenes have Si=Si distances roughly 0.1 Å longer than the average. (2) The disilenes generally have a slightly trans-bent structure, consistent with theoretical calculations on $H_2Si{=}SiH_2$,[4,38] although some of them ((*E*)-**3**, **2**, (*E*)-**4**, **22**) are planar or almost planar. The planarity of the disilene framework does not appear to correlate directly with the size and nature of the substituents, suggesting rather flat potential energy surfaces for bending. (3) The twist angle γ (see Table II) between the two planes of the silicon and attached carbon or silicon atoms is mostly 0° or close to 0°, but substantially larger in some disilenes (**1** · C_7H_8 **22**, (*E*)- and (*Z*)-**27**), again suggesting a flat potential energy surface for twisting.

The classic disilene **1** is unusual in that it exists in at least three crystalline modifications: orange and yellow unsolvated forms and a yellow toluene solvate (Fig. 2). The orange polymorph has a helical conformation in which all of the mesityl substituents are twisted in the same direction; thus molecules of **1** in this form are chiral.[51] The toluene solvate has an unusual conformation in which two mesityl rings cis to each other are nearly coplanar with the Si=Si bond, while the other two cis mesityl groups are nearly orthogonal.[41] The structure of the yellow unsolvated form is not yet known. Because of the flat potential surface for the Si=Si

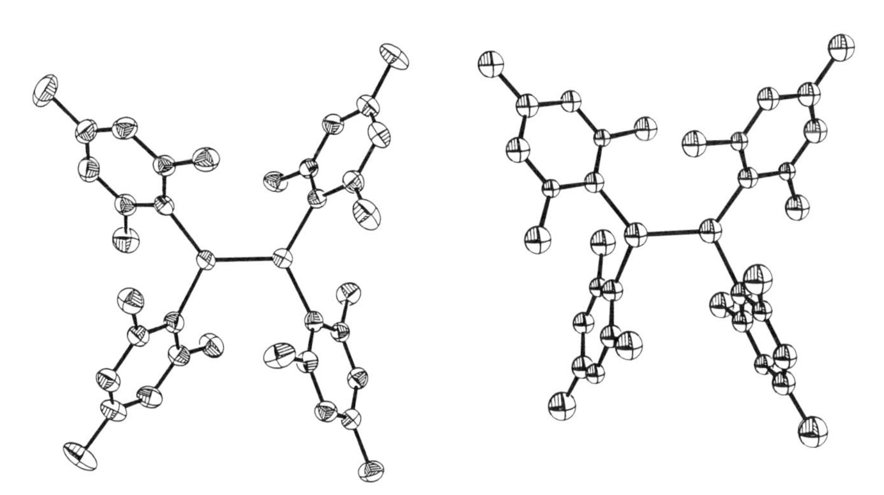

FIG. 2. Thermal ellipsoid diagrams for X-ray crystal structures of **1**. Left, orange unsolvated form; right, yellow toluene-containing crystals (**1** · C_7H_8). The toluene, not shown, occupies a position between the two mesityl rings which are nearly orthogonal to the Si=Si bond.[41,51]

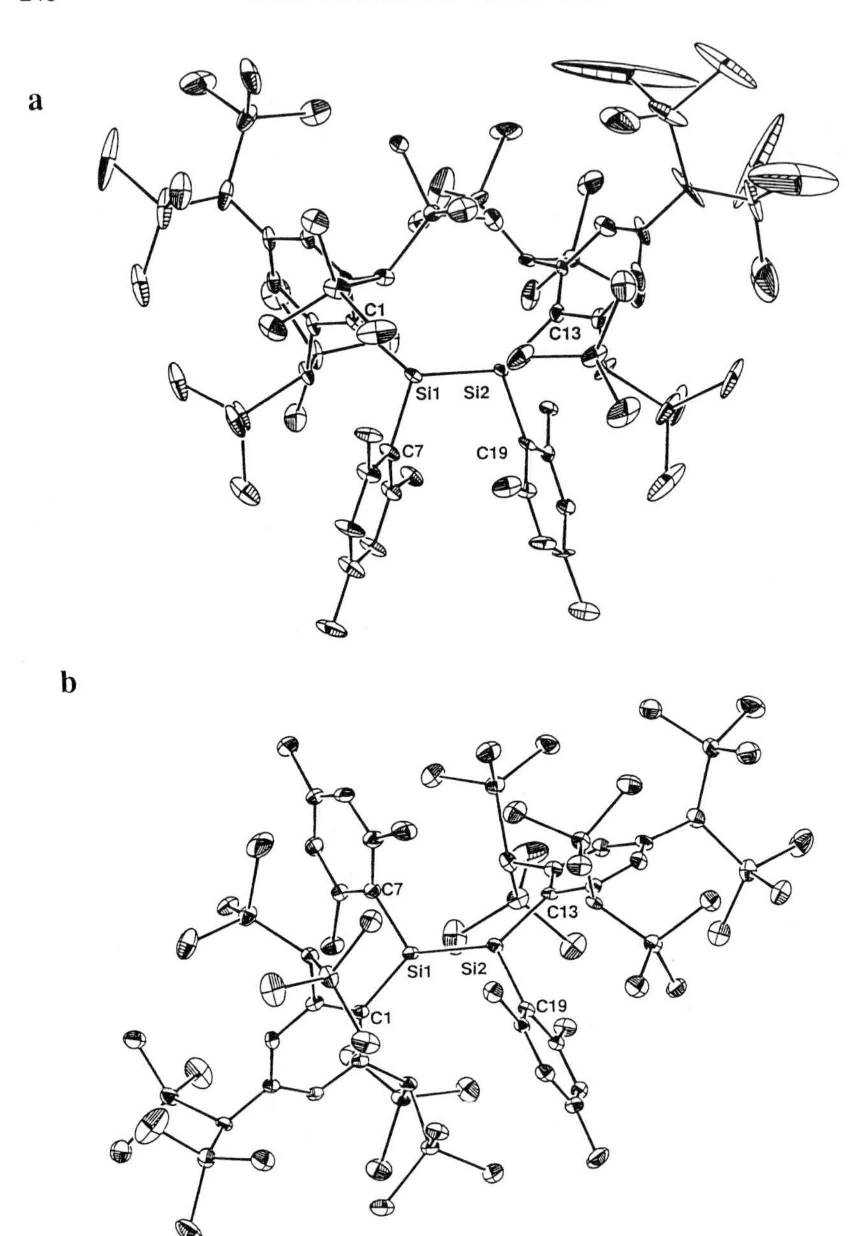

FIG. 3. Thermal ellipsoid diagrams for the *Z* (a) and *E* (b) isomers of **27**, 30% probability level.[24]

bond, it is likely that polymorphic forms will be found for other disilenes in the future.

The structural features of (*E*)- and (*Z*)-**27**[24] merit special comments for two reasons: (1) they are the only pair of *E*- and *Z*-isomers for which the X-ray structures have been analyzed (Fig. 3); (2) they are probably the most sterically congested disilenes so far reported. The Si═Si bond lengths of the **27** isomers, 2.195 and 2.228 Å, are much larger than those of other disilenes having carbon substituents on the silicon atoms, which vary within a range of 2.138–2.160 Å. Both (*E*)- and (*Z*)-**27** are substantially twisted (γ = 8.7° or 14°) as well as trans-bent (δ = 7.6–14.6°). Highly sterically hindered alkenes, such as compounds **33**, **34**, and **35**, relieve their strain mainly by twisting along the C═C axis (28.6° for **33**, 17° for **34**, 37.5° for **35**) rather than by pyramidalization or elongation of the C═C bond (elongations are 1.5% for **33**, 0.5% for **34**, 2.2% for **35**, compared to the normal C═C length of 1.337 Å).[52,53] In contrast, (*Z*)- and (*E*)-**27** relieve their strain by changing the bond angles of Si—Si—C_{Tbt} and Si—Si—C_{Mes} as well as lengthening the Si═Si bond (the degree of lengthening is 2.2% for (*Z*)-**27** and 3.8% for (*E*)-**27**, compared to the mean value of the other carbon-substituted Si═Si bond lengths (2.147 Å)). These characteristic deformations of **27** indicate the "soft" nature of the Si═Si bond in comparison with the C═C bond.

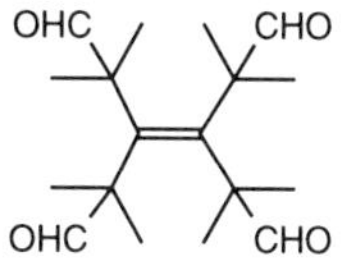

33 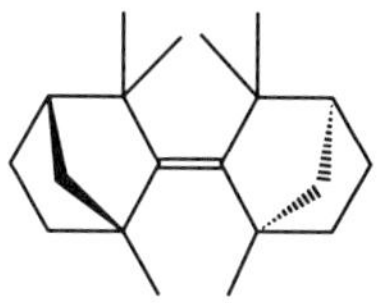 **34** 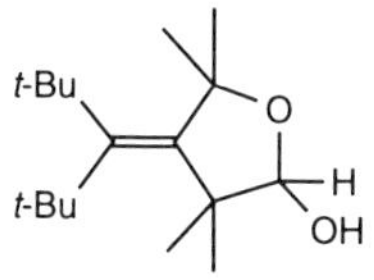 **35**

VII

IONIZATION POTENTIALS, OXIDATION, AND REDUCTION

Very recently, photoelectron spectra have been recorded for disilenes **1**, **3**, and **25**.[54] Consistent with the electronic spectra, the lowest ionizations were assigned to the Si═Si π-orbital. The He(I) photoelectron spectrum for **1** is shown in Fig. 4. The low-energy ionization appears at 7.53 eV. The more intense, higher energy bands were shown to be due to ionizations from mesityl π-orbitals by comparison with the photoelectron spectra of model compounds without a Si═Si bond.

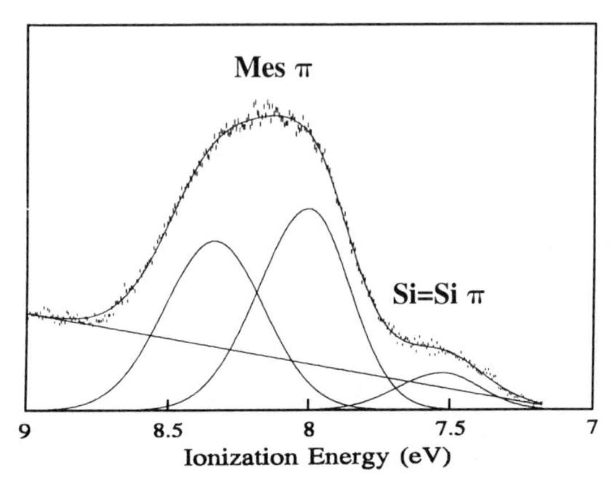

FIG. 4. Helium(I) photoelectron spectrum of **1**, showing ionization of Si=Si π-electrons at 7.53 eV.[54]

The Si=Si π-ionization energies observed were 7.85 eV for **3**, 7.53 eV for **1**, and 7.24 eV for **25**. The decrease from **3** to **1** can be explained by mixing of the mesityl and Si=Si π-orbitals in **1**, pushing the Si=Si π-orbital up in energy. In **3**, this mixing is inhibited by the orthogonal arrangement of the aromatic rings and the Si=Si π-bond, as mentioned above. The very low ionization energy in **25** may reflect weakening or distortion of the Si=Si π-bond in this highly hindered disilene (the structure is unknown).

It is interesting to compare these photoelectron results with solution oxidation and reduction measurements carried out earlier by cyclic voltammetry.[55] Since both oxidation and reduction were irreversible, true oxidation and reduction potentials could not be obtained. Peak potentials were measured, however, and should show any important trends. Compounds studied included three tetraaryldisilenes (**1**, **11**, and **12**) in which aromatic Si—Si conjugation is possible and two dialkyldimesityldisilenes (**3** and **4**) in which the orthogonal arrangement should prevent such orbital mixing. Little difference was found among the peak potentials for oxidation, which all fell in the region +0.38 to +0.54 V vs. SCE. However, significant differences were found in the reduction peak potentials: −2.12, −2.03, and −2.20 V for **1**, **11**, and **12**, compared to −2.66 and −2.64 V for **3** and **4**. Here, the orbital mixing between aryl π*- and Si=Si π*-orbitals works in the opposite sense, pushing the Si=Si π*-orbital downward so that reduction of the tetraaryldisilenes is more favorable.

The photoelectron spectra of **3** and **25** (but not **1**) showed evidence of thermolytic dissociation of the disilenes into silylenes in the spectrometer. This type of dissociation will be treated in Section VIII.B.

VIII

REACTIONS

Many of the fundamental reaction patterns of disilenes, except for thermal dissociation into silylenes, have been described in the previous reviews,[2–5] but ongoing studies on the reactivity of disilenes continue to reveal a rich and diverse chemistry.

A. E–Z *Isomerization*

Information on the E–Z isomerization of the disilenes sheds light on the nature of the chemical bonding of the silicon–silicon double bond. Kinetic parameters, which have been reported for cis–trans isomerization of five disilenes, are listed in Table IV.

The activation enthalpies for cis–trans isomerization in these disilenes, which are considered to be a measure of the Si=Si π-bond strength, range from 24.7 to 30.6 kcal mol^{-1}. These values are in good agreement with the π-bond strength (22–28 kcal mol^{-1}) in $H_2Si{=}SiH_2$ predicted by *ab initio* calculations,[38,56] indicating that π-bonding in disilenes is somewhat weaker than the C=C π-bond in such similar olefins as stilbene, in which E_a (Z to E) is 43 kcal mol^{-1}.[57] The rotational barriers in olefins are known to depend greatly upon substitution, ranging from 65 to less than 20 kcal mol^{-1} and usually decreasing with increasing steric bulk of the substituents.[58] It appears that the situation is more complicated in disilenes. The same trend is emerging in dialkyldiaryldisilenes, as exemplified by the t-butyl vs. 1-adamantyl derivatives (30.6 for **3** and 28.1 kcal mol^{-1} for **4**).

TABLE IV

ACTIVATION PARAMETERS FOR CIS–TRANS ISOMERIZATION IN DISILENES

	E_a (kcal mol^{-1})	log$A$	$\Delta H^{\neq}$ (kcal mol^{-1})	$\Delta S^{\neq}$ (cal mol^{-1} K^{-1})	Reference
(Z)**3**→(E)-**3**	31.3	15.7	30.6	11.0	8
			23	$-8 \sim -12$	4
(E)-**3**→(Z)-**3**			28	$-13 \sim -17$	4
(Z)-**7**→(E)-**7**	25.4	13.2	24.7	-0.3	8
(Z)-**4**→(E)-**4**	28.8 ± 0.6	14.2	28.1 ± 0.6	5.4 ± 1.8	9
(E)-**4**→(Z)-**4**			27	-4	9
(Z)-**20**→(E)-**20**			25.6 ± 0.8	-5.0 ± 2.5	23b
(E)-**20**→(Z)-**20**			26.7 ± 0.8	-2.4 ± 2.5	23b
(Z)-**26**→(E)-**26**			26.3 ± 0.5	-4.0 ± 2.0	23b
(E)-**26**→(Z)-**26**			27.0 ± 0.5	-2.4 ± 2.0	23b

However, the reverse trend is observed, albeit to a minor extent, in the case of tetraaryldisilenes, with the $\Delta H^{\neq}$ of the more hindered 2,6-diisopropylphenyl derivative (26.3 and 27.0 kcal mol^{-1}) being larger than that of the less hindered 2,6-diethylphenyl derivative (25.6 and 26.7 kcal mol^{-1}). Although the reason for the difference between these two types of disilenes remains to be studied, it is probably due to the complexity of conformational equilibria in disilenes owing to the soft nature of the Si═Si bond.

For the amino derivative **7** ($\Delta H^{\neq}$ = 24.7 kcal mol^{-1}), it is likely that both electronic and steric effects are important.

More recently, a new mode of cis–trans isomerization of a disilene has been suggested for the extremely hindered disilene **27**. As will be detailed in Section VIII.B, **27** undergoes thermal dissociation into the corresponding silylenes. Monitoring the thermolysis of (*Z*)-**27** at 50°C by ^{1}H and ^{29}Si NMR reveals a competitive formation of the isomerized (*E*)-**27** and benzosilacyclobutene **37**, which is most likely formed by intramolecular insertion of silylene **36** into the C—H bond of the *o*-bis(trimethylsilyl)-methyl group (Scheme 3).[22,59] This suggests the possible occurrence of cis–trans isomerization via a dissociation–association mechanism.

B. *Thermal Dissociation into Silylenes*

As shown in Scheme 3, disilene **27** undergoes thermal dissociation into silylene **36** under very mild conditions (about 50°C).[22,59] This represents the first example of such dissociation. The formation of **36** was confirmed

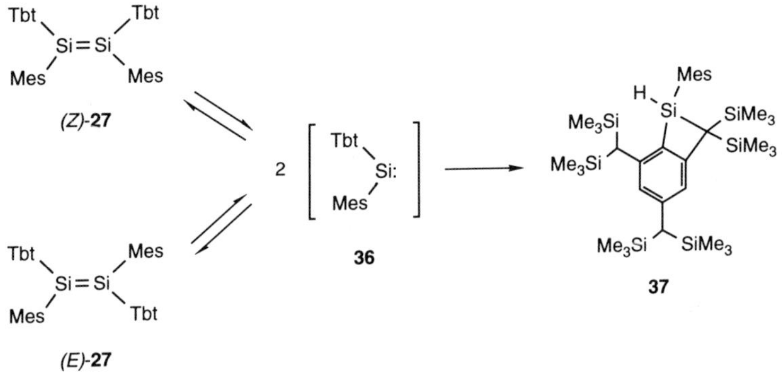

Scheme 3

SCHEME 4

by the reactions with methanol, triethylsilane, and 2,3-dimethyl-1,3-butadiene, all of which are well known trapping reagents for a silylene (Scheme 4). Silanol **38** is most likely produced by the hydrolysis of a silacyclopropane, the initial [1 + 2] cycloadduct of **36** with the diene. The unusually facile dissociation of **27** into silylene **36** is apparently due to the severe steric congestion in **27** caused by the extremely bulky Tbt group, which is manifested in the unusually long Si═Si bond distance (see Section VI). Activation parameters for the thermal dissociation are shown in Table V.

The other disilenes are rather thermally stable (see Section IV.A.), and it has been experimentally established that upon heating they undergo *E–Z* isomerization without breaking of the Si—Si σ-bond.[60,61] It is therefore of interest to compare the $\Delta G^{\neq}$ values of the thermal dissociation of **27** with those of the cis–trans isomerization of structurally similar disilenes **20** and **26** at the same temperature. The $\Delta G^{\neq}$ values at 50°C obtained from the $\Delta H^{\neq}$ and $\Delta S^{\neq}$ values in Tables IV and V are listed in Table VI. Since the $\Delta G^{\neq}$ values for cis–trans isomerization for the 2,6-diisopropylphenyl derivative **26** are almost similar or slightly larger than those for the 2,6-diethylphenyl derivative **20**, the $\Delta G^{\neq}$ values for (*Z*)-**27**/(*E*)-**27** isomerization are reasonably estimated to be similar to or slightly larger than those for the 2,6-diisopropylphenyl derivative **26**, which are much larger than

TABLE V

ACTIVATION PARAMETERS FOR THE THERMAL DISSOCIATION OF **27** INTO **36**

	(*Z*)-**27**	(*E*)-**27**
E_a (kcal mol^{-1})	26.16 ± 0.37	25.61 ± 0.64
log*A*	15.04 ± 0.23	13.54 ± 0.38
$\Delta H^{\neq}$ (kcal mol^{-1})	25.45 ± 0.34	24.97 ± 0.64
$\Delta S^{\neq}$ (cal mol^{-1} K^{-1})	7.97 ± 0.97	1.3 ± 1.7

TABLE VI

$\Delta G^{\neq}$ VALUES FOR THERMAL CIS–TRANS ISOMERIZATION AND DISSOCIATION OF DISILENES[a]

	(E)-**20**→(Z)-**20**	(Z)-**20**→(E)-**20**	(E)-**26**→(Z)-**26**	(Z)-**26**→(E)-**26**	(Z)-**27**→**36**	(E)-**27**→**36**
$\Delta G^{\neq}$ (kcal mol^{-1})	27.2	27.5	27.6	27.8	22.9	24.6

[a] $\Delta G^{\neq}$ values are referred to those at 323 K.

the observed $\Delta G^{\neq}$ values for the dissociation of **27**. This suggests that for **27**, dissociation is a much faster process than bond rotation, and that cis–trans isomerization probably proceeds by the former route.

This silylene formation from **27** under mild conditions permits the synthesis of a variety of interesting carbo- and heterocycles, most of which are new types of compounds. The results are summarized in Schemes 5 and 6. The reactions with benzene and naphthalene represent the first examples of [2+1] cycloadditions of a silylene with aromatic C=C double bonds.[59,62a] The reactions with carbon disulfide and isocyanide (Scheme 6) are also of great interest because of their unusual reaction patterns.[62b]

The dissociation into silylenes observed in photoelectron spectroscopy for **3** and **25** has been described in Section VII. In addition, thermal dissociation of **25** has recently been noted.[63] This requires a somewhat higher temperature than that for dissociation of **27**. At 100°C, **25** shows no reactivity toward Et_3SiH or (remarkably) toward EtOH, but at 120°C it reacts with both substrates to give silylene trapping products (Scheme

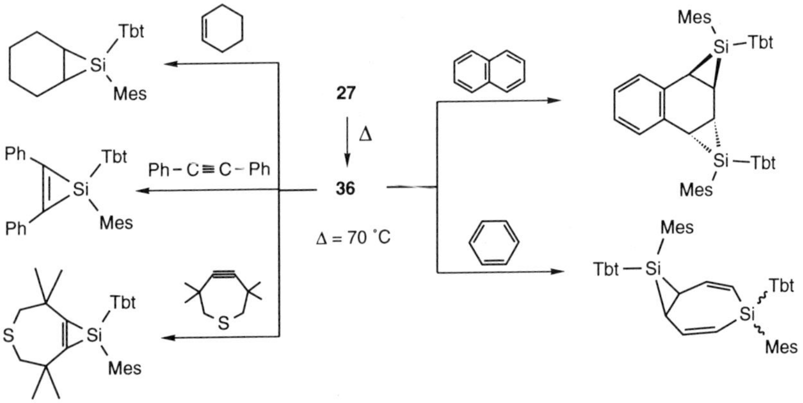

SCHEME 5

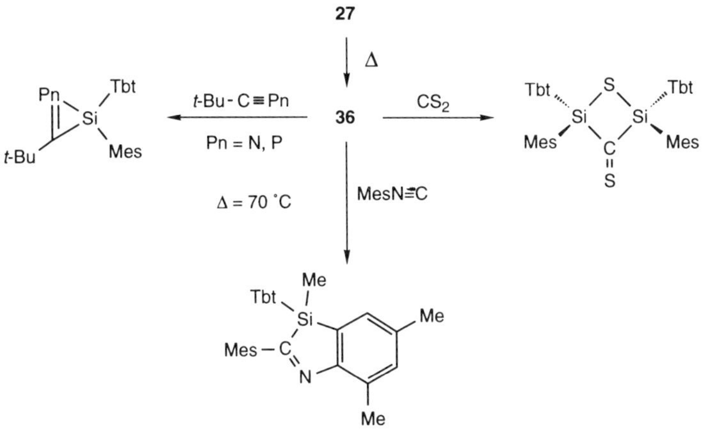

SCHEME 6

7). Earlier, it had been shown that **25** reacts with CH_3OH under photolysis to give similar silylene trapping products; this reaction may involve photochemical dissociation of **25** into silylenes.[23a]

These recent observations suggest that thermal dissociation of disilenes into silylenes may be a more general phenomenon than has previously been thought. Additional examples are likely to be reported in the future.

C. *1,2-Migration of Aryl Groups*

Thermal 1,2-diaryl rearrangement in disilenes was first demonstrated for tetraaryldisilene **13**, which was found to give a mixture of (*Z*)- and (*E*)-**5**.[64] More recently, kinetic data, as well as the scope and limitation of the 1,2-rearrangement, have been reported (Scheme 8).[11,12] Similar rearrangement was also observed for 2,6-dimethyl-4-*t*-butylphenyl derivative **14** to give a mixture of (*Z*)- and (*E*)-**6**, but not for disilenes Mes(R)Si=SiMes(R) (**3**: R = *t*-Bu; **7**: R = $N(SiMe_3)_2$; **8**: R = Tip). These re-

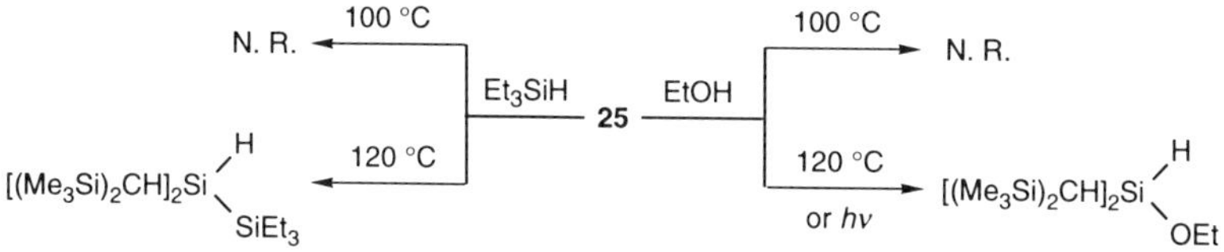

SCHEME 7

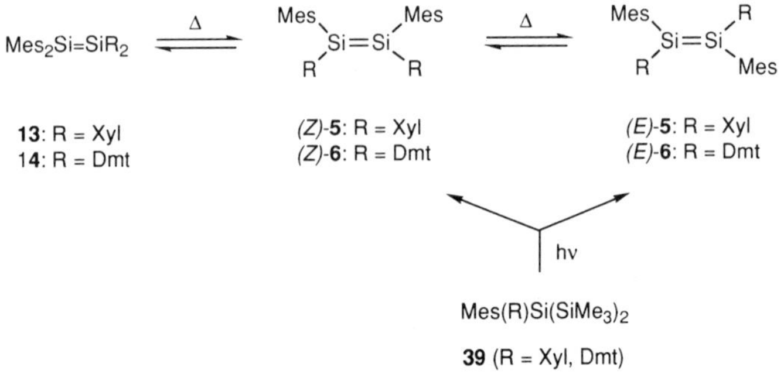

$Mes_2Si{=}SiR_2$ ⇌ (Δ) (Mes)(R)Si=Si(Mes)(R) ⇌ (Δ) (Mes)(R)Si=Si(R)(Mes)

13: R = Xyl
14: R = Dmt

(*Z*)-**5**: R = Xyl
(*Z*)-**6**: R = Dmt

(*E*)-**5**: R = Xyl
(*E*)-**6**: R = Dmt

hν

$Mes(R)Si(SiMe_3)_2$

39 (R = Xyl, Dmt)

SCHEME 8

arrangements can be attained from the opposite direction by using photolysis of trisilanes **39**. In this case, a mixture of *E*- and *Z*-isomers of **5** or **6** is formed first, and then **13** or **14** is observed after a delay. A kinetic study on **13** and (*E*/*Z*)-**5** has revealed that the reaction is first-order, with the activation parameters of $\Delta H^{\neq} = 15 \pm 2$ kcal mol^{-1} and $\Delta S^{\neq} = -36 \pm 4$ eu for the reaction of **13** to **5** and $\Delta H^{\neq} = 14 \pm 2$ kcal mol^{-1} and $\Delta S^{\neq} = -37 \pm 4$ eu for the reaction of **5** to **13**. In view of the fact that the reaction is intramolecular and takes place only for tetraaryldisilenes, a concerted dyotropic mechanism[65] proceeding through a bicyclobutane-like transition state or intermediate was proposed. The mechanism involving such an ordered transition state is consistent with the large negative entropy of activation observed for the reaction.

Alkyl groups are known to be less mobile than aryls in rearrangement, so the lack of rearrangement in **3** is not surprising. The nonoccurrence of rearrangement for **8** is most likely due to the steric bulk of the Tip group, which renders the migration difficult. On the other hand degenerate 1,2-aryl rearrangements probably occur in symmetric tetraaryldisilenes like **1**, but are not observed since no net chemical change takes place.

D. *1,2-Additions*

Disilenes readily add halogens[14,66] and active hydrogen compounds (HX), such as hydrogen halides,[63,66] alcohols, and water,[27,63] as well as hydride reagents, such as tin hydride and lithium aluminum hydride.[66] These reactions are summarized in Scheme 9. The reaction of the stereoisomeric disilene (*E*)-**3** with hydrogen chloride and alcohols led to a mixture of *E*- and *Z*-isomers, but the reaction with chlorine gave only one of the two possible stereoisomers, thus indicating that the former two reactions proceed stepwise while the latter occurs without Si—Si rotation.

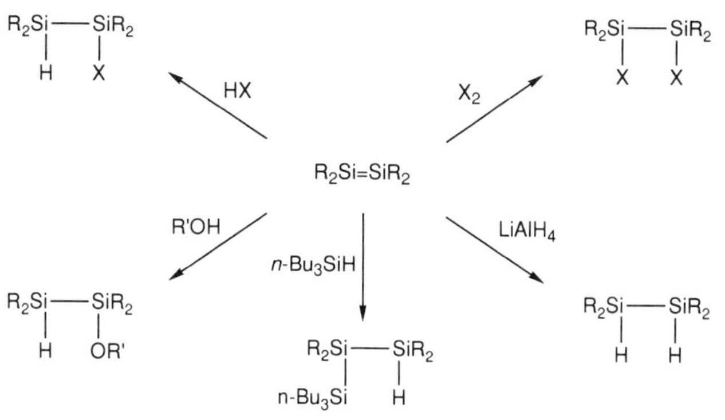

$R_2Si{=}SiR_2$ + HX → $R_2Si(H){-}SiR_2(X)$

$R_2Si{=}SiR_2$ + X_2 → $R_2Si(X){-}SiR_2(X)$

$R_2Si{=}SiR_2$ + R'OH → $R_2Si(H){-}SiR_2(OR')$

$R_2Si{=}SiR_2$ + n-Bu_3SiH → $R_2Si(n\text{-}Bu_3Si){-}SiR_2(H)$

$R_2Si{=}SiR_2$ + $LiAlH_4$ → $R_2Si(H){-}SiR_2(H)$

SCHEME 9

Recently, however, Sekiguchi *et al.* reported that the transient disilene (E)- and (Z)-PhMeSi=SiMePh reacted with alcohols in a syn fashion with high diastereoselectivity, the extent of which depended on the concentration and steric bulk of alcohols used.[31] These facts suggest that in the reaction with a sterically hindered disilene like (Z)-**3**, the addition of alcohols might also proceed diastereoselectively under appropriate reaction conditions.

E. *[2 + 2] Cycloadditions*

Disilenes are very reactive toward a variety of unsaturated bonds, such as C=C, C≡C, C=N, C≡N, C=O, C=S, N=N, and N=O, giving interesting four-membered ring compounds that are otherwise difficult to synthesize.[3b] Reactions using stable tetraaryl- or dialkyldiaryldisilenes and marginally stable tetra-*t*-butyldisilene (**21**), which is generated by photolysis of the corresponding cyclotrisilane, are summarized in Schemes 10 and 11, respectively. Some of them merit further comments.

Disilene **1** reacts not only with aldehydes and ketones but also with some esters (Eq. 10)[68] and acid chlorides (Eqs. 11 and 12).[69] In the reaction

$$\mathbf{1} + \text{2-furyl–OCOCH}_3 \longrightarrow \text{2-furyl–O–}\overset{|}{C}(CH_3)\text{ in ring } [C{-}O{-}SiMes_2{-}SiMes_2] \quad (10)$$

$$\mathbf{1} + C_3F_7COCl \longrightarrow \text{ring } [Mes_2Si{-}SiMes_2{-}O{-}C(C_3F_7)(Cl)] \quad (11)$$

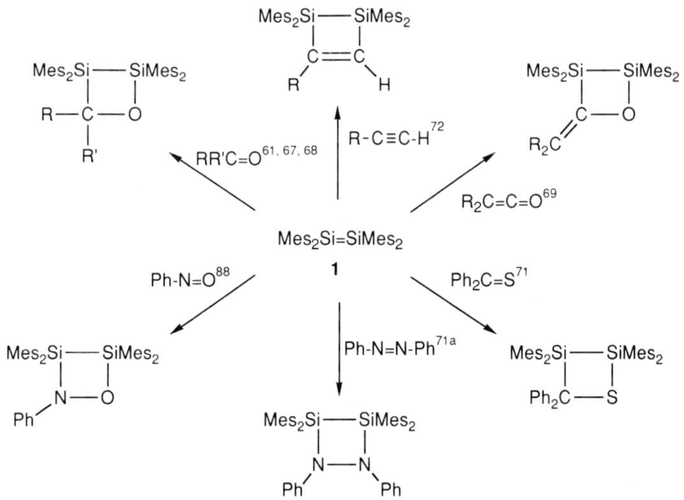

SCHEME 10

of Eq. (12), the substituted benzoyl chloride acts as a 4π-system, including 2π-electrons of the aromatic ring. The 1,2,3-oxadisiletanes **40** obtained from **21** and acetone or benzophenone undergo further rearrangement, giving the corresponding silyl vinyl ethers. (Eq. 13).[70]

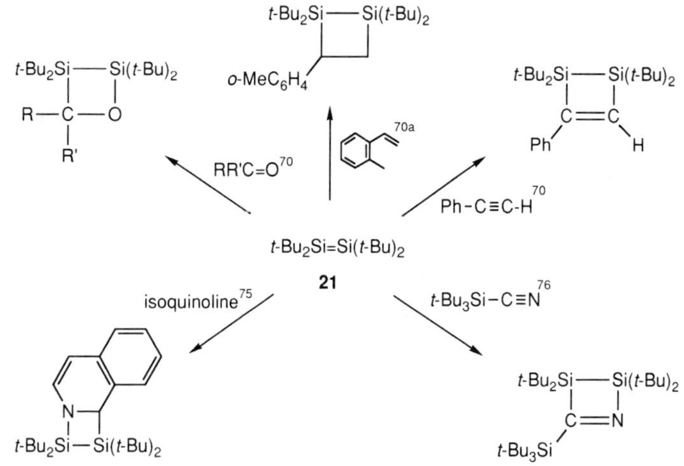

SCHEME 11

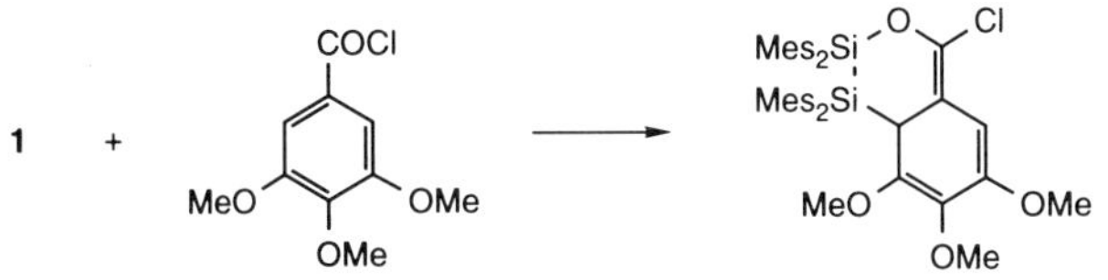

(12)

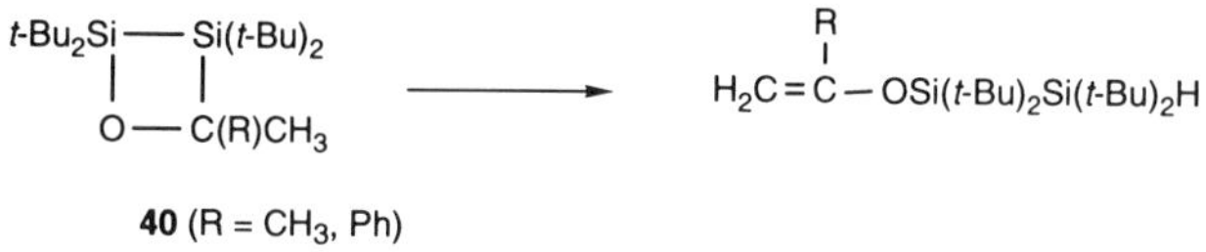

(13)

40 ($R = CH_3$, Ph)

The adducts **41** from **1** and ketones or thiobenzophenone undergo interesting photochemical cycloreversion to afford a silanone or silanethione intermediate **42** in addition to silene **43**; both of these intermediates are trapped by ethanol, as shown in Eq. (14).[68,71] In the reaction with the thiobenzophenone adduct **41** (R = Ph, X = S), the intermediate silene **43** (R = Ph) was detected by ^{29}Si NMR.[71]

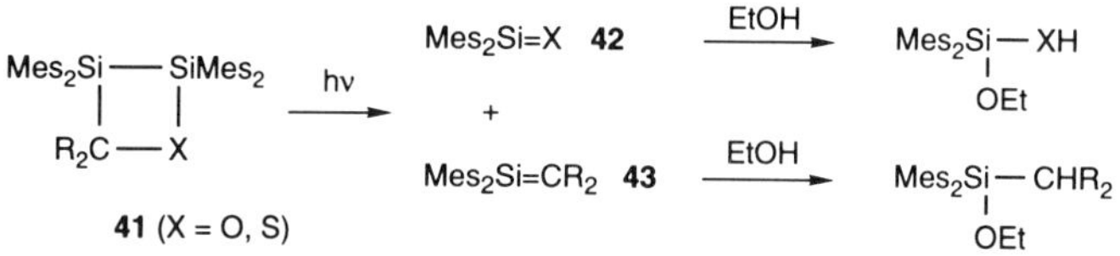

(14)

41 (X = O, S)

In the reaction with acetylenic compounds (PhC≡CH or EtOC≡CH), (*E*)-**3** afforded 1 : 1 stereoisomeric products, indicating that the reaction proceeds stepwise via a diradical or dipolar intermediate.[72]

It is noteworthy that **21** can also react with the C═C bonds in pyridine[73,74] and isobenzofuran[73] skeletons (Eqs. 15–17); by contrast, **1** does not react with alkenes and conjugated dienes.[68]

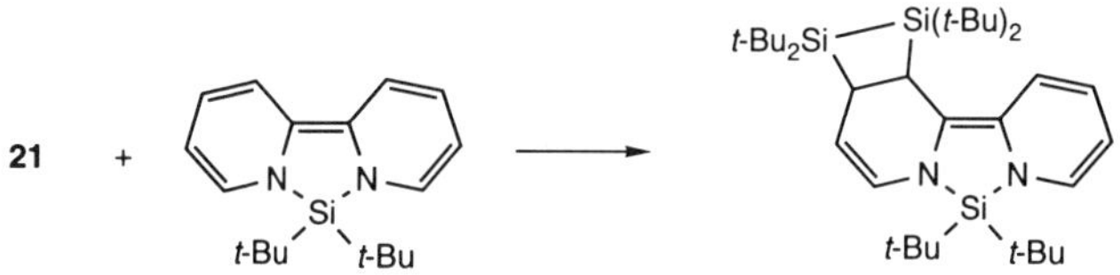

(15)

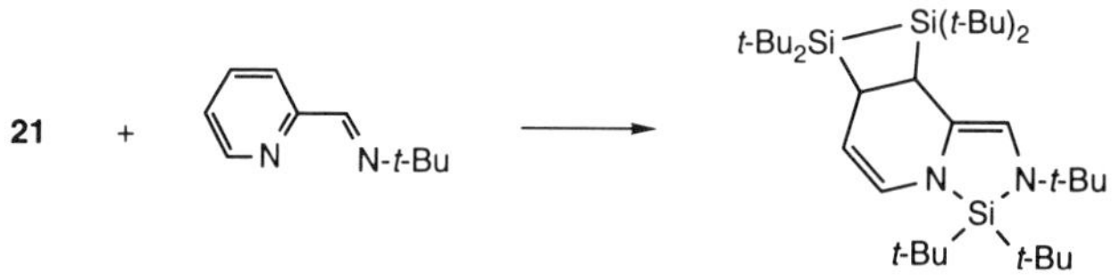

(16)

21 + (1,3-diphenylisobenzofuran: Ph, O, Ph) ⟶ $t\text{-}Bu_2Si$, $Si(t\text{-}Bu)_2$, Ph, O, Ph (17)

Disilene **21** can react with the C=N bond in a cyclic imine [73] and cumulenes such as carbodiimide[75] and keteneimine[74] (Eqs. 18–21). The reactions with cumulenes are not straightforward, although the initial products are most likely [2 + 2] cycloadducts.

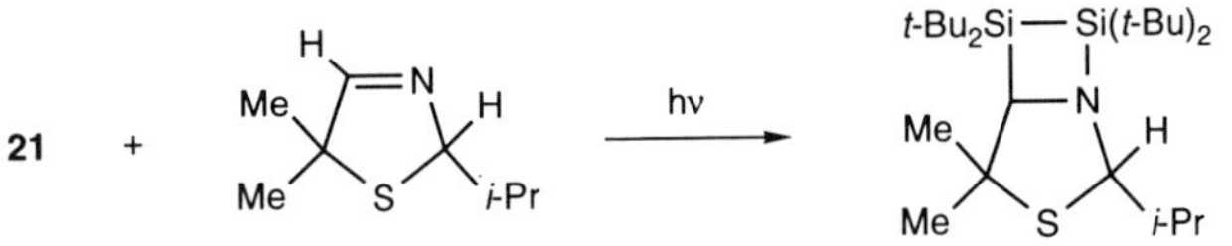

21 + (H, N, H, Me, Me, S, i-Pr) —hν→ $t\text{-}Bu_2Si—Si(t\text{-}Bu)_2$, N, Me, Me, S, H, i-Pr (18)

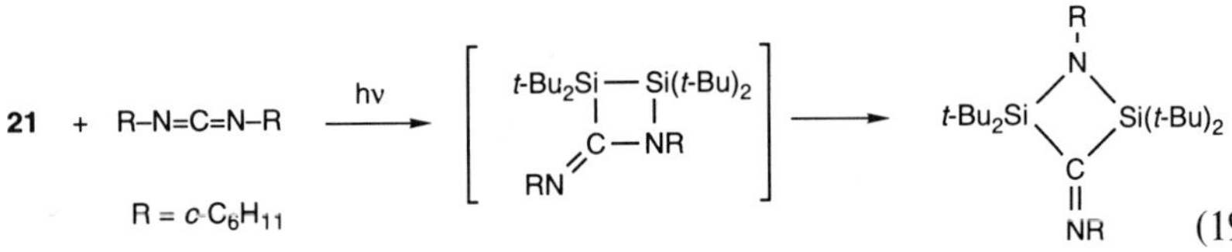

21 + R–N=C=N–R —hν→ [$t\text{-}Bu_2Si—Si(t\text{-}Bu)_2$, C—NR, RN] ⟶ R, N, $t\text{-}Bu_2Si$, $Si(t\text{-}Bu)_2$, C, NR (19)

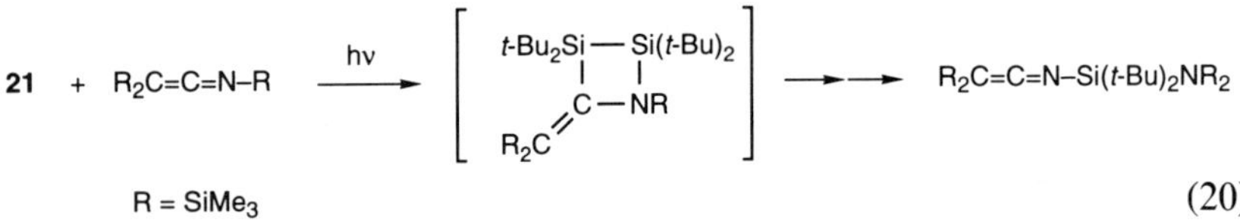

$R = c\text{-}C_6H_{11}$

21 + $R_2C{=}C{=}N{-}R$ —hν→ [$t\text{-}Bu_2Si—Si(t\text{-}Bu)_2$, C—NR, R_2C] ⟶⟶ $R_2C{=}C{=}N{-}Si(t\text{-}Bu)_2NR_2$

$R = SiMe_3$ (20)

21 + $Me_2C{=}C{=}N{-}Ph$ —hν→ [$t\text{-}Bu_2Si—Si(t\text{-}Bu)_2$, C—N, Me_2C, H] ⟶ $t\text{-}Bu_2Si—Si(t\text{-}Bu)_2$, H–C—N, Me_2C

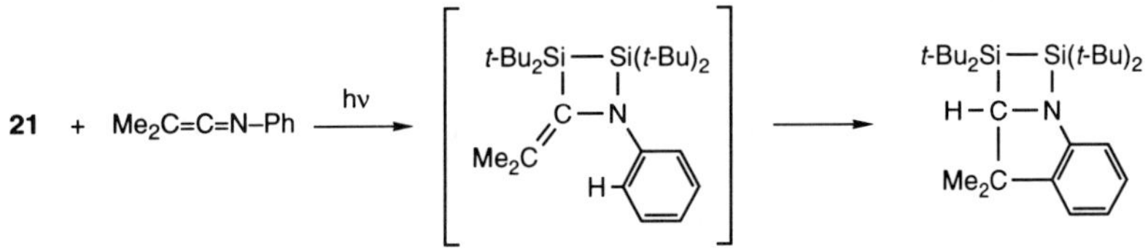

(21)

In contrast to a simple [2 + 2] cycloaddition with a nitrile, **21** undergoes an interesting reaction with an isocyanate involving multiple cleavage and recoupling of bonds to give a five-membered ring compound (Eq. 22).[76]

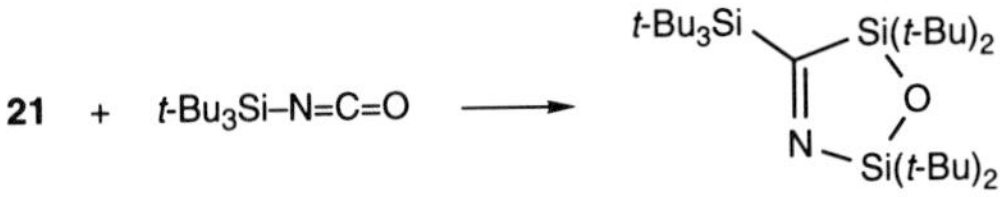

21 + $t\text{-}Bu_3Si{-}N{=}C{=}O$ ⟶ $t\text{-}Bu_3Si$, $Si(t\text{-}Bu)_2$, O, N, $Si(t\text{-}Bu)_2$ (22)

F. *[2 + 1] Cycloadditions*

Disilenes react with various types of reagents to afford novel three-membered cyclic compounds that are otherwise inaccessible. Even though some are not mechanistically [2 + 1] cycloadditions, all the reactions in which three-membered rings are formed from disilenes are summarized in Scheme 12.

While the reaction of **11** with diazomethane gives disilacyclopropane **44**,[77] that of **1** with *p*-tolyldiazomethane affords an azadisilirane derivative **45**[78] via [2 + 1] cycloaddition at the terminal nitrogen of the diazomethane. Similar [2 + 1] cycloadditions also take place in the reactions with aliphatic azides to give **46**,[82] while aromatic azides product [2 + 3] cycloadducts **47**.[79] Both **46** and **47** give the corresponding azadisiliranes **46′** and **46″** upon heating.[79] Disilenes **1** and **2** are readily oxidized by N_2O[80] and *m*-chloroperbenzoic acid (MCPBA)[7] to give the corresponding oxadisiliranes **48** and **49**, respectively (for oxidation with molecular oxygen, see Section VIII.H). Reactions of **1** with elemental chalcogens proceed at room temperature to give the corresponding chalcogenadisiliranes **50**,[81] **51**,[82] and **52**.[82] Cycloaddition of **1** with an isonitrile, a carbene-like reagent, affords **53**.[83]

Thiadisiliranes are also formed in high yield in the reaction of disilenes with episulfides. Thus **1** with cyclohexene sulfide gave **50**[84]; and the reac-

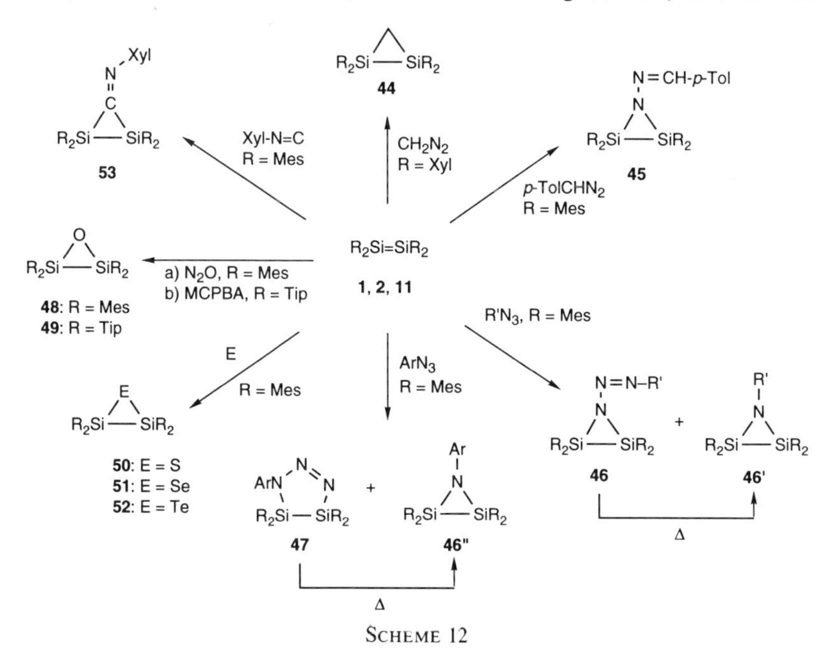

SCHEME 12

SCHEME 13

tions of (E)-**10**, (Z)-**10**, and an E/Z mixture of **3** with propylene sulfide or ethylene sulfide all yielded the corresponding thiadisiliranes.[85]

From X-ray crystal structures of the products, the reactions of these stereoisomeric disilenes with episulfides, and with sulfur, were shown to proceed with retention of configuration at silicon. These findings suggest that the reaction proceeds in a concerted fashion, through intermediates or transition states involving tetracoordination for the sulfur atom being transferred (Scheme 13). Similar intermediates are believed to occur in other sulfur-transfer reactions.[86]

The structures of the three-membered ring compounds obtained in these reactions are of great interest. The pertinent structural parameters are listed in Table VII. The column headed ΣSi gives the sum of the bond

TABLE VII

STRUCTURAL PARAMETERS OF THREE-MEMBERED RING COMPOUNDS DERIVED FROM DISILENES

Compound	E	R	d(Si—Si) (Å)	ΣSi (deg.)	Reference
48	O	Mes	2.227	360.0	80
50	S	Mes	2.289	357.3, 357.5	81
51	Se	Mes	2.303	355.6, 356.0	82
52	Te	Mes	2.337	355.3, 355.8	82
44	CH_2	Mes	2.272	357.4, 357.7	77
53	C=NXyl	Mes	2.328	357.5, 354.9	83
46′ (R′ = $SiMe_3$)	$NSiMe_3$	Mes	2.232	357.6, 358.1	79
49	O	Tip	2.254	359.7, 359.8	7

angles around the silicon, excluding the E atom(s). Most of these compounds have Si—Si distances shorter than the normal single bond length of 2.34–2.35 Å, as well as near-planar arrangements of R, R, and Si substituents around the silicon. The extreme is reached for disilaoxirane **48**, which has ΣSi = 360° and a Si—Si distance of only 2.227 Å, which is closer to a normal double bond than to the single bond length. These unusual structures suggest that some π-bonding may remain between the silicon atoms in **48**, and perhaps in other molecules in this series as well.[87] The nature of the chemical bonding in molecules such as **48** is however still controversial.

G. *Other Cycloadditions*

As mentioned in Section VIII.F (Scheme 12), **1** reacts with aromatic azides to give [3 + 2] cycloadducts **47**. Sulfonyl and phosphoryl azides, however, react with **1** in a different manner to afford five-membered ring compounds (Eqs. 23 and 24).[79]

1 + p-TolSO_2N_3 ⟶ [five-membered ring: p-Tol(O)S=N–Mes_2Si—$SiMes_2$–O] (23)

1 + $Ph_2P(O)N_3$ ⟶ [five-membered ring: Ph_2P=N–Mes_2Si—$SiMes_2$–O] (24)

Reaction of **1** with nitrobenzene gives dioxazadisilolidine **55**. Since an intermediate having a single ^{29}Si NMR peak appears at low temperature, the reaction most likely proceeds via the initial formation of [3 + 2] cycloadduct **54** (Eq. 25).[88]

1 + $PhNO_2$ ⟶ [Ph–N(O–Mes_2Si—$SiMes_2$–O)] **54** ⟶ O–N(Ph), Mes_2Si, $SiMes_2$, O **55** (25)

Although disilenes do not undergo Diels–Alder reaction with 1,3-dienes,[68] they react with heterodienes like benzil,[67,89] acylimines,[90] and 1,4-diazabutadienes[91] to give [4 + 2] cycloadducts (Eqs. 26–28).

1 or **21** + PhC(O)C(O)Ph ⟶ R_2Si—SiR_2 (O, O; Ph, Ph) (26)

21 + PhC(O)C(=NMe)Ph ⟶ t-Bu_2Si—Si(t-Bu)$_2$ (O, N–Me; Ph, Ph) (27)

21 + RN=CH–CH=NR ⟶ t-Bu_2Si—Si(t-Bu)$_2$ (R–N, N–R) (28)

Reactions of **1** with epoxides involve some cycloaddition products, and thus will be treated here. Such reactions are quite complicated and have been studied in some depth.[84,92] With cyclohexene oxide, **1** yields the disilaoxirane **48**, cyclohexene, and the silyl enol ether **56** (Eq. 29). With (*E*)- and (*Z*)-stilbene oxides (Eq. 30) the products include **48**, (*E*)- and (*Z*)-stilbenes, the *E*- and *Z*-isomers of silyl enol ether **57**, and only one (trans) stereoisomer of the five-membered ring compound **58**. The products have been rationalized in terms of the mechanism detailed in Scheme 14, involving a ring-opened zwitterionic intermediate, allowing for carbon–carbon bond rotation and the observed stereochemistry.

1 + cyclohexene oxide ⟶ $OSiMes_2SiMes_2H$ (**56**) + Mes_2Si—$SiMes_2$ (O) (**48**) (29)

1 + Ph–epoxide–Ph ⟶ Ph, $OSiMes_2SiMes_2H$, Ph, H (**57**) + Ph, O, $SiMes_2$, Ph, $SiMes_2$ (**58**) + Mes_2Si—$SiMes_2$ (O) (**48**) (30)

H. *Oxidation*

As described in Section IV.A, disilenes are usually highly reactive toward oxygen as expected from their low oxidation potentials[55] even when they have sterically hindering groups.[7,93,94] Solid disilenes react with atmospheric oxygen to afford 1,3-cyclodisiloxanes (Scheme 15). Reaction of disilenes in solution with molecular oxygen leads to 1,2-disiladioxetanes **59** as the major products, accompanied at room temperature by a smaller

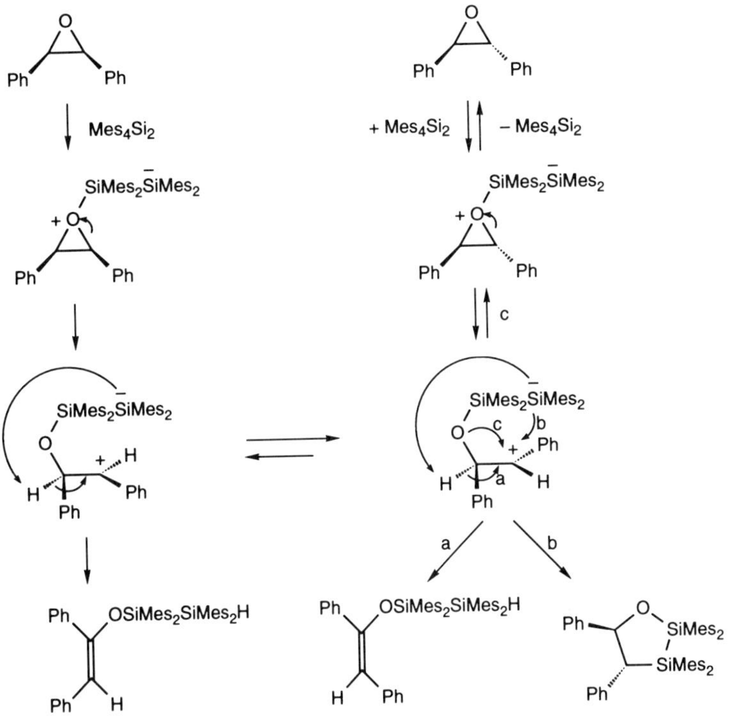

SCHEME 14

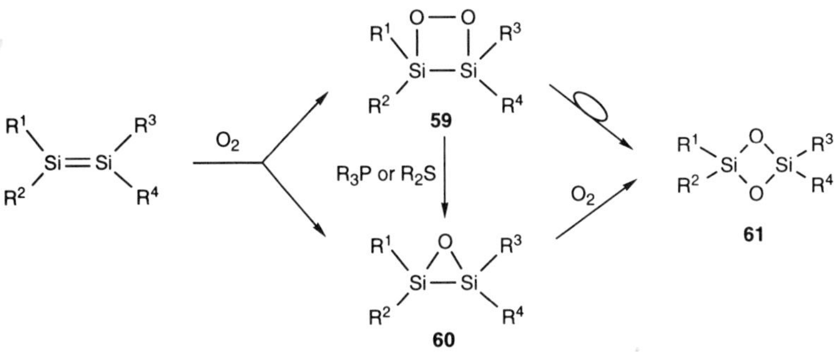

59a, **60a**, **61a**: $R^1 = R^3 = Mes$, $R^2 = R^4 = t\text{-}Bu$

SCHEME 15

amount of disilaoxiranes **60** (see also **48** and **49** in Scheme 12) which further react with oxygen to give **61**. Compounds **59** undergo a quantitative rearrangement to **61**, either thermally or photochemically. Partial deoxygenation of **59** proceeds in the presence of phosphines or sulfides to form **60**.[93]

The stereochemistry of dioxygen oxidation has been studied for the di-*t*-butyldimesityldisilene (**3**).[93] Oxidation of (*E*)-**3** produced (*E*)-**59a**, (*E*)-**60a**, and ultimately (*E*)-**61a** exclusively, showing that all the steps in the sequence are stereospecific with retention of configuration. (Similar oxidation of a mixture of (*E*)- and (*Z*)-**3** gave isomeric mixtures of **59a**, **60a**, and **61a** having the same proportions of stereoisomers as in the starting material.) Oxidation of **3** was found to be stereospecific both in solution and in the solid state.

Retention of configuration is expected for conversion of **3** to **59a**. It is curious, however, for the rearrangement of **59a** to **60a**, since the most obvious pathway—opening of the O—O bond followed by rotation and reclosing—should lead to inversion of configuration. Double isotopic labeling with ^{18}O was used to show that the rearrangement of **59a** to **61a**, as well as the initial oxidation of **3** to **59a**, is fully intramolecular.

The rearrangement of **59a** to **61a** in the solid state could be observed by single-crystal X-ray analysis. Figure 5 shows a thermal ellipsoid plot for (*E*)-**59a** partially rearranged (20.3%) to (*E*)-**61a**. The population of oxygens in the 1,2-dioxetane positions is reduced by 20%, and 20% population of oxygens is found in the rearranged 1,3-positions. Beyond about 25% rearrangement the crystal lattice is destroyed, so the process can no longer be followed by X-ray crystallography.

The X-ray results show that in rearrangement in the solid, one oxygen

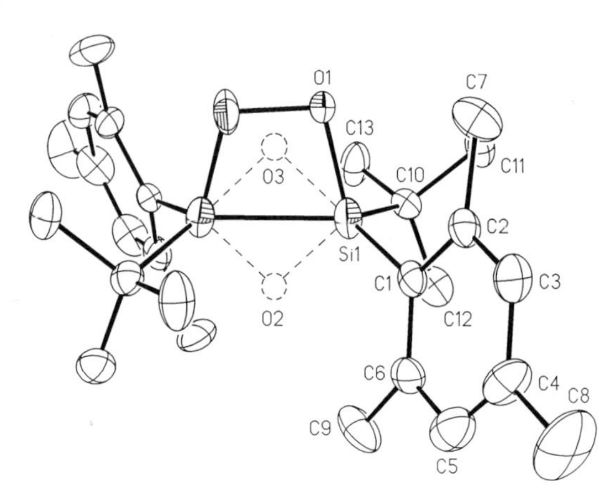

FIG. 5. Thermal ellipsoid diagram for X-ray structure of **59a** partially (20.3%) rearranged to **61a**.

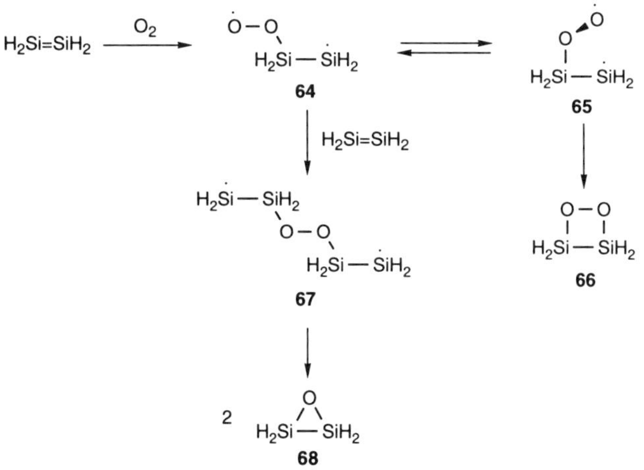

SCHEME 16

atom shifts only a short distance, but the other must move across the Si—Si bond, a distance of nearly 4 Å. The mechanism of this remarkable rearrangement is not understood in detail.

The early stages in the oxidation of disilene have been treated theoretically for the parent molecule $H_2Si{=}SiH_2$.[95] The first intermediate along the reaction coordinate is the open-chain trans diradical **64** (Scheme 16), which is in equilibrium with a gauche form, **65**. From the latter, closure to the 1,2-dioxetane **66** would probably be rapid. The open-chain form can react with a second molecule of disilene to give the diradical **67**, which could collapse into two molecules of the disilaoxirane **68**. If similar steps are followed in the oxidation of **3**, they must be quite rapid, since the relative configuration at the silicon atoms is maintained in both products, **59a** and **61a**.[93]

The dioxygen oxidation of **2** takes a slightly different course.[7,96] The initial product is again the 1,2-dioxetane, part of which rearranges to give the 1,3-dioxetane **63** while a hydrogen-transfer reaction gives **62** (Eq. 31). Transfer of hydrogen from an isopropyl group to silicon is probably favored by the steric bulk of the substituents in the 1,2-dioxetane. The air oxidation of **25** is also abnormal; the major product is the disilaoxirane.[23]

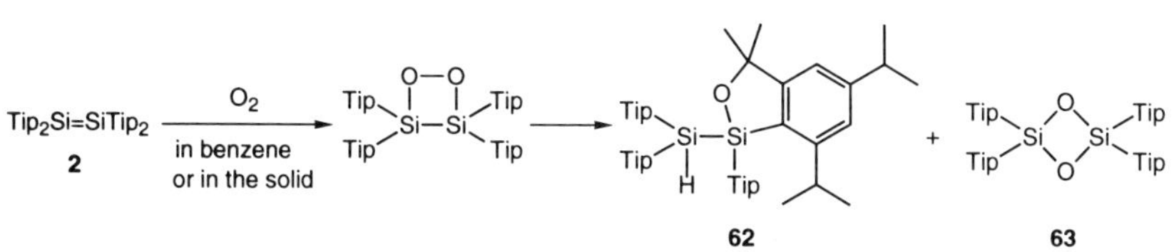

(31)

The highly hindered disilene **27** is very stable toward oxygen in solution; **27** in benzene can survive bubbling of oxygen at room temperature, and the oxidation is completed only after about 10 h at 50°C. This stability is surprising considering that **4**, the most stable of the disilenes, undergoes oxidation at −78°C within several minutes by oxygen bubbling. The reaction is stereospecific, as with the other disilenes; (*E*)- and (*Z*)-**27** give the corresponding (*E*)- and (*Z*)-1,3-cyclodisiloxanes, respectively (Eq. 32).[24]

$$\underset{\mathbf{27}}{\text{Tbt(Mes)Si=Si(Mes)Tbt}} \xrightarrow[\text{in benzene or in the solid}]{O_2} \text{Tbt(Mes)Si}\langle O_2 \rangle\text{Si(Mes)Tbt} \quad (32)$$

The 1,3-dioxetanes **61**, the first cyclodisiloxanes, are of special interest; cyclosiloxanes with larger ring size are critical intermediates in the industrial manufacture of silicon polymers. In addition, compounds **61** have unusual structures, summarized in a recent publication.[94] These compounds exhibit short Si—Si separations of about 2.39 Å, only slightly longer than the normal Si—Si single bond length. This feature has stirred controversy concerning the possible existence of cross-ring bonding in **61**. The nature of bonding in these molecules is still being debated in the literature.[87,97]

I. *Reactions with P_4 and As_4*

Quite unusual reactions take place between disilenes and the active forms of the group 15 elements, P_4 and As_4. In nearly all reactions of disilenes, the Si—Si π-bond is broken in the first step, leaving the σ bond between the silicons intact. It was therefore a surprise when the products isolated from the reaction of disilenes **1**, **11**, and **12** with P_4 were found to have the [1.1.0]bicyclobutane structure **69a–c**.[98,99] In these butterfly-shaped compounds the silicon atoms are completely separated, although a P—P σ-bond remains (Eq. 33).

$$P_4 + R_2Si{=}SiR_2 \xrightarrow[C_6H_5CH_3]{40\ °C} \text{R}_2\text{Si}(\text{P–P})\text{SiR}_2 \quad (33)$$

69a, R = Mes
69b, R = Xyl
69c, R = Dmt

The products are, in turn, starting materials for a rich chemistry, only superficially explored to date. The phosphorus atoms serve as Lewis bases toward metals. Coordination complexes with one or two tungsten atoms have been isolated (Eq. 34). They also readily undergo insertion reactions

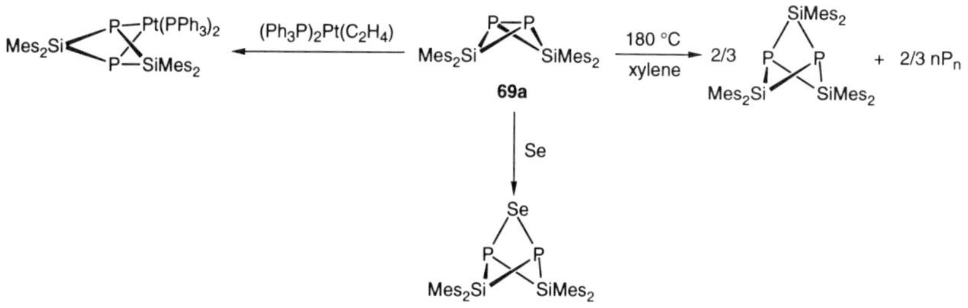

SCHEME 17

into the P—P bond to give unusual asterane-like products, as shown for **69a** in Scheme 17.[100]

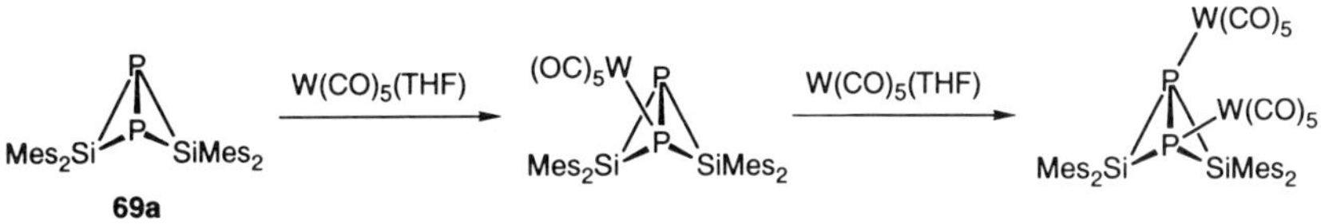

(34)

The highly hindered disilene **2** did not react with white phosphorus, even under forcing conditions. With disilene **3**, which is more hindered than **1** but less so than **2**, the reaction with P_4 was more complicated. It proceeded slowly, producing small amounts of both stereoisomers of the bicyclobutane compounds **70** and **70′**. The major product, however, was a more complex compound containing four phosphorus and four silicon atoms, also obtained as a mixture of two stereoisomers. Two-dimensional ^{31}P NMR spectroscopy established the probable structures to be **71**.[98]

Compounds **71** are precursors to the bicyclobutanes, as shown in Scheme 18; heating of the reaction mixture led to their disappearance, with formation of **70** and **70′**. At the same time rearrangement of **70** to the exo–exo isomer **70′** took place. This appears to be the only reaction of disilenes known to occur with inversion of configuration at silicon.

The reaction wtih As_4, generated by vaporization of metallic arsenic into cold toluene, has so far been reported only for **1**.[101] Two principal products (79%) were formed in 1 : 3 ratio: the bicyclobutane **72** and the remarkable compound **73** (Scheme 19). The structure of the latter was established by X-ray crystallography. It appears to be an intermediate on the reaction pathway, since long heating at 95°C converted it completely to **72**.

A common pathway has been tentatively put forward to account for both the arsenic and phosphorus reaction products (Scheme 20). Breaking of two bonds to one of the atoms in the P_4 or As_4 tetrahedron would lead

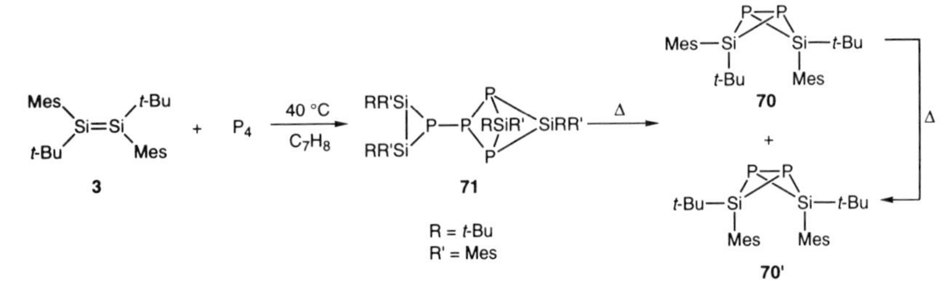

Mes t-Bu Si=Si t-Bu Mes
3
+ P_4
40 °C
C_7H_8
RR'Si RR'Si P—P RSiR' SiRR' P P
71
R = t-Bu
R' = Mes
Δ
Mes—Si P—P Si—t-Bu t-Bu Mes
70
+
t-Bu—Si P—P Si—t-Bu Mes Mes
70'
Δ

SCHEME 18

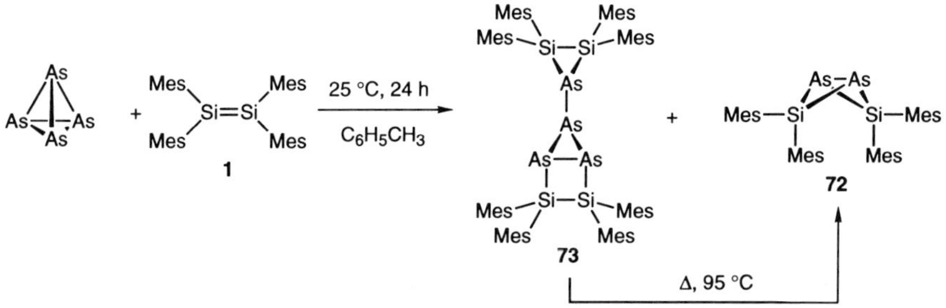

As As As As
+
Mes Mes Si=Si Mes Mes
1
25 °C, 24 h
$C_6H_5CH_3$
Mes Mes Mes Mes Si—Si As As As—As Si—Si Mes Mes Mes Mes
73
+
As—As Mes—Si Si—Mes Mes Mes
72
Δ, 95 °C

SCHEME 19

M M M M
+
Mes Mes Si=Si Mes Mes
1
Mes Mes Mes Mes Si—Si M M M—M Si—Si Mes Mes Mes Mes
74
Mes Mes Mes Mes Si—Si M M M—M Mes—Si Si—Mes Mes Mes
75
Mes Mes Mes Mes Si—Si M M
+
M—M Mes—Si Si—Mes Mes Mes

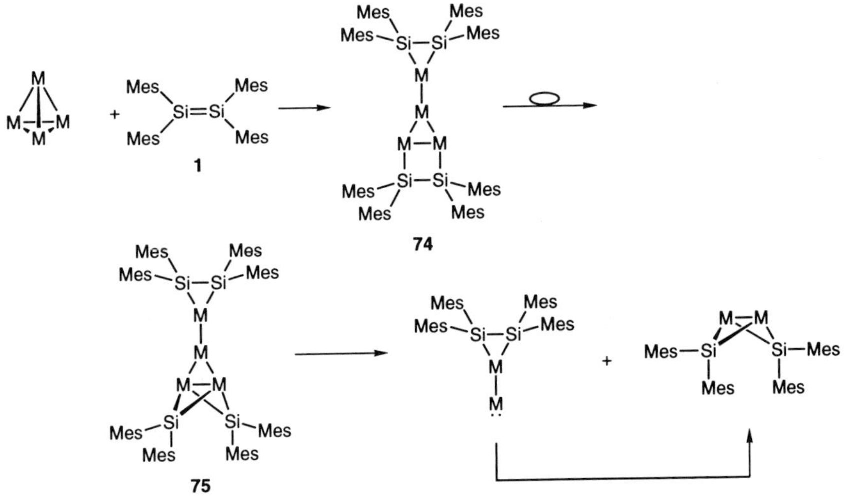

SCHEME 20

to an intermediate that could be trapped to give **74**. The latter could rearrange intramolecularly to the more open structure **75**, which could fragment into two bicyclobutane molecules. This common pathway would require that **74** be, for some reason, stable for arsenic but not for phosphorus.

J. *Disilene–Metal Complexes*

Reactions of disilenes to form transition metal complexes have hardly been investigated; complexes are reported only for platinum. An example is shown in Eq. (35). The Pt complex **76** was identified spectroscopically and from trapping products, but it was a minor product in the reaction and crystals could not be obtained.[102]

$$(Ph_3P)_2Pt(C_2H_4) \quad + \quad \mathbf{1} \longrightarrow (Ph_3P)_2Pt(SiMes_2{-}SiMes_2)\ \ \mathbf{76} \tag{35}$$

The chemistry of silylene–metal complexes has developed in quite another direction, however, from reactions of disilyl–metal complexes, leading to complexes of otherwise unstable disilenes such as $Me_2Si{=}SiMe_2$. Molybdenum and tungsten complexes have been particularly well investigated by Berry and co-workers,[103] and platinum complexes have also been isolated.[104] Readers interested in this field are directed to a 1992 review of silylene, silene, and disilene–metal complexes.[105]

IX

CONCLUSION

At this time, although details remain to be elucidated, it seems that most of the main kinds of reactions of disilenes, have been discovered. These have led to numerous new classes of cyclic compounds, some with unprecedented and beautiful structures. In addition, the chemical bonding in disilenes has received serious study and is at least partially understood.

What developments are likely in the future? Now that many reactions are known, the investigation of reaction mechanisms of disilenes is just beginning. Much is likely to be learned about details of reaction pathways. Further studies, both experimental and theoretical, should lead to a more complete understanding of the chemical bonding in disilenes. Vibrational spectra and electronic excited states of disilenes have received very little attention as yet, and almost nothing is known of disilene photochemistry. All are likely to be explored further in years to come.

Finally, the chemical properties of the new structures available from disilene addition reactions have hardly been touched. In future developments, theoretical and experimental studies are likely to proceed together and complement one another, as they have from the beginning days of this research.

Acknowledgments

We are indebted to Drs. N. Tokitoh and H. Suzuki of the University of Tokyo for their valuable assistance in preparing this manuscript for publication. R.W. thanks Tokohashi University of Technology for a Visiting Professorship which made writing this chapter possible.

References

(1) West, R.; Fink, M. J.; Michl, J. *Science* **1981,** *214*, 1343.
(2) (a) Raabe, G.; Michl, J. *Chem. Rev.* **1985,** *85*, 419. (b) Raabe, G.; Michl, J. In *The Chemistry of Organic Silicon Compounds*; Patai, S.; Rappoport, Z., Eds.; Wiley: New York, 1989, p. 1015.
(3) (a) West, R. *Angew. Chem., Int. Ed. Engl.* **1987,** *26*, 1201. (b) West, R. In *The Chemistry of Inorganic Ring Compounds*; Steudel, R., Ed.; Elsevier: Amsterdam, 1992; p. 35.
(4) Tsumuraya, T.; Batcheller, S. A.; Masamune, S. *Angew. Chem., Int. Ed. Engl.* **1991,** *30*, 902.
(5) Weidenbruch, M. *Coord. Chem. Rev.* **1994,** *130*, 275.
(6) Tan, R. P.; Gillette, G. R.; Yokelson, H. B.; West, R. *Inorg. Synth.* **1992,** *29*, 19.
(7) Millevolte, A. J.; Powell, D. R.; Johnson, S. G.; West, R. *Organometallics* **1992,** *11*, 1091.
(8) Michalczyk, M. J.; West, R.; Michl, J. *Organometallics* **1985,** *4*, 826.
(9) Shepherd, B. D.; Powell, D. R.; West, R. *Organometallics* **1989,** *8*, 2664.
(10) Archibald, R. S.; van den Winkel, Y.; Millevolte, A. J.; Desper, J. M.; West, R. *Organometallics* **1992,** *11*, 3276.
(11) Yokelson, H. B.; Siegel, D. A.; Millevolte, A. J.; Maxka, J.; West, R. *Organometallics* **1990,** *9*, 1005.
(12) Archibald, R. S.; van den Winkel, Y.; Powell, D. R.; West, R. *J. Organomet. Chem.* **1993,** *446*, 67.
(13) Sekiguchi, A.; Hagiwara, K.; Ando, W. *Chem. Lett.* **1987,** 209.
(14) Masamune, S.; Hanzawa, Y.; Murakami, S.; Bally, T.; Blount, J. F. *J. Am. Chem. Soc.* **1982,** *104*, 1150.
(15) (a) Masamune, S.; Murakami, S.; Snow, J. T.; Tobita, H.; Williams, D. J. *Organometallics* **1984,** *3*, 333. (b) Murakami, S.; Collins, S.; Masamune, S. *Tetrahedron Lett.* **1984,** *25*, 2131. (c) Batcheller, S. A. Ph.D. Dissertation, Massachusetts Institute of Technology, Cambridge, MA, 1989. (d) Murakami, S. Ph.D. Dissertation, University of Tokyo, 1985.
(16) Masamune, S.; Tobita, H.; Murakami, S. *J. Am. Chem. Soc.* **1983,** *105*, 6524.
(17) (a) Watanabe, H.; Okawa, T.; Kato, M.; Nagai, Y. *J. Chem. Soc., Chem. Commun.* **1983,** 781. (b) Watanabe, H.; Kato, M.; Okawa, T.; Nagai, Y.; Goto, M. *J. Organomet. Chem.* **1984,** *271*, 225.

(18) (a) Schäfer, A.; Weidenbruch, M.; Peters, K.; von Schnering, H. G. *Angew. Chem., Int. Ed. Engl.* **1984,** *23*, 302. (b) Boudjouk, P.; Samaraweera, U.; Sooriyakumaran, R.; Chrusciel, J.; Anderson, K. R. *Angew. Chem., Int. Ed. Engl.* **1988,** *27*, 1355.
(19) Matsumoto, H.; Sakamoto, A.; Nagai, Y. *J. Chem. Soc., Chem. Commun.* **1986,** 1768.
(20) Watanabe, H.; Takeuchi, K.; Fukawa, N.; Kato, M.; Goto, M.; Nagai, Y. *Chem. Lett.* **1987,** 1341.
(21) Kira, M.; Maruyama, T.; Kabuto, C.; Ebata, K.; Sakurai, H. *Angew. Chem., Int. Ed. Engl.* **1994,** *33*, 1489.
(22) Tokitoh, N.; Suzuki, H.; Okazaki, R.; Ogawa, K. *J. Am. Chem. Soc.* **1993,** *115*, 10428.
(23) (a) Masamune, S.; Eriyama, Y.; Kawase, T. *Angew. Chem., Int. Ed. Engl.* **1987,** *26*, 584. (b) Batcheller, S. A.; Tsumuraya, T.; Tempkin, O.; Davis, W. M.; Masamune, S. *J. Am. Chem. Soc.* **1990,** *112*, 9394.
(24) Suzuki, H.; Tokitoh, N.; Okazaki, R.; Harada, J.; Ogawa, K.; Tomoda, S.; Goto, M. *Organometallics* **1995,** *14*, 1016.
(25) Watanabe, H.; Takeuchi, K.; Nakajima, K.; Nagai, Y.; Goto, M. *Chem. Lett.* **1988,** 1343.
(26) Weidenbruch, M.; Pellmann, A.; Pan, Y.; Pohl, S.; Saak, W.; Marsmann, H. *J. Organomet. Chem.* **1993,** *450*, 67.
(27) Roark, D. N.; Peddle, G. J. D. *J. Am. Chem. Soc.* **1972,** *94*, 5837.
(28) Masamune, S.; Murakami, S.; Tobita, H. *Organometallics* **1983,** *2*, 1464.
(29) Sakurai, H.; Nakadaira, Y.; Kobayashi, T. *J. Am. Chem. Soc.* **1979,** *101*, 487.
(30) Sakurai, H.; Nakadaira, Y.; Sakaba, H. *Organometallics* **1984,** *3*, 1484. Sakurai, H.; Sakamoto, K.; Kira, M. *Chem. Lett.* **1984,** 1379.
(31) Sekiguchi, A.; Maruki, I.; Sakurai, H. *J. Am. Chem. Soc.* **1993,** *115*, 11460.
(32) Gillette, G. R.; Noren, G.; West, R. *Organometallics* **1989,** *8*, 487. **1990,** *9*, 2925.
(33) Ando, W.; Hagiwara, K.; Sekiguchi, A. Sakakibara, A.; Yoshida, H. *Organometallics* **1988,** *7*, 558.
(34) Pearsall, M.-A.; West, R. *J. Am. Chem. Soc.* **1988,** *110*, 7228.
(35) Jutzi, P.; Holtmann, U.; Bögge, H.; Müller, A. *J. Chem. Soc., Chem. Commun.* **1988,** 305.
(36) Maxka, J.; Apeloig, Y. *J. Chem. Soc., Chem. Commun.* **1990,** 737.
(37) Apeloig, Y. In *The Chemistry of Organosilicon Compounds;* Patai, S.; Rappoport, Z., Eds.; Wiley: New York, 1989; p. 57.
(38) Jacobsen, H.; Ziegler, T. *J. Am. Chem. Soc.* **1994,** *116,* 3667.
(39) Krogh-Jespersen, K. *J. Am. Chem. Soc.* **1985,** *107*, 537.
(40) Karni, M.; Apeloig, Y. *J. Am. Chem. Soc.* **1990,** *112*, 8584.
(41) Fink, M. J.; Michalczyk, M. J.; Haller, K. J.; West, R.; Michl, J. *Organometallics* **1984,** *3*, 793.
(42) Zilm, K. W.; Fink, M. J.; Grant, D. M.; West, R.; Michl, J. *Organometallics* **1983,** *2*, 193.
(43) Cavalieri, J. D.; West, R.; DuChamp, J. C.; Zilm, K. W. *Phosphorus, Sulfur, Silicon* **1994,** *92*, 213. See also Cavalieri, J. D. Ph.D. Thesis, University of Wisconsin-Madison, 1993.
(44) Fleischer, U.; Schindler, M.; Kutzelnigg, W. *J. Chem. Phys.* **1987,** *86*, 6337.
(45) Tossell, J. A.; Lazzeretti, P. *Chem. Phys. Lett.* **1986,** *128*, 420.
(46) Apeloig, Y.; West, R. unpublished data.
(47) Hinton, J. F.; Guthrie, P. L.; Palay, P.; Wolinski, K. *J. Magn. Reson. A* **1993,** *103*, 188.
(48) Yokelson, H. B.; Millevolte, A. J.; Adams, B. R.; West, R. *J. Am. Chem. Soc.* **1987,** *109*, 4116.
(49) Galasso, V.; Fronzoni, G. *Chem. Phys.* **1986,** *103*, 29.
(50) Fink, M. Ph.D. Thesis, University of Wisconsin–Madison, **1983**.

(51) Shepherd, B. D.; Campana, C. F.; West, R. *Heteroatom. Chem.* **1990,** *1*, 1.
(52) Krebs, A.; Kaletta, B.; Nickel, W.-U.; Rüger, W.; Tikwe, L. *Tetrahedron* **1986,** *42*, 1693.
(53) Garratt, P. J.; Payne, D.; Tocher, D. A. *J. Org. Chem.* **1990,** *55*, 1909. Brooks, P. R.; Bishop, R.; Counter, J. A.; Tiekink, E. R. T. *J. Org. Chem.* **1994,** *59*, 1365.
(54) Guhn, N. F.; Lichtenberger, D.; Spitzner, H.; West, R. unpublished studies. Guhn, N. F. Ph.D. Thesis, University of Arizona, Tucson, 1994.
(55) Shepherd, B. D.; West, R. *Chem. Lett.* **1988,** 183.
(56) (a) Schmidt, M. W.; Truong, P. N.; Gordon, M. S. *J. Am. Chem. Soc.* **1987,** *109*, 5217. (b) Kutzelnigg, W. *Angew. Chem., Int. Ed. Engl.* **1984,** *23*, 272. (c) Olbrich, G.; Potzinger, P.; Reimann, B.; Walsh, R. *Organometallics* **1984,** *3*, 1267.
(57) Kistiakowski, G. B.; Smith, W. R. *J. Am. Chem. Soc.* **1934,** *56*, 638.
(58) (a) Sandstrom, J. *Top. Stereochem.* **1983,** *14*, 83. (b) Saltiel, J.; Charlton, J. L. In *Rearrangements in Ground and Excited States*; de Mayo, P., Ed.; Academic Press: New York, 1980; Vol. 3, pp. 25–89 (Vol. 42 of *Organic Chemistry, A Series of Monographs*, Wasserman, H. H., Ed.). (c) Gano, J. E.; Lenoir, D.; Park, B.-S.; Roesner, R. A. *J. Org. Chem.* **1987,** *52*, 5636. (d) Sakurai, H.; Ebata, K.; Kabuto, C.; Nakadaira, Y. *Chem. Lett.* **1987,** 301. (e) Shvo, Y. *Tetrahedron Lett.* **1968,** 5923.
(59) Suzuki, H.; Tokitoh, N.; Okazaki, R. *Bull. Chem. Soc. Jpn.*, **1995,** *68*, 2471.
(60) Michalczyk, M. J.; West, R.; Michl, J. *Organometallics* **1985,** *4*, 826.
(61) Fink, M. J.; DeYoung, D. J.; West, R.; Michl, J. *J. Am. Chem. Soc.* **1983,** *105*, 1070.
(62) (a) Suzuki, H.; Tokitoh, N.; Okazaki, R. *J. Am. Chem. Soc.* **1994,** *116*, 11572. (b) Suzuki, H.; Tokitoh, N.; Okazaki, R. unpublished results.
(63) Spitzner, H.; West R. unpublished results.
(64) Yokelson, H. B.; Maxka, J.; Siegel, D. A.; West, R. *J. Am. Chem. Soc.* **1986,** *108*, 4239.
(65) Reetz, M. *Adv. Organomet. Chem.* **1977,** *16*, 33.
(66) DeYoung, D. J.; Fink, M. J.; Michl, J.; West, R. *Main Group Met. Chem.* **1987,** *1*, 19.
(67) (a) Boudjouk, P.; Han, B.-H.; Anderson, K. R. *J. Am. Chem. Soc.* **1982,** *104*, 4992. (b) Boudjouk, P. *Nachr. Chem., Tech. Lab.* **1983,** *31*, 798.
(68) Fanta, A. D.; DeYoung, D. J.; Belzner, J.; West, R. *Organometallics* **1991,** *10*, 3466.
(69) Fanta, A. D.; Belzner, J.; Powell, D. R.; West, R. *Organometallics* **1993,** *12*, 2177.
(70) Schäfer, A.; Weidenbruch, M.; Pohl, S. *J. Organomet. Chem.* **1985,** *282*, 305.
(70a) Weidenbruch, M.; Kroke, E.; Marsmann, H.; Pohl, S.; Saak, W. *J. Chem. Soc., Chem. Commun.* **1994,** 1233.
(71) Kabeta, K.; Powell, D. R.; Hanson, J.; West, R. *Organometallics* **1991,** *10*, 827.
(71a) Sakakibara, A.; Kabe, Y.; Shimizu, T.; Ando, W. *J. Chem. Soc., Chem. Commun.* **1991,** 43.
(72) DeYoung, D. J.; West, R. *Chem. Lett.* **1986,** 883.
(73) Weidenbruch, M.; Flintjer, B.; Pohl, S.; Haase, D.; Martens, J. *J. Organomet. Chem.* **1988,** *338*, C1.
(74) Weidenbruch, M.; Lesch, A.; Peters, K.; von Schnering, H. G. *J. Organomet. Chem.* **1992,** *423*, 329.
(75) Weidenbruch, M.; Lesch, A.; Peters, K.; von Schnering, H. G. *Chem. Ber.* **1990,** *123*, 1795.
(76) Weidenbruch, M.; Flintjer, B.; Pohl, S.; Saak, W. *Angew. Chem., Int. Ed. Engl.* **1989,** *28*, 95.
(77) Masamune, S.; Murakami, S.; Tobita, H.; Williams, D. J. *J. Am. Chem. Soc.* **1983,** *105*, 7776.
(78) Piana, H.; Schubert, U. *J. Organomet. Chem.* **1988,** *348*, C19.
(79) Gillette, G. R.; West, R. *J. Organomet. Chem.* **1990,** *394*, 45.

(80) Yokelson, H. B.; Millevolte, A. J.; Gillette, G. R.; West, R. *J. Am. Chem. Soc.* **1987,** *109*, 6865.

(81) West, R.; DeYoung, D. J.; Haller, K. J. *J. Am. Chem. Soc.* **1985,** *107*, 4942.

(82) Tan, R. P.; Gillette, G. R.; Powell, D. R.; West, R. *Organometallics* **1991,** *10*, 546.

(83) Yokelson, H. B.; Millevolte, A. J.; Haller, K. J.; West, R. *J. Chem. Soc., Chem. Commun.* **1987,** 1605.

(84) Mangette, J. E.; Powell, D. R.; West, R. *J. Chem. Soc., Chem. Commun.* **1993,** 1348.

(85) Mangette, J. E.; Powell, D. R.; West, R. *Organometallics* **1995,** *14*, 3551.

(86) Steudel, R. *Top. Curr. Chem.* **1982,** *102*, 149. Laitinen, R. S.; Pakkanen, T. A.; Steudel, R., *J. Am. Chem. Soc.* **1987,** *109*, 710.

(87) Grev, R. S.; Schaefer, H. F. *J. Am. Chem. Soc.* **1987,** *109*, 6577.

(88) Gillette, G. R.; Maxka, J.; West, R. *Angew. Chem., Int. Ed. Engl.* **1989,** *28*, 54.

(89) Weidenbruch, M.; Schäfer, A.; Thom, K. L. *Z. Naturforsch. B: Anorg. Chem., Org. Chem.* **1983,** *38B*, 1695.

(90) Weidenbruch, M.; Piel, H.; Lesch, A.; Peters, K.; von Schnering, H. G. *J. Organomet. Chem.* **1993,** *454*, 35.

(91) Weidenbruch, M.; Lesch, A.; Peters, K. *J. Organomet. Chem.* **1991,** *407*, 31.

(92) Mangette, J. E.; Powel, D. R.; West, R. *Organometallics* **1994,** *13*, 4097.

(93) McKillop, K. L.; Gillette, G. R.; Powell, D. R.; West, R. *J. Am. Chem. Soc.* **1992,** *114*, 5203.

(94) Sohn, H.; Tan, R. P.; Powell, D. R.; West, R. *Organometallics* **1994,** *13*, 1390.

(95) McKillop, K. L.; West, R.; Clark, T.; Hoffman, H. *Z. Naturforsch., B: Anorg. Chem., Org. Chem.* **1994,** *49B*, 1737.

(96) Watanabe, H.; Takeuchi, K.; Nakajima, K.; Nagai, Y.; Goto, M. *Chem. Lett.* **1988,** 1343; but cf. ref. 7.

(97) Gordon, M. S.; Puckwood, T. J.; Carroll, M. T., Boatz, J. A., *J. Phys. Chem.* **1991,** *95*, 4332. Liang, C.; Allen, L. C. *J. Am. Chem. Soc.* **1991,** *113*, 1878.

(98) Fanta, A. D.; Tan, R. P.; Comerlato, N. M.; Driess, M.; Powell, D. R.; West, R. *Inorg. Chim. Acta* **1992,** *198–200*, 733.

(99) Fanta, A. D.; Driess, M.; Powell, D. R.; West, R. *Angew. Chem., Int. Ed. Engl.* **1989,** *28*, 1038.

(100) Fanta, A. D.; Driess, M.; Powell, D. R.; West, R. *J. Am. Chem. Soc.* **1991,** *113*, 7806.

(101) Tan, R. P.; Comerlato, N. M.; Powell, D. R.; West, R. *Angew. Chem., Int. Ed. Engl.* **1992,** *31*, 1217.

(102) Pham, E. K.; West, R. *Organometallics* **1990,** *9*, 1517.

(103) (a) Hong, D.; Damrauer, N. H.; Carroll, P. J.; Berry, D. H. *Organometallics* **1993,** *12*, 3698. (b) Berry, D. H.; Chey, J.; Zipin, H. S.; Carroll, P. J. *Polyhedron* **1991,** *10*, 1189. *J. Am. Chem. Soc.* **1990,** *112*, 452.

(104) Heyn, R. H.; Tilley, T. D. *J. Am. Chem. Soc.* **1992,** *114*, 1917.

(105) Lickiss, P. D. *Chem. Soc. Rev.* **1992,** *21*, 271.

Stable Doubly Bonded Compounds of Germanium and Tin

K. M. BAINES and W. G. STIBBS

Department of Chemistry
University of Western Ontario
London, Ontario, Canada N6A 5B7

I
INTRODUCTION

Doubly bonded compounds of the heavier group 14 elements, i.e., germanium and tin, are now well known. For doubly bonded germanium compounds, many reviews have been published. The earliest review, published in 1982 by J. Satgé, is an excellent summation of the investigations on reactions involving transient doubly bonded germanium compounds.[1] The results in this area of chemistry from the Toulouse group were updated by Satgé in 1990.[2] A more comprehensive review on the subject of doubly bonded germanium compounds also appeared in the same year.[3] Finally, the most recent review to appear, from the French group, summarizes results in the area of stable doubly bonded germanium compounds.[4] In addition, N. Wiberg summarized his group's results on both unsaturated silicon and germanium compounds in 1984, and S. Masamune reviewed the field of doubly bonded silicon, germanium, and tin compounds in 1991.[5] Stable doubly bonded compounds of the heavier group 14 elements have also been reviewed along with doubly bonded compounds of group 15 elements.[6] The theoretical aspects of multiply bonded silicon and germanium compounds have been summarized in an excellent review by R. Grev.[7] Finally, in a special issue of *Main Group Metal Chemistry*, the invited lectures of the "Ge-Sn-Pb Tokyo '93" conference were published, and several of these concerned aspects of doubly bonded germanium and tin compounds.[8–10] The purpose of this review is to update the previously published work. Although the last review on doubly bonded germanium compounds was published in 1994, the literature covered was from mid- to late 1992. The literature covered in this review extends to the end of 1994. No attempt is made to summarize the early work in this area; the reader is referred to the previously mentioned works for details. The current review details the physical properties, synthesis, and reactivity of *stable* unsaturated germanium and tin compounds. By *stable,* we mean compounds that are capable of existing at room temperature in solution. After careful consideration of the method of presentation, it was decided to approach the material from a reactivity viewpoint rather than to divide it on the basis of the classes of unsaturated compounds. Thus, the synthesis of the various derivatives is divided up into general synthetic methods, and the reactivity of the unsaturated species is classified according to the type of reactions—addition, reduction, and rearrangements. It is hoped that this will provide a new view of the overall field, thereby generating some new perspectives in this exciting field of chemistry.

II

GERMANIUM

A. *Physical Properties*

1. *Structure Determinations*

Several compounds containing a tricoordinate germanium atom have had their structures determined by X-ray crystallography; these are listed in Table I. The fold angle θ is defined as the angle between the Ge—E vector and the GeL_2 plane. The twist angle τ is defined as the angle between the GeL_2 and the EL_2 planes about the Ge—E vector.

Some general trends are worthy of note. Except for the digermenes, the tricoordinate germanium atom is trigonal-planar. Although the degree of shortening of the Ge═E double bond length varies from 4 to 10%, the majority are shortened from 7 to 9% relative to the average Ge—E single

TABLE I

SELECTED BOND LENGTHS AND ANGLES IN DOUBLY BONDED GERMANIUM COMPOUNDS

Entry	Compound[a]	Ge=E (Å)	Angles (deg.)				Ref.
			C—Ge—C	C—Ge—E	Fold	Twist	
1	Mes_2Ge=fluorenylidene	1.803(4)			0	6	11
2	$[(Me_3Si)_2N]_2Ge$=C(B-t-Bu)$_2$C$(SiMe_3)_2$	1.827(4)	N—Ge—N 110.5(2)	N—Ge—E 125.3(2) 124.1(2)	1.7	36	12
3	$[(Me_3Si)_2CH]_2Ge$=$Ge[CH(SiMe_3)_2]_2$	2.347(2)	112.5(2)	113.7(3) 122.3(2)	32	0	14
4	Dep_2Ge=$GeDep_2$	2.213(2)	115.4(2)	118.7(1) 124.3(1)	12	10	13

TABLE I (*continued*)

Entry	Compound[a]	Ge=E (Å)	Angles (deg.)				Ref.
			C—Ge—C	C—Ge—E	Fold	Twist	
5	$Dip(Mes)Ge{=}Ge(Dip)Mes$	2.301(1)	109.9(2)	111.6(2) 124.0(2)	36	7	15
6	$[(Me_3Si)_2CH]_2Ge{=}N{-}Si(t\text{-}Bu)_2N_3$	1.704(5)	122.9(3)	121.8(3) 115.2(3)			16, 17
7	$[[(Me_3Si)_2CH]_2Ge{=}N{-}SiMes_2]_2$	1.681(8)	116.3(5)	119.1(5) 124.6(4)			16
8	$Me_3Si{-}N{=}Ge$ cage with $Si(Me)$, $N(t\text{-}Bu)$ groups	1.688(9)	N—Ge—N 108.3(3) N→Ge—N 78.5(3) 79.5(3)	N—Ge=N 125.8(4) 123.6(4) N→Ge=N 122.6(4)			18

9	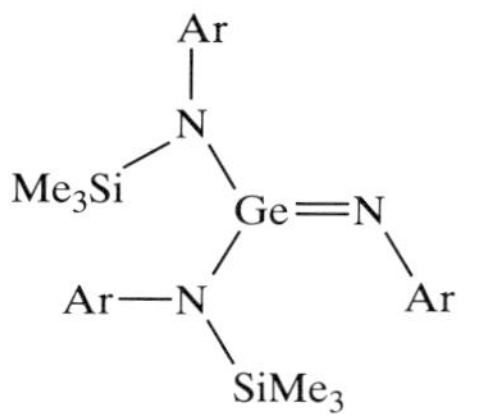 Ar N Me_3Si Ge=N Ar—N Ar $SiMe_3$	Ar = Dip 1.703(2) Ar = Mes 1.691(3)	Ar = Dip 120.7(1) Ar = Mes 115.4(1)	Ar = Dip 108.4(1) 130.9(1) Ar = Mes 110.6(1) 134.0(1)		25
10	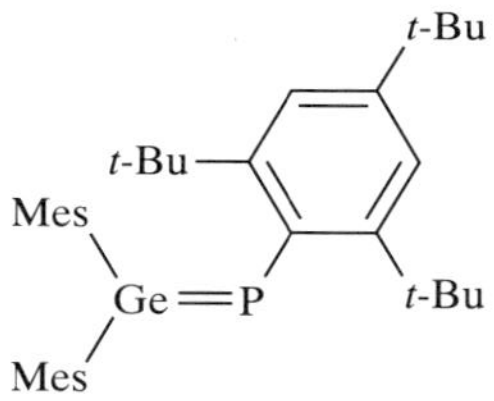 *t*-Bu *t*-Bu Mes Ge=P *t*-Bu Mes	2.138(3)	112.9(4)	111.8(3) 135.2(4)	cis: 10 trans: 13	19
11	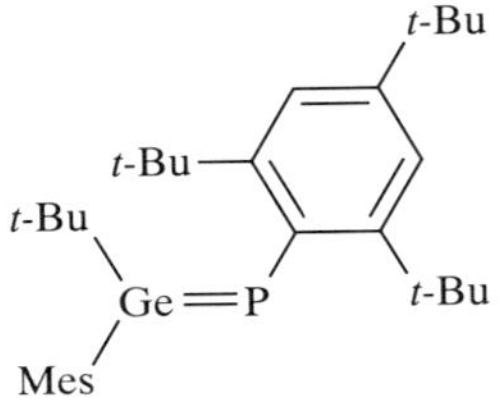 *t*-Bu *t*-Bu *t*-Bu Ge=P *t*-Bu Mes	2.144(3)	110.9(5)	114.4(3) 134.7(4)		20
12	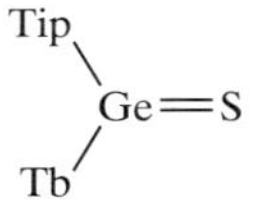 Tip Ge=S Tb	2.049(3)	118.4(4)	124.8(3) 116.2(3)		21

TABLE I (*continued*)

Entry	Compound[a]	Ge=E (Å)	Angles (deg.) C—Ge—C	C—Ge—E	Fold	Twist	Ref.
13	t-Bu, N, S, Ge, Me, Si, t-Bu, N, Si, N, t-Bu, Me, N, t-Bu	2.063(3)	N—Ge—N 110.6(3) N→Ge—N 79.8(2) 79.5(2)	N—Ge=S 121.4(2) 123.0(2) N→Ge=N 128.9(2)			23
14	Tip, Ge=Se, Tb	2.180(2)	119.1(4)	126.4(3) 113.8(3)			22
15	N, E, N, Ge, N, N	E = S 2.110(2) E = Se 2.247(1) E = Te 2.466(1)					24

[a] Mes = 2,4,6-trimethylphenyl; Dep = 2,6-diethylphenyl; Dip = 2,6-diisopropylphenyl; Tip = 2,4,6-triisopropylphenyl; Tb = 2,4,6-tris[bis(trimethylsilyl)methyl]phenyl.

bond length. This is slightly smaller than the degree of shortening observed for C—C to C═C (12%), but not surprising considering the relative weakness of Ge—E π-bond strengths.

The molecular structures of two germenes have been determined (entries **1** and **2**, Table I) by X-ray methods. In both, the germanium atom is essentially trigonal-planar and there is a shortening of 8–9% (Ge—C 1.98 Å) of the Ge—C bond. In the fluorenylidene-substituted germene shown below, 16 atoms (those of the fluorenylidene group, the germanium atom, and the ipso carbon atoms of the mesityl groups) all lie in the same plane, which is consistent with stabilization of the Ge═C by electronic interactions between Ge═C and the fluorenylidene moiety.[11]

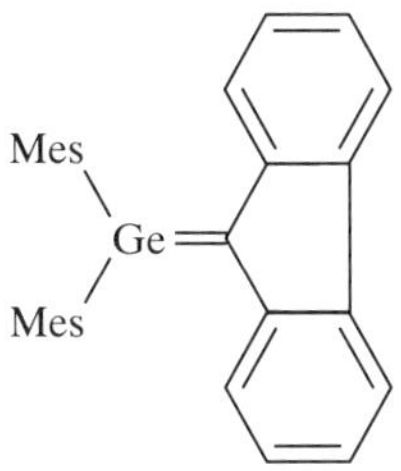

Mes = 2,4,6-trimethylphenyl

From the chemical shifts of the carbon atom of the double bond and the directly substituted boron atoms, the bis[bis(trimethylsilyl)amino]-substituted germene, shown below, appears to have significant ylide character.[12] This molecule is also significantly twisted by 36° about the Ge—E axis.

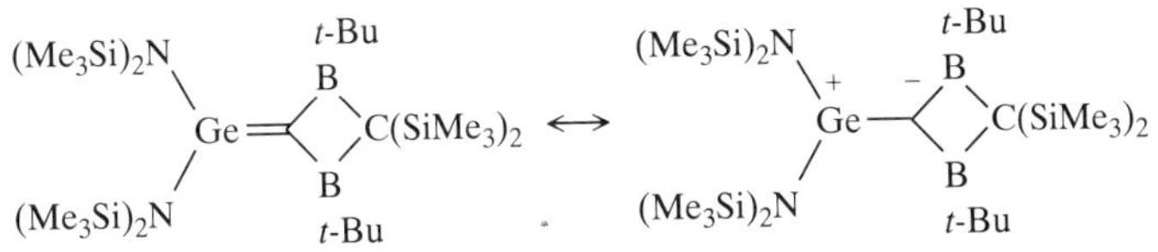

The molecular structures of three digermenes, RR′Ge═GeRR′, have been determined (entries **3** (R = R′ = $CH(SiMe_3)_2$), **4** (R = R′ = 2,6-diethylphenyl), and **5** ((*Z*)-R = mesityl, R′ = 2,6-diisopropylphenyl); Table I). All three have trans-bent double bonds. The trans-bent geometry has been calculated to have the lowest energy of all the Ge_2H_4 isomers and lies 3.2 kcal/mol below the planar form.[7] The fold angle (as defined above) appears to increase with steric congestion and, as the germanium atom becomes more pyramidalized, the degree of bond shortening decreases (9% for RR′Ge═GeRR′ (R = R′ = 2,6-diethylphenyl)[13] and 4% for RR′

Ge=GeRR′ (R = R′ = $CH(SiMe_3)_2$)[14]; Ge—Ge 2.44 Å). The digermene with the longest Ge=Ge double bond length (entry **3**, (R = R′ = CH$(SiMe_3)_2$) dissociates readily in solution to two germylenes, although this does not appear to be related to the conformation of the digermene since the bulky substituted (*Z*)-1,2-bis(2,6-diisopropylphenyl)-1,2-dimesityldigermene does not appear to dissociate under the same conditions.[15] The small twist observed about the Ge—E vector also may relieve some steric strain.

The structures of five germanimines have been examined by X-ray crystallography (entries **6–9**, Table I). In each case, the germanium atom is trigonal-planar and an 8–10% shortening of the Ge—N bond is observed. In entries **6**, **7**, and **9**, the germanimine unit is almost planar, as would be expected from a normal p_π–p_π double bond. The Ge—N—L (L = silyl ligand) angle of entries **6** and **7** is enlarged to 136–137° from the expected 120°.[16,17] The base-stabilized germanimine (entry **8**, Table I) has an interesting structure[18]: N^2, N^3, and N^5 lie in nearly the same plane (the sum of the angles about Ge is 357.7°) and N^1 forms an angle of 57° to this plane. The Ge—N^5 distance is shortened to the same extent as the germanimine structures in entries **6**, **7** and **9** (9%); however, the plane in which the N^2, N^3, and N^5 atoms lie is orthogonal to the Ge, N^5, and Si plane, thus ruling out a conventional p_π–p_π double bond. Like that of the *N*-silylated germanimines in entries **6** and **7**, the Ge—N—Si angle is enlarged to 151.2(7)°. The structure suggests that the compound is best described as a hybrid of the following resonance contributors, although the Ge—N—Si angle widening may simply be the result of steric interactions.

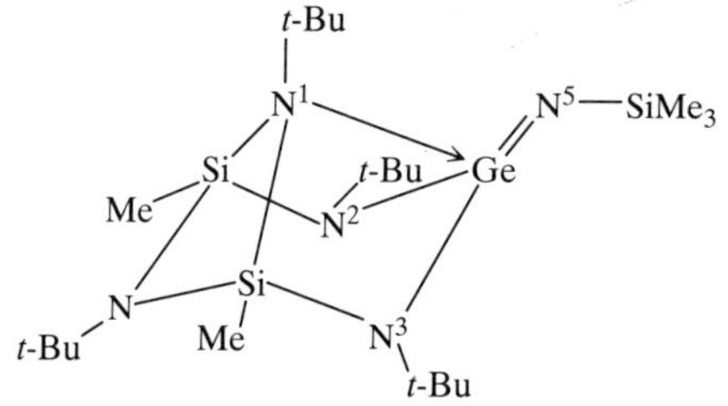

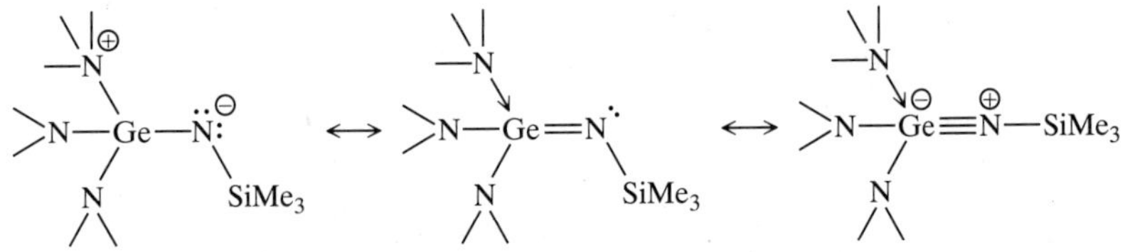

The molecular structures of germaphosphenes (entries **10** and **11**, Table I) show much the same characteristics as germanimines.[19,20] They are planar about the germanium atom and the Ge=P double bond is about 8% shorter than its single-bond analog. The angle at phosphorus is smaller than in germanimines (102.6–107.5°), reflecting the increased *p*-character used in bonding to the phosphorus. The same trend has been observed on going from diimines to diphosphenes to diarsenes.

One example of an uncoordinated diarylgermanethione,[21] with a germanium–sulfur double bond, and one example of an uncoordinated diarylgermaneselone,[22] with a germanium–selenium double bond, have been synthesized (entries **12** and **14**, respectively; Table I). Both compounds are kinetically stabilized by the extremely bulky 2,4,6-tris[bis(trimethylsilyl)-methyl]phenyl and the 2,4,6-triisopropylphenyl groups. Again, the germanium is trigonal-planar in each case. The Ge—S bond is shortened by 8% to 2.049(3) Å and the Ge—Se bond is shortened by approximately 7% to 2.180(2) Å. A base-stabilized germanethione (entry **13**, Table I) has also been described[23]:

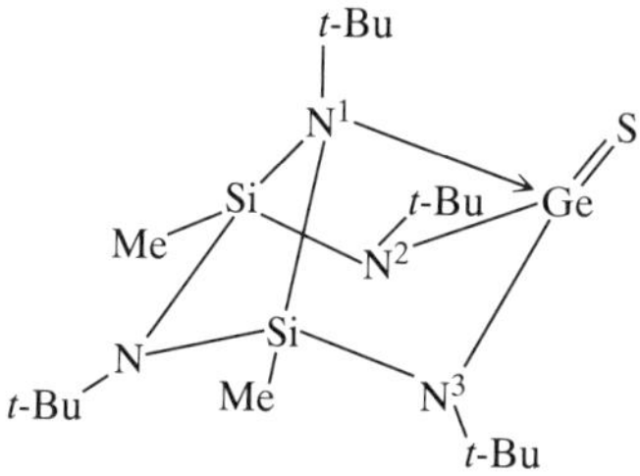

The germanium atom is four-coordinate and is best described as trigonal-planar with a fourth ligand: the sum of the angles about germanium (N^2, N^3, and S) is 355° and the Ge—N^1 bond length (2.050 Å) is significantly longer than the Ge—N^2 or —N^3 bond lengths (av. 1.884 Å). The Ge—S bond length (2.063(3) Å) is comparable to that in the diarylgermanethione (entry **12**, Table I). Some interesting terminal chalcogenido complexes (E = S, Se, and Te) of germanium containing the octamethyldibenzotetraaza[14]annulene ligand have recently been synthesized (entry **15**, Table I).[24] Examination of the Ge—E bond distance indicates that the bonding of the Ge—E moiety is best described as a hybrid of the Ge=E and Ge^+—E^- resonance structures.

2. *NMR Spectroscopy*

Although no ^{73}Ge NMR spectra of stable doubly bonded germanium compounds have been recorded, NMR chemical shift data for the heteroatom of the double bond are available; these are listed in Table II. The chemical shift of the sp^2-hybridized carbon atom of dimesitylneopentylgermene (entry **1**, Table II), without any polar substituents, falls at δ 124.2.[26] The chemical shifts of the boron-substituted sp^2-hybridized carbon atoms of two stable germenes (entry **2**, Table II) fall in the upfield region of the

TABLE II

NMR SPECTRAL DATA OF DOUBLY BONDED GERMANIUM COMPOUNDS

Entry	Compound[a]	Chemical shift, δ	Solvent: standard	Reference
1	$Mes_2Ge{=}CHCH_2$-*t*-Bu	^{13}C: 124.20	C_7D_8	26
2	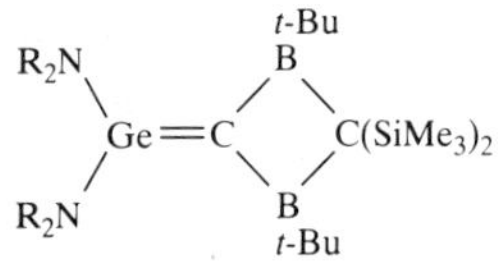 $(R_2N)_2Ge{=}C$ in a four-membered ring $C[B(t\text{-}Bu)]_2C(SiMe_3)_2$	$NR_2 = N(SiMe_3)_2$ ^{13}C: 115 NR_2 = —N(*t*-Bu)$SiMe_2$N(*t*-Bu)— ^{13}C: 93	DME-d_{10}	12
3	(R_2CH)-*t*-BuGe=CR_2 (CR_2 = fluorenylidene)	^{13}C: 79.80	C_6D_6	28
4	$Mes_2Ge{=}SiMes_2$	^{29}Si: 80.6	C_7D_8: $SiMe_4$	30
5	$[Mes(Me_3Si)N]_2Ge{=}NMes$	^{15}N: −295, −193 (ratio 2 : 1)	THF: $MeNO_2$	25
6	$Mes_2Ge{=}P$—Ar	^{31}P: 175.4	C_6D_6: H_3PO_4	32
7	$Mes_2Ge{=}P$—Tip	^{31}P: 145.3	hexane/Et_2O: H_3PO_4	35
8	(*Z*)-Mes(*t*-Bu)Ge=P—Ar	^{31}P: 169.2	pentane: H_3PO_4	20
9	(*E*)-Mes(*t*-Bu)Ge=P—Ar	^{31}P: 157.4	pentane: H_3PO_4	20
10	*t*-$Bu_2Ge{=}P$—Ar	^{31}P: 156.6	not given: H_3PO_4	20
11	Tip(Tb)Ge=Se	^{77}Se: 940.6	C_6D_6: $SeMe_2$	22

[a] Mes = 2,4,6-trimethylphenyl; Ar = 2,4,6-tri-*t*-butylphenyl; Tip = 2,4,6-triisopropylphenyl; Tb = 2,4,6-tris[bis(trimethylsilyl)methyl]phenyl.

typical range of the analogous carbon atoms in alkenes (80–150 ppm).[27] The shielded carbon atoms are undoubtedly the result of an important contribution from the zwitterionic structure shown below to the overall structure.

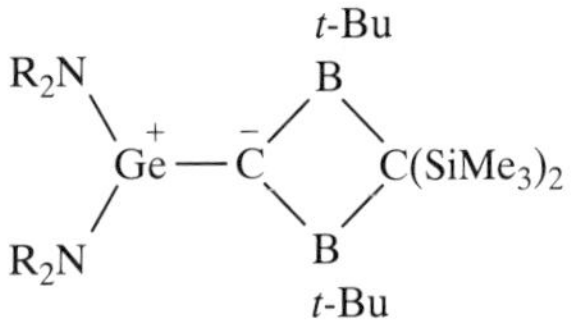

$$(R_2N)_2\overset{+}{Ge}—\overset{-}{C}\langle[B(t\text{-}Bu)]_2\rangle C(SiMe_3)_2$$

It is not clear why the sp^2-hybridized carbon atom of a fluorenylidene germene (entry **3**, Table II) is so shielded (δ 79.8), unless this chemical shift is from the spectrum of the THF adduct.[28] The chemical shift of the silicon atom in the germasilene (entry **4**, Table II) and the phosphorus atom in germaphosphenes (entries **6–10**, Table II) is deshielded compared to the corresponding silicon analogs. For example, for $Mes_2M{=}SiMes_2$

with M = Si, δ ^{29}Si = 63.6[29] and with M = Ge, δ ^{29}Si = 80.6[30]; while for $Mes_2M{=}P{-}Ar$ (Ar = 2,4,6-tri-*t*-butylphenyl) with M = Si, δ ^{31}P = 136.0[31] and with M = Ge, δ ^{31}P = 175.4.[32] The same trend, the shift to lower field upon substitution with germanium, has also been noted in phenylated propane analogs of silicon and germanium and has been reasonably ascribed to the greater electronegativity of germanium compared to silicon.[33] There is one reported ^{15}N chemical shift of a germanimine (entry **5**, Table II). The value of δ −295 is comparable to the corresponding value found for a silanimine (t-$Bu_2Si{=}NSi$-t-Bu_3; δ ^{14}N −230[34]), although not to lower field; however, the nature of the substituents on the double bond is quite different, and therefore the chemical shifts are not directly comparable. The chemical shift of the selenium atom in the germaneselone (entry **11**, Table II)[22] is the first reported of its kind for heavier group 14 elements; like dialkyl selenoketones (δ = 1600–2200),[22] the signal is shifted downfield.

The chemical shifts of the phosphorus atoms in germaphosphenes is temperature-dependent. The value shifts to higher fields at low temperature and to lower fields at high temperature.[20]

3. *IR/Raman Spectroscopy*

Relatively few infrared/Raman data have been obtained for doubly bonded germanium compounds. $Mes_2Ge{=}CR_2$ (CR_2 = fluorenylidene) has a valence vibration at 988 cm^{-1} which has been assigned to the Ge—C double bond.[11] A series of *N*-trifluoromethyl-substituted germanimines have been characterized by infrared spectroscopy.[36] The Ge=N stretching vibrations of $CF_3N{=}GeXX'$ occur at 1030 cm^{-1} for X = X′ = H; 1028 cm^{-1} for X = H, X′ = $ON(CF_3)_2$; 1070 cm^{-1} for X = X′ = $ON(CF_3)_2$; 1057 cm^{-1} for X = H, X′ = Cl; and 1076 cm^{-1} for X = X′ = Cl. Germaphosphenes have a characteristic Raman stretching frequency at approximately 500 cm^{-1} (MesRGe=PAr; Ar = 2,4,6-tri-*t*-butylphenyl; R = Mes, 503 cm^{-1} [19]; R = *t*-Bu, 501.5 cm^{-1} [20]). The stretching vibrations of the Ge=S and Ge=Se bonds in the kinetically stabilized germanethione[21] and the analogous germaneselone[22] (RR′Ge=E; E = S, Se; R = 2,4,6-triisopropylphenyl, R′ = 2,4,6-tris[bis(trimethylsilyl)methyl]phenyl) have been observed at 521 and 382 cm^{-1}, respectively.

4. *Ultraviolet/Visible Spectroscopy*

Germenes vary from pale yellow ([—B-*t*-Bu($Me_3Si)_2$C(B-*t*-Bu)—]C=Ge[—N-*t*-Bu($SiMe_2$)N-*t*-Bu—][12]) to orange-red ((R_2CH)-*t*-Bu-Ge=CR_2 (CR_2 = fluorenylidene)[28]). Tetraaryldigermenes and -germasilenes are yellow (λ_{max} 408–418 nm), germaphosphenes are orange, the germanethione is orange-yellow,[21] and the germaneselone is red.[22] Table

TABLE III

ULTRAVIOLET/VISIBLE DATA OF DOUBLY BONDED GERMANIUM COMPOUNDS

Entry	Compound[a]	λ_{max} (nm) (log ε)	Solvent	Ref.
1	$Dmp_2Ge{=}GeDmp_2$	408 (3.9)	cyclohexane	37
2	$Mes_2Ge{=}GeMes_2$	410	cyclohexane	38
3	$Dep_2Ge{=}GeDep_2$	263 (4.1), 412 (3.9)	hexane	13
4	$Dip_2Ge{=}GeDip_2$	273 (4.1), 418 (4.6)	methylcyclohexane	39
5	$Tip_2Ge{=}GeTip_2$	279 (4.1), 305 (4.0), 418 (4.5)	methylcyclohexane	39
6	(*Z*)-DipMesGe=GeDipMes	280 (4.1), 412 (4.6)	methylcyclohexane	15
7	DepMesGe=GeDepMes	412 (3.6–4.0)	—	5b
8	*t*-BuMesGe=Ge-*t*-BuMes	378	—	5b
9	$Mes_2Ge{=}SiMes_2$	414	cyclohexane	30
10	$Mes_2Ge{=}N$– (thiophene ring) MeO_2C S	261, 325	cyclohexane	40
11	$Mes_2Ge{=}N$– (benzene ring) MeO_2C	261 (4.8), 282 (4.5), 343 (4.4), 391 (4.1)	cyclohexane	41
12	$[(Me_3Si)_2N]_2Ge{=}N{-}N{=}C(CO_2Me)_2$	502 (1.7)	diethyl ether	42
13	Tip(Tb)Ge=S	229 (4.7), 340 (3.7), 450 (2)	hexane	21
14	Tip(Tb)Ge=Se	519 (2.1)	—	22

[a] Dmp = 2,6-dimethylphenyl; Mes = 2,4,6-trimethylphenyl; Dep = 2,6-diethylphenyl; Dip = 2,6-diisopropylphenyl; Tip = 2,4,6-triisopropylphenyl; Tb = 2,4,6-tris[bis(trimethylsilyl)methyl]phenyl.

III[37–42] lists the electronic absorption data for doubly bonded germanium compounds.

5. *Double Bond Strengths*

With the availability of stable geometric isomers of doubly bonded germanium compounds, experimental determinations of the π-bond strength can be made. The enthalpy of activation for double bond isomerization in Mes(Tip)Ge=Ge(Tip)Mes (Tip = 2,4,6-triisopropylphenyl) has been determined: for the *Z–E* conversion, 22.2 ± .3 kcal/mol and for the *E–Z* conversion, 20.0 ± 0.3 kcal/mol.[15] These values agree well with recent theoretical estimations.[7] The isomerization barrier in germaphosphenes has also been directly determined from the interconversion of (Z)- and (E)-Mes(t-Bu)Ge=P—Ar (Ar = 2,4,6-tri-t-butylphenyl) by NMR spectroscopy. A value of 22.3 kcal/mol was found.[20]

B. *Synthetic Methods*

1. *Thermal or Photochemical Generation*

The photolysis of hexaarylcyclotrigermanes was used to synthesize the first stable digermenes. The photolyses are generally carried out at 254 nm in hydrocarbon solvents (e.g., cyclohexane or 3-methylpentane). Presumably, two equivalents of the cyclotrigermane form three equivalents of the digermene, the third equivalent being formed by dimerization of the diarylgermylene.

$$2\ \mathrm{Ar_2Ge}\overset{\mathrm{GeAr_2}}{\triangle}\mathrm{GeAr_2} \xrightarrow{h\nu} 3\ \mathrm{Ar_2Ge{=}GeAr_2} \qquad (1)$$

Ar = 2,6-dimethylphenyl,[37] 2,6-diethylphenyl,[13] 2,4,6-trimethylphenyl[43]

Depending on the substituent, thermolysis of the three-membered ring may also yield the digermene, generally as an intermediate. Thermolysis of hexamesitylcyclotrigermane has been a particularly useful route to tetramesityldigermene.[38]

Tetramesitylgermasilene has been obtained by an analogous route. Thermolysis[44] or photolysis[30] (at 254 or 350 nm) of hexamesitylsiladigermirane ($SiGe_2Mes_6$) occurs regioselectively to give the germasilene (and the germylene) as an intermediate. The germasilene is stable in solution at low temperature.

Closely related to the generation of digermenes from the photolysis or thermolysis of cyclotrigermanes is the photolysis of diarylbis(trimethylsilyl) germanes to give the diarylgermylene, which dimerizes to give the corresponding digermene. The photolyses are generally carried out at 254 nm in hydrocarbon solvents (e.g., 3-methylpentane or methylcyclohexane).

$$\mathrm{ArAr'Ge(SiMe_3)_2} \xrightarrow{h\nu} \mathrm{ArAr'Ge{:}} + \mathrm{Me_3SiSiMe_3} \xrightarrow{\times 2} \mathrm{ArAr'Ge{=}GeArAr'} \qquad (2)$$

Ar = Ar′ = 2,6-dimethylphenyl[45]

Ar = Ar′ = 2,4,6-trimethylphenyl[45–47]

Ar = Ar′ = 2,6-diethylphenyl[45,48]

Ar = 2,4,6-trimethylphenyl, Ar′ = t-butyl[45,47]

Ar = Ar′ = 2,4,6-triisopropylphenyl[45,46]

2. *Dehydrohalogenation*

Dehydrohalogenation appears to be the most general route for the synthesis of doubly bonded germanium compounds. This route has been used extensively for the formation of germenes and germaphosphenes, and less often for the synthesis of germanimines. The use of fluorine as the halogen is preferred, although chlorine has also been employed. *t*-Butyllithium is the most commonly used base because of its low nucleophilicity, but high basicity. The compounds synthesized by this route are summarized in Table IV.

TABLE IV

DEHYDROHALOGENATION REACTIONS OF HALOGERMANES

$$R^1\text{—}Ge(R^2)(X)\text{—}E(R^3)(H)\text{—}R^4 \xrightarrow{\text{base}} (R^1)(R^2)Ge\text{=}E(R^3)(R^4)$$

Entry	R^1	R^2	X	E	R^3	R^4	Base	Ref.
1	Mes[a]	Mes	F	C	fluorenylidene		*t*-BuLi	51
2	Bis[b]	Bis	F	C	fluorenylidene		*t*-BuLi	52
3	Bis	Mes	F	C	fluorenylidene		*t*-BuLi	52
4	fluorenyl	fluorenyl	F	C	fluorenylidene		*t*-BuLi	28
5	fluorenyl	*t*-Bu	Cl	C	fluorenylidene		*t*-BuLi	28
6	Mes	Mes	F	P	Ar[c]	—	*t*-BuLi	32
7	Mes	Mes	Cl	P	Ar	—	DBU	32
8	Mes	Mes	Cl	P	Ar	—	$Me_3P\text{=}CH_2$	32
9[d]	Mes	*t*-Bu	F	P	Ar	—	*t*-BuLi	20
10	*t*-Bu	*t*-Bu	F	P	Ar	—	*t*-BuLi	20
11	Mes	Mes	Cl	N	2-($C(=O)OMe$)phenyl	—	*t*-BuLi	41, 53
12	Mes	Mes	F	N	2-($C(=O)NMe_2$)phenyl	—	*t*-BuLi	41, 53
13	Mes	Mes	Cl	N	2-($C(=O)OMe$)-3-thienyl	—	*t*-BuLi	40

[a] Mes = 2,4,6-trimethylphenyl; [b]Bis = bis(trimethylsilyl)methyl; [c]Ar = 2,4,6-tri-*t*-butylphenyl; [d]*E* and *Z* isomers formed.

Closely related to this method is the synthesis of dimesitylneopentylgermene by the addition of *t*-BuLi to a fluorovinylgermane.[26] Presumably, the reaction occurs via an addition–elimination mechanism.

$$\mathrm{Mes_2GeF{-}CH{=}CH_2} + t\text{-}\mathrm{BuLi} \rightarrow \mathrm{Mes_2Ge{=}CHCH_2}\text{-}t\text{-}\mathrm{Bu} \tag{3}$$

Also related is the synthesis of 2,2-dimesityl-1-(2,4,6-triisopropylphenyl)germaphosphene by the reaction between difluorodimesitylgermane and the appropriate dilithiophosphide.[35]

$$\mathrm{Mes_2GeF_2} + \mathrm{TipPLi_2} \rightarrow \mathrm{Mes_2Ge{=}PTip} \tag{4}$$

Attempts to form the fluorophosphinogermane as the precursor to the germaphosphene failed; the reaction between Mes_2GeF_2 and TipPHLi gave the diphosphinogermane as the major product.

Attempts to synthesize a digermadiene or a germasilene from the dehydrohalogenation of $(MesGeFCHR_2)_2$[49] or $Ar_2GeHSiClMes_2$[50] (CHR_2 = fluorenylidene; Ar = Mes or Tip), respectively, by treatment with t-BuLi were unsuccessful.

3. *Coupling of Germylenes with Various Reagents*

Although digermenes are readily available from the coupling of transient germylenes (Section II.B.1.), heteronuclear doubly bonded germanium systems are more accessible from the reactions of stable diaza- or bis[bis(trimethylsilyl)methyl]germylenes with various reagents. Coupling of stable diazagermylenes with an electrophilic diboryl-substituted carbene yielded some of the first examples of stable germenes (Eq. 5),[12]

$$(\mathrm{Me_3Si})_2\mathrm{C{-}B}(t\text{-}\mathrm{Bu})\text{(}\mathrm{C{=}B{-}}t\text{-}\mathrm{Bu}\text{)} \longrightarrow \left[(\mathrm{Me_3Si})_2\mathrm{C}(\mathrm{B}\text{-}t\text{-}\mathrm{Bu})_2\mathrm{C{:}}\right] \xrightarrow{\mathrm{GeR_2}} (\mathrm{Me_3Si})_2\mathrm{C}(\mathrm{B}\text{-}t\text{-}\mathrm{Bu})_2\mathrm{C{=}GeR_2} \tag{5}$$

R = $N(SiMe_3)_2$; R_2 = N-*t*-BuSiMe$_2$N-*t*-bu

and coupling of the bis[bis(trimethylsilyl)amino]germylene with a diazo compound yielded the first stable germanimine (Eq. 6).[54]

$$[(\mathrm{Me_3Si})_2\mathrm{N}]_2\mathrm{Ge{:}} + \mathrm{N_2C(COOMe)_2} \rightarrow [(\mathrm{Me_3Si})_2\mathrm{N}]_2\mathrm{Ge{=}N{-}N{=}C(COOMe)_2} \tag{6}$$

An apparently general route to germanimines involves the reaction of diazagermylenes with either aryl[25] or silyl[23,55] azides. A particularly interesting example is the reaction of two equivalents of bis[bis(trimethylsilyl)methyl]germylene with silyldiazides to give bis(germanimines).[16] Diazagermylenes have also been reacted with elemental sulfur,[23,24] selenium,[24] or tellurium (with PMe_3)[24] to give base-stabilized germanethiones, -selones, and -telones, respectively.

4. *Miscellaneous*

Attempts to form cyclotrigermanes (and, subsequently, digermenes; see Section II.B.1) from the reductive cyclization of diaryldichlorogermanes and lithium naphthalenide (LiNp) with particularly bulky substituents on the aryl groups (i.e., isopropyl) led, instead, directly to the tetraaryldigermenes by reductive coupling.

$$ArAr'GeCl_2 + LiNp \rightarrow ArAr'Ge{=}GeArAr' \tag{7}$$

Ar = Ar′ = 2,4,6-triisopropylphenyl[39]

Ar = Ar′ = 2,6-diisopropylphenyl[39]

Ar = 2,4,6-triisopropylphenyl, Ar′ = 2,4,6-trimethylphenyl[15]

A novel synthesis of a germanimine involves the reaction between germane and CF_3NO.[36] The *N*-trifluoromethylgermanimine was subsequently treated with $(CF_3)_2NO$ to give the mono- and the bis[bis(trifluoromethyl)nitroxy]-substituted derivatives. These derivatives were subsequently treated with hydrogen chloride to give the mono- or the dichloro derivatives, respectively.

$$GeH_4 + CF_3NO \rightarrow CF_3N{=}GeH_2 \tag{8}$$

The first examples of a diarylgermanethione[21] or -germaneselone[22] not stabilized by interaction with a Lewis base were synthesized by the desulfurization or deselenization of the corresponding 1,2,3,4,5-tetrathia- or -selenagermolanes using triphenylphosphine.

$$\text{Tb(Tip)Ge(Y}_4\text{)} \xrightarrow[-3\,Y{=}PPh_3]{+3\,PPh_3} \text{Tb(Tip)Ge{=}Y} \tag{9}$$

Y = S, Se
Tb = 2,4,6-tris[bis(trimethylsilyl)methyl]phenyl

C. *Reactivity*

1. *Complex Formation*

Germenes readily form complexes with such Lewis bases as diethyl ether,[11,51] tetrahydrofuran,[52] and triethylamine.[51] The germanimine[18] and the germanethione,[23] synthesized by Veith, are intramolecularly coordinated by a nitrogen Lewis base. The coordination of this base is believed to contribute to the overall stability of these unsaturated compounds. The dimesitylgermanimine derivatives of methyl anthranilate,[41,53] *N,N*-dimethylanthranilamide[41,53] and 3-amino-2-(carbomethoxy)thiophene,[40] most probably are stabilized to some extent by coordination of the carbonyl oxygen atom to the germanium atom.

Lewis base

$$\mathrm{R_2Ge{=}ER(R)}$$

2. *Addition of Protic Reagents*

The addition of the oxygen–hydrogen bond across the double bond of unsaturated germanium compounds is perhaps the most important and general reaction this class of compounds can undergo. The reaction usually occurs cleanly and in high yield and is taken as solid evidence for the existence of a doubly bonded compound. The most commonly used alcohol is methanol, although ethanol and water are also employed for this purpose. Methanol has been added to germenes,[26,28,51,52] germasilenes,[44] digermenes,[13,37,48] germanimines,[40,41] (ethanol[41,53,56] and *t*-butanol[56] have also been added to germanimines), and germaphosphenes[32]; and water has been added to germenes,[51] germasilenes,[57] digermenes,[57] germanimines,[40,41,53] germaphosphenes,[20,32] and a germanethione.[21]

$$\mathrm{R_2Ge{=}ER'_{(2)}} \xrightarrow{\mathrm{R''OH}} \mathrm{R_2Ge(OR'')-ER'_{(2)}H} \tag{10}$$

R" = Me, Et, H

With one exception, the addition of water and alcohols to the unsaturated germanium compounds is regioselective with formation of the germanium–oxygen bond. Only in the addition of water to tetramesitylgermasilene were two regioisomers obtained in a 7 : 1 ratio favoring the Si—O isomer.[57] The factors that control the regioselectivity of this reaction are not well understood. Because of the lack of appropriate precursors, little is known of the stereochemistry of the addition: the addition of water to (*E*)-Mes(*t*-Bu)Ge=P—Ar (Ar = 2,4,6-tri-*t*-butylphenyl) gave a mixture of diastereomers.[20] More experiments are needed to make any definitive statements.

The addition of hydrogen halides, most commonly hydrogen chloride, to a variety of unsaturated germanium compounds has been examined.

With one exception (the addition of HCl to tetramesitylgermasilene), the addition is regioselective, giving the halogermane. In the case of the germasilene, a 7:1 mixture of regioisomers was obtained, the chlorogermane being the major isomer.[58] Only one example of addition of a hydrogen halide to a stereochemically pure unsaturated germanium compound was examined: the reaction between hydrogen fluoride and (*E*)-(*t*-Bu)Mes Ge=PAr (Ar = 2,4,6-tri-*t*-butylphenyl) gave a mixture of diastereomers.[20]

$$R_2Ge{=}ER'_{(2)} \xrightarrow[\text{(or } Et_3N\cdot HX)]{HX} \underset{X}{\underset{|}{R_2Ge}}-\underset{H}{\underset{|}{ER'_{(2)}}} \tag{11}$$

X = Cl; ER′ = fluorenylidene; R = 2,4,6-trimethylphenyl[26]
X = Cl; E = Si; R = R′ = 2,4,6-trimethylphenyl[58]
X = Cl; E = N; R = 2,4,6-trimethylphenyl; R′ = 2-(carbomethoxy)phenyl[41]
X = Cl; E = N; R = 2,4,6-trimethylphenyl; R′ = 2-(carbomethoxy)thiophen-3-yl[40]
X = Cl; E = N, R = $(MeSi)_2(t\text{-}Bu)_4$; R′ = trimethylsilyl[56]
X = Cl; E = P; R = 2,4,6-trimethylphenyl; R′ = 2,4,6-tri-*t*-butylphenyl[32]
X = F; E = P; R = 2,4,6-trimethylphenyl; R′ = 2,4,6-tri-*t*-butylphenyl[20]

There are a few isolated cases of the addition of amines, thiols, carboxylic acids, and a phosphorus ylide to doubly bonded germanium compounds. Again, the reactions are regioselective, with the nucleophilic portion of the weak acid adding to the germanium and the proton adding to the heteroatom.

$$Mes_2Ge{=}ER_{(2)} + AH \longrightarrow \underset{A}{\underset{|}{Mes_2Ge}}-\underset{H}{\underset{|}{ER_{(2)}}} \tag{12}$$

A = EtS; ER_2 = fluorenylidene[51]
A = NMe_2; E = N; R = 2-(carbomethoxy)thiophen-3-yl[40]
A = *i*PrS; E = P; R = 2,4,6-tri-*t*-butylphenyl[59]
A = NHPh; E = P; R = 2,4,6-tri-*t*-butylphenyl[59]
A = CH_3COO; E = P; R = 2,4,6-tri-*t*-butylphenyl[59]
A = Me_3PCH; E = P; R = 2,4,6-tri-*t*-butylphenyl[59]

3. *Addition of Halogens, Haloalkanes, and Disulfides*

The addition of bromine or iodine to $Mes_2Ge{=}PAr$ (Ar = 2,4,6-tri-*t*-butylphenyl) gave two major products: the vicinal dihalide and the (halogermyl)phosphine.[59]

$$Mes_2Ge{=}PAr + X_2 \longrightarrow Mes_2Ge(X){-}PAr(X) + Mes_2Ge(X){-}PAr(H) \quad (13)$$

$$X = Br, I$$

Carbon tetrachloride reacts with the same germaphosphene to yield dichlorodimesitylgermane and the *P*-aryldichlorophosphaalkene.[59] Presumably, in the formation of these products the first step is addition across the carbon–chlorine bond to give the chlorogermane, which decomposes to the observed products.

Addition across the carbon–halogen bond has been observed in the reaction between a base-stabilized germanimine or germanethione and methyl iodide.[56]

$$(MeSi)_2(t\text{-}BuN)_4Ge{=}E(R) + CH_3I \longrightarrow (MeSi)_2(t\text{-}BuN)_4Ge(CH_3{-}E(R))(I) \quad (14)$$

$$ER = NSiMe_3 \text{ or } E = S$$

On the other hand, the reaction between chloroform and dimesitylgermanimine derivatives of methyl anthranilate[41] and *N*,*N*-dimethylanthranilamide[53] resulted in addition across the carbon–hydrogen bond with formation of the secondary amine. The greater acidity of the hydrogen in $CHCl_3$ and/or the greater basicity in these germanimine derivatives may partially explain the differences in reactivities.

The sulfur–sulfur bond of dimethyl disulfide adds across the double bond of both $Mes_2Ge{=}PAr$ (Ar = 2,4,6-tri-*t*-butylphenyl)[59] and $Mes_2Ge{=}CR_2$ (CR_2 = fluorenylidene)[51] possibly via a radical mechanism.

4. *Addition of Organometallic Reagents*

Methyllithium has been added to the base-stabilized germanimine $(MeSi)_2(t\text{-}BuN)_4Ge{=}NSiMe_3$[56] and to $Mes_2Ge{=}PAr$ (Ar = 2,4,6-tri-*t*-butylphenyl),[59] *t*-butyllithium has been added to $Mes_2Ge{=}CHCH_2\text{-}t\text{-}Bu$[26] and $Mes_2Ge{=}PAr$,[59] and MeMgI has been added to the same germaphosphene.[59] The reactions are quenched with a protic agent such as an alcohol or water. The alkyl portion of the organometallic reagent always adds to the germanium atom.

5. *Addition of π-Bonded Reagents*

a. *Addition of Diazo Compounds and Azides.* The addition of diazo compounds and azides to germenes and digermenes has been examined; the two doubly bonded germanium compounds exhibit interesting differences in their reactivity.

The addition of diazomethane to $Mes_2Ge{=}CR_2$ (CR_2 = fluorenylidene) yielded a stable 4-germa-1-pyrazoline (see Scheme 1).[60,61] Care must be taken to avoid traces of base in the reaction mixture: in an excess of diazomethane, a tautomer of the pyrazoline, the 4-germa-2-pyrazoline, is formed; and in an excess of the germene, two equivalents of the germene add to diazomethane.[61] Thermolysis or photolysis of the 4-germa-1-pyrazoline aparently gives an intermediate germirane that decomposes to 9-methylenefluorene and dimesitylgermylene.[60,61] The germylene was trapped with 2,3-dimethylbutadiene or methanol. In contrast, the addition of 9-diazofluorene or diphenyldiazomethane gave a transient germanimine, apparently via nucleophilic attack by the terminal nitrogen of the diazo compound instead of the diazo carbon atom.[62] This is readily explained by the greater steric bulk at the carbon in these substituted diazoalkanes. The transient germanimine underwent dimerization to give a cyclodigermazane (see Scheme 2).

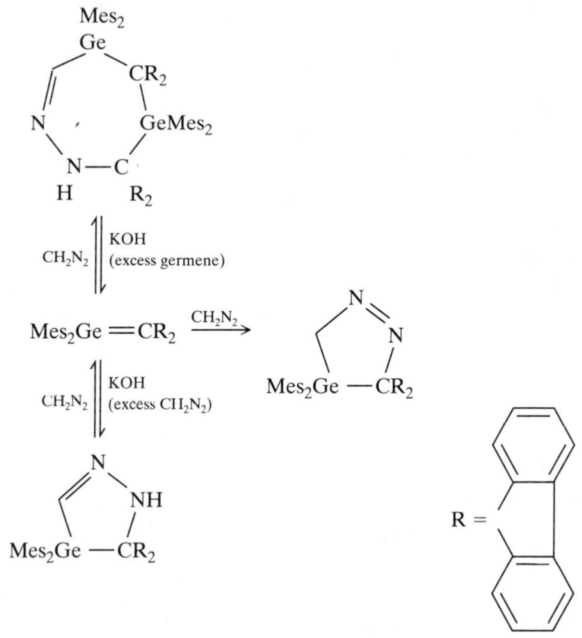

SCHEME 1

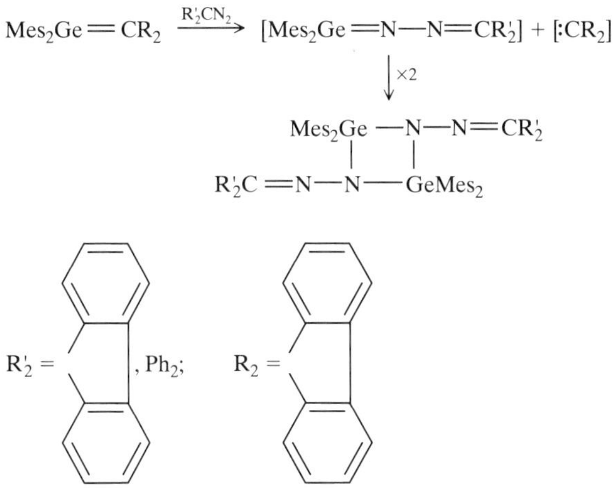

$$Mes_2Ge{=}CR_2 \xrightarrow{R'_2CN_2} [Mes_2Ge{=}N{-}N{=}CR'_2] + [{:}CR_2]$$

$$\downarrow \times 2$$

$$\begin{array}{ccc} Mes_2Ge & -N- & N{=}CR'_2 \\ | & | & \\ R'_2C{=}N{-}N & - & GeMes_2 \end{array}$$

$R'_2 =$ (fluorenylidene), Ph_2; $R_2 =$ (fluorenylidene)

SCHEME 2

The addition of diazomethane to tetrakis(2,6-diethylphenyl)digermene results in the formation of the stable digermirane, presumably via the digermapyrazoline intermediate.[63–65] The analogous reaction takes place upon treatment of tetrakis(2,4,6-triisopropylphenyl)digermene with diazomethane, albeit in low yield.[65] Similarly, the reaction of tetrakis(2,6-diethylphenyl)digermene[63–65] or tetramesityldigermene[65] with phenyl azide gives the digermaaziridine.

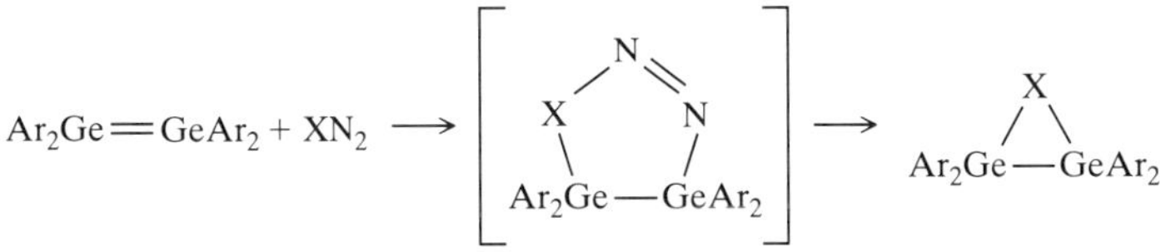

$$Ar_2Ge{=}GeAr_2 + XN_2 \longrightarrow \left[\begin{array}{c} X{-}N{=}N \\ | \quad\quad | \\ Ar_2Ge{-}GeAr_2 \end{array} \right] \longrightarrow \begin{array}{c} X \\ / \ \backslash \\ Ar_2Ge{-}GeAr_2 \end{array} \tag{15}$$

X = CH_2; Ar = 2,6-diethylphenyl
X = CH_2; Ar = 2,4,6-triisopropylphenyl
X = NPh; Ar = 2,6-diethylphenyl
X = NPh; Ar = 2,4,6-trimethylphenyl

b. *Addition of Carbonyl Compounds and Their Derivatives.* Four different modes of reactivity have been observed between carbonyl compounds and their derivatives and doubly bonded germanium compounds. A formal [2 + 2] reaction can occur between the carbonyl compound and the unsaturated germanium compound; with α,β-unsaturated ketones and aldehydes, both [2 + 2] and [2 + 4] products are observed; and finally, with germenes that have an allylic-type hydrogen, ene reactions, where both the germene or the carbonyl compound can act as the ene component, have been observed (see Scheme 3).

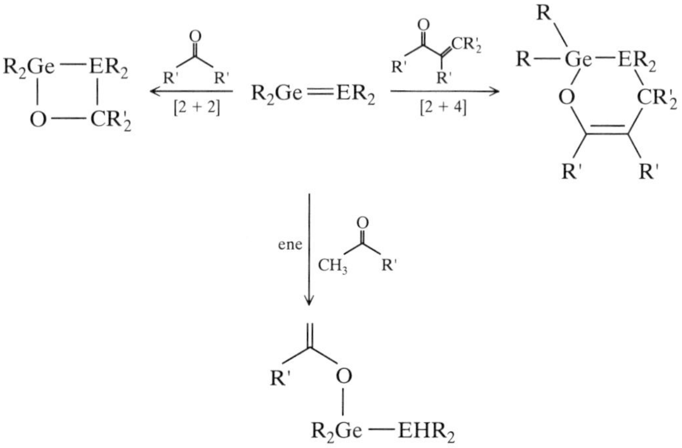

$$R_2Ge{=}ER_2 \xrightarrow[{[2+4]}]{R'C(O)C(R')=CR'_2} \text{germaoxacyclohexene};\quad R_2Ge{=}ER_2 \xrightarrow[{[2+2]}]{R'C(O)R'} \text{2-germaoxetane};\quad R_2Ge{=}ER_2 \xrightarrow{\text{ene},\ CH_3C(O)R'} R'C(=CH_2)O{-}R_2Ge{-}EHR_2$$

SCHEME 3

The addition of a variety of carbonyl compounds to dimesitylfluorenylidenegermane has been investigated. Aryl-substituted aldehydes and ketones (namely, benzaldehyde and benzophenone) add in a [2 + 2] manner to give the 2-germaoxetanes regioselectively, and acetone adds in an ene-type reaction, again regioselectively, with germanium–oxygen bond formation.[66] On the other hand, α,β-unsaturated aldehydes and ketones (namely, acrolein, methacrolein, crotonaldehyde, methyl vinyl ketone, methyl 2-methyl-1-propenyl ketone, and tetraphenylcyclopentadienone) yield the germaoxacyclohexene exclusively via a formal [2 + 4] cycloaddition, also regioselectively, with formation of the germanium–oxygen bond.[67] 1,1-Dimesitylneopentylgermene shows a greater versatility in its behavior toward carbonyl compounds.[68] With methyl vinyl ketone, again, only the germaoxacyclohexene is formed regioselectively, and with ace-

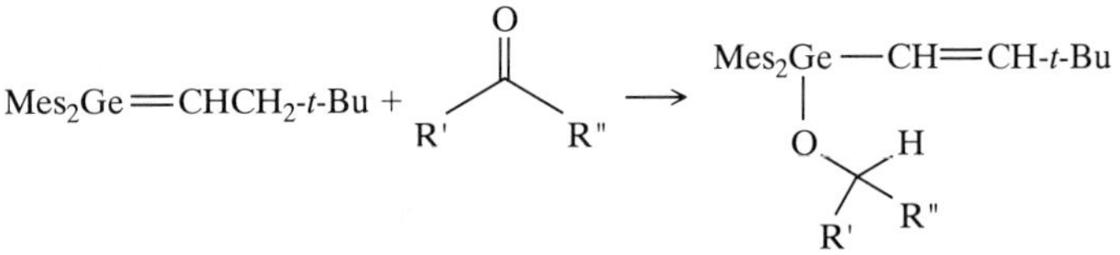

$$Mes_2Ge{=}CHCH_2\text{-}t\text{-}Bu + R'C(O)R'' \longrightarrow Mes_2Ge(CH{=}CH\text{-}t\text{-}Bu){-}O{-}CH(R')(R'') \tag{16}$$

tone or acetaldehyde, the germene also acts as an enophile in an ene-type reaction. However, with benzaldehyde, benzophenone, and acrolein, the carbonyl compound acts as the enophile, and the germene as the ene component. With acrolein, a germaoxacyclohexene is also obtained in about equal yield. Pearson's theory has been used to explain the predominance of [2 + 4]-type cycloaddition products in the reactions of germenes with α,β-unsaturated carbonyl compounds.

The addition of carbonyl compounds to germasilenes and digermenes has not been as extensively investigated as additions to germenes. Tetramesitylgermasilene adds benzaldehyde, pivalaldehyde, and acetone regioselectively in a [2 + 2] cycloaddition to give only the 2,3-silagermaoxetanes.[69] An ene-type reaction with acetone was observed neither in this case nor in the addition of acetone to tetrakis(2,6-diethylphenyl)digermene, where only the digermaoxetane was obtained.[63] Tetramesityldigermene added formaldehyde,[38,70] thiobenzophenone,[38,70] and adamantanethione,[71] again in a [2 + 2] manner. Methyl vinyl ketone represents an interesting example of the addition reactions of carbonyl compounds to germasilenes.[58] In this case, both regioisomers of the [2 + 2] addition reaction were obtained, as well as one regioisomer (most probably with a silicon–oxygen bond) of a silagermaoxacyclohexene from a [2 + 4] reaction.

The base-stabilized germanimine $(Me_2Si)_2(N\text{-}t\text{-}Bu)_4Ge{=}N(SiMe_3)$ does not react with benzophenone, even upon heating.[72] Apparently, no other reactions between stable germanimines and simple carbonyl compounds have been investigated; however, the reaction with a diketone, 3,5-di-*t*-butyl-*ortho*-quinone, has. Both a [2 + 2] and a [2 + 4] cycloaddition reaction have been postulated; the initially formed adducts are unstable and decompose (see Scheme 4).

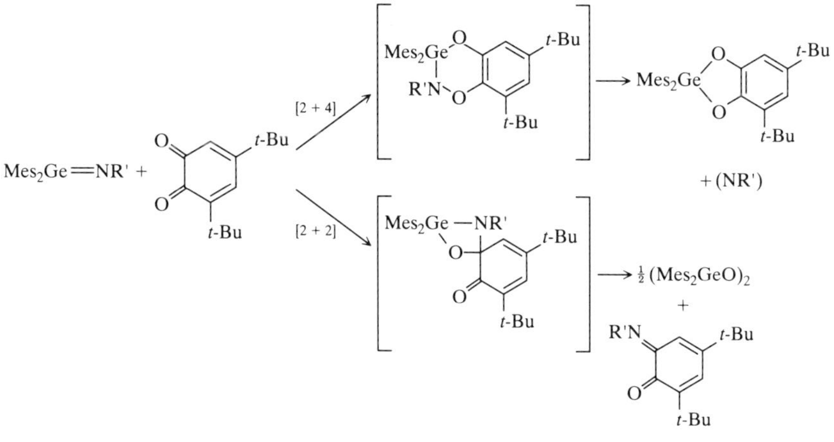

[2 + 4]: R' = 2-(carbomethoxy)phenyl,[53] 2-(carbo-*N*,*N*-dimethylamino)phenyl[53]
2-(carbomethoxy)thiophen-3-yl[40]
[2 + 2]: R' = 2-(carbomethoxy)phenyl,[53] 2-(carbo-*N*,*N*-dimethylamino)phenyl[53]

SCHEME 4

The addition of benzaldehyde[35] and a variety of α,β-unsaturated aldehydes and ketones (namely, acrolein, methacrolein, crotonaldehyde, 3-methyl-2-butenal, methyl vinyl ketone, methyl 2-methyl-1-propenyl ketone, and methyl 1-methylethenyl ketone)[73] to $Mes_2Ge{=}PAr$ (Ar = 2,4,6-triisopropylphenyl or 2,4,6-tri-*t*-butylphenyl) has been investigated. All additions are regioselective with formation of the germanium–oxygen bond. With either germaphosphene, benzaldehyde gives a [2 + 2] adduct, methyl vinyl ketone gives a [2 + 4]-type adduct, and methyl 2-methyl-1-propenyl ketone gives an ene adduct. With methyl 2-methyl-1-propenyl ketone, a six-membered ring is formed in both cases, but when the aryl group on phosphorus is 2,4,6-tri-*t*-butylphenyl, minor amounts of an ene product are also obtained. With acrolein, methacrolein, and crotonaldehyde, the [2 + 4]-type adducts are the major products with both germaphosphenes. With the bulkier 2,4,6-tri-*t*-butylphenyl group on phosphorus, minor amounts of the germaoxaphosphetanes are formed in reaction with methacrolein and crotonaldehyde. However, with 3-methyl-2-butenal, the [2 + 2] product predominates with both germaphosphenes; and with the less bulky 2,4,6-triisopropylphenyl group, some of the four-membered ring is still obtained. All of the addition reactions to germaphosphenes are regioselective, with formation of the germanium–oxygen bond. The product distribution has been rationalized, for the most part, on steric grounds.

Related to the addition reactions of carbonyl compounds is the addition of phenyl isothiocyanate to a stabilized germanethione[21] and germaneselone,[22] and carbon disulfide to the base-stabilized germanimine $(Me_2Si)_2(N\text{-}t\text{-}Bu)_4Ge{=}N(SiMe_3)$.[72] A regioselective [2 + 2] addition is observed across the carbon–sulfur double bond with formation of the germanium–sulfur bond. In the case of the germanimine, the four-membered ring is not stable and undergoes a retro-[2 + 2] reaction to give the stable germanethione and trimethylsilyl isothiocyanate.[72]

$$(Tb)(Tip)Ge{=}E + PhNCS \longrightarrow (Tb)(Tip)Ge\overline{-E-C(=N-Ph)-S-} \quad (17)$$

E = S, Se

c. *Addition of Unsaturated Nitrogen Compounds.* Dimesitylfluorenylidenegermane has been reacted with a variety of unsaturated nitrogen compounds.[74] The reactions are summarized in Scheme 5. Analogous reactions of other unsaturated germanium compounds have not been investigated. With benzophenonimine, insertion into the NH bond is observed,

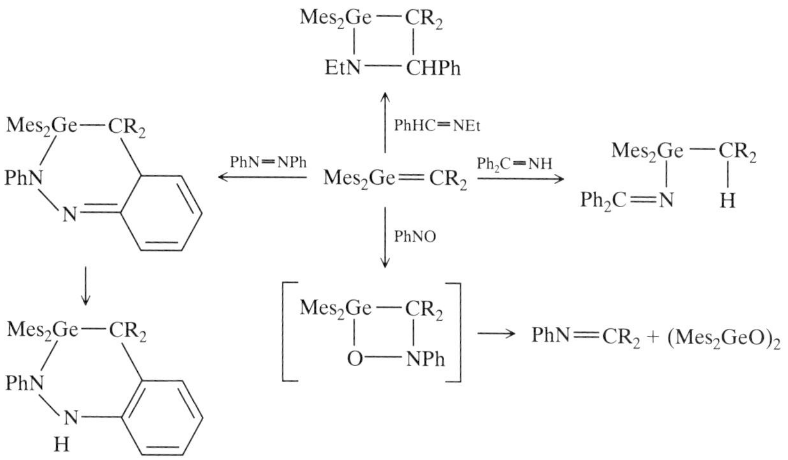

$Mes_2Ge{=}CR_2 \xrightarrow{PhHC{=}NEt}$ $Mes_2Ge{-}CR_2$ / EtN—CHPh

$Mes_2Ge{=}CR_2 \xrightarrow{Ph_2C{=}NH}$ $Mes_2Ge{-}CR_2$ / $Ph_2C{=}N$ H

$Mes_2Ge{=}CR_2 \xrightarrow{PhN{=}NPh}$ $Mes_2Ge{-}CR_2$ / PhN / N= (six-membered ring) $\rightarrow$ $Mes_2Ge{-}CR_2$ / PhN / N—H (aromatized)

$Mes_2Ge{=}CR_2 \xrightarrow{PhNO}$ [$Mes_2Ge{-}CR_2$ / O—NPh] $\longrightarrow PhN{=}CR_2 + (Mes_2GeO)_2$

CR_2 = fluorenylidene

SCHEME 5

whereas with *N*-ethylbenzaldimine, a [2 + 2] cycloaddition takes place regioselectively to afford the 2-germaazetidine. Presumably, both reactions begin with nucleophilic attack of the nitrogen at the unsaturated germanium center followed by transfer of a proton or cyclization. However, intermediate formation of the [2 + 4] adduct, where the imine acts as the 4π-component, followed by rearrangement cannot be excluded. In contrast, the reaction of the germene with azobenzene yields the [2 + 4] adduct exclusively. The initially formed six-membered ring undergoes rearrangement at room temperature to give the aromatized product. Finally, with nitrosobenzene, phenylfluorenylidenamine and 2,2,4,4-tetramesityldigerma-1,3-dioxetane were isolated, presumably the products of a retro-[2 + 2] reaction of the initially formed [2 + 2] cycloadduct. Again, intermediate formation of the [2 + 4] product cannot be ruled out.

d. *Addition of Alkynes.* Tetramesityldigermene,[38,70] tetrakis(2,6-diethylphenyl)digermene,[63] and tetramesitylgermasilene[75] add phenylacetylene in a formal [2 + 2]-cycloaddition reaction to give dimetallacyclobutenes. The addition to tetramesitylgermasilene is completely regioselective.

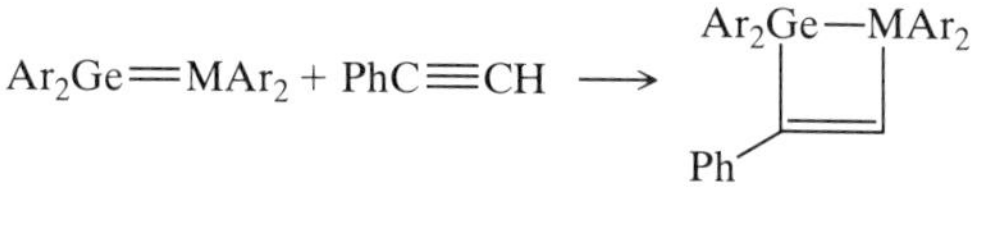

$$Ar_2Ge{=}MAr_2 + PhC{\equiv}CH \longrightarrow \begin{array}{c} Ar_2Ge{-}MAr_2 \\ | \quad\quad | \\ Ph{-}C{=}CH \end{array} \qquad (18)$$

An analogous reaction has been proposed to take place between acetylene and tetramesityldigermene during the thermolysis of hexamesitylcyclotrigermane and acetylene in the presence of $Pd(PPh_3)_4$.[76] Quite a different mode of reactivity is observed between stabilized germanimines and germaphosphenes and phenylacetylene. In these cases, the terminal carbon–hydrogen bond of the alkyne adds across the double bond, with the *sp*-hybridized carbon becoming attached to the germanium atom. The regiochemistry of the addition is readily explained in terms of the polarity of the double bond; however, the difference in the reactivity between germasilenes and digermenes relative to the germanimines and germaphosphenes is not easily explained. Perhaps it is a reflection of the greater zwitterionic character of germanimines and germaphosphenes.

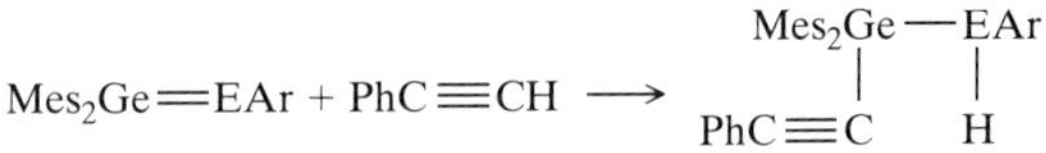

$$Mes_2Ge{=}EAr + PhC{\equiv}CH \longrightarrow Mes_2Ge(C{\equiv}CPh){-}E(H)Ar \qquad (19)$$

E = N; Ar = 2-(carbomethoxy)phenyl[41] or 2-(carbo-*N*-*N*-dimethylamino)phenyl[41,53]
E = P; Ar = 2,4,6-tri-*t*-butylphenyl[59]

A rather unusual cycloaddition reaction takes place between dimesitylfluorenylidenegermane and *t*-butylphosphaalkyne. This reaction could begin with a formal [2 + 4] cycloaddition between the germene and the alkyne, with the germene acting as the 4π-component.[77]

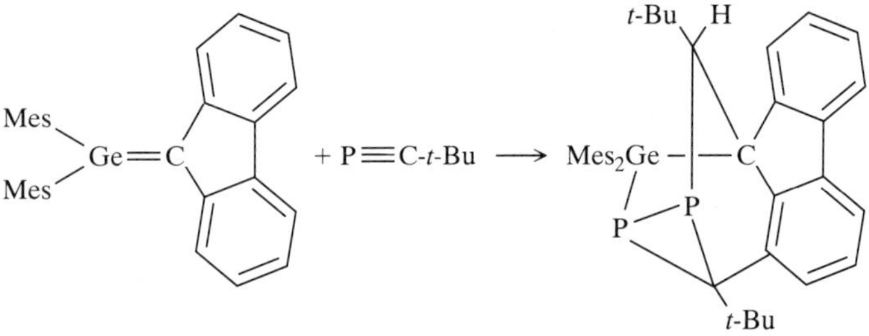

$$Mes_2Ge{=}C(\text{fluorenylidene}) + P{\equiv}C\text{-}t\text{-Bu} \longrightarrow \text{product (Mes}_2\text{Ge, 2 P, 2 } t\text{-Bu, H)} \qquad (20)$$

e. *Addition of Nitrones and Nitrile Oxides.* Stable germenes, germanimines, and germaphosphenes react regioselectivly in a [2 + 3] cycloaddition with nitrones, forming the strong germanium–oxygen bond.

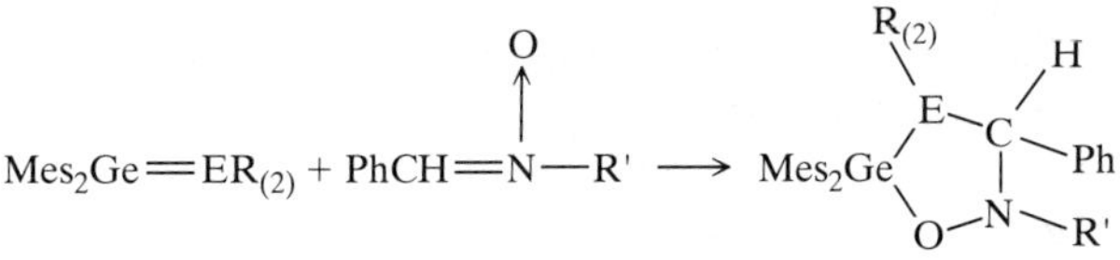

$$Mes_2Ge{=}ER_{(2)} + PhCH{=}N(\rightarrow O){-}R' \longrightarrow Mes_2Ge{-}E(R_{(2)}){-}C(H)(Ph){-}N(R'){-}O{-}(\text{ring}) \qquad (21)$$

ER_2 = fluorenylidene; R' = *t*-butyl[51]
E = N; R = 2-(carbomethoxy)phenyl; R' = *t*-butyl,[41,53] phenyl[41]
E = N; R = 2-(carbo-*N*-*N*-dimethylamino)phenyl; R' = *t*-butyl,[53] phenyl[41]
E = P; R = 2,4,6-triisopropylphenyl; R' = *t*-butyl[35] (two disastereomers formed)
E = P; R = 2,4,6-tri-*t*-butylphenyl; R' = *t*-butyl[35] (one disastereomer detected)

The addition of mesityl nitrile oxide to stable derivatives of a germanethione[21] and a germaneselone[22] is closely related to the addition of nitrones. The reaction is regioselective again, with formation of the germanium–oxygen bond.

$$\text{Tb(Tip)Ge=E} + \text{MesCNO} \longrightarrow \text{Tb(Tip)}\overline{\text{Ge–E–C(Mes)=N–O}} \qquad (22)$$

E = S, Se

f. *Addition of Dienes.* 2,3-Dimethylbutadiene appears to be the only diene that has been investigated in reactions with doubly bonded germanium compounds. With germenes, a germanethione, and a germaneselone, only the six-membered ring adduct, presumably from a formal [2 + 4] cycloaddition, is formed. 2,3-Dimethylbutadiene does not appear to undergo a [2 + 4] cycloaddition reaction with germasilenes,[57] digermenes,[15,48] or base-stabilized germanimines.[72] Although a cycloaddition product between tetramesityldigermene and 2,3-dimethylbutadiene was reported,[38] it was later recognized that an erroneous assignment of the structure had been made.[57,70]

$$R^1R^2\text{Ge=E}(R_2) + \text{CH}_2\text{=C(CH}_3\text{)C(CH}_3\text{)=CH}_2 \longrightarrow R^1R^2\overline{\text{Ge–CH}_2\text{C(CH}_3\text{)=C(CH}_3\text{)CH}_2\text{–E}(R_2)} \qquad (23)$$

ER_2 = fluorenylidene; $R^1 = R^2$ = 2,4,6-trimethylphenyl[51]
ER_2 = fluorenylidene; $R^1 = R^2$ = fluorenyl[28]
E = S; R^1 = 2,4,6-tris[bis(trimethylsilyl)methyl]phenyl, R^2 = 2,4,6-triisopropylphenyl[21]
E = Se; R^1 = 2,4,6-tris[bis(trimethylsilyl)methyl]phenyl, R^2 = 2,4,6-triisopropylphenyl[22]

6. *Reaction with Reagents Containing Group 16 Elements*

Only oxygenation reactions of digermenes (specifically, those of tetrakis(2,6-diethylphenyl)- and tetrakis(2,6-diisopropylphenyl)digermene) have been investigated.[78] Three different ring systems are accessible, depending on the reagent employed (see Scheme 6). Treatment of the digermene with oxygen in the cold or at room temperature gave the 1,2-digermadioxetane. Upon heating, this compound isomerizes by insertion of the oxygen–oxygen bond into a benzylic carbon–hydrogen bond. Photolysis also results in isomerization of the 1,2-digermadioxetane ring system; however, under these conditions, the 1,3-digermadioxetane is formed. The 1,3-digermadioxetane can also be obtained directly by oxygenation of the digermene using dimethyl sulfoxide (DMSO) or *N*-methylmorpholine-*N*-oxide. Alternatively, the 1,3-digermadioxetane can be synthesized by treatment of the digermene with nitrous oxide to give a digermaoxirane; subsequent treatment of the oxirane with DMSO or *N*-methylmorpholine-*N*-oxide yields the 1,3-digermadioxetane. These reactions are summarized in Scheme 6 for Ar = 2,6-diethylphenyl. Most of these reactions have been carried out with tetrakis(2,6-diisopropylphenyl) digermene as well.

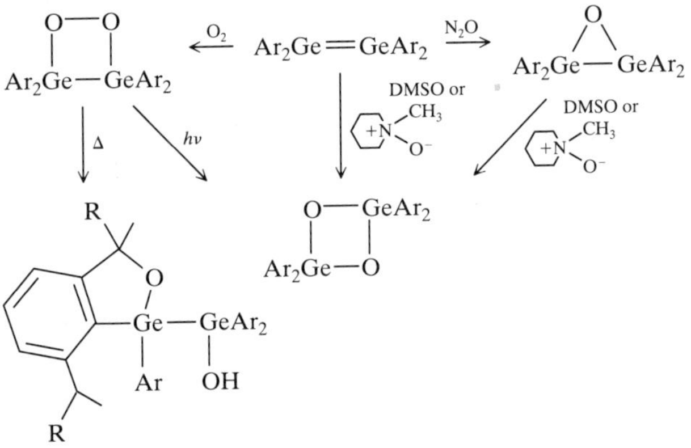

O_2 ; $Ar_2Ge{=}GeAr_2$; N_2O ; DMSO or CH_3 $+N$ O^- ; Δ ; $h\nu$; $Ar_2Ge{-}GeAr_2$; R ; O ; Ge ; $GeAr_2$; Ar ; OH ; $Ar_2Ge{-}O$

Ar = 2,6-diethylphenyl, R = H

SCHEME 6

Thia-, selena-, and telluradigermiranes have been synthesized by the addition of sulfur, selenium, or tellurium, respectively, to digermenes.

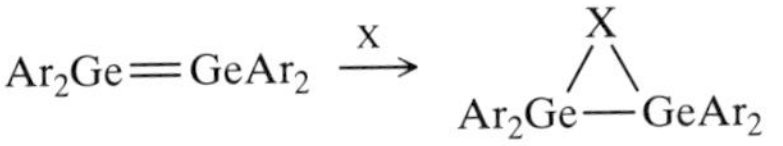

$$Ar_2Ge{=}GeAr_2 \xrightarrow{X} Ar_2Ge(X)GeAr_2 \tag{24}$$

Ar = 2,6-diethylphenyl; X = Se,[63] Te[79]
Ar = 2,4,6-trimethylphenyl; X = S (S_8),[43] Se,[43] Te[79]

An analogous reaction has been carried out with germaphosphenes.[80]

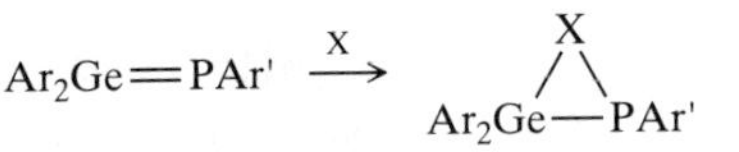

$$Ar_2Ge{=}PAr' \xrightarrow{X} Ar_2Ge(X)PAr' \tag{25}$$

Ar = 2,4,6-trimethylphenyl; Ar' = 2,4,6-tri-*t*-butylphenyl; X = S (S_8), Se

7. *Reduction Reactions*

Germenes and germaphosphenes can be reduced with lithium aluminum hydride to give the corresponding germanes. This reaction apparently has not been examined for other unsaturated germanium derivatives.

$$R_2Ge{=}ER'_{(2)} \xrightarrow{LiAlH_4} \underset{H}{R_2Ge}{-}\underset{H}{ER'_{(2)}} \qquad (26)$$

R = 2,4,6-trimethylphenyl; ER'_2 = fluorenylidene[11]
E = C; R = 2,4,6-trimethylphenyl; R'_2 = hydrogen, neopentyl[26]
E = P; R = 2,4,6-trimethylphenyl; R' = 2,4,6-tri-*t*-butylphenyl[59]
E = P; R = *t*-butyl; R' = 2,4,6-tri-*t*-butylphenyl[20]

When E = P, R = 2,4,6-trimethylphenyl, and R′ = 2,4,6-tri-*t*-butylphenyl (Eq. 26), the same reduction can be accomplished using $BH_3 \cdot SMe_2$.[59] The borane addition product is also obtained ($Mes_2GeH{-}PArBH_2$).

A rather unusual reduction reaction was observed upon treatment of tetrakis(2,6-diisopropylphenyl)digermene with excess lithium naphthalenide. Cleavage of an aryl–germanium bond occurs to give a digermenyllithium reagent, which was trapped by methanol.[39] The generality of this reaction has yet to be demonstrated. Similar treatment of tetrakis(2,6-diethylphenyl)digermene led to products that were converted upon methanolysis to the corresponding diaryl- and tetraaryldigermane.

$$Dip_2Ge{=}Ge(Dip)_2 \xrightarrow{LiNp} Dip_2Ge{=}Ge(Dip)(Li(DME)_2) \xrightarrow{MeOH} Dip_2Ge(H){-}Ge(Dip)(H){-}OMe$$

Dip = 2,6-diisopropylphenyl
DME = 1,2-oxyethanedimeth (27)

8. *Rearrangements*

Tetramesitylgermasilene[30] and -digermene[81] undergo a facile (at or below room temperature) 1,2-mesityl shift to give the corresponding silyl- or germylgermylene, respectively. The germylene has been trapped with triethylsilane[30,81] and 2,3-dimethylbutadiene.[57,70] In the case of the germasilene, the shift is completely regioselective to give only the germylene.

$$Mes_2Ge{=}MMes_2 \rightarrow Mes\ddot{G}e{-}MMes_3$$
$$M = Si, Ge \qquad (28)$$

It is interesting that the aryl groups in tetrakis(2,6-diethylphenyl)digermene do not shift under the same conditions or upon thermolysis, presumably because of steric hindrance. However, upon photolysis, tetrakis(2,6-diethylphenyl)digermene dissociates to the diarylgermylene.[48]

Thermolysis of 1-(2,4,6-tri-*t*-butylphenyl)-2,2-dimesitylgermaphosphene gives a germaphosphetene, presumably via a radical mechanism.[82]

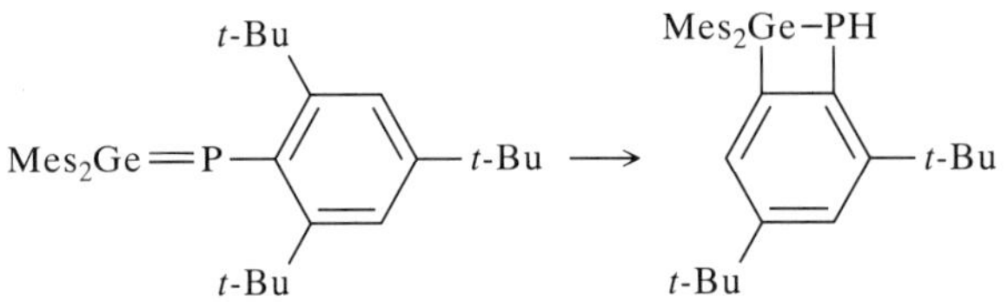

$$Mes_2Ge{=}P\text{-}C_6H_2(t\text{-}Bu)_3 \longrightarrow \text{(Mes}_2\text{Ge–PH benzocyclic product with two } t\text{-Bu)} \quad (29)$$

III
TIN

A. Physical Properties

1. *Structure Determinations*

Only a few structures of tricoordinate tin compounds have been determined crystallographically. The first was that of the distannene $[Sn[CH(SiMe_3)_2]_2]_2$ (entry **1**, Table V), initially reported in 1976.[14,83] The compound was considered to be the dimer of the dialkylstannylene, as the compound was found to exist as the stannylene in solution. The centrosymmetric dimer had a Sn—Sn bond length of 2.768(1) Å (versus 2.780(4) Å for Sn_2Ph_6)[84] with the geminal alkyl groups in a trans-bent arrangement. The sum of angles at Sn is 342°. There is a fold angle of 32°, but no twist of the SnC_2 planes aboiut the Sn—Sn axis. The structure of the distannene can be explained in terms of a weak bent double bond formed from the stannylene by overlap of the filled sp_xp_y-nonbonding orbital of one monomer unit with the vacant p_z-orbital of the other.

The stannene (entry **2**, Table V) shown below has been examined by X-ray crystallography.[85] The length of the bond between the tin atom and the tricoordinated carbon atom was found to be 2.025(4) Å, significantly shorter (6%) than the bonds to the tetracoordinated carbon atoms (2.152(5) and 2.172(4) Å) and in reasonable agreement with that calculated for the parent compound $H_2C{=}SnH_2$ (1.982 Å).[86] The average twist angle about the Sn=C bond is 61° and the fold angle is 5°. Thc bonds between the boron atoms and the tricoordinate carbon atom are short (1.510(6) and 1.494(7) Å), indicating a significant contribution from the ylide form shown below.

The stannaketenimine (entry **3**, Table V) shown below was found to have a bent structure with an angle at the dicoordinate carbon atom of 154°.[87] The coordination geometry of the tin atom is distorted owing to the short contact with one of the fluorine atoms of the trifluoromethyl substituents. The bond between the tin atom and the doubly bonded carbon

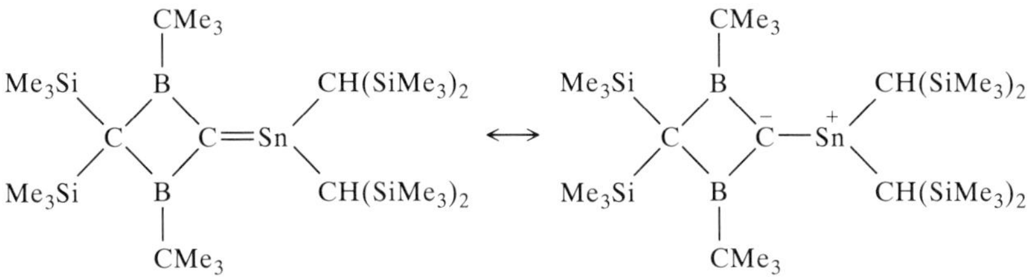

TABLE V

SELECTED BOND LENGTHS AND ANGLES IN TIN DOUBLY BONDED COMPOUNDS

Entry	Compound[a]	Sn=E (Å)	Angles (deg.)				Ref.
			C—Sn—C	C—Sn—E	Fold	Twist	
1	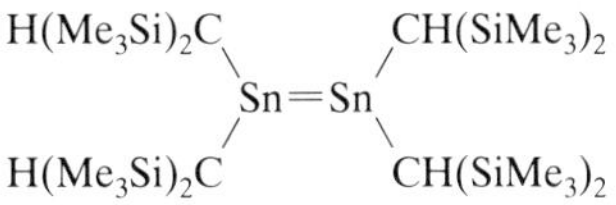 $[H(Me_3Si)_2C]_2Sn=Sn[CH(SiMe_3)_2]_2$	2.768(1)	109.2(2)	112.0(1)	41	0	83, 14
2	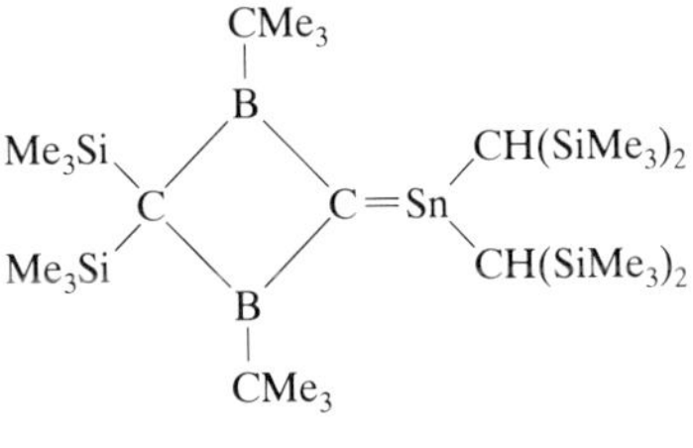 $(Me_3Si)_2C(BCMe_3)_2C=Sn[CH(SiMe_3)_2]_2$	2.025(4)	104.8(2)	129.2(2) 125.6(2)	5	61	85
3	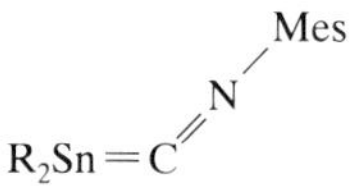 $R_2Sn=C=N-Mes$	2.397(3)	102.6(1)	104.9(1) 83.4(1)			87
4	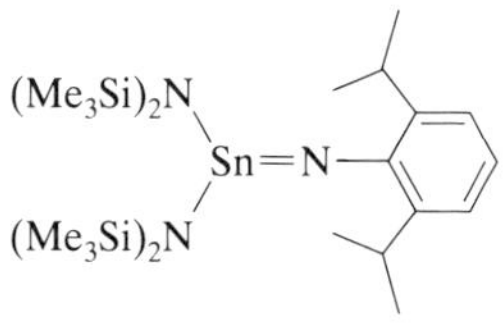 $[(Me_3Si)_2N]_2Sn=N-(2,6\text{-}iPr_2C_6H_3)$	1.921(2)	N—Sn—N 115.8(1)	N—Sn—E 114.3(1) 129.1(1)			88

[a] R = 2,4,6-tris(trifluoromethyl)phenyl; Mes = 2,4,6-trimethylphenyl.

atom (2.397(3) Å) is significantly longer than the distances between the tin atom and the ipso carbon atoms (2.306(2) and 2.314(3) Å).

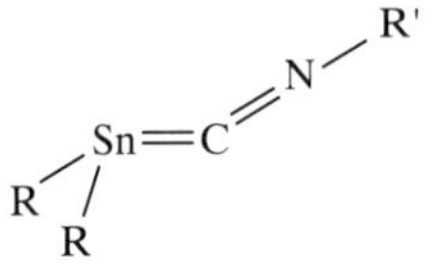

$R_2Sn{=}C{=}N{-}R'$

R = 2,4,6-$(CF_3)_3C_6H_2$
R' = 2,4,6-$(CH_3)_3C_6H_2$

The only reported structure of a stannanimine (entry **4**, Table V), shown below, was found to have a trigonal-planar arrangement of the three nitrogen atoms about the tin atom.[88] The Sn═N double bond is significantly shorter than the two Sn—N single bonds (1.921(2) versus 2.015(2) and 2.030(3) Å).

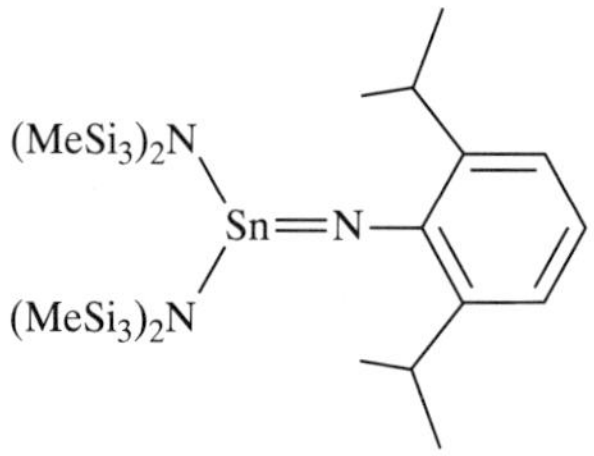

$[(MeSi_3)_2N]_2Sn{=}N{-}$(2,6-diisopropylphenyl)

2. *NMR Spectroscopy*

The structures of doubly bonded compounds containing tin readily lend themselves to study by multinuclear NMR spectroscopy, as tin has two NMR active isotopes: ^{117}Sn and ^{119}Sn. The chemical shifts of doubly bonded tin species in the ^{119}Sn NMR spectrum cover a broad range from $\delta = -150$ to $+835$ (Table VI); however, the majority of the signals occur at low field: $\delta = 400$ or above.

The stannenes (entry **1**, Table VI) have the expected low-field ^{119}Sn chemical shifts, the difference of almost 200 ppm presumably arising from the different nature of the ligands attached to the tin atom. The shielding of the boron atom, as indicated by its chemical shift, suggests a negative π-charge at the boron atoms as a result of a considerable contribution from the ylide resonance structure ($\delta^{11}B = 64$ for entry **1a** and 50 for entry **1b**, Table VI).[85] The ^{119}Sn NMR signal of the solvated stannene (entry **2**, Table VI) is shifted to higher field ($\delta = 288$) because of complexation of the tin atom with a solvent molecule.[89] The chemical shifts of the tricoordinated carbon atoms occur between $\delta = 91$ (entry **1b**, Table VI) and $\delta = 142$ (entry **1a**, Table VI).

The stannaketenimine (entry **3**, Table VI) has a ^{119}Sn resonance at high field ($\delta = -150$), which can be explained by the distortion of the molecule

TABLE VI

^{119}Sn NMR CHEMICAL SHIFTS OF DOUBLY BONDED TIN COMPOUNDS

Entry	Compound	^{119}Sn Chemical shift, δ	Solvent: standard	Ref.
1	CMe_3 B Me_3Si C $C{=}SnR_2$ Me_3Si B CMe_3 a: $R = CH(SiMe_3)_2$ b: $R_2 = [—N(t\text{-}Bu)(SiMe_2)(t\text{-}Bu)N—]$	a: 835 b: 647	not reported	85
2	$R_2(solvent)Sn{=}CR'_2$ $R = 2,4,6\text{-}(i\text{-}Pr)_3C_6H_2$; R'_2C = fluorenylidene a: solvent = diethyl ether b: solvent = tetrahydrofuran	a: 288.0 b: 287.3	C_6D_6	89
3	$R_2Sn{=}C{=}N{-}R'$ $R = 2,4,6\text{-}(CF_3)_3C_6H_2$; $R' = 2,4,6\text{-}(CH_3)_3C_6H_2$	−150	C_6D_6/toluene	87
4	$R_2Sn{=}SnR_2$ a: $R = 2,4,6\text{-}(i\text{-}Pr)_3C_6H_2$ b: $R = CH(SiMe_3)_2$	a: 427.3 or 410 b: 692 (solid state), 740 and 725 (solution)	a: methylcyclohexane-d_{14}: Me_4Sn b: toluene	a: 90 b: 91
5	$[(Me_3Si)_2N]_2Sn{=}NR$ $R = 2,4,6\text{-}(iPr)_3C_6H_2$	−3.5	C_7D_8: Me_4Sn	88
6	$R_2Sn{=}PR'$ a: $R = CH(SiMe_3)_2$; $R' = 2,4,6\text{-}(t\text{-}Bu)_3C_6H_2$ b: $R = 2,4,6\text{-}(i\text{-}Pr)_3C_6H_2$; $R' = 2,4,6\text{-}(t\text{-}Bu)_3C_6H_2$	a: 658.3 b: 499.5	a: Me_4Sn b: not reported	a: 92 b: 93, 94

from the linear idealized ketenimine structure and the elongation of the Sn=C double bond.[87]

The ^{119}Sn chemical shift of $Ar_2Sn{=}SnAr_2$ (entry **4a**, Table VI) is at low field and the tin–tin coupling satellites ($^1J(^{119}Sn{-}^{117}Sn)$ = 2930 Hz) are of the relative peak intensities expected only for a structure containing a tin–tin bond.[90] The room-temperature ^{119}Sn NMR spectrum of $[(Me_3Si)_2CH]_2Sn{=}Sn[CH(SiMe_3)_2]_2$ was recorded in the solid state. As in solution, the compound exists as the stannylene; the low-temperature solution ^{119}Sn NMR spectrum of this compound showed two signals at $\delta = 740$ and 725, which were assigned to distannenes from the temperature dependence of the solution ^{13}C shift of the methine carbons.[91]

The only reported ^{119}Sn NMR spectrum of a stannanimine (entry **5**, Table VI) shows a single signal at the surprisingly high field of $\delta = -3.5$, for which there is no obvious explanation. The other spectroscopic data, ^{1}H, ^{13}C, and ^{29}Si NMR spectra, are consistent with the stannanimine structure.[88]

The stannaphosphenes (entry **6**, Table VI) display low-field chemical shifts in both the ^{31}P and ^{119}Sn NMR spectra. The ^{31}P NMR shifts are δ 204.7[92] and 170.7[93,94] for entries **6a** and **6b**, respectively; the ^{31}P chemical shift of (2,4,6-triisopropylphenyl)$_2$Sn=PCH(SiMe$_3$)$_2$ has also been recorded (δ 168.7).[8] The ^{31}P chemical shift of entry **6b** was found to be temperature-dependent; shifting to higher fields with decreasing temperature, a similar observation has also been made for germaphosphenes. The tin–phosphorus coupling is characteristic of these compounds as it is much larger than that for singly bonded tin–phosphorus compounds: $^{1}J(P{-}^{117}Sn) = 2191$ Hz and $^{1}J(P{-}^{119}Sn) = 2295$ Hz for entry **6a** and $^{1}J(P{-}^{117}Sn) = 2110$ Hz and $^{1}J(P{-}^{119}Sn) = 2208$ Hz for entry **6b**, compared to values of ~1000 for singly bonded species.

3. *IR Spectroscopy*

The only doubly bonded tin compound for which the IR spectrum has been reported is the stannaketenimine $[2,4,6\text{-}(CF_3)_3C_6H_2]_2Sn{=}C{=}N[2,4,6\text{-}(CH_3)_3C_6H_2)]$. The C—N stretching vibration (2166 cm^{-1}) is shifted relative to that of mesityl isocyanide (2118 cm^{-1}); this phenomenon is also observed for isocyanide–transition-metal complexes.[87]

4. *UV Spectroscopy*

Compounds containing doubly bonded tin are highly colored due to a strong absorption in the visible spectrum (π–π^* transition of the double bond). The extinction coefficients of the distannene $[[(Me_3Si)_2CH]_2Sn]_2$ (entry **3**, Table VII) were found to depend on concentration and temperature[5]: that for the absorption at 332 nm increases with dilution while that for the 495 nm absorption decreases. This suggests that the absorptions arise from two different species, presumably the distannene and the stannylene. By comparison with the other absorption maxima of doubly bonded tin compounds, the absorption at 495 nm is probably that of the distannene.

TABLE VII

UV SPECTROSCOPY DATA FOR DOUBLY BONDED TIN COMPOUNDS

Entry	Compound[a]	λ_{max} (nm)	log ε	Solvent	Ref.
1	$Tip_2(Et_2O)Sn{=}CR_2$ CR_2 = fluorenylidene	542	not determined	Et_2O	89
2	$Tip_2Sn{=}SnTip_2$	494	4.59	methylcyclohexane	90
3	$[(Me_3Si)_2CH]_2Sn{=}Sn[CH(SiMe_3)_2]_2$	495, 332	temperature- and concentration-dependent	hexane	5, 83
4	$[(Me_3Si)_2CH]_2Sn{=}PAr$ Ar = tri-*t*-butylphenyl	460, 350	not reported	THF, pentane	92
5	$Tb(Tip)Sn{=}S$	473	not reported	hexane	95

[a] Tip = 2,4,6-triisopropylphenyl; Tb = 2,4,6-tris[bis(trimethylsilyl)methyl]phenyl.

B. *Synthetic Methods*

1. *Thermal or Photochemical Generation*

The low-temperature photolysis of hexakis(2,4,6-triisopropylphenyl) cyclotristannane leads quantitatively to the corresponding distannene (Eq. 30).[90] The distannene was found to be in equilibrium with the cyclotristannane over a wide range of temperatures. The simplest explanation for this is that the stannylene is formed as an intermediate from either the cyclotristannane or the distannene, but quickly adds to the distannene or dimerizes. This equilibrium can be followed by ^{119}Sn NMR spectroscopy at elevated temperatures.[96]

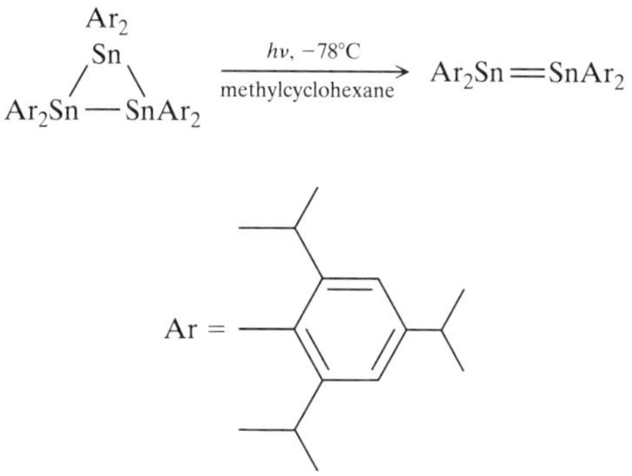

$$\text{cyclo-}(Ar_2Sn)_3 \xrightarrow[\text{methylcyclohexane}]{h\nu,\ -78^{\circ}C} Ar_2Sn{=}SnAr_2 \qquad (30)$$

Ar = 2,4,6-triisopropylphenyl

Stannanethiones and stannaneselones can be generated by the thermal retrocycloaddition of trichalcogenastannolanes or the dechalcogenation of tetrachalcogenastannolanes in the presence of triphenylphosphine

(Scheme 7).[97,98] These doubly bonded species can be identified from their trapping reactions.

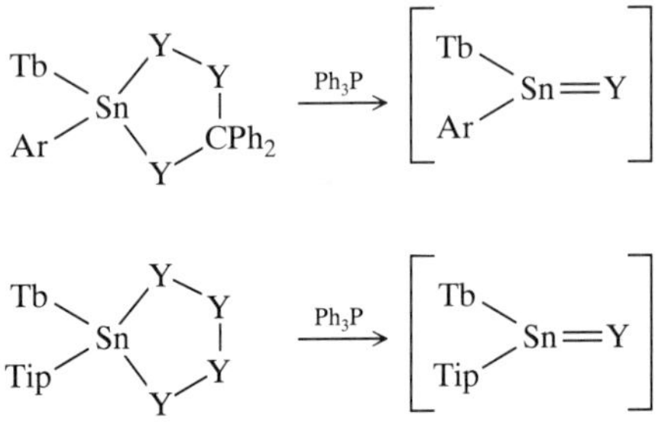

$$\text{Tb(Ar)Sn}\langle \text{Y–Y–CPh}_2\text{–Y} \rangle \xrightarrow{Ph_3P} \left[\text{Tb(Ar)Sn=Y} \right]$$

$$\text{Tb(Tip)Sn}\langle \text{Y–Y–Y–Y} \rangle \xrightarrow{Ph_3P} \left[\text{Tb(Tip)Sn=Y} \right]$$

Y = S, Se; Ar = Tip, Mes

SCHEME 7

The thermolysis of a linear (fluorostannyl)silylphosphine has been reported to give the corresponding stannaphosphene.[8] This reaction can be considered to be a defluorosilylation and, as such, is analogous to the dehydrohalogenation reactions discussed below.

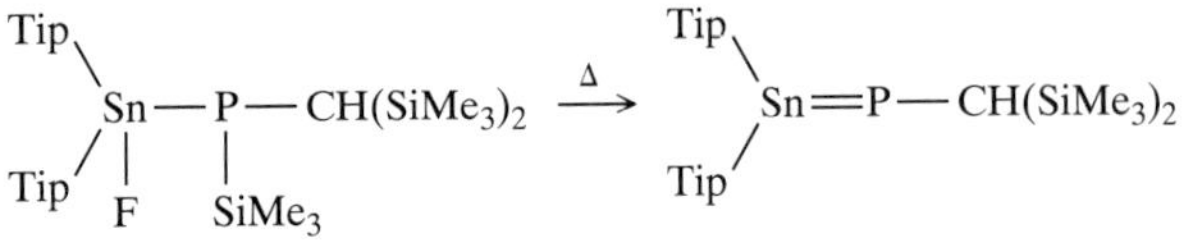

$$Tip_2Sn(F)\text{—}P(SiMe_3)\text{—}CH(SiMe_3)_2 \xrightarrow{\Delta} Tip_2Sn\text{=}P\text{—}CH(SiMe_3)_2 \quad (31)$$

2. *Dehydrohalogenation*

The dehydrohalogenation by organolithium reagents of the appropriate halo derivatives is a very useful route to unsaturated main group species. The use of fluorine as the halogen and a bulky organolithium as reagent generally gives the best results, as side reactions, such as direct alkylation of the metal by the organolithium compound or lithium–halogen exchange, are minimized. Three stannenes have been synthesized using this method.

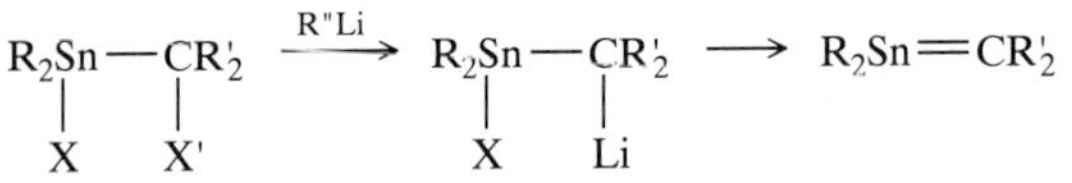

$$R_2Sn(X)\text{—}CR'_2(X') \xrightarrow{R''Li} R_2Sn(X)\text{—}CR'_2(Li) \longrightarrow R_2Sn\text{=}CR'_2 \quad (32)$$

Stannene 1: R = $CH(SiMe_3)_2$; CR'_2 = fluorenylidene; X = F; X' = H; R" = *t*-butyl
Stannene 2: R = 2,4,6-triisopropylphenyl; CR'_2 = fluorenylidene; X = F; X' = H; R" = *t*-butyl
Stannene 3: R = methyl; R' = $SiMe_3$; X = X' = Br; R" = phenyl

Stannene 1 was found to be an extremely air-sensitive compound, which was identified only by its trapping reactions.[99] Replacement of the bis(trimethylsilyl)methyl groups with the bulky aromatic group 2,4,6-triisopropylphenyl led to the isolation of the stable stannene 2 as its diethyl ether complex.[89] Stannene 3 is short-lived at low temperatures and readily gives the head-to-tail dimer.[100]

Dehydrofluorination has also been utilized for the synthesis of stable stannaphosphenes. The stannaphosphene with bis(trimethylsilyl)methyl groups attached to tin could not be obtained in pure form because of its high reactivity[92]; replacing these alkyl groups with the bulky aryl substituent 2,4,6-triisopropylphenyl eliminated this problem and resulted in a stannaphosphene that is still air- and moisture-sensitive but stable under an inert atmosphere for prolonged periods of time and in solution at 60°C.[93,94]

$$\underset{\text{F}\quad\ \text{H}}{R_2Sn{-}PAr} \xrightarrow{t\text{-BuLi}} \underset{\text{F}\quad\ \text{Li}}{R_2Sn{-}PAr} \longrightarrow R_2Sn{=}PAr \tag{33}$$

Ar = 2,4,6-tri-*t*-butylphenyl; R = $(Me_3Si)_2CH$, 2,4,6-triisopropylphenyl

3. *Coupling of Stannylenes with Various Reagents*

The synthesis of doubly bonded tin compounds by the coupling of stannylenes, however useful, is limited by the need for a stable stannylene and often a second divalent species (for example, a carbene or isonitrile). The simplest example of this reaction is the formation of tetrakis[bis(trimethylsilyl)methyl]distannene from two molecules of the corresponding stannylene,[83] with which it is in equilibrium in solution as evidenced by NMR spectroscopy.[91]

Two stannenes have been synthesized by the reaction of a stannylene with a boranediylborirane (Eq. 34).[85] The boranediylborirane has been shown to react toward suitable reagents as though it were the carbene,[101] which is only slightly higher in energy than the boranediylborirane.[102] The reaction occurs at room temperature in pentane solution. The resulting stannene has a considerable contribution from the ylide resonance structures. The carbene arising from the boranediylborirane is extremely electrophilic, and therefore the stannenes can be considered formally to be adducts of the stannylene as Lewis base and the carbene.

$$(Me_3Si)_2C(B\text{-}t\text{-Bu})C{=}B\text{-}t\text{-Bu} \longrightarrow \left[(Me_3Si)_2C(B\text{-}t\text{-Bu})_2C{:}\right] \xrightarrow{SnR_2} (Me_3Si)_2C(B\text{-}t\text{-Bu})_2C{=}SnR_2 \tag{34}$$

R = $CH(SiMe_3)_2$; R_2 = N-*t*-BuSiMe$_2$N-*t*-Bu

A stannaketenimine has been synthesized by the low-temperature reaction of a diarylstannylene with mesityl isocyanide.[87] The stannaketenimine is thermally stable, crystals being obtained by sublimation at 40°C/0.01 mmHg. In this case, the stannylene acts as the Lewis acid for formation of the adduct and, along with the stannenes described above, covers the full range of bonding modes available to the stannenes.

$$R_2Sn + :C{=}N{-}Mes \rightleftharpoons R_2Sn{=}C{=}N{-}Mes \quad (35)$$

$$R = 2,4,6\text{-}(CF_3)_3C_6H_2$$

A stannanimine has been prepared from a highly hindered diazastannylene and 2,6-diisopropylphenylazide.[88] The stannanimine is stable in solution at −30°C, but rearranges within two weeks by the intramolecular addition of a C—H bond of one of the isopropyl groups across the Sn=N bond.

$$[(Me_3Si)_2N]_2Sn + N_3R \xrightarrow[4\,h,\ -N_2]{\text{hexane},\ -30°C} [(Me_3Si)_2N]_2Sn{=}NR \quad (36)$$

$$R = 2,6\text{-diisopropylphenyl}$$

A stable diarylstannanethione has been synthesized by the reaction of an extremely bulky stannylene with either styrene episulfide[95] or less than one equivalent of elemental sulfur.[9] If the 2,4,6-tris[bis(trimethylsilyl) methyl]phenyl group (Tb) is replaced by the smaller mesityl group, no monomeric products are observed; thus the stability of the stannanethione is attributed to the demand of the substituents.

$$(Tb)(Tip)Sn: \xrightarrow[\text{or } \frac{1}{8} S_8]{\text{Ph-episulfide}} (Tb)(Tip)Sn{=}S \quad (37)$$

C. *Reactivity*

2. *Addition of Protic Reagents*

Generally, protic reagents add across the double bonds of unsaturated tin species, the nucleophilic portion of the reagent becomes attached to the tin atom, and the proton shifts to the other element of the double bond. Examples of reactions of this type are given in Table VIII,[103,104] according to the equation given there.

In contrast, $[2,4,6\text{-}(CF_3)_3C_6H_2]_2Sn{=}C{=}NMes$ reacts with *t*-butyl alco-

TABLE VIII

REACTIONS OF DOUBLY BONDED TIN COMPOUNDS WITH PROTIC REAGENTS

$$R^1R^2Sn{=}ER^3R^4 \xrightarrow{AH} R^2{-}Sn(R^1)(A){-}E(R^3)(H){-}R^4$$

Entry	R_1	R_2	E	R^3	R^4	A	Ref.
1	Bis[a]	Bis	C	fluorenylidene		HO	99
2	Tip[b]	Tip	C	fluorenylidene		MeO	89
3	$(Me_3Si)_2N$	$(Me_3Si)_2N$	N	Dip[c]	—	DipNH	103
4	Bis	Bis	P	Ar[d]	—	HO	104
5	Bis	Bis	P	Ar	—	MeO	92, 104
6	Bis	Bis	P	Ar	—	PhNH	104
7	Bis	Bis	P	Ar	—	*i*PrS	104
8	Bis	Bis	P	Ar	—	Cl	92, 104
9	Tip	Tip	P	Ar	—	HO	93
10	Tip	Tip	P	Ar	—	MeO	93
11	Tb[e]	Tip	S	—	—	HO	9, 97

[a] Bis = bis(trimethylsilyl)methyl; [b]Tip = 2,4,6-triisopropylphenyl; [c]Dip = 2,6-diisopropylphenyl; [d]Ar = 2,4,6-tri-*t*-butylphenyl; [e]Tb = 2,4,6-tris[bis(trimethylsilyl)methyl]phenyl.

hol to give $(t\text{-BuO})_2Sn$ and mesityl isocyanide.[87] The reaction of the stannaketenimine as the stannylene is not unexpected, considering the strength of the tin–carbon bond, which is weaker than a tin–carbon single bond (dissociation enthalpy $\Delta H^\circ = 29.6 \pm 0.4$ kJ mol^{-1}, dissociation entropy $\Delta S^\circ = 90 \pm 1$ JK^{-1} mol^{-1}).

2. *Addition of Organometallic Reagents*

In general alkyllithiums and Grignard reagents add across the double bond of compounds containing doubly bonded tin.[100,104] However, in the reaction of $Bis_2Sn{=}CR_2$ (Bis = bis(trimethylsilyl)methyl) with *t*-BuLi (see Scheme 8), *t*-BuLi apparently acts as a single electron-transfer reagent to give the stannyl radical anion which abstracts a hydrogen from the solvent, resulting in the formation of a tin hydride anionic intermediate.[99] Examples of these reactions are shown in Scheme 8. The resulting metal species can be quenched with a protic reagent (H_2O, D_2O, MeOH) or an alkyl halide.[99,100,103,104]

A stannanethione and stannaneselone were found to react with $Os_3(CO)_{12}$ to give a mixture of two (sulfur) or three (selenium) metal–chal-

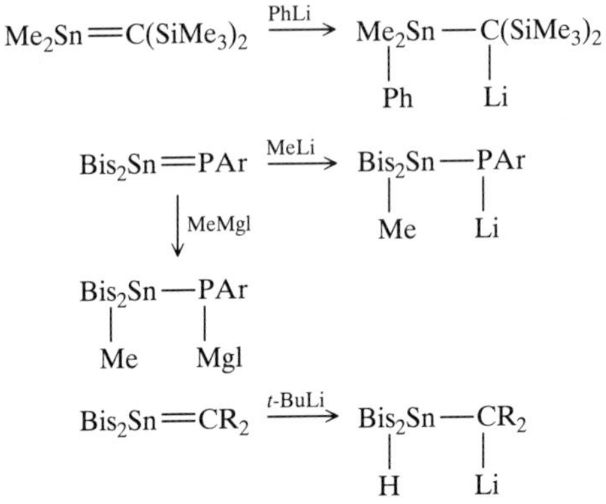

$$Me_2Sn{=}C(SiMe_3)_2 \xrightarrow{PhLi} Me_2Sn(Ph){-}C(SiMe_3)_2(Li)$$

$$Bis_2Sn{=}PAr \xrightarrow{MeLi} Bis_2Sn(Me){-}PAr(Li)$$

$$Bis_2Sn{=}PAr \xrightarrow{MeMgI} Bis_2Sn(Me){-}PAr(MgI)$$

$$Bis_2Sn{=}CR_2 \xrightarrow{t\text{-}BuLi} Bis_2Sn(H){-}CR_2(Li)$$

Ar = 2,4,6-tri-*t*-butylphenyl; CR_2 = fluorenylidene

SCHEME 8

cogen mixed clusters which included the first examples of transition-metal complexes bridged by stannanethione and stannaneselone.[105]

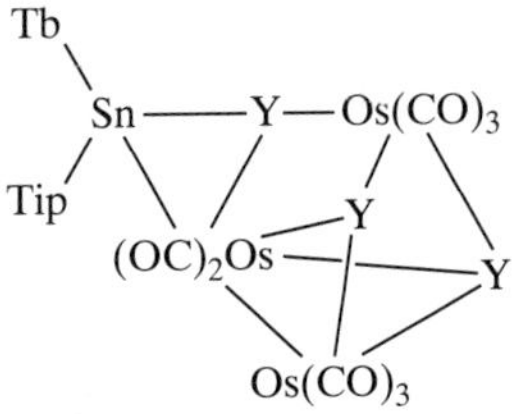

Tb, Tip–Sn–Y–$Os(CO)_3$; $(OC)_2Os$; $Os(CO)_3$; Y; Y

3. *Addition of π-Bonded Reagents*

a. *Dimerization.* Stannenes,[8,100] stannanimines,[100] stannaphosphenes,[8] stannanethiones,[95,97,106] and stannaneselones[97,106] all undergo dimerization by [2 + 2] cycloaddition. In all cases the addition takes place in a head-to-tail manner, as depicted below.

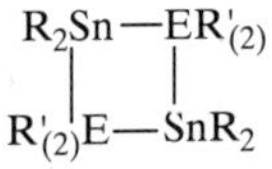

$$\begin{array}{c} R_2Sn{-}ER'_{(2)} \\ | \quad\quad | \\ R'_{(2)}E{-}SnR_2 \end{array}$$

Dimerization of the stannanethiones and stannaneselones with two different substituents resulted in a mixture of the cis and trans isomers.

b. *Addition of Carbonyl Compounds and Their Derivatives.* Stannenes,[8] stannanimines,[103] and stannaphosphenes[8] react with benzaldehyde and benzophenone in a [2 + 2] manner; the regiochemistry is such that,

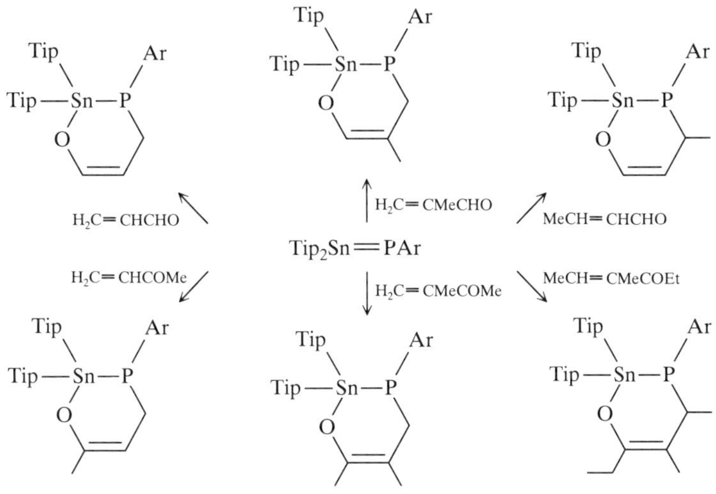

Ar = 2,4,6-tri-*t*-butylphenyl

SCHEME 9

as in all additions of carbonyl compounds to doubly bonded tin species, a Sn—O bond is formed rather than a Sn—C bond. The stannene $Tip_2Sn{=}CR_2$ (CR_2 = fluorenylidene) undergoes a [2 + 2] cycloaddition with benzaldehyde and benzophenone, but the benzaldehyde adduct dissociates to the stannanone and alkene.[8] The stannanimine $[(Me_3Si)_2N]_2Sn{=}NDip$ also reacts in a [2 + 2] fashion with the C=O bond of 2,6-diisopropylphenyl isocyanate rather than adding to the C=N bond or undergoing a [2 + 3] addition.[103]

A stannaphosphene, $Tip_2Sn{=}PAr$, was reacted with a series of α,β-unsaturated aldehydes and ketones and, in all cases, was found to undergo exclusive [2 + 4] cycloaddition, leading to the corresponding six-membered ring stannaoxaphosphorinenes (Scheme 9).[115] An ene reaction is observed between the stannaphosphene and acetone.[8]

c. *Addition of Alkynes.* Tetrakis(2,4,6-triisopropylphenyl)distannene reacts with phenylacetylene to give a 1,2-distannacyclobutene derivative. No reaction occurs with 1,3-dienes or diketones (cyclic or open-chain).[107,108]

Distannenes are in equilibrium with the corresponding stannylenes in solution; therefore, reaction with alkynes can lead either to stannacyclopropene or 1,2-distannacyclobut-3-ene derivatives. Reaction of tetrakis(2,4,6-triisopropylphenyl)- or tetrakis[bis(trimethylsilyl)methyl]distannene with a ring-strained cycloheptyne gave the stannacyclopropene derivatives,[109] whereas reaction of tetrakis[bis(trimethylsilyl)methyl]distannene with cyclooctyne gave the 1,2-distannacyclobut-3-ene derivative.[110] This can be explained in terms of the competing mechanisms shown in Scheme 10.

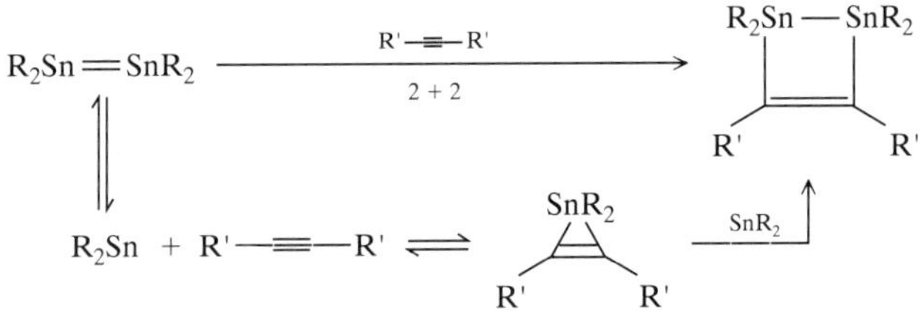

$R_2Sn{=}SnR_2$ —(R'—≡—R', 2 + 2)→ $R_2Sn—SnR_2$ (1,2-distannacyclobutene, R', R')

$R_2Sn{=}SnR_2$ ⇌ R_2Sn + R'—≡—R' ⇌ stannacyclopropene (SnR_2; R', R') —(SnR_2)→ distannacyclobutene

SCHEME 10

The [2 + 2] addition of tetrakis[bis(trimethylsilyl)methyl]distannene to *t*-BuC≡P gives a phosphadistannacyclobutene; no spectroscopic evidence for the presence of a phosphastannacyclopropene was found, suggesting that thc reaction proceeds via a [2 + 2] mechanism rather than stannylene insertions.[111] A stannanethione was found to react with thiocumulenes such as carbon disulfide and phenyl isothiocyanate, giving the [2 + 2] cycloadducts; the regiochemistry is such that the sulfur atom of the thiocumulene is bonded to the tin atom exclusively.[9,95]

d. *Addition of Azides and Nitrile Oxides.* A stannene was found to react with azido(di-*t*-butyl)methylsilane resulting in the [2 + 3] adduct (Scheme 11). The cycloadduct undergoes [2 + 3] cycloreversion at elevated temperature to give a stannanimine with the loss of bis(trimethylsilyl)diazomethane; in the presence of excess *t*-Bu_2MeSiN_3, the [2 + 3] cycloadduct of the stannanimine is formed.[100]

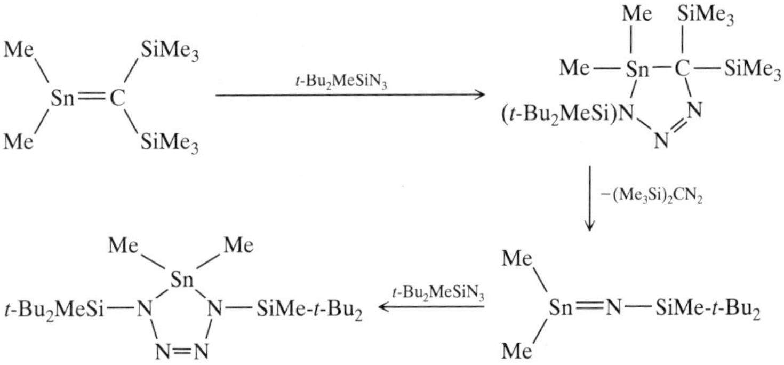

$Me_2Sn{=}C(SiMe_3)_2$ —(*t*-Bu_2MeSiN_3)→ cycloadduct: Me—Sn(Me)—C(SiMe₃)(SiMe₃), (*t*-Bu_2MeSi)N, N=N ring

↓ $-(Me_3Si)_2CN_2$

$Me_2Sn{=}N—SiMe$-*t*-Bu_2 —(*t*-Bu_2MeSiN_3)→ *t*-Bu_2MeSi—N(SnMe_2)N—SiMe-*t*-Bu_2 (ring with N=N)

SCHEME 11

The stannanimine $[(Me_3Si)_2N]_2Sn{=}NDip$(Dip = 2,6-diisopropylphenyl), undergoes a [2 + 3] cycloaddition with benzonitrile oxide depicted below,[103] with 2,6-diethylphenyl azide, a stannatetrazole is formed.[88]

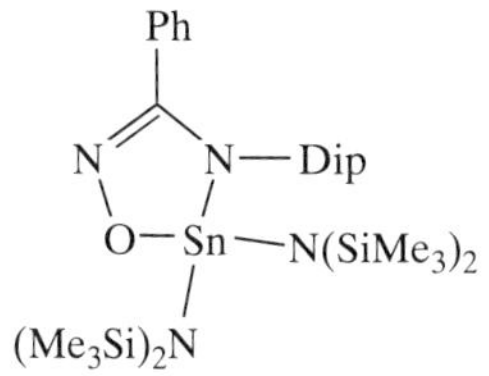

Ph
N N—Dip
O—Sn—$N(SiMe_3)_2$
$(Me_3Si)_2N$

A stannanethione and stannaneselone were found to give [2 + 3] cycloadducts with mesityl nitrile oxide.[9]

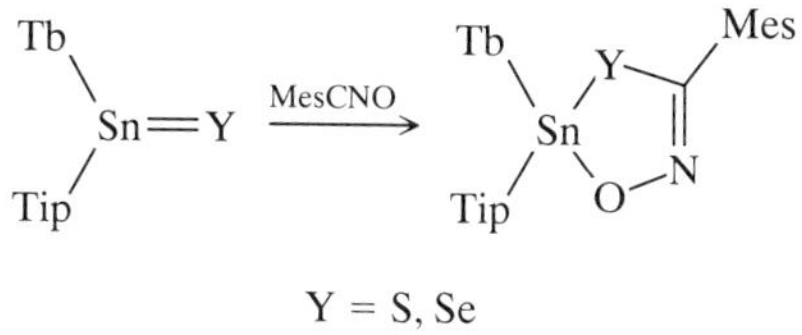

Tb(Tip)Sn=Y —MesCNO→ Tb(Tip)Sn(Y–C(Mes)=N–O) (38)

Y = S, Se

e. *Addition of Dienes* The stannaethene $Me_2Sn{=}C(SiMe_3)_2$ undergoes a [2+4] cycloaddition with 2,3-dimethylbutadiene, in contrast to the related stannanimine $Me_2Sn{=}N(Si\text{-}t\text{-}Bu_2Me)$, which undergoes an ene reaction with the same diene (Scheme 12).[100] Another stannanimine, $[(Me_3Si)_2N]_2Sn{=}NDip$, undergoes a [2+4] cycloaddition reaction with acrolein ($CH_2{=}CHCH{=}O$), but an ene reaction with methyl vinyl ketone.[103]

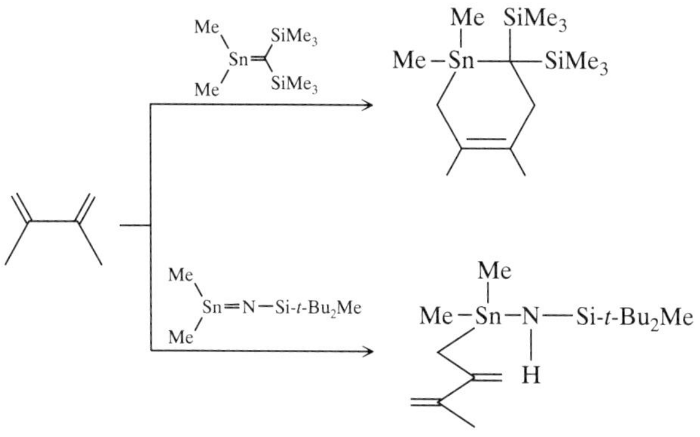

$Me_2Sn{=}C(SiMe_3)_2$ → Me, Me–Sn, Me, $SiMe_3$, $SiMe_3$ (cyclohexene)

$Me_2Sn{=}N{-}Si\text{-}t\text{-}Bu_2Me$ → Me–Sn(Me)–N(H)—$Si\text{-}t\text{-}Bu_2Me$

SCHEME 12

The only example of a [2+4] addition of a stannanethione is that between TbTipSn=S and 2,3-dimethylbutadiene.[106]

4. *Reactions with Reagents Containing Group 16 Elements*

Whereas several reports of doubly bonded tin compounds' being sensitive to decomposition in air are to be found in the literature, there is only one instance of a controlled oxidation. Tetrakis[bis(trimethylsilyl)methyl]-distannene was treated with Me_3NO to give the corresponding cyclo-1,3-distannoxane.[112]

Tetrakis(2,4,6-triisopropylphenyl)distannene reacts with selenium and tellurium in a stepwise manner, giving first the chalcogenadistannirane and subsequently the 1,3-dichalcogenadistannetane. The reaction of the distannene with elemental sulfur proceeds differently, leading to a mixture of the 1,3-dithiadistannetane and the 1,2-dithiadistannetane. The thiadistannirane is easily accessible by treatment of the distannene with methylthiirane; an excess of the thiirane leads exclusively to the 1,3-dithiadistannetane (see Scheme 13).[107,113,114]

A stannaphosphene, $[(Me_3Si)_2CH]_2Sn{=}PAr$, reacts with selenium in a way similar to that described above to give a transient stannaselenaphosphirane, which decomposes after one hour at room temperature to give three main products: diphosphene, a 1,3-diselenastannetane, and the expected four-membered ring. This can be explained in terms of the decomposition of the stannaselenaphosphirane as shown in Scheme 14.[104] Dimethyl disulfide adds across the double bond of the stannaphosphene, leading to $Bis_2(MeS)Sn{-}P(SMe)Ar$, which decomposes at room temperature to give the diphosphene and bis(bis(trimethylsilyl)methyl)di(methylthio)stannane.[104]

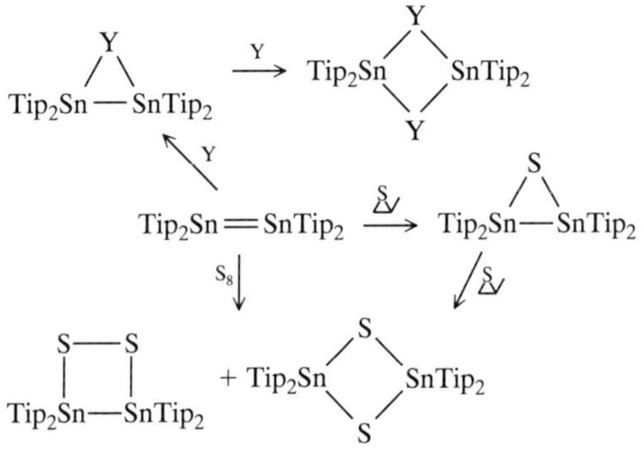

Y = Se, Te

SCHEME 13

$$Bis_2Sn{=}PAr \xrightarrow{Se} \underset{Bis_2Sn-PAr}{\overset{Se}{\triangle}} \longrightarrow [Bis_2Sn{=}Se] + [ArP\!:]$$

$$\underset{Se}{\downarrow} \quad Bis_2Sn(\mu\text{-}Se)_2PAr \qquad \underset{\times 2}{\downarrow} \quad (Bis_2SnSe)_2 \qquad \underset{\times 2}{\downarrow} \quad ArP{=}PAr$$

Ar = 2,4,6-tri-*t*-butylphenyl

SCHEME 14

A stannanethione and stannaneselone have been trapped by their reaction with styrene oxide.[97,106] The resulting oxathiastannolane and oxaselenastannolane were obtained as mixtures of the possible regio- and stereoisomers.

$$Tb(Tip)Sn{=}Y \xrightarrow{\text{styrene oxide}} Tb(Tip)Sn(Y\text{-}CH_2\text{-}CHPh\text{-}O) \qquad (39)$$

Y = S, Se

5. *Reduction Reactions and Rearrangements*

The reduction of a stannaphosphene with lithium aluminum hydride has been reported to give the stannylphosphine.[92] An excess of hydride leads to cleavage of the phosphorus–tin single bond.[104]

$$Bis_2Sn{=}PAr \xrightarrow{LiAlH_4} Bis_2Sn(H){-}P(H)Ar \xrightarrow{LiAlH_4} Bis_2SnH_2 + ArPH \qquad (40)$$

Ar = 2,4,6-tri-*t*-butylphenyl

A stannanimine was found to rearrange in solution by intramolecular addition of the C—H bond of one of the isopropyl groups across the Sn=N bond.[88]

$$[(Me_3Si)_2N]_2Sn{=}NDip \xrightarrow{>-30^\circ C,\ <0^\circ C} [(Me_3Si)_2N]_2Sn(CMe_2\text{-}C_6H_3(iPr)\text{-}NH) \qquad (41)$$

IV

SUMMARY

After just over a decade of research on the synthesis and chemistry of doubly bonded germanium and tin compounds, certain trends in synthetic methods and reactivity are emerging. There are currently three general routes for the synthesis of compounds of this type: (1) thermal or photochemical cleavage of a cyclotrimetallane, (2) dehydrohalogenation, and (3) coupling of a stable germylene or stannylene with a carbene (or carbene equivalent), a diazo compound, or an azide. In the addition of σ-bonded reagents, the reactions are remarkably regioselective in almost all cases. The addition of some π-bonded reagents provides some interesting insights into the nature of the doubly bonded compounds. For example, in the addition of phenylacetylene, a formal [2+2] cycloaddition is observed in some cases (i.e., digermenes and germasilenes); in others, addition across the acetylenic C—H bond is observed (i.e., germanimines and germaphosphenes), suggesting a greater zwitterionic character of the germanimines (not unexpected) and germaphosphenes (somewhat surprising) compared to digermenes and germasilenes. At the present time, not enough data are available to compare the reactivity of the different classes of unsaturated compounds with the same reagent although, clearly, such a comparison would offer greater understanding of the nature of this class of compounds as a whole. In addition, almost no mechanistic studies have been carried out on the addition reactions of unsaturated germanium and tin compounds. There is much to be done in this area of chemistry.

Acknowledgments

Useful comments from Professor C.J. Willis of the Department of Chemistry, The University of Western Ontario are gratefully acknowledged. Assistance in the form of a grant to K.M.B. from the NSERC (Canada) and a NATO Postdoctoral fellowship to W.G.S. is also acknowledged.

References

(1) Satgé, J. *Adv. Organomet. Chem.* **1982,** *21*, 241.
(2) Satgé, J. *J. Organomet. Chem.* **1990,** *400*, 121.
(3) Barrau, J.; Escudié, J.; Satgé, J. *Chem. Rev.* **1990,** *90*, 283.
(4) Escudié, J.; Couret, C.; Ranaivonjatovo, H.; Satgé, J. *Coord. Chem. Rev.* **1994,** *130*, 427.
(5) (a) Wiberg, N. *J. Organomet. Chem.* **1984,** *273*, 141. (b) Tsumuraya, T.; Batcheller, S. A.; Masamune, S. *Angew. Chem., Int. Ed. Engl.* **1991,** *30*, 902.

(6) Cowley, A. H.; Norman, N. C. *Prog. Inorg. Chem.* **1986,** *34*, 1. Norman, N. C. *Polyhedron* **1993,** *12*, 2431.
(7) Grev, R. S. *Adv. Organomet. Chem.* **1991,** *33*, 125.
(8) Escudié, J.; Couret, C.; Ranaivonjatovo, H.; Anselme, G.; Delpon-Lacaze, G.; Rodi, A. K.; Satgé, J. *Main Group Met. Chem.* **1994,** *17*, 33.
(9) Tokitoh, N.; Matsuhashi, Y.; Shibata, K.; Matsumoto, T.; Suzuki, H.; Saito, M.; Manmaru, K.; Okazaki, R. *Main Group Met. Chem.* **1994,** *17*, 55.
(10) Ando, W.; Kabe, Y.; Nami, C. *Main Group Met. Chem.* **1994,** *17*, 209.
(11) Lazraq, M.; Escudié, J.; Couret, C.; Satgé, J.; Dräger, M.; Dammel, R. *Angew. Chem., Int. Ed. Engl.* **1988,** *27*, 828.
(12) Meyer, H.; Baum, G.; Massa, W.; Berndt, A. *Angew. Chem., Int. Ed. Engl.* **1987,** *26*, 798.
(13) Snow, J. T.; Murakami, S.; Masamune, S.; Williams, D. J. *Tetrahedron Lett.* **1984,** *25*, 4191.
(14) Goldberg, D. E.; Hitchcock, P. B.; Lappert, M. F.; Thomas, K. M.; Thorne, A. J.; Fjeldberg, T.; Haaland, A.; Schilling, B. E. R. *J. Chem. Soc., Dalton Trans.* **1986,** 2387.
(15) Batcheller, S. A.; Tsumuraya, T.; Tempkin, O.; Davis, W. M.; Masamune, S. *J. Am. Chem. Soc.* **1990,** *112*, 9394.
(16) Ando, W.; Ohtaki, T.; Kabe, Y. *Organometallics* **1994,** *13*, 434.
(17) Ohtaki, T.; Kabe, Y.; Ando, W. *Heteroatom. Chem.* **1994,** *5*, 313.
(18) Veith, M.; Becker, S.; Huch, V. *Angew. Chem., Int. Ed. Engl.* **1990,** *29*, 216.
(19) Dräger, M.; Escudié, J.; Couret, C.; Ranaivonjatovo, H.; Satgé, J. *Organometallics* **1988,** *7*, 1010.
(20) Ranaivonjatovo, H.; Escudié, J.; Couret, C.; Satgé, J.; Dräger, M. *New J. Chem.* **1989,** *13*, 389.
(21) Tokitoh, N.; Matsumoto, T.; Manmaru, K.; Okazaki, R. *J. Am. Chem. Soc.* **1993,** *115*, 8855.
(22) Matsumoto, T.; Tokitoh, N.; Okazaki, R. *Angew. Chem., Int. Ed. Engl.* **1994,** *33*, 2316.
(23) Veith, M.; Becker, S.; Huch, V. *Angew. Chem., Int. Ed. Engl.* **1989,** *28*, 1237.
(24) Kuchta, M. C.; Parkin, G. *J. Chem. Soc., Chem. Commun.* **1994,** 1351.
(25) Meller, A.; Ossig, G.; Maringgele, W.; Stalk, D.; Herbst-Irmer, R.; Freitag, S.; Sheldrick, G. M. *J. Chem. Soc., Chem. Commun.* **1991,** 1123.
(26) Couret, C.; Escudié, J.; Deplon-Lacaze, G.; Satgé, J. *Organometallics* **1992,** *11*, 3176.
(27) Levy, G. C.; Nelson, G. L. *Carbon-13 Nuclear Magnetic Resonance for Organic Chemists;* Wiley (Interscience): New York, 1972; p. 59.
(28) Anselme, G.; Escudié, J.; Couret, C.; Satgé, J. *J. Organomet. Chem.* **1991,** *403*, 93.
(29) West, R.; Fink, M.; Michl, J. *Science* **1981,** *214*, 1343.
(30) Baines, K. M.; Cooke, J. A. *Organometallics* **1992,** *11*, 3487.
(31) Smit, C. N.; Bickelhaupt, F. *Organometallics* **1987,** *6*, 1156.
(32) Escudié, J.; Couret, C.; Satgé, J.; Adrianarison, M.; Andriamizaka, J.-D. *J. Am. Chem. Soc.* **1985,** *107*, 3378.
(33) Charissé, M.; Mathes, M.; Simon, D.; Dräger, M. *J. Organomet. Chem.* **1993,** *445*, 39.
(34) Wiberg, N.; Schurz, K.; Reber, G.; Müller, G. *J. Chem. Soc., Chem. Commun.* **1986,** 591.
(35) Ranaivonjatovo, H.; Escudié, J.; Couret, C.; Satgé, J. *J. Organomet. Chem.* **1991,** *415*, 327.
(36) Ang, H. G.; Lee, F. K. *J. Chem. Soc., Chem. Commun.* **1989,** 310.
(37) Masamune, S.; Hanzawa, Y.; Williams, D. J. *J. Am. Chem. Soc.* **1982,** *104*, 6136.
(38) Ando, W.; Tsumuraya, T. *J. Chem. Soc., Chem. Commun.* **1989,** 770.
(39) Park, J.; Batcheller, S. A.; Masamune, S. *J. Organomet. Chem.* **1989,** *367*, 39.

(40) Rivière-Baudet, M.; Morère, A. *J. Organomet. Chem.* **1992,** *431*, 17.
(41) Rivère-Baudet, M.; Khallaayoun, A.; Satgé, J. *Organometallics* **1993,** *12*, 1003.
(42) Glidewell, C.; Lloyd, D.; Lumbard, K. W.; McKechnie, J. S.; Hursthouse, M. B.; Short, R. L. *J. Chem. Soc., Dalton Trans.* **1987,** 2981.
(43) Tsumuraya, T.; Sato, S.; Ando, W. *Organometallics* **1988,** *7*, 2015.
(44) Baines, K. M.; Cooke, J. A. *Organometallics* **1991,** *10*, 3419.
(45) Ando, W.; Itoh, H.; Tsumuraya, T. *Organometallics* **1989,** *8*, 2759.
(46) Ando, W.; Itoh, H.; Tsumuraya, T.; Yoshida, H. *Organometallics* **1988,** *7*, 1880.
(47) Ando, W.; Tsumuraya, T.; Sekiguchi, A. *Chem. Lett.* **1987,** 317.
(48) Collins, S.; Murakami, S.; Snow, J. T.; Masamune, S. *Tetrahedron Lett.* **1985,** *26*, 1281.
(49) Chaubon-Deredempt, M. A.; Escudié, J.; Couret, C. *J. Organomet. Chem.* **1994,** *467*, 37.
(50) Baines, K. M., Groh, R. J.; Joseph, B.; Parshotam, U. R. *Organometallics* **1992,** *11*, 2176.
(51) Couret, C.; Escudié, J.; Satgé, J.; Lazraq, M. *J. Am. Chem. Soc.* **1987,** *109*, 4411.
(52) Lazraq, M.; Couret, C.; Escudié, J.; Satgé, J.; Soufiaoui, M. *Polyhedron* **1991,** *10*, 1153.
(53) Rivière-Baudet, M.; Khallaayoun, A.; Satgé, J. *J. Organomet. Chem.* **1993,** *462*, 89.
(54) Glidewell, C.; Lloyd, D.; Lumbard, K. W.; McKechnie, J. S. *Tetrahedron Lett.* **1987,** *28*, 345.
(55) Pfeiffer, J.; Maringgele, W.; Noltemeyer, M.; Meller, A. *Chem. Ber.* **1989,** *122*, 245.
(56) Veith, M.; Detemple, A.; Huch, V. *Chem. Ber.* **1991,** *124*, 1135.
(57) Baines, K. M.; Cooke, J. A.; Dixon, C. E.; Liu, H. W.; Netherton, M. R. *Organometallics* **1994,** *13*, 631.
(58) Cooke, J. A. Ph.D. Thesis, University of Western Ontario, 1994.
(59) Escudié, J.; Couret, C.; Andrianarison, M.; Satgé, J. *J. Am. Chem. Soc.* **1987,** *109*, 386.
(60) Egorov, M. P.; Kolesnikov, S. P.; Nefedov, O. M.; Krebs, A. *J. Organomet. Chem.* **1989,** *375*, C5.
(61) Lazraq, M.; Escudié, J.; Couret, C.; Satgé, J.; Soufiaoui, M. *J. Organomet. Chem.* **1990,** *397*, 1.
(62) Lazraq, M.; Couret, C.; Declercq, J. P.; Dubourg, A.; Escudié, J.; Rivière-Baudet, M. *Organometallics* **1990,** *9*, 845.
(63) Batcheller, S. A.; Masamune, S. *Tetrahedron Lett.* **1988,** *29*, 3383.
(64) Ando, W.; Tsumuraya, T. *Organometallics* **1988,** *7*, 1882.
(65) Tsumuraya, T.; Sato, S.; Ando, W. *Organometallics* **1990,** *9*, 2061.
(66) Lazraq, M.; Couret, C.; Escudié, J.; Satgé, J.; Dräger, M. *Organometallics* **1991,** *10*, 1771.
(67) Lazraq, M.; Escudié, J.; Couret, C.; Satgé, J.; Soufiaoui, M. *Organometallics* **1992,** *11*, 555.
(68) Deplon-Lacaze, G.; Couret, C.; Escudié, J.; Satgé, J. *Main Group Met. Chem.* **1993,** *16*, 419.
(69) Baines, K. M.; Cooke, J. A.; Vittal, J. J. *Heteroatom. Chem.* **1994,** *5*, 293.
(70) Tsumuraya, T.; Kabe, Y.; Ando, W. *J. Organomet. Chem.* **1994,** *482*, 131.
(71) Tsumuraya, T.; Sato, S.; Ando, W. *Organometallics* **1989,** *8*, 161.
(72) Veith, M.; Detemple, A. *Phosphorus, Sulfur, Silicon* **1992,** *65*, 17.
(73) Ranaivonjatovo, H.; Escudié, J.; Couret, C.; Declercq, J.-P.; Dubourg, A.; Satgé, J. *Organometallics* **1993,** *12*, 1674.
(74) Lazraq, M.; Escudié, J.; Couret, C.; Satgé, J.; Soufiaoui, M. *Organometallics* **1991,** *10*, 1140.
(75) Liu, H. W. M.Sc. Thesis, University of Western Ontario, 1994.
(76) Tsumuraya, T.; Ando, W. *Organometallics* **1990,** *9*, 869.

(77) Lazraq, M.; Escudié, J.; Couret, C.; Bergsträsser, U.; Regitz, M. *J. Chem. Soc., Chem. Commun.* **1993,** 569.
(78) Masamune, S.; Batcheller, S. A.; Park, J.; Davis, W. M. *J. Am. Chem. Soc.* **1989,** *111*, 1888.
(79) Tsumuraya, T.; Kabe, Y.; Ando, W. *J. Chem. Soc., Chem. Commun.* **1990,** 1159.
(80) Andrianarison, M.; Couret, C.; Declercq, J.-P.; Dubourg, A.; Escudié, J.; Ranaivonjatovo, H.; Satgé, J. *Organometallics* **1988,** *7*, 1545.
(81) Baines, K. M.; Cooke, J. A.; Vittal, J. J. *J. Chem. Soc., Chem. Commun.* **1992,** 1484.
(82) Andrainarison, M.; Couret, C.; Declerq, J.-P.; Dubourg, A.; Escudié, J.; Satgé, J. *J. Chem. Soc., Chem. Commun.* **1987,** 921.
(83) Davidson, P. J.; Harris, D. H.; Lappert, M. F. *J. Chem. Soc., Dalton Trans.* **1976,** 2268.
(84) Haupt, H. J.; Huber, F.; Preut, H. *Z. Anorg. Allg. Chem.* **1973,** *396*, 81.
(85) Meyer, H.; Baum, G.; Massa, W.; Berger, S.; Berndt, A. *Angew. Chem., Int. Ed. Engl.* **1987,** *26*, 546. *Pure Appl. Chem.* **1987,** *59*, 1011.
(86) Dobbs, K. D.; Hehre, W. J. *Organometallics* **1986,** *5*, 2057.
(87) Grützmacher, H.; Freitag, S.; Herbst-Irmer, R.; Sheldrick, G. S. *Angew. Chem., Int. Ed. Engl.* **1992,** *31*, 437.
(88) Ossig, G.; Meller, A.; Freitag, S.; Herbst-Irmer, R. *J. Chem. Soc., Chem. Commun.* **1993,** 497.
(89) Anselme, G.; Ranaivonjatovo, H.; Escudié, J.; Couret, C.; Satgé, J. *Organometallics* **1992,** *11*, 2748.
(90) Masumune, S.; Sita, L. R. *J. Am. Chem. Soc.* **1985,** *107*, 6390.
(91) Zilm, K. W.; Lawless, G. A.; Merrill, R. M.; Millar, J. M.; Webb, G. G. *J. Am. Chem. Soc.* **1987,** *109*, 7236.
(92) Couret, C.; Escudié, J.; Satgé, J.; Rahaarinirina, A.; Andriamizaka, J. D. *J. Am. Chem. Soc.* **1985,** *107*, 8280.
(93) Ranaivonjatovo, H.; Escudié, J.; Couret, C.; Satgé, J. *J. Chem. Soc., Chem. Commun.* **1992,** 1047.
(94) Ranaivonjatovo, H.; Escudié, J.; Couret, C.; Satgé, J. *Phosphorus, Sulfur, Silicon* **1993,** *76*, 61.
(95) Tokitoh, N.; Saito, M.; Okazaki, R. *J. Am. Chem. Soc.* **1993,** *115*, 2065.
(96) Weidenbruch, M.; Schäfer, A.; Kilian, H.; Pohl, S.; Saak, W.; Marsmann, H. *Chem. Ber.* **1992,** *125*, 563.
(97) Tokitoh, N.; Matsuhashi, Y.; Goto, M.; Okazaki, R. *Chem. Lett.* **1992,** 1595.
(98) Matsuhashi, Y.; Tokitoh, N.; Okazaki, R. *Organometallics* **1993,** *12*, 2573.
(99) Anselme, G.; Couret, C.; Escudié, J.; Richelme, S.; Satgé, J. *J. Organomet. Chem.* **1991,** *418*, 321.
(100) Wiberg, N.; Vasisht, S.-K. *Angew. Chem., Int. Ed. Engl.* **1991,** *30*, 93.
(101) Wehrman, R.; Klusik, H.; Berndt, A. *Angew. Chem., Int. Ed. Engl.* **1984,** *23*, 826.
(102) Budzelaar, P. H. M.; Schleyer, P. von R.; Krogh-Jespersen, K. *Angew. Chem., Int. Ed. Engl.* **1984,** *23*, 825.
(103) Ossig, G.; Meller, A.; Freitag, S.; Herbst-Irmer, R.; Sheldrick, G. M. *Chem. Ber.* **1993,** *126*, 2247.
(104) Escudié, J.; Couret, C.; Raharinirina, A.; Satgé, J. *New J. Chem.* **1987,** *11*, 627.
(105) Tokitoh, N.; Matsuhashi, Y.; Okazaki, R. *Organometallics* **1993,** *12*, 2894.
(106) Matsuhashi, Y.; Tokitoh, N.; Okazaki, R. *Organometallics* **1993,** *12*, 2573.
(107) Schäfer, A.; Weidenbruch, M. *Phosphorus, Sulfur, Silicon* **1992,** *65*, 13.
(108) Weidenbruch, M.; Schäfer, A.; Kilian, H.; Pohl, S.; Saak, W.; Marsmann, H. *Chem. Ber.* **1992,** *125*, 563.

(109) Sita, L. R.; Bickerstaff, R. D. *J. Am. Chem. Soc.* **1988,** *110*, 5208. *Phosphorus, Sulfur Silicon Relat. Elem.* **1989,** *41*, 31.
(110) Sita, L. R.; Kinoshita, I.; Lee, S. P. *Organometallics* **1990,** *9*, 1644.
(111) Cowley, A. H.; Hall, S. W.; Nunn, C. M.; Power, J. M. *Angew. Chem., Int. Ed. Engl.* **1988,** *27*, 838.
(112) Edelman, M. A.; Hitchcock, P. B. *J. Chem. Soc., Chem. Commun.* **1990,** 1116.
(113) Schäfer, A.; Weidenbruch, M.; Saak, W.; Pohl, S.; Marsmann, H. *Angew. Chem., Int. Ed. Engl.* **1991,** *30*, 962.
(114) Schäfer, A.; Weidenbruch, M.; Saak, W.; Pohl, S.; Marsmann, H. *Angew. Chem., Int. Ed. Engl.* **1991,** *30*, 834.
(115) Kandri-Rodi, A.; Ranaivonjatovo, H.; Escudié, J. *Organometallics* **1994,** *13*, 2787.

Diheteroferrocenes and Related Derivatives of the Group 15 Elements: Arsenic, Antimony, and Bismuth

ARTHUR J. ASHE III and SALEEM AL-AHMAD

Department of Chemistry
The University of Michigan
Ann Arbor, Michigan

I
INTRODUCTION

The heterobenzenes of the group 15 elements (**1–5**) comprise a series in which elements of an entire column of the periodic table have been incorporated into aromatic rings. The comparative study of this series has been extremely valuable for evaluating p–p π-bonding between carbon and the heavier elements.[1] However, the heterobenzene series has two important limitations. Only arsabenzene has a well developed aromatic chemistry. Moreover, stibabenzene and particularly bismabenzene are so labile that it has been difficult to obtain derivatives stable enough for study.

The diheteroferrocenes (**6–10**) form a series analogous to that of the corresponding heterobenzenes. The diheteroferrocene series partially circumvents the limitations of the heterobenzene series and provides additional context for p–p π-bonding between carbon and the group 15 elements. Since stable derivatives have now been prepared for the entire series, it seems appropriate to review the chemistry of the heavier diheteroferrocenes and closely related η^5-heterolyl transition metal complexes (**13–15**, **18–20**). The extensive work on phosphaferrocenes (**7**, **18**) has been the subject of a prior review by Mathey and is mentioned only selectively here.[2] Similarly, azaferrocenes (**6**, **11**) are largely omitted from this review.[3] We have attempted to cover the published literature through

1994, although some unpublished material from the authors' laboratory is also included.

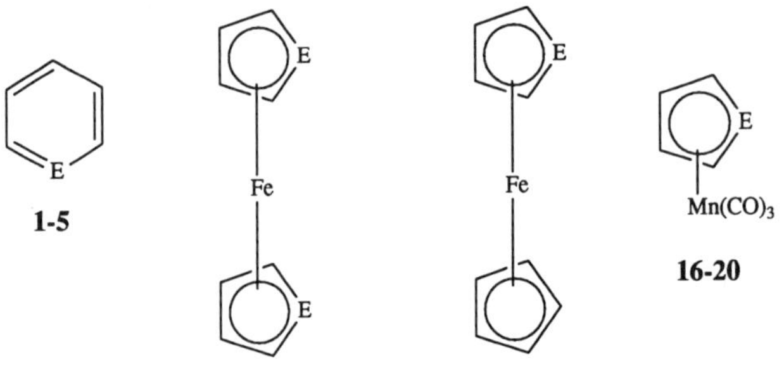

1-5 **6-10** **11-15** **16-20**

1,6,11,16, E=N
2,7,12,17, E=P
3,8,13,18, E=As
4,9,14,19, E=Sb
5,10,15,20, E=Bi

II

SYNTHESIS

Invariably, 1,1′-diheteroferrocenes (**23**) have been prepared from the corresponding 1-phenylheteroles (**21**). The reaction of **21** with lithium in THF (Scheme 1) affords a solution of phenyllithium and the lithium heterocyclopentadienide **22**. After the phenyllithium is removed by reac-

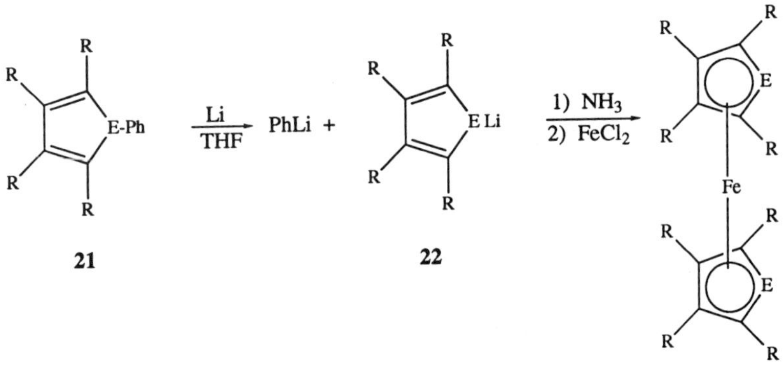

Scheme 1

tion with excess NH_3 or other reagent, addition of $FeCl_2$ affords the desired 1,1′-diheteroferrocene (**23**) in 30–70% yield.

Thus the synthesis of diheteroferrocenes depends on the availability of the corresponding heteroles, which have been prepared by just a few routes. 1-Phenyl-2,5-disubstituted arsoles are prepared generally from the corresponding 1,3-diynes by the base-catalyzed addition of phenylarsine. In this manner, Märkl and Hauptmann have prepared 1-phenyl-2,5-dimethylarsole (**25**) in 30% yield.[4] Subsequently, Mathey *et al.* used this arsole to prepare the first diarsaferrocene **26**.[5] (see Scheme 2.)

The Märkl route has not been applied to the synthesis of the heavier heteroles owing to the great difficulties of handling very labile phenylstibine[6] and phenylbismuthine.[7] However, Sb- and Bi-heteroles are available by an alternative route using organotin chemistry. Thermal hydrostannation of 2,4-hexadiyne (**24**) with dibutyltin dihydride affords the stannole **27** in 15–20% yield.[8] Although the yield is poor, the ready availability of the diyne allows multigram quantities of **27** to be prepared easily. As shown in Scheme 3, direct exchange of **27** with phenylarsenic dichloride or phenylantimony dichloride gives the corresponding heteroles **25**[9] and **28**,[8] respectively, in 65% yield. Stibole **28** was converted to tetramethyl-1,1′-distibaferrocene (**29**) in 50% yield.[8,10]

The attempted exchange reaction between **27** and phenylbismuth diiodide gave only intractable products. Apparently the bismole **31** is unstable under these Lewis acidic conditions. However, iodination of **27** gave **30** which, on lithiation followed by reaction with phenylbismuth diiodide, afforded the desired 1-phenyl-2,5-dimethylbismole (**31**),[11] which could be

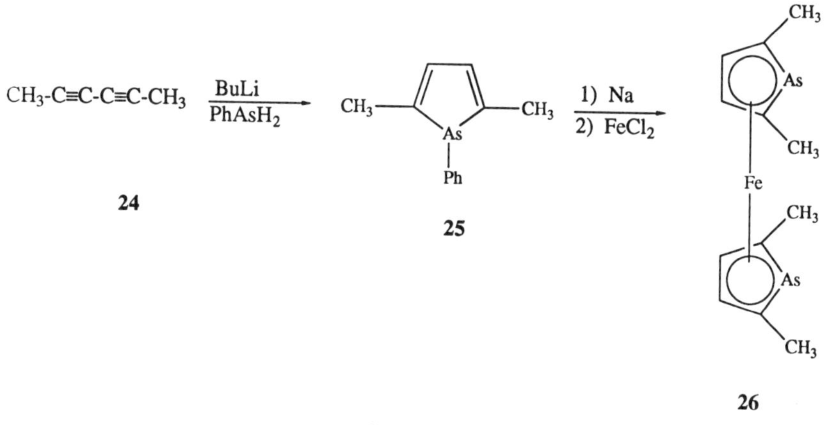

CH_3-C≡C-C≡C-CH_3 (**24**) —BuLi, $PhAsH_2$→ **25** (As–Ph, 2,5-CH_3) —1) Na, 2) $FeCl_2$→ **26**

SCHEME 2

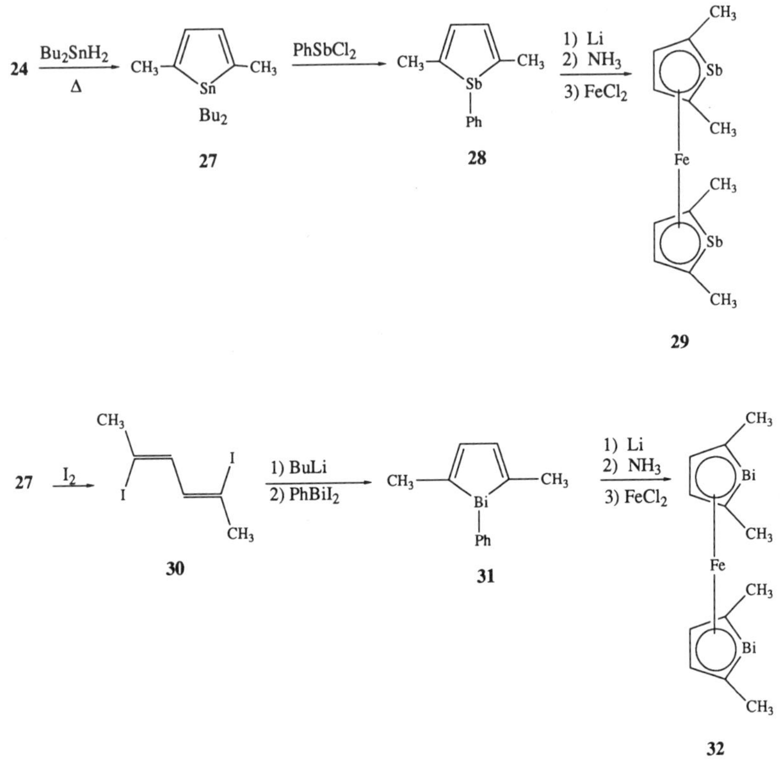

SCHEME 3

converted to the 2,2′,5,5′-tetramethyl-1,1′-dibismaferrocene (**32**) in 37% yield.[12]

There is a good synthesis of the *C*-unsubstituted 1-phenylarsole (**34**). The reaction of dilithiophenylarsine with the readily available mixture of isomeric 1,4-dichloro-1,3-butadienes (**33**) gives **34** in 25% yield.[13] The arsole **34** is easily converted to 1,1′-diarsaferrocene (**8**), while 1,1′-diphosphaferrocene (**7**) can be prepared analogously, as shown in Scheme 4. The lack of availability and low stability of phenylstibine and phenylbismuthine limit extension to preparation of the heavier diheteroferrocenes.

However, the reaction of (1*Z*,3*Z*)-1,4-dilithio-1,3-butadiene (**36**) with phenylantimony dichloride (Scheme 5) affords 1-phenylstibole (**37**),[14] which has been converted to 1,1′-distibaferrocene (**9**).[15] Unfortunately, there is no totally satisfactory synthetic route to dilithio compound **36**. Sn/Li exchange of (1*Z*,3*Z*)-1,4-bis(trimethylstannyl)-1,3-butadiene (**38**) with methyllithium gives a good yield of **36**, but the preparation of **38** is

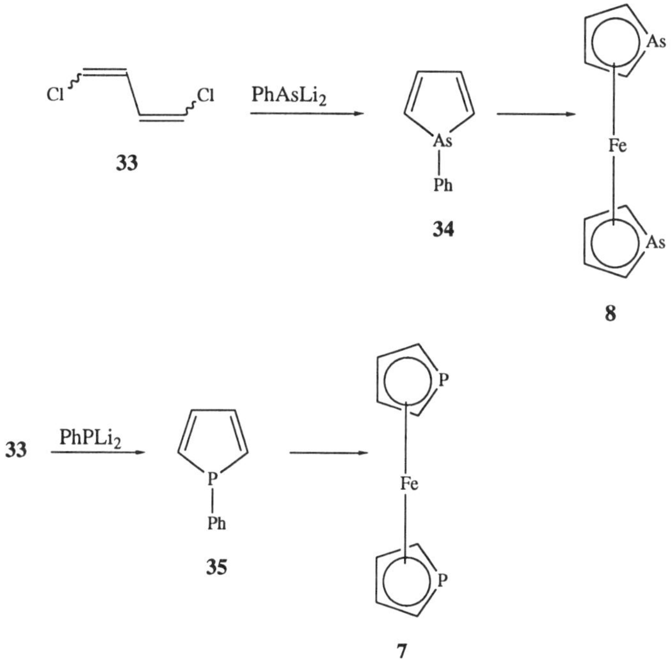

Cl Cl PhAsLi$_2$ As Ph As Fe As

33 34 8

33 PhPLi$_2$ P Ph P Fe P

35 7

SCHEME 4

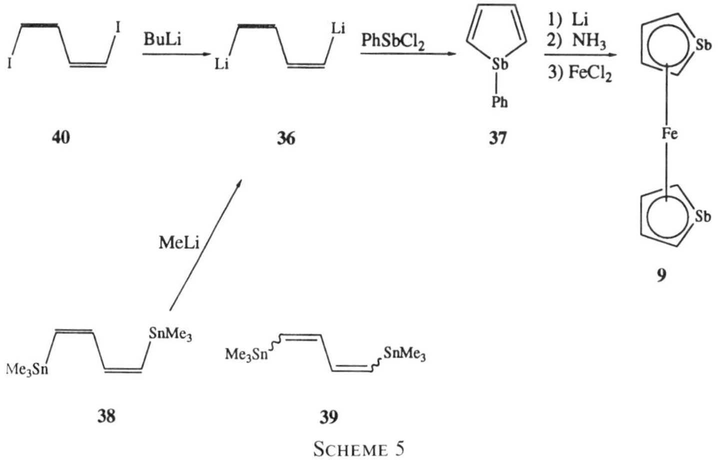

I I BuLi Li Li PhSbCl$_2$ Sb Ph 1) Li 2) NH_3 3) $FeCl_2$ Sb Fe Sb

40 36 37 9

MeLi

Me_3Sn $SnMe_3$ Me_3Sn $SnMe_3$

38 39

SCHEME 5

difficult.[16] Mixed isomers of 1,4-bis(trimethylstannyl)-1,3-butadiene (**39**) undergo concomitant isomerization and Sn/Li exchange with conversion to **36** in moderate yields.[16] However, this reaction has not yet been optimized.

Tetrasubstituted heteroles are generally obtained from the corresponding organozirconium compounds. Thus 2-butyne (**40**) reacts with zirconocene dichloride and magnesium amalgam to give zirconole **41** (Scheme 6).[17] Direct exchange of **41** with $AsCl_3$ affords 1-chloro-2,3,4,5-tetramethylarsole (**42**), which gives the 1-phenylarsole (**43**) on treatment with phenyllithium.[18] 1-Phenylarsole (**43**) can be converted to octamethyldiarsaferrocene (**44**) in the usual manner.[19] A slightly modified approach has been used to prepare phenylstiboles and bismoles. Iodination of **41** gives **45**. Lithiation of **45** followed by reaction with phenylantimony dichloride or phenylbismuth diiodide affords 2,3,4,5-tetramethyl-1-phenylstibole (**46**) or 2,3,4,5-tetramethyl-1-phenylbismole (**47**), respectively.[19,20] Conversion of **46** and **47** to the corresponding octamethyl-1,1′-diheteroferrocenes **48** and **49** is routine.

CH_3-C≡C-CH_3 (**40**) → ($ZrCp_2Cl_2$, Mg(Hg)) → **41** → ($AsCl_3$) → **42** → (PhLi) → **43** → **44**

41 → (I_2) → **45** → ($PhEX_2$) → **46** E = Sb, **47** E = Bi → **48** E = Sb, **49** E = Bi

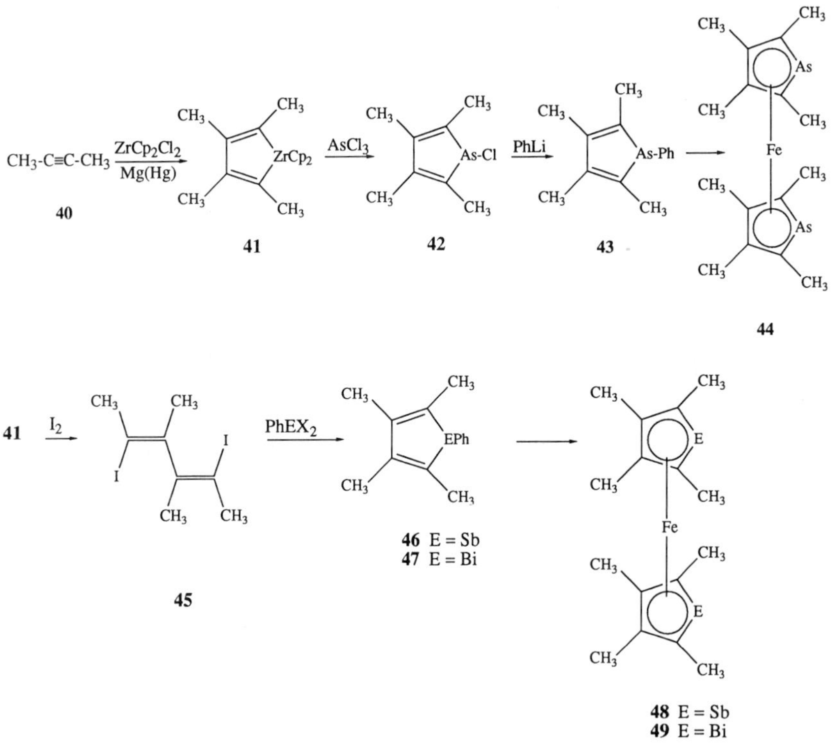

SCHEME 6

Addition of zirconocene to unsymmetric acetylenes is often regiospecific. Thus the reaction of 1-trimethylsilyl-1-propyne (**50**) with zirconocene dichloride and magnesium amalgam gives the zirconocycle **51**, where the large trimethylsilyl groups are adjacent to zirconium.[21] Iodination of **51** affords diiodide **52**, which has been converted to the corresponding distibaferrocene **55** and dibismaferrocene **56**.

This sequence, which is shown in Scheme 7, may also be adapted to prepare 3,4-disubstituted heteroles and hence to preparation of the

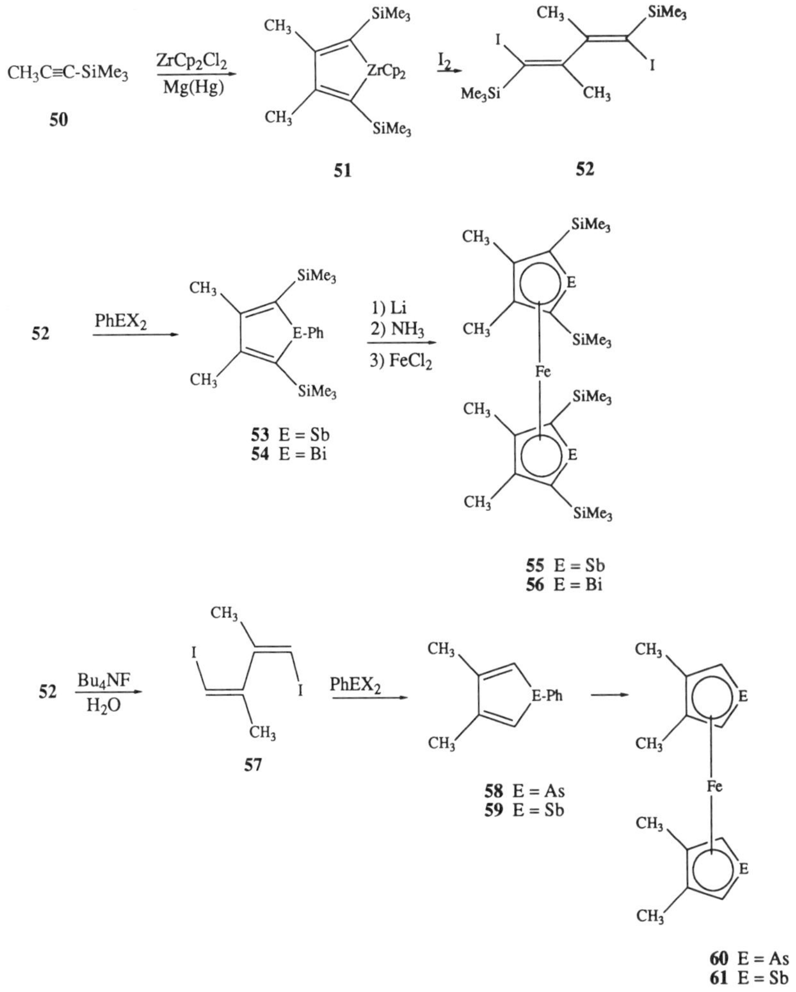

SCHEME 7

corresponding tetrasubstituted diheteroferrocenes. Desilylation of **52** with Bu_4NF and water gives **57** in high yield. Lithiation of **57** followed by reaction with appropriate phenylpnictogen dihalide affords the corresponding 3,4-dimethyl-1-phenylheterole (**58** and **59**). Conversion of **58** and **59** to 3,3′,4,4′-tetramethyl-1,1′-diarsaferrocene (**60**) and 3,3′,4,4′-tetramethyl-1,1′,distibaferrocene (**61**), respectively, can be accomplished in the usual manner.[22]

Although a number of monophosphaferrocenes have been prepared, only a few monoarsa-, monostiba-, and monobismaferrocenes have been reported. Tetraphenylarsolyl derivatives were the first to be prepared. As shown in Scheme 8, they are obtained ultimately from the addition of lithium to tolane (**62**), which gives 1,4-dilithio-1,2,3,4-tetraphenylbutadiene (**63**).[23] The reaction of **63** with $AsCl_3$ gives 1-chloro-2,3,4,5-tetraphenylarsole (**64**). Reaction of **64** with $Fe(CO)_2CpNa$ affords σ-complex **65**, which gives 2,3,4,5-tetraphenylarsaferrocene (**66**) on pyrolysis.[24] Similarly, the reaction of 2,5-dimethyl-1-phenylarsole (**25**) with potassium followed by $Fe(CO)_2CpI$ gave the analogous σ-complex **67**, which afforded 2,5-dimethylarsaferrocene (**68**) on pyrolysis.[5]

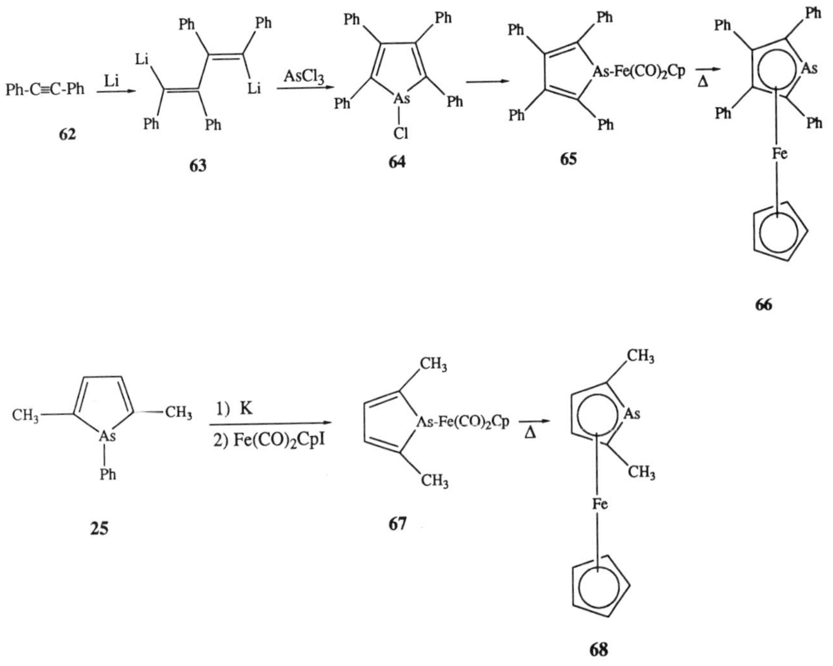

SCHEME 8

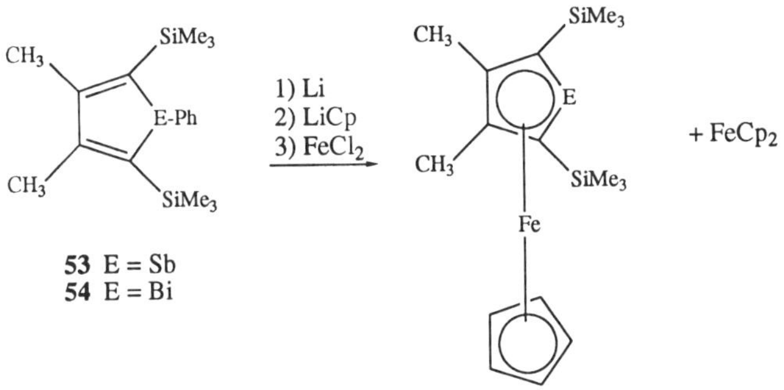

1) Li
2) LiCp
3) $FeCl_2$

$+ FeCp_2$

53 E = Sb
54 E = Bi

69 E = Sb
70 E = Bi

SCHEME 9

The reaction of the crowded stibole **53** and bismole **54** with lithium followed by addition of lithium cyclopentadienide (Scheme 9) gives mixtures of the corresponding monostibaferrocene **69** and monobismaferrocene **70**, respectively, with ferrocene.[25] The more volatile ferrocene is easily removed by sublimation. Rather small amounts of the crowded diheteroferrocenes **55** or **56** are formed in this reaction.

Finally, (η^5-heterolyl)$Mn(CO)_3$ complexes, the heterocymantrenes, have been prepared for the complete family of group 15 elements from N to Bi. As shown in Scheme 10, 2,5-dimethylarsacymantrene (**71**) is obtained directly from 1-phenyl-2,5-dimethylarsole (**25**) in 50% yield by heating with $Mn_2(CO)_{10}$.[26] The tetraphenylarsacymantrene (**74**) is obtained by pyrolysis of the corresponding σ-complex **72**. Rhenium complex **75** is obtained by an analogous reaction. The 2,5-dimethylstibacymantrene (**76**)[8] and 2,5-dimethylbismacymantrene (**78**)[27] were obtained by similar routes.

III

STRUCTURES

X-Ray structures have been determined for at least one derivative for each of the group 15 1,1′-diheteroferrocenes. These compounds include dibismaferrocene **32**[12]; distibaferrocenes **29**,[10] **48**,[19] and **55**[21a]; diarsaferrocenes **26**[25] and **44**[19]; and diphosphaferrocene **79**.[28] In addition, structures[29] have been determined for the closely related bismaferrocene **70**,[30] phosphaferrocenes **80** and **81**,[31] arsacymantrene **74**,[32] and phosphacymantrene

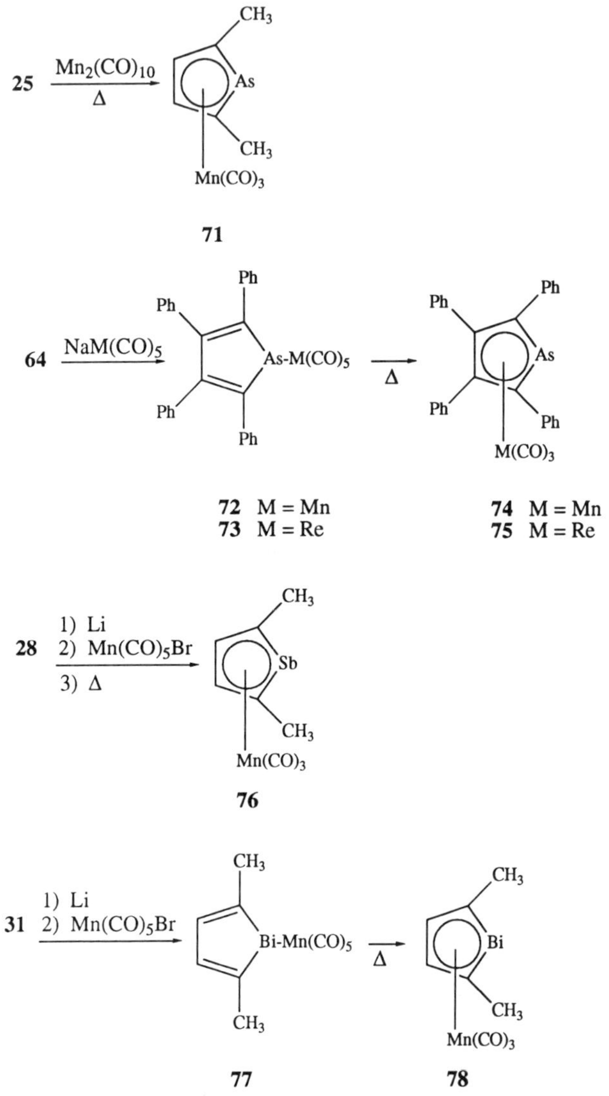

25 $\xrightarrow[\Delta]{Mn_2(CO)_{10}}$ 71

64 $\xrightarrow{NaM(CO)_5}$ 72 M = Mn, 73 M = Re $\xrightarrow{\Delta}$ 74 M = Mn, 75 M = Re

28 $\xrightarrow[\text{3) } \Delta]{\text{1) Li; 2) } Mn(CO)_5Br}$ 76

31 $\xrightarrow{\text{1) Li; 2) } Mn(CO)_5Br}$ 77 $\xrightarrow{\Delta}$ 78

SCHEME 10

82.[33] Structural data on lithiotetramethylarsolide **83**[18] and lithiotetramethylphospholide **84**[34] also make useful comparisons.

Selected structural parameters of the family of diheteroferrocenes are compared in Tables I and II, while typical structures, those of **44** and **48**, are illustrated in Figs. 1 and 2. Table III lists comparable structural data

TABLE I

INTRA-RING BOND LENGTHS AND BOND ANGLES OF 1,1′-DIHETEROFERROCENES

Compound	Bond lengths (Å)			Bond angles (deg.)			Ref.
	$d(EC\alpha)$	$d(C\alpha C\beta)$	$d(C\beta C\beta)$	$\langle C\alpha EC\alpha$	$\langle EC\alpha C\beta$	$\langle C\alpha C\beta C\beta$	
$(C_4H_2\alpha\text{-}Me_2Bi)_2Fe$ (**32**)	2.22	1.40	1.43	76.4	113	118	12
$(C_4H_2\alpha\text{-}Me_2Sb)_2Fe$ (**29**)	2.11	1.41	1.43	79.5	113	117	10
$(C_4Me_4Sb)_2Fe$ (**48**)	2.11	1.42	1.44	79.0	114	116	19
$(C_4\alpha\text{-}(SiMe_3)_2\beta\text{-}Me_2Sb)_2Fe$ (**55**)	2.11	1.44	1.43	81.0	111	117	21
$(C_4H_2\alpha\text{-}Me_2As)_2Fe$ (**26**)	1.90	1.40	1.42	85.8	112	115	25
$(C_4Me_4As)_2Fe$ (**44**)	1.90	1.42	1.45	85.8	114	114	19
$(C_4H_4\beta\text{-}Me_2P)Fe$ (**79**)	1.76	1.42	1.42	88.2	115	111	30

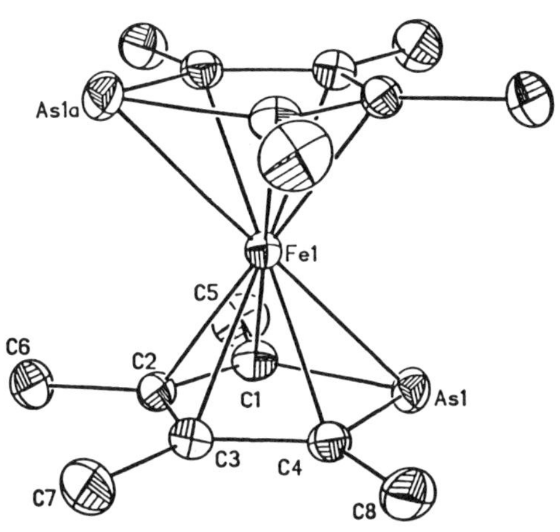

FIG. 1. ORTEP drawing of the solid-state structure of 2,2′,3,3′,4,4′,5,5′-octamethyl-1,1′-diarsaferrocene (**44**). [Reprinted with permission from Ashe *et al.*[19] Copyright 1994 American Chemical Society.]

TABLE II

SELECTED DISTANCES BETWEEN HETEROATOMS, IRON, AND RINGS OF 1,1′-DIHETEROFERROCENES

Compound	Distances (Å)						Ref.
	E—Fe	Cα—Fe	Cβ—Fe	Fe—PL(CαCβCβCα)[a]	E—PL(CαCβCβCα)	E···E	
$(C_4H_2\alpha\text{-}MeBi)_2Fe$ (**32**)	2.68	2.12	2.05	1.60	0.28	3.69	12
$(C_4H_2\alpha\text{-}Me_2Sb)_2Fe$ (**29**)	2.57	2.13	2.07	1.63	0.18	3.68	10
$(C_4Me_4Sb)_2Fe$ (**48**)	2.59	2.11	2.08	1.62	0.23	3.58	19
$(C_4\alpha\text{-}(SiMe_3)_2\beta\text{-}Me_2Sb)_2$ Fe (**55**)	2.57	2.14	2.10	1.63	0.23	—	21
$(C_4H_2\alpha\text{-}Me_2As)_2Fe$ (**26**)	2.40	2.11	2.05	1.65	0.09	3.44	25
$(C_4Me_4As)_2Fe$ (**44**)	2.40	2,11	2.08	1.65	0.09	—	19
$(C_4H_2\beta\text{-}Me_2P)_2Fe$ (**79**)	2.28	2.06	2.07	1.64	0.04	—	30

[a] PL(CαCβCβCα) is the plane through the ring-carbon atoms.

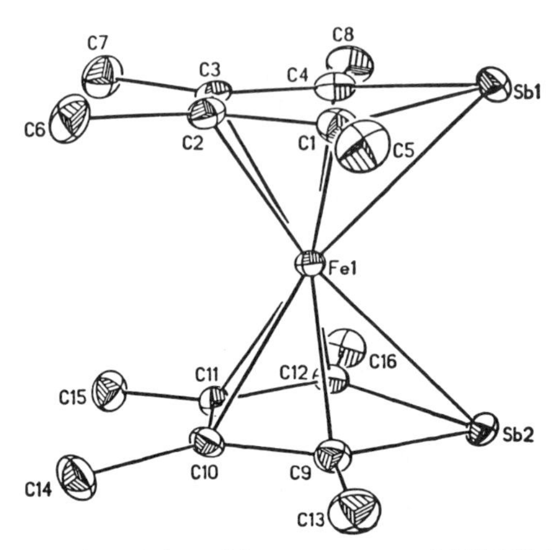

FIG. 2. ORTEP drawing of the solid-state structure of 2,2′,3,3′,4,4′,5,5′-octamethyl-1,1′-distibaferrocene (**48**). [Reprinted with permission from Ashe *et al.*[19] Copyright 1994 American Chemical Society.]

for heterobenzenes **2**,[35–37] **3**,[37–40] and **4**[40,41] and for heteroles **85**,[42] **86**,[9] **87**,[43] and **88**.[12]

In all cases, the 1,1′-diheteroferrocenes have metallocene-type structures in which the Fe atom is sandwiched between two symmetric η^5-heterolyl rings. Although the heterolyl rings are not planar, the four C

TABLE III
SELECTED DISTANCES AND ANGLES FOR HETEROBENZENES AND HETEROLES

Compound	Distances (Å)				⟨CαECα (deg.)	Ref.
	Cα—E	Cα—Cβ	Cβ—Cγ	E—PL(CαCβCβCα)[a]		
C_5H_5Bi (**5**)	[2.15][b]	—	—	—	—	—
C_5H_5Sb (**4**)	2.05	1.40	1.39	0	93	40, 41
C_5H_5As (**3**)	1.85	1.39	1.40	0	97	37–40
C_5H_5P (**2**)	1.73	1.41	1.38	0	101	35–37
$(C_4H_2\alpha\text{-}Me_2Bi)_2$ (**88**)	2.24	1.35	1.44	0.06	79.3	12
$(C_4H_2\alpha\text{-}Me_2Sb)_2$ (**87**)	2.14	1.32	1.46	0.08	81.5	43
$(C_4H_2\alpha\text{-}Me_2As)_2$ (**86**)	1.94	1.34	1.44	0.17	87.1	9
$C_4H_4PCH_2Ph$ (**85**)	1.78	1.34	1.44	0.21	90.7	42

[a] PL(CαCβCβCα) is the plane through the ring-carbon atoms.
[b] Estimated value.

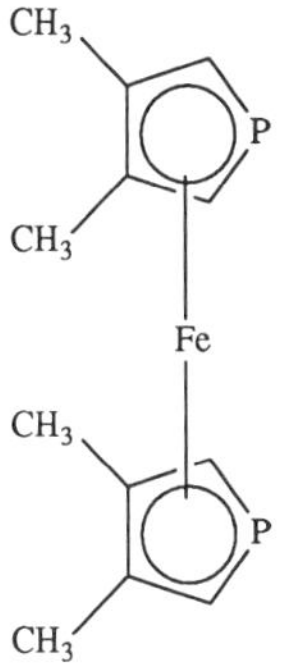

CH_3 CH_3 P Fe CH_3 CH_3 P

79

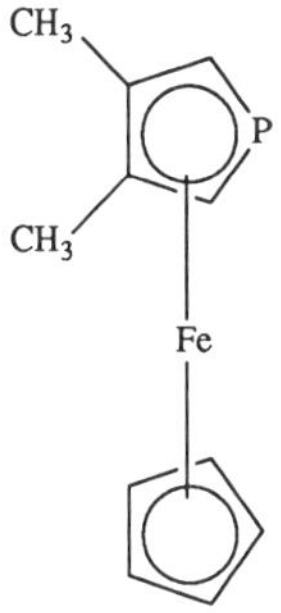

CH_3 CH_3 P Fe

80

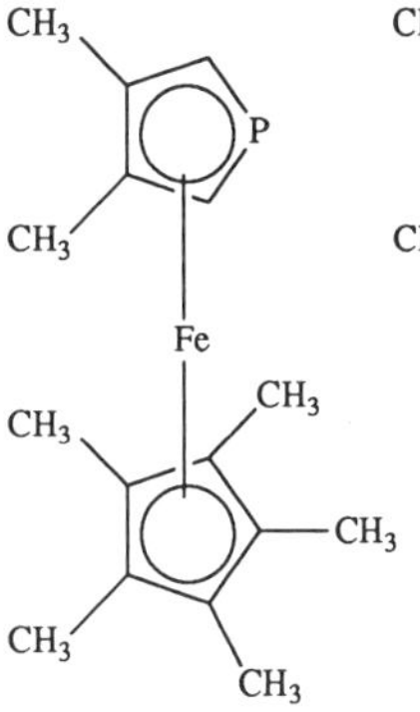

CH_3 CH_3 P Fe CH_3 CH_3 CH_3 CH_3 CH_3

81

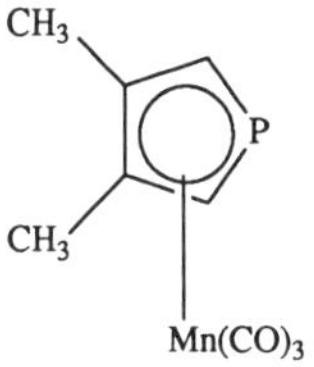

CH_3 CH_3 P $Mn(CO)_3$

82

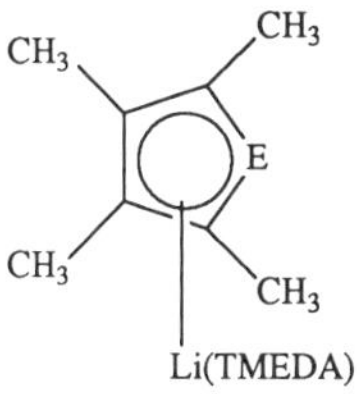

CH_3 CH_3 E CH_3 CH_3 Li(TMEDA)

83 E = As
84 E = P

PCH_2Ph

85

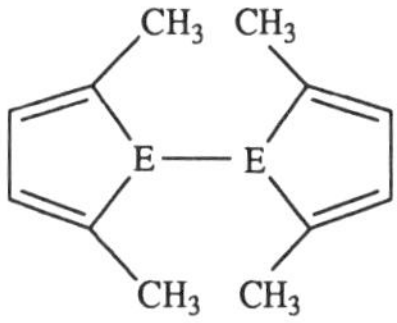

CH_3 CH_3 E — E CH_3 CH_3

86 E = As
87 E = Sb
88 E = Bi

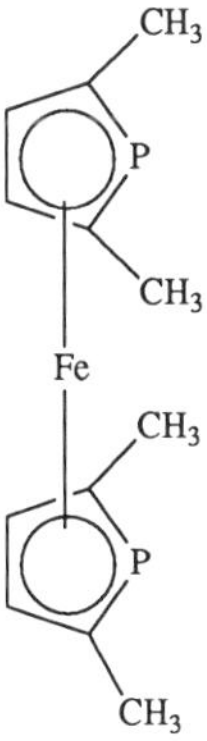

CH_3 P CH_3 Fe CH_3 P CH_3

89

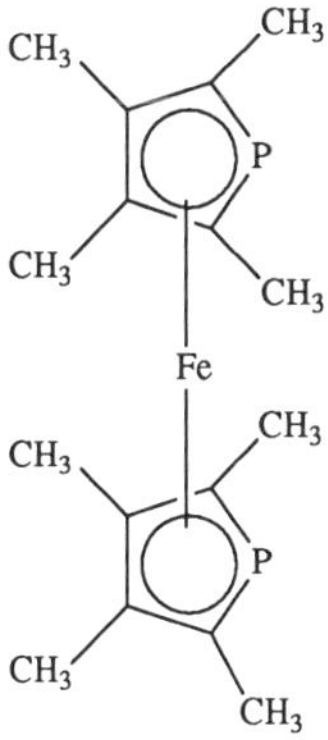

CH_3 CH_3 P CH_3 CH_3 Fe CH_3 CH_3 P CH_3 CH_3

90

atoms of each ring are coplanar. These two planes are either parallel or very close to parallel above and below the Fe atom. Since the C—Fe distances do not vary greatly (2.05–2.14 Å), the C-atom planes are necessarily slightly closer to the Fe atoms for the larger rings. In every case, the heteroatom (E) is displaced out of its C-ring plane away from Fe. The heteroatom displacement is progressive in the series P $<$ As $<$ Sb $<$ Bi. It is probable that this distortion is a consequence of the need to accommodate π-bonding to the C atoms and the increasingly large heteroatoms simultaneously. In the uncomplexed heteroles the distortion from planarity actually decreases with atomic number P $>$ As $>$ Sb $>$ Bi (see Table III).

The intra-ring C—C bond distances of the complexed heterolyl rings of the 1,1′-diheteroferrocenes vary over a small range (1.40–1.45 Å). In no individual case does the Cα—Cβ bond length differ significantly from the Cβ—Cβ bond length. The typical value (1.42 Å) is comparable with the C—C bond length of ferrocene (1.41 Å). In contrast, the C—C bonds of the nonaromatic uncomplexed group 15 heteroles show the expected variation between the Cα—Cβ double bonds (1.32–1.35 Å) and the Cβ-—Cβ single bonds (1.44–1.46 Å). The C—E bond distances of the diheteroferrocenes are much longer than the C—C bond distances. They range from 2.22 Å for the Bi—C bond of **32** to 1.78 Å for the P—C bond of **79**. In all cases, the C—E bonds are 0.02–0.04 Å shorter than the C—E single bond distances of corresponding uncomplexed heteroles. Therefore, the intra-ring bond distances of the diheteroferrocenes indicate that the heterolyl rings are π-complexed aromatic rings.

It is useful to compare the C—E π-bonding of the diheteroferrocenes with that of the corresponding heterobenzenes. The C—P π-bond (1.76 Å) of diphosphaferrocene (**79**) is longer than the C—P bond (1.73 Å) of phosphabenzene (**2**). Apparently this difference is a consequence of metal coordination, since (η^6-2,4,6-triphenylphosphabenzene)$Cr(CO)_3$[44] has an identical 1.76-Å C—P bond distance. The C—As π-bond (1.85 Å) of the uncoordinated arsabenzene (**3**) is 0.05 Å shorter than those of diarsaferrocenes **26** and **44**, while the C—Sb π-bond (2.05 Å) of stibabenzene (**4**) is 0.07 Å shorter than the average C—Sb bond of the distibaferrocenes **29**, **48**, and **55**. If the structural relationship between the lighter diheteroferrocenes and heterobenzenes can be extrapolated, we may anticipate that the C—Bi π-bond of bismabenzene will be close to 2.15 Å.

The heterolyl rings of the diheteroferrocenes are rather distorted pentagons. The CEC bond angles are much smaller than the 108° found for ferrocenes. The CEC bond angle decreases as the atomic number of E increases, from 88.2° for **79** to 76.4° for **32**. Since this trend of decreasing bond angles about E is shown by the corresponding heteroles and acyclic group 15 compounds, it is considered the normal pattern. The smaller CEC bond angles mean that the ECαCβ and the CαCβCβ bond angles must exceed 108°. However, this change probably results in a decrease in strain since the C(sp^2) bond angles more closely approach their ideal value of 120°. The combination of decreasing CEC bond angles and increasing C—E bond lengths serves to minimize structural changes in the carbocyclic portion of the ring. Thus the very large heteroatoms seem to be easily accommodated in the heterolyl rings.

IV

CONFORMATIONS

It is of interest that several of the diheteroferrocenes populate conformers with short interannular separations between the heteroatoms. The various conformations differ by the relative orientation between the two rings. It is convenient to define the conformational variable θ as the dihedral angle between the planes normal to each ring that contain both iron and the heteroatom. When the heteroatoms are eclipsed (C_{2v} symmetry) as in Fig. 2, $\theta = 0°$. When they are anti (C_{2h} symmetry) as in Fig. 1, $\theta = 180°$. The intermediate gauche conformations (C_2 symmetry) have $0° > \theta > 180°$. Crystalline diphosphaferrocene **79** adopts a gauche conformation with $\theta \sim 140°$. Tetramethyldiarsaferrocene **26** adopts a C_{2v} conformation, while octamethyldiarsaferrocene **44** has a C_{2h} conformation. On the other hand, distibaferrocenes **29** and **48** and dibismaferrocene **32** all crystallize in the C_{2v} eclipsed conformation. However, the very crowded distibaferrocene **55** adopts a gauche conformation ($\theta \sim 70°$), which allows the four bulky trimethylsilyl groups to stagger in a gear-like manner. Distibaferrocene (**55**) closely resembles 1,1′,3,3′-tetrakis(trimethylsilyl)ferrocene.[21b]

The eclipsed conformations of **29**, **48**, and **32** bring the pairs of Sb and Bi atoms on opposing rings very close (3.58–3.69 Å). These distances are significantly shorter than van der Waals radius separation (4.4 Å for Sb · · · Sb and 4.6 Å for Bi · · · Bi). Thus a direct secondary bonding between the pnictogen atoms seems likely.

Both the π-bonding of the heterolyl rings and the highly acute CEC bond angles imply that the pnictogen atoms use their ultimate p-orbitals for bonding while the lone pair is largely s-hybridized. In the C_{2v} conformations, these p-orbitals are approximately aligned and their inter-ring overlap is significant. Thus the conditions exist for appreciable inter-ring bonding.

An intermolecular analog of this inter-ring pnictogen bonding is found in the thermochromic distibines and dibismuthines.[45] For example, 2,2′,5,5′-tetramethylbistibole (**87**) crystallizes so that the Sb atoms are aligned in chains with close contacts between the Sb atoms of adjacent molecules. It seems more than coincidental that the intermolecular contacts in **87** occur at virtually the same distance as the intramolecular contacts in **29**. See Figure 3.

A similar relationship exists between the short inter-ring Bi · · · Bi contact (3.69 Å) in **32** and the short intermolecular Bi · · · Bi distance (3.66 Å) in bibismole **88** and other thermochromic dibismuthines. However, no analogy can be drawn with the conformation of diarsaferrocene **26** and the solid-state packing of solid 2,2′,5,5′-tetramethylbiarsole (**86**), since **86** and diarsines in general do not crystallize in forms with short intermolecular As · · · As contacts. Similarly, the observation that octamethyldiarsaferrocene does not have a short inter-ring As · · · As contact suggests that As · · · As secondary bonding is very weak.

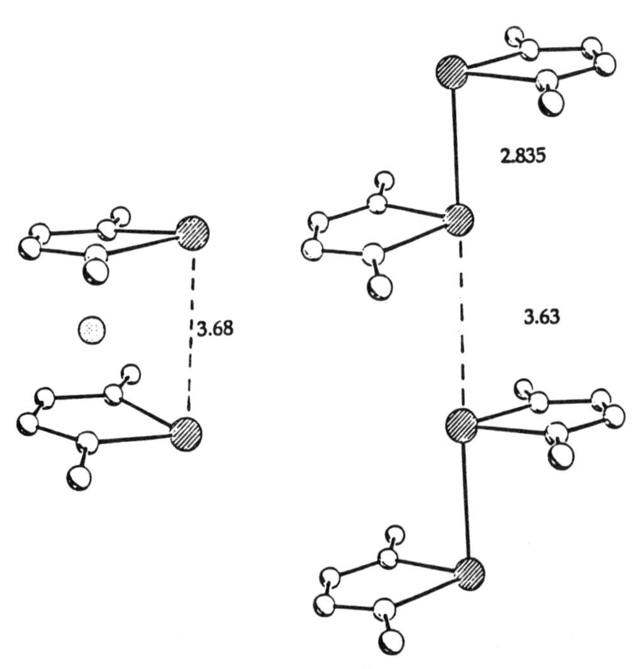

FIG. 3. Comparison of the inter-ring distance (Å) in **29** with the intermolecular Sb · · · Sb interaction in **87**. [Reprinted with permission from Ashe *et al.*[10] Copyright 1991 American Chemical Society.]

The conformational properties of various 1,1′-diheteroferrocenes (**7–10**) have been the subject of three computational studies using extended Hückel methods.[19,46,47] 1,1′-Diphosphaferrocene has also been studied using the Fenske–Hall approach.[48] and an MS Xα method.[46] Where they overlap, the four treatments are in reasonable qualitative agreement.

The total energies of **7–10** were determined as a function of the angle θ. Setting the energy of the C_{2h} conformation ($\theta = 180°$) of each diheteroferrocene equal to zero allows a comparison of the relative conformational energies as a function of θ. These data are plotted in Fig. 4.

For all the diheteroferrocenes, the conformation energies change relatively little as θ varies from 180 to 100°. However, the conformational energies diverge markedly as θ changes from 100 to 0°. The eclipsed conformation of diphosphaferrocene is destablized by 9.6 kcal/mol. The energy of diarsaferrocene changes little, but distibaferrocene and dibismaferrocene are stabilized by 3 and 7 kcal/mol, respectively.

Thus the eclipsed conformations become more favorable as the heteroatoms become larger. However, if the overlap integral involving the two heteroatoms is arbitrarily set at zero, the stabilization of the C_{2v} conformations vanishes and the C_{2h} conformations become the global minimum for

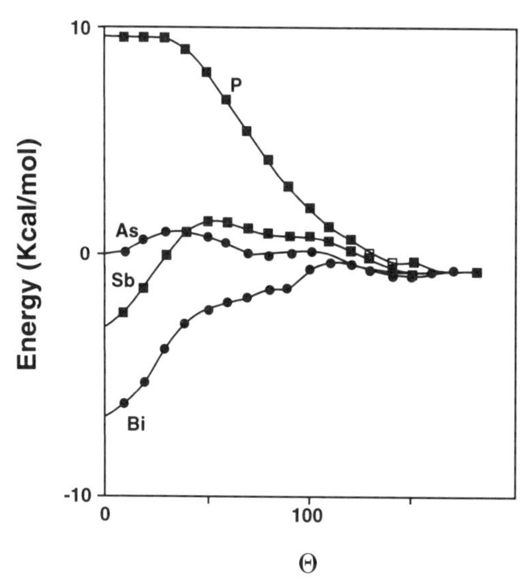

FIG. 4. Relative energies of the 1,1′-diheteroferrocenes $(C_4H_4E)_2Fe$, where E = P, As, Sb, Bi, as a function of the angle θ. [Reprinted with permission from Ashe *et al.*[19] Copyright 1994 American Chemical Society.]

7–10. Thus the calculations indicate that the direct through-space bonding between the pnictogen atom *p*-orbitals controls the conformations of the heavier diheteroferrocenes.

V
SPECTRA

The ^{1}H- and ^{13}C-NMR chemical shift values of these series of polymethyl-1,1′-diheteroferrocenes are collected in Tables IV[49] and V,[49] respectively. In all cases, the signals show upfield shifts relative to their corresponding uncomplexed heterole precursors. Although most of the chemical shift values are unexceptional, the bismuth compounds do show rather low-field signals. In each series the ^{13}C-NMR shifts show progressive shifts to lower field with increasing atomic number of the pnictogen atom. However, similar effects are observed for both aromatic, e.g., **2–5**, and nonaromatic, e.g., **86–88**, organopnictogen compounds.

The mass spectra of the 1,1′-diheteroferrocenes are unexceptional. The molecular ion is always prominent and is usually the base peak.

TABLE IV

^{1}H NMR Chemical Shift Values of the Polymethyl-1,1′-Diheteroferrocenes

Compound	δ (ppm)				Ref.
	α—CH	α—Me	β—CH	β—Me	
$(C_4H_2\alpha\text{-}Me_2E)_2Fe$					
89, E = P	—	1.80	4.90	—	22
26, E = As	—	1.90	5.12	—	22, 26
29, E = Sb	—	1.84	5.63	—	10
32, E = Bi[a]	—	2.17	7.15	—	12
$(C_4H_2\beta\text{-}Me_2E)_2Fe$					
79, E = P	3.71	—	—	2.11	2
60, E = As	4.32	—	—	2.05	22
61, E = Sb	3.90	—	—	2.01	22
$(C_4Me_4E)_2Fe$					
90, E = P[a]	—	1.63	—	1.96	49
44, E = As[a]	—	1.73[b]	—	1.64[b]	19
48, E = Sb[a]	—	1.65[b]	—	1.50[b]	19
49, E = Bi[a]	—	1.50[b]	—	1.86[b]	22

[a] Solvent is C_6D_6; in all other cases it is $CDCl_3$.
[b] Relative assignment uncertain.

TABLE V

^{13}C NMR CHEMICAL SHIFT VALUES OF THE POLYMETHYL-1,1′-DIHETEROFERROCENES

Compound	δ (ppm)				Ref.
	α—C	α—Me	β—C	β—Me	
$(C_4H_4\alpha\text{-}Me_2E)_2Fe$					
89, E = P	97.9	16.3	84.4	—	22
26, E = As	108.6	19.0	86.5	—	22, 26
29, E = Sb	113.6	23.0	93.5	—	10
32, E = Bi[a]	n.o.[b]	29.1	109.6	—	12
$(C_4H_2\beta\text{-}Me_2E)_2Fe$					
79, E = P	82.1	—	97.1	16.1	2
60, E = As	89.8	—	98.5	16.8	22
61, E = Sb	86.4	—	104.8	18.4	22
$(C_4Me_4E)_2Fe$					
90, E = P[a]	94.0	14.1	95.1	12.7	49
44, E = As[a]	103.8[c]	17.0[c]	97.2[c]	13.9[c]	19
48, E = Sb[a]	105.6[c]	20.9[c]	103.9[c]	14.8[c]	19
49, E = Bi[a]	n.o.	26.1[c]	117.9	18.9[c]	22

[a] Solvent is C_6D_6; in all other cases it is $CDCl_3$.
[b] Not observed.
[c] Relative assignment of α and β is uncertain.

VI

ELECTROCHEMISTRY

The redox chemistry of several phosphaferrocenes,[31,50] 1,1′-diarsaferrocene (**7**),[13] the complete series of 2,2′,5,5′-tetramethyl-1,1′-diheteroferrocenes (**89**, **26**, **29**, **32**),[22] and octamethyl-1,1′-diheteroferrocenes (**90**, **44**, **48**, **49**)[22] has been investigated by cyclic voltammetry. These compounds undergo quasi-reversible one-electron oxidations (0/+) to their radical cations and irreversible one-electron reductions (0/−) to their radical anions. The data are summarized in Table VI.

In each series the oxidations show monotonic cathodic shifts, while the reductions show monotonic anodic shifts as the heteroatom increases in atomic number. Thus it is easier both to oxidize and reduce the heavier diheteroferrocenes.

For the oxidations, the cathodic shifts (P→As→Sb→Bi) are nearly the same for the tetramethyl series and the octamethyl series. Since the shifts are independent of the degree of methyl substitution, these shifts can be used to estimate the $E_{1/2}$ values for the series of unsubstituted diheteroferrocenes from the experimental value for 1,1′-diarsaferrocene.

TABLE VI

SELECTED ELECTROCHEMICAL DATA FOR 1,1′-DIHETEROFERROCENES[a,b]

Compound	$E_{1/2}(+/0)$ (volts)[c]	$E_{1/2}(0/-)$ (volts)[d]
$(C_4H_4E)_2Fe$		
Ferrocene, E = CH	0.490	—
7, E = P	[0.882][e,f]	—
8, E = As	0.730	−2.190
9, E = Sb	[0.453][e]	—
10, E = Bi	[0.190][e]	—
$(C_4H_2\alpha\text{-}Me_2E)_2Fe$		
89, E = P	0.722	−2.352
26, E = As	0.578	−2.258
29, E = Sb	0.293	−2.036
32, E = Bi	0.030	−2.004
$(C_4Me_4E)_4Fe$		
90, E = P	0.565	−2.507
44, E = As	0.405	−2.381
48, E = Sb	0.141	−2.176
49, E = Bi	−0.133	−2.140

[a] Ashe *et al.*[22]

[b] In DME/(nBu$_4$NClO$_4$) (0.1 *M*) at glassy carbon vs. SCE, $T = -40°C$, $v = 100$ mV s^{-1}.

[c] Quasi-reversible.

[d] Irreversible.

[e] Estimated.

[f] Also experimentally measured using other conditions. See Elschenbroich *et al.*[51]

Diphosphaferrocene[50] and diarsaferrocene[13] are harder to oxidize than ferrocene. Distibaferrocene is only slightly easier to oxidize than ferrocene, while dibismaferrocene is easier to oxidize.

The MO treatments of the diheteroferrocenes indicate that the HOMO is effectively an Fe(d_{z^2}) orbital. The variation in the $E_{1/2}$ (0/+) value should measure the ability of the heterolyl ligand to withdraw electron density from Fe. Thus the phospholyl and arsolyl ligands are more effective at withdrawing electron density from Fe than are the Cp ligands of ferrocene. Apparently, the effective π-electronegativities of P and As are greater than those for C. A similar conclusion was reached by comparison of the redox behavior of bis(η^6-phosphabenzene)metal and bis(η^6-arsabenzene)-metal complexes with bis(benzene)metal complexes.[52] A study of the MCD spectra of heterobenzenes **2**, **3**, and **4** also indicated a high π-electronegativity of the heteroatoms.[53]

VII

ELECTROPHILIC AROMATIC SUBSTITUTION

Phosphaferrocenes have a well developed aromatic chemistry which has been explored largely by the Mathey group.[2] Although less work has been done on the arsenic compounds, the chemistry appears to be very similar. Both 3,3′,4,4′-tetramethyl-1,1′-diphosphaferrocene (**79**)[28] and 3,3′,4,4′-tetramethyl-1,1′-diarsaferrocene (**60**)[22] can be acetylated to afford good yields of the monoacetyl derivatives **91** and **92**, respectively (Scheme 11). Under more severe conditions with excess acetylating agent, each mono adduct affords mixtures of two diastereomeric diadducts **93**, **94** and **95**, **96**, respectively. In each case the meso and racemic isomers could be separated by column chromatography, but the spectra did not allow relative identification.

As shown in Scheme 12, acetylation of 1-phosphaferrocene (**12**) gives a mixture of 85% 2-acetyl-1-phosphaferrocene (**97**) and 15% 3-acetyl-1-phosphaferrocene (**98**).[54] In contrast, the acetylation of 1,1′-diarsaferrocene (**8**) gives **99** as the exclusive monoacetylation product.[13] Further acetylation of **99** affords a mixture of the two diastereomeric products **100** and **101**.[55] Apparently the α position is more reactive in both the P and As series, but the relative rates for α- vs. β-substitution are greater in the As series. If the more reactive α position is blocked, substitution can take place at the β position in both series. Thus the acetylation of 2,2′,5,5′-tetraphenyl-1,1′-diphosphaferrocene (**102**) gives the 3-substituted product **103** as well as substantial amounts of phenyl-substituted product

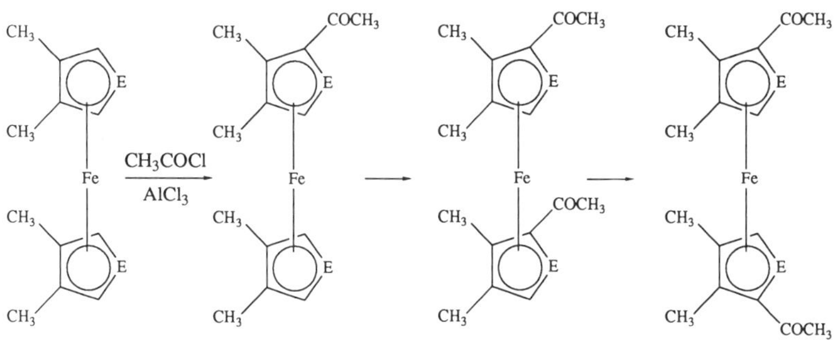

CH_3COCl / $AlCl_3$

79 E = P
60 E = As

91
92

93
95

94
96

SCHEME 11

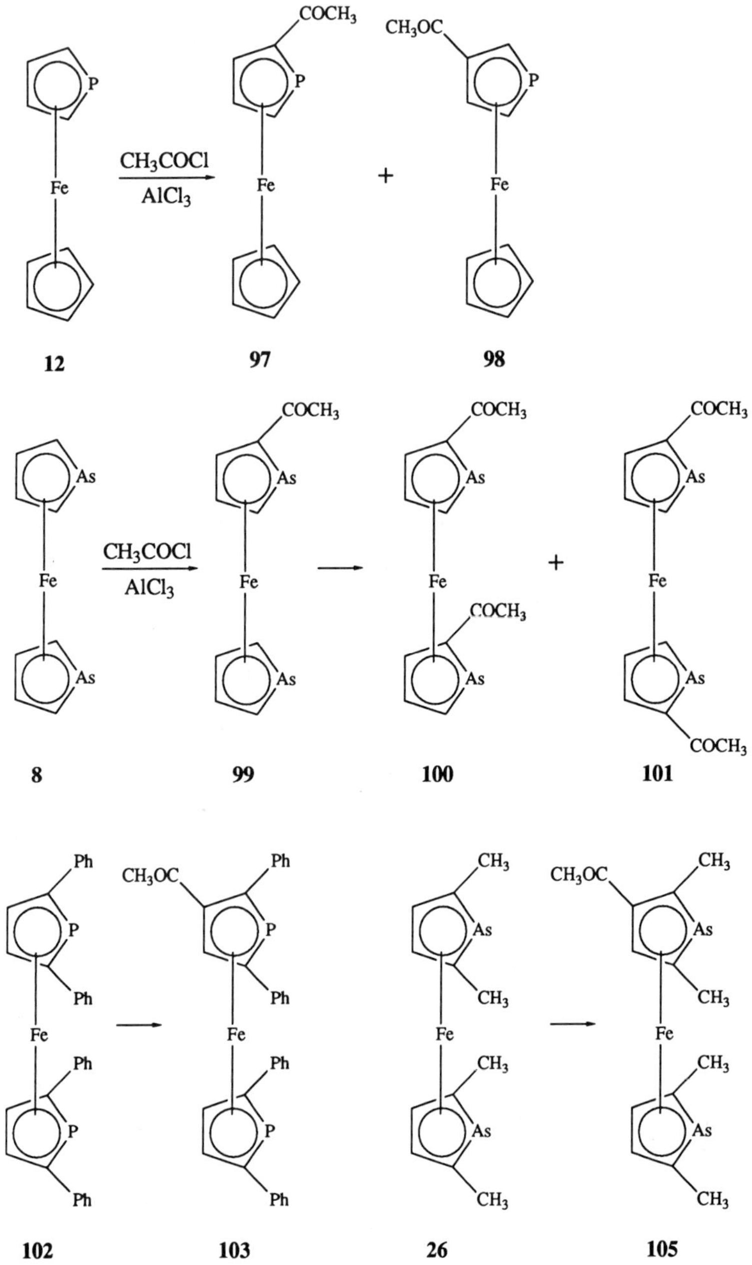

CH_3COCl / $AlCl_3$

12 97 98

CH_3COCl / $AlCl_3$

8 99 100 101

102 103 26 105

SCHEME 12

104.[56] Acetylation of 2,2′,5,5′-tetramethyl-1,1′-diarsaferrocene (**26**) gives the 3-acetylated product **105**.

The greater reactivity of the α positions of phosphaferrocenes has been rationalized on the basis of charge-density effects.[48] Alternatively, it has been argued that the π-orbitals centered on Cα are more easily perturbed by attacking electrophiles.[46] We suggest that the lower reactivity of the β position may reflect a lower stability of the intermediate for substitution. Perhaps the cross-conjugated heterodiene–iron complex, which must be an intermediate for β-substitution, is higher in energy than the terminally substituted heterodiene–iron complex, which is the intermediate for α-substitution. See Scheme 13.

Attempted acetylation of 2,2′,5,5′-tetramethyl-1,1′-distibaferrocene (**29**) or 3,3′,4,4′-tetramethyl-1,1′-distibaferrocene (**61**) led to destruction of the ring system and the formation of intractable products. Acid-catalyzed H/D isotopic exchange (Scheme 14) is the simplest electrophilic aromatic substitution. Both 1,1′-diphosphaferrocene[55] (**7**) and 1,1′-diarsaferrocene (**8**)[13] undergo rapid exchange at the α positions when treated

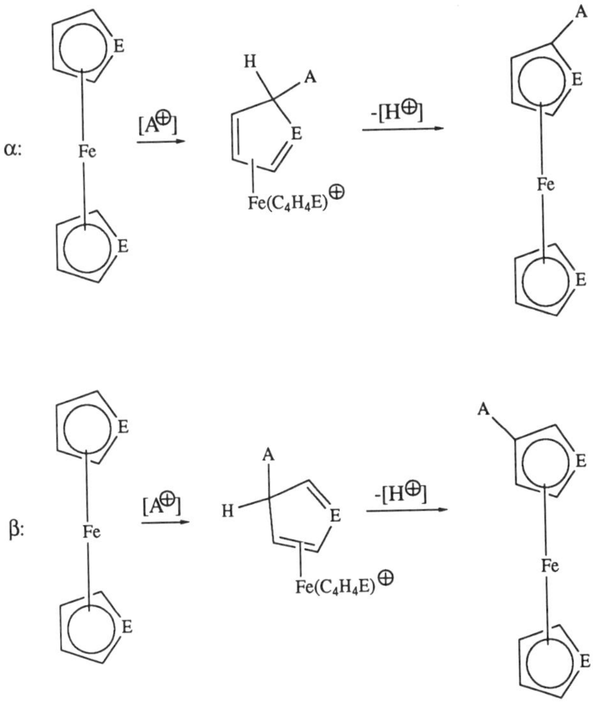

Scheme 13

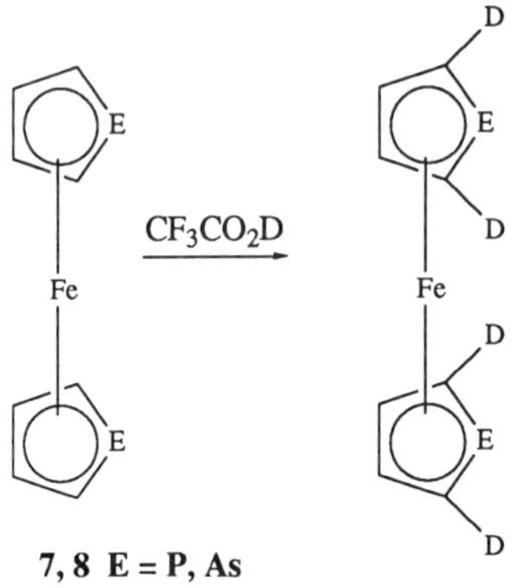

CF_3CO_2D

7, 8 E = P, As

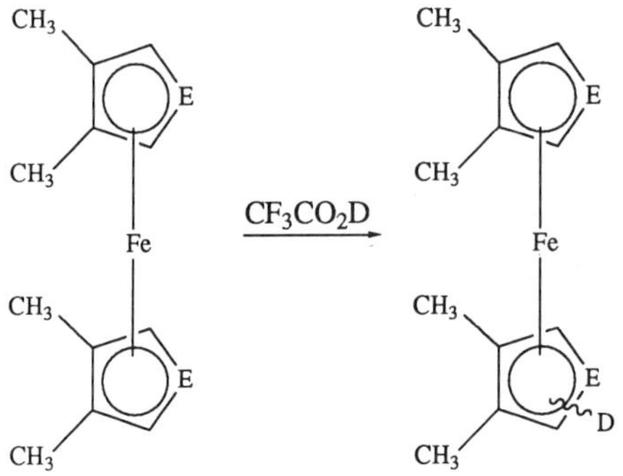

CF_3CO_2D

60, 61, 79, 106
E = As, Sb, P, CH
Relative Rates: Sb>As>CH>P

SCHEME 14

with trifluoroacetic acid-*d*, at −20°. Heating either **7** or **8** in trifluoracetic acid-*d*, allowed partial exchange at the β position. Qualitatively, **8** undergoes exchange faster than ferrocene, while **7** reacts more slowly. Similarly, it has been found that **60** undergoes exchange faster than 1,1′,2,2′-tetramethylferrocene (**106**) but **79** reacts more slowly.[22] Although 3,3′,4,4′-tetramethyl-1,1′-distibaferrocene (**61**) is destroyed by treatment with acid, H/D exchange has been observed in dilute solutions in toluene. Exchange of **61** is faster than **60**.[22] The higher reactivity of the heavier diheteroferrocenes is consistent with their lower oxidation potentials.

VIII

HETEROCYMANTRENES AND RELATED COMPLEXES

(η^5-2,5-Dimethylheterolyl)Mn$(CO)_3$ complexes (**107**, **108**,[57] **71**,[26] **76**,[8] **78**[27]) have been reported for all the group 15 elements. Simlarly, (η^5-2,3,4,5-tetraphenylphospholyl and -arsolyl)Mn$(CO)_3$ and -Re$(CO)_3$ derivatives have been prepared.[24] The ^{1}H-NMR and ^{13}C-NMR spectra of these complexes correspond closely to those of the 1,1′-diheteroferrocenes. Comparison of the carbonyl stretching frequencies obtained from the IR spectra show that the ν(CO) decreases with atomic number of the heteroatom, N > P > As > Sb ~ Bi (see Table VII). These data suggest that the heterolyl ligands become better π-donors as the atomic numbers of the heteroatoms increase. A similar conclusion has been drawn for the heterolyl ligands of the 1,1′-diheteroferrocenes from the observed decrease in the electrochemical $E_{1/2}$ (0/+) values with increasing atomic number.[22]

2,5-Dimethylarsacymantrene (**71**) undergoes Friedel–Crafts acetylation at the β position to give adduct **110**.[26] By analogy to the heteroferrocene series, it is presumbed that the α position would be more reactive; however, no α-unsubstituted arsacymantrenes have been reported. For comparison, 3,4-dimethylphosphacymantrene (**111**) is clearly acylated at the α position. See Scheme 15.

TABLE VII

CARBONYL STRETCHING FREQUENCIES OF THE COMPLEXES (η^5-HETEROLYL)Mn$(CO)_3$ FROM IR SPECTRA

Complex	ν(CO) (cm^{-1})	Reference
$Mn(CO)_3(C_4H_2\alpha\text{-}Me_2N)^a$ (**107**)	2035, 1958, 1950	57
$Mn(CO)_3(C_4H_2\alpha\text{-}Me_2P)^a$ (**108**)	2023, 1954, 1951	57
$Mn(CO)_3(C_4Ph_4P)^b$ (**109**)	2034, 1969, 1955	24
$Mn(CO)_3(C_4H_2\alpha\text{-}Me_2As)^a$ (**71**)	2024, 1952, 1946	57
$Mn(CO)_3(C_4H_2\alpha\text{-}Me_2As)^c$ (**71**)	2020, 1948, 1942	26
$Mn(CO)_3(C_4Ph_4As)^b$ (**74**)	2024, 1960, 1944	24
$Mn(CO)_3(C_4H_2\alpha\text{-}Me_2Sb)^b$ (**76**)	2010,[e] 1948	8
$Mn(CO)_3(C_4H_2\alpha\text{-}Me_2Bi)^d$ (**78**)	2011, 1942, 1937	27

[a] Decalin.
[b] Cyclohexane.
[c] Hexadecane.
[d] Hexane.
[e] Misquoted as 2100 cm^{-1} in Ashe and Diephouse.

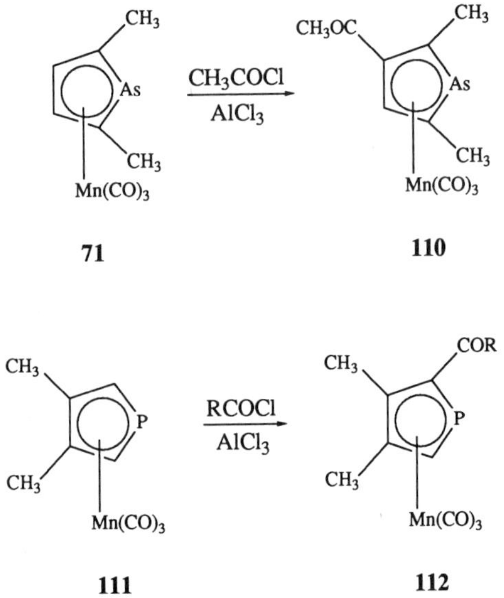

CH_3COCl / $AlCl_3$

71 → **110**

RCOCl / $AlCl_3$

111 → **112**

SCHEME 15

Tetraphenylarsacymantrene (**74**) undergoes several CO ligand displacements without disruption of the arsolyl ligand (Scheme 16).[24] It is reported that these displacements take place more readily and in higher yield than those for the corresponding pyrolyl derivatives. However, **108** and **71** are found to be much less reactive than **107** toward nonphotochemical CO displacement by phosphines.[37]

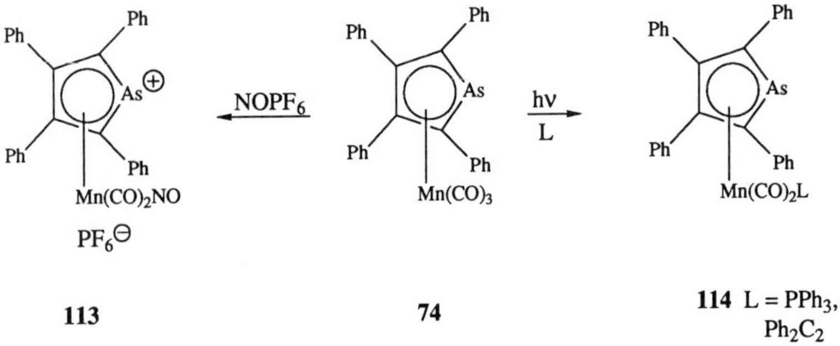

$NOPF_6$; hν / L

113 ← **74** → **114** L = PPh_3, Ph_2C_2

SCHEME 16

IX

CONCLUDING REMARKS

Many of the properties of the group 15 element diheteroferrocences are very similar to ferrocenes and other metallocenes. It seems justified to regard the diheteroferrocenes as perturbed ferrocenes just as we regard the group 15 heterobenzenes as perturbed benzenes. Thus, it is very clear that the elements phosphorus, arsenic, antimony, and bismuth can take part in π-bonding in a manner similar to carbon.

ACKNOWLEDGMENTS

We are especially grateful to our able co-workers whose names are mentioned in the appropriate references. The Research Corporation, the donors of Petroleum Research Fund of the American Chemical Society, and the National Science Foundation are gratefully acknowledged for their financial support.

REFERENCES

(1) Ashe, A. J., III. *Acc. Chem. Res.* **1978,** *11*, 153. Ashe, A. J., III. *Top. Curr. Chem.* **1982,** *105*, 125.
(2) Mathey, F. *Nouv. J. Chim.* **1987,** *11*, 585.
(3) Kuhn, N.; Jendral, K.; Stubenrauch, S.; Mynott, R. *Inorg. Chim. Acta* **1993,** *206*, 1 and the earlier papers in this series.
(4) Märkl, G.; Hauptmann, H. *Tetrahedron Lett.* **1968,** 3257.
(5) Thiollet, G.; Mathey, F.; Poilblanc, R. *Inorg. Chim. Acta* **1979,** *32*, L67.
(6) Wieber, M. *Gmelin Handbook of Inorganic Chemistry*, 8th ed. 1981; Part 2, p. 91.
(7) Phenylbismuthine has not been prepared. For a bismuth hydride, see: Amberger, E. *Chem. Ber.* **1961,** *94*, 1447.
(8) Ashe, A. J., III; Diephouse, T. R. *J. Organomet. Chem.* **1980,** *202*, C95.
(9) Ashe, A. J., III; Butler, W. M.; Diephouse, T. R. *Organometallics* **1983,** *2*, 1005.
(10) Ashe, A. J., III; Diephouse, T. R.; Kampf, J. W.; Al-Taweel, S. M. *Organometallics* **1991,** *10*, 2068.
(11) Ashe, A. J., III; Drone, F. J. *Organometallics* **1984,** *3*, 495.
(12) Ashe, A. J., III; Kampf, J. W.; Puranik, D. B.; Al-Taweel, S. M. *Organometallics,* **1992,** *11*, 2743.
(13) Ashe, A. J., III; Mahmoud, S.; Elschenbroich, C.; Wünsch, M. *Angew. Chem., Int. Ed. Engl.* **1987,** *26*, 229.
(14) Ashe, A. J., III; Drone, F. J. *Organometallics* **1985,** *4*, 1478.
(15) Ashe, A. J., III; Al-Taweel, S. M. unpublished work.
(16) Ashe, A. J., III; Mahmoud, S. *Organometallics* **1988,** *7*, 1878.
(17) Fagan, P. J.; Nugent, W. A. *J. Am. Chem. Soc.* **1988,** *110*, 2310. Fagan, P. J.; Nugent, W. A.; Calabrese, J. C. *J. Am. Chem. Soc.* **1994,** *116*, 1880.
(18) Sendlinger, S. C.; Haggerty, B. S.; Rheingold, A. L.; Theopold, K. H. *Chem. Ber.* **1991,** *124*, 2453.
(19) Ashe, A. J., III; Kampf, J. W.; Pilotek, S.; Rousseau, R. *Organometallics* **1994,** *13*, 4067.

(20) Also see: Spence, R. E. v. H.; Hsu, D. P.; Buchwald, S. L. *Organometallics* **1992,** *11*, 3492.
(21) (a) Ashe, A. J., III; Kampf, J. W.; Al-Taweel, S. M. *Organometallics* **1992,** *11*, 1491. (b) Okuda, J.; Herdtweck, E. *J. Organomet. Chem.* **1989,** *373*, 99 **1990,** *385*, C39.
(22) Ashe, A. J., III; Al-Ahmad, S.; Pilotek, S.; Puranik, D. B.; Elschenbroich, C.; Behrendt, A. *Organometallics* **1995,** *14*, 2689.
(23) Leavitt, F. C.; Manuel, T. A.; Johnson, F.; Matternas, L. U.; Lehman, D. S. *J. Am. Chem. Soc.* **1960,** *82*, 5099. Braye, E. H.; Hübel, W.; Caplier, I. *J. Am. Chem. Soc.* **1961,** *83*, 4406.
(24) Abel, E. W.; Towers, C. *J. Chem. Soc., Dalton Trans.* **1979,** *814*. Abel, E. W.; Clark, N.; Towers, C. *J. Chem. Soc., Dalton Trans.* **1979,** 1552.
(25) Chiche, L.; Galy, J.; Thiollet, G.; Mathey, F. *Acta Crystallogr., Sect. B: Struct. Crystallogr. Cryst. Chem.* **1980,** *B36*, 1344.
(26) Thiollet, G.; Poilblanc, R.; Voigt, D., Mathey, F. *Inorg. Chim. Acta* **1978,** *30*, L294.
(27) Ashe, A. J., III; Kampf, F. W.; Puranik, D. B. *J. Organomet. Chem.* **1993,** *447*, 197.
(28) De Lauzon, G.; Deschamps, B.; Fischer, J.; Mathey, F.; Mitschler, A. *J. Am. Chem. Soc.* **1980,** *102*, 994.
(29) Mathey, F.; Mitschler, A.; Weiss, R. *J. Am. Chem. Soc.* **1977,** *99*, 3537.
(30) Ashe, A. J., III; Kampf, J. W.; Al-Taweel, S. M. *J. Am. Chem. Soc.* **1992,** *114*, 372.
(31) Roman, E.; Leiva, A. M.; Casasempere, M. A.; Charrier, C.; Mathey, F.; Garland, M. T.; Le Marouille, J.-Y. *J. Organomet. Chem.* **1986,** *309*, 323.
(32) Abel, E. W.; Nowell, I. W.; Modinos, A. G. J.; Towers, C. *J. Chem. Soc., Chem. Commun.* **1973,** 258.
(33) Breque, A.; Mathey, F.; Santini, C. *J. Organomet. Chem.* **1979,** *165*, 129.
(34) Douglas, T.; Theopold, K. H. *Angew. Chem., Int. Ed. Engl.* **1989,** *28*, 1367.
(35) Kuczkowski, R. L.; Ashe, A. J., III. *J. Mol. Spectrosc.* **1972,** *42*, 457.
(36) Wong, T. C.; Bartell, L. S. *J. Chem. Phys.* **1974,** *61*, 2840.
(37) Wong, T. C.; Ashe, A. J., III. *J. Mol. Struct.* **1978,** *48*, 219.
(38) Lattimer, R. P.; Kuczkowski, R. L.; Ashe, A. J., III; Meinzer, A. L. *J. Mol. Spectrosc.* **1975,** *57*, 428.
(39) Wong, T. C.; Ashe, A. J., III; Bartell, L. S. *J. Mol. Struct.* **1975,** *25*, 65.
(40) Wong, T. C.; Ferguson, M. G.; Ashe, A. J., III. *J. Mol. Struct.* **1979,** *52*, 231.
(41) Fong, G.; Kuczkowski, R. L.; Ashe, A. J., III. *J. Mol. Spectrosc.* **1978,** *70*, 197.
(42) Coggon, P.; Engel, J. F.; McPhail, A. T.; Quin, L. D. *J. Am. Chem. Soc.* **1970,** *92*, 5779. Coggon, P.; McPhail, A. T. *J. Chem. Soc., Dalton Trans.* **1973,** 1888.
(43) Ashe, A. J., III; Butler, W.; Diephouse, T. R. *J. Am. Chem. Soc.* **1981,** *103*, 207.
(44) Vahrenkamp, H.; Nöth, H. *Chem. Ber.* **1972,** *105*, 1148.
(45) Ashe, A. J., III. *Adv. Organomet. Chem.* **1990,** *30*, 77.
(46) Guimon, C.; Gonbeau, D.; Pfister-Guillouzo, G.; de Lauzon, G.; Mathey, F. *Chem. Phys. Lett.* **1984,** *104*, 560.
(47) Su, M.-D.; Chu, S.-Y. *J. Phys. Chem.* **1989,** *93*, 6043.
(48) Kostic, N. M.; Fenske, R. F. *Organometallics* **1983,** *2*, 1008.
(49) Nief, F.; Mathey, F.; Richard, L.; Robert, F. *Organometallics* **1988,** *7*, 921.
(50) Lemoine, P.; Gross, M., Braunstein, P., Mathey, F.; Deschamps, B.; Nelson, J. H. *Organometallics* **1984,** *3*, 1303.
(51) Elschenbroich, C.; Nowotny, M.; Metz, B.; Massa, W.; Granlich, J., Biehler, K.; Sauer, W. *Angew. Chem., Int. Ed. Engl.* **1991,** *30*, 547. Elschenbroich, C.; Bär, F.; Bilger, E.; Mahrwald, D.; Nowotny, M.; Metz, B. *Organometallics* **1993,** *12*, 3373.
(52) Elschenbroich, C.; Kroker, J.; Massa, W.; Wünsch, M.; Ashe, A. J., III. *Angew. Chem., Int. Ed. Engl.* **1986,** *25*, 571.

(53) Waluk, J.; Klein, H.-P.; Ashe, A. J., III; Michl, J. *Organometallics* **1989,** *8*, 2804.
(54) Mathey, F. *J. Organomet. Chem.* **1977,** *139, 77.*
(55) Al-Taweel, S. A. M. Ph.D. Dissertation, University of Michigan, Ann Arbor, 1988.
(56) Roberts, R. M. G.; Wells, A. S. *Inorg. Chim. Acta* **1987,** *130*, 93.
(57) Kershner, D. L.; Basolo, F. *J. Am. Chem. Soc.* **1987,** *109*, 7396.

Boron–Carbon Multiple Bonds

JOHN J. EISCH

Department of Chemistry
State University of New York at Binghamton
Binghamton, New York

I
INTRODUCTION

A. *Historical Background*

Experimental evidence for multiple bonding between boron and its main-group neighbors first stemmed from the pioneering work of Alfred Stock on the reaction of boron hydride with ammonia.[1] Especially significant was the isolation of the colorless liquid $B_3N_3H_6$ (**1**) by Stock and Pohland[2] in 1926, as physicochemical measurements have been able to establish that **1** consists of a planar, hexagonal array of alternating B and N centers having the same B—N distances of 1.44 Å. Since such a separation is intermediate between that typical for ordinary B—N (1.54 Å) and B═N (1.36 Å) bonds, the structure of the ring system, termed borazole or borazine, has been depicted in terms of resonance structures **1a–1c** (formal charges omitted). Because of further remarkable similarities in physical properties, as well as in bonding, to those of benzene, **1** has been labeled as "inorganic benzene."[3]

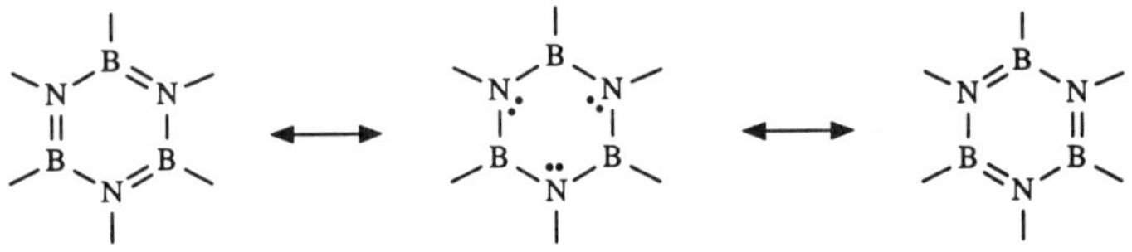

1a **1b** **1c**

Pertinent to multiple bonding between boron and carbon are the extensive studies of Michael Dewar's students in the early 1960s on the synthesis, structure, and aromaticity of 1,2-azaboraarenes, such as the 1,2-azaborine (**2**), the 1-aza-2-boranaphthalene (**3**), and the 9-aza-10-boraphenanthrene (**4**).[4] In order to account for the aromatic character of such rings, as deduced from physical properties (such as NMR data) and chemical stability, it is clearly necessary for π-bonding between B and C centers to be significant.[5] Although some observations on the possible aromaticity of boracyclopolyenes have also been published in the 1960s,[5] little definitive information has become available until the last decade.

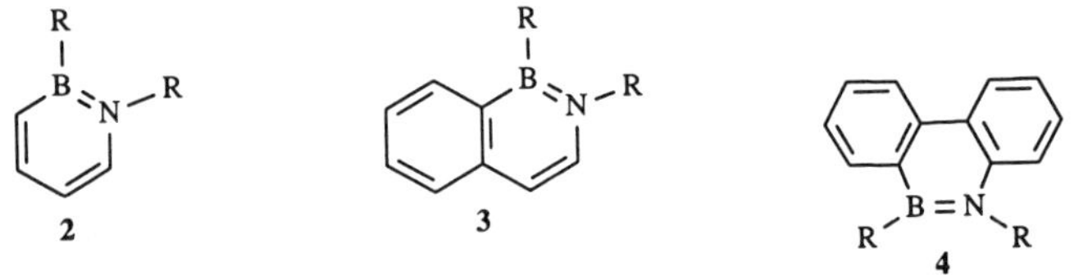

2 **3** **4**

The question of multiple bonding between boron and carbon when there is no cyclic aromatic character to be gained, as for example in vinylboranes (structures **5**a and **5**b), is more difficult to decide.

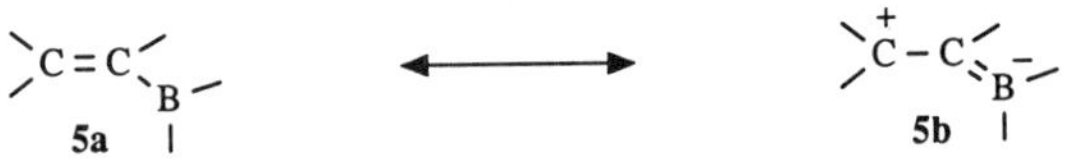

5a **5b**

But the discovery of the carboranes in the early 1960s revealed that bonding possibilities other than simple σ- or π-bonds between B and C centers were necessary to understand the structure of such compounds as 1,5-dicarba-closo-pentaborane(5) [**6** in Eq. (1)], which is obtained in low yield in an electrical discharge.[6] Ordinary valence conventions cannot account for the bonding of boron to five other atoms, and hence the concept of electron-deficient bonding must be invoked for boron. Although carbon seems to adhere to normal tetravalence, again it should be remembered

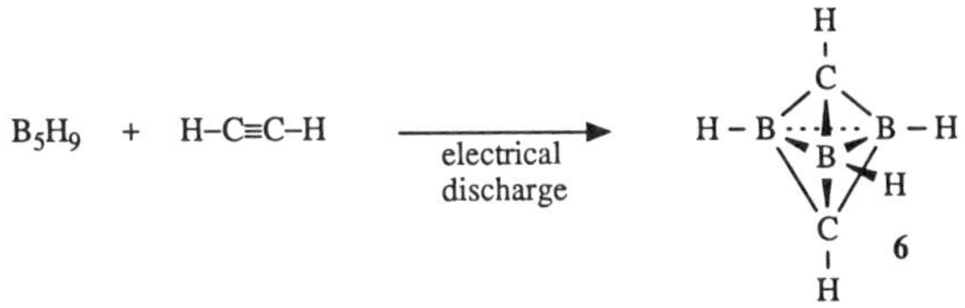

$$B_5H_9 + H{-}C{\equiv}C{-}H \xrightarrow[\text{discharge}]{\text{electrical}} \mathbf{6} \qquad (1)$$

that such B—C bonds as indicated in **6** are not ordinary two-electron bonds.[7] The existence of carboranes should alert chemists to the possibility that bond-carbon multiple bonding could involve both ordinary π-bonding of linear or planar systems and electron-deficient, three-center B—C—B bonding leading to three-dimensional clusters.

B. *Scope of This Survey in the Light of Previous Treatments*

This review will restrict itself to boron–carbon multiple bonding in carbon-rich systems, as encountered in organic chemistry, and leave the clusters of carboranes rich in boron to the proper purview of the inorganic chemist. Insofar as such three-dimensional clusters are considered at all in these review, interest will focus on the carbon-rich carboranes and the effect of ring size and substituents, both on boron and carbon, in determining the point of equilibrium between the cyclic organoborane and the isomeric carborane cluster. A typical significant example would be the potential interconversion of the 1,4-dibora-2,5-cyclohexadiene system (**7**) and the 2,3,4,5-tetracarbahexaborane(6) system (**8**) as a function of substituents R (Eq. 2).

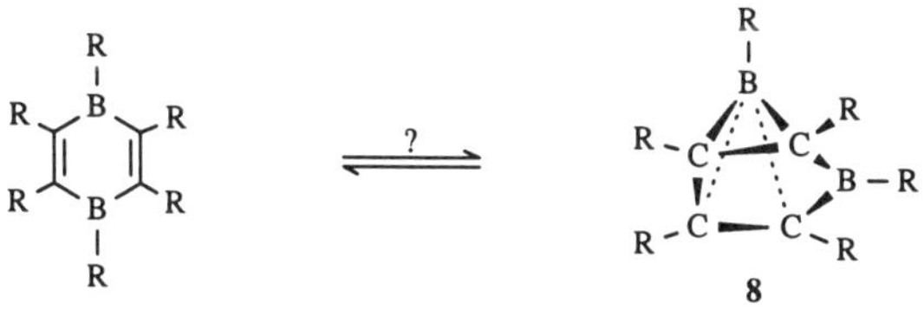

$$\mathbf{7} \overset{?}{\rightleftharpoons} \mathbf{8} \qquad (2)$$

The central problem for both cyclic and open-chain α,β-unsaturated boranes, such as 1-alkynyl- and 1-alkenylboranes, will be to employ experimental data discriminatingly and critically to assess what, if any, the π-bonding component of the actual C—B bond might be [*cf.* resonance structures **5a** and **5b**]. Where such unsaturation is geometrically constrained in a boracycloalkene (**9**) or in a boracyclopolyene (**10** or **11**), the fostering of boron-$2p_z$/carbon $2p_z$ orbital overlap may lead to enhanced B—C π-bonding, which may be stabilizing or destabilizing of the π-electron system, depending upon whether allylic-like or aromatic delocalization ensues (**9** and **10**) or whether Hückel antiaromaticity results (**11**).

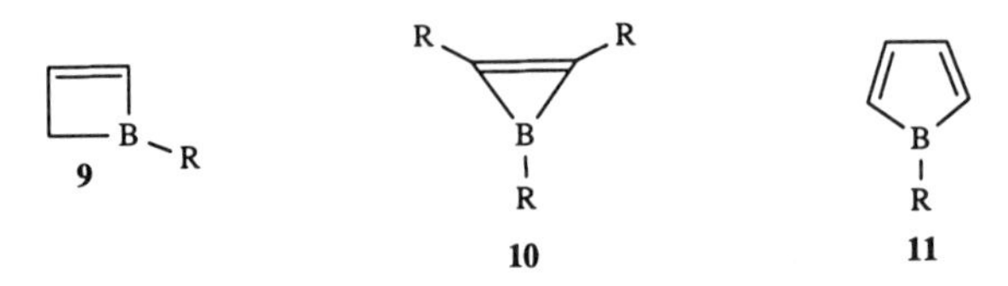

Two other types of π-interaction between boron and carbon centers that merit consideration are σ-bond hyperconjugation (**12a** and **12b**) and homoallylic delocalization (**13a**, **13b**, and **13c**, $n = 1,2, \ldots$), respectively.

Because of the necessity of spatially appropriate overlap of the $2p_z$-orbital on boron with the adjacent sp^3-hybridized carbon orbital (**12**) or with the π-MO of the double bond (**13**), any such π-bonding should be most sensitive to molecular conformation. Hence, the occurrence of such π-bonding might be more evident in certain highly rigid cyclic structures, such as **14** and **15**.

The assessment of π-bonding between boron and carbon in such structures as the foregoing has been the subject of several reviews, which have appeared periodically since the early 1960s.[8–10] Generally, a wide array of physical measurements, especially spectroscopic and diffraction data, has been scrutinized in these surveys, either to allow or to dismiss the importance of multiple B—C bonding. As will be seen, such judgments are not always easy or clear-cut, and the reality of this bonding will probably be, in part, abidingly controversial.[10]

C. *Theoretical Treatments of Boron–Carbon π-Bonding*

Consideration of the covalent radii and electronegativities (Allred–Rochow) of those first-row elements forming covalent bonds reveals a closer similarity in both these parameters for boron and carbon than for any other neighboring pair, except for that of nitrogen and oxygen (Table I). Substitution of boron for carbon in any hydrocarbon chain or ring would therefore cause only a modest perturbation in the covalent bonding. If one further substitutes a boron center for carbon while maintaining an isoelectronic configuration, then the appropriate substitution is sp^2-hybridized boron for a carbocationic center. Illustrative of this *Gedanken-experiment* are the transformations of the allyl (**16**) and cyclopropenyl (**18**) cations into a vinylborane (**17**) and a borirene (**19**), respectively (Eqs. 3 and 4).

$$H_2C{=}CH{-}\overset{+}{C}H_2 \ (\mathbf{16}) \longrightarrow H_2C{=}CH{-}BH_2 \ (\mathbf{17}) \quad (3)$$

$$\text{cyclo-}[HC{=}CH{-}\overset{+}{C}H] \ (\mathbf{18}) \longrightarrow \text{cyclo-}[HC{=}CH{-}BH] \ (\mathbf{19}) \quad (4)$$

Since carbon and nitrogen centers, with covalent radial and electronegativity differences of $\Delta r = 0.07$ and $\Delta\chi = 0.57$, are known to form π-bonds in C═N and C≡N linkages, some degree of π-bonding between neighboring sp^2-hybridized carbon and boron centers—as in **17** and **19** and as suggested by resonance structures **5a** and **5b**—has long been assumed.[9] With reference to C⩵B bonding, the covalent radial and electronegativity differences between carbon and boron centers are even smaller: $\Delta r = 0.04$ and $\Delta\chi = 0.49$. To this circumstantial evidence favoring

TABLE I

COVALENT RADII AND ELECTRONEGATIVITIES (ALLRED–ROCHOW) OF CARBON'S FIRST-ROW NEIGHBORS IN THE PERIODIC TABLE

Element	Li	Be	B	C	N	O	F
Covalent radius (Å)	1.23	0.89	0.81	0.77	0.70	0.66	0.64
χ_{AR}	0.97	1.47	2.01	2.50	3.07	3.50	4.10

boron–carbon multiple bonding can be added the experimental evidence gathered on the borazine system (**1**; *vide supra*), where the existence of multiple bonding between boron and nitrogen ($\Delta\chi = 0.11$ and $\Delta\chi = 1.06$) has been established beyond reasonable doubt. In light of these considerations, the abiding question with such boranes as **17** and **19** has not been whether boron–carbon multiple bonding *exists*, but rather, how significantly such π-bonding *affects* the structure and reactions of **17** and **19**. From this perspective, the only compelling and decisive answer to the question of boron–carbon π-bonding must come from experimental evidence that has been critically evaluated.

In setting the stage for such π-bonding in structures **17** and **19**, one essential geometric requirement must be met: the adjacent $2p$-orbitals on carbon and on boron must lie in the same plane for maximal overlap. This criterion is met of necessity in **19** because of the ring constraint, but not in **17** where rotation about the B—C bond can reduce π-overlap. Indeed, this structural change provides an experimental test of such π-bonding: to the extent that there is B—C π-bonding in **17**, rotation about the B—C bond will be restricted. This restriction may also be evident in the difference of the NMR signals for the two hydrogens at low temperatures.[11]

Theoretical treatments of B—C π-bonding of varying degrees of sophistication and refinement have been carried out since 1960 in order to estimate changes in π-electron density for such linear (**17**) and cyclic (**19**) systems, to calculate B—C bond rotational barriers and electronic spectra, to compute NMR chemical shifts, to assess delocalization energies for such organoboranes, and to determine potential energy surfaces for organoborane rearrangements. These approximate molecular orbital calculations (or, as some chemists prefer, computations) range all the way from the simple Hückel LCAO–MO method[12,13] through extended Hückel,[14] LCAO–SCF,[15] CNDO,[16,17] INDO[18] methods on to state-of-the-art *ab initio* MO calculations employing geometries fully optimized with the 6-31G* basis set.[19–21]

From the most humble, qualitative resonance depiction given by **5a** and **5b** to the advanced *ab initio* calculations of the activation energies of [1,3]-sigmatropic rearrangements of allylboranes[21] (Eq. 5),

$$\mathrm{H{-}C^{*}H{=}CH{-}CH_2{-}BR_2} \xrightarrow{\mathrm{H}^{\ddagger}} \mathrm{R_2B{-}C^{*}H_2{-}CH{=}CH{-}H} \qquad (5)$$

all of these theoretical models have heuristic value: they account for an experimental observation in a qualitative or quantitative manner *or* they predict a given experimental outcome in a general or precise way. Since practitioners of *ab initio* methods claim to provide quantitative estimates not only of ground-state geometries and relative energies of isomeric organoborane structures, but also of dynamic chemical and physical processes, their results merit not only a respectful hearing but also an equally impartial critique. Such calculations must not be granted sovereignty merely by virtue of the elegance of their mathematical techniques; rather, they must be judged only by their success in reproducing experimental results. Despite statements to the contrary,[21] theoreticians are less often leaders, and more often stimulating companions, of organoboron experimentalists. Experience has taught that neither one can move too far ahead of the other without getting lost.

D. *Experimental Criteria for the Importance of Boron–Carbon π-Bonding*

Being simply a kind of electron delocalization, boron–carbon π-bonding should manifest itself both in the structure and the chemical properties of organoboron compounds. The most useful experimental observations for the theoretician are naturally quantitative; thus, structural data from a variety of diffraction, spectroscopic, and spectrometric sources as well as chemical reaction information, such as kinetics, stereo- and regioselectivity, and detectable reaction intermediates, should be sought out as the proving ground for boron–carbon π-bonding.

1. *Structural Criteria*

In deciding on the contribution of π-bonding between boron and carbon centers, it is essential to distinguish between an inductive effect (*I*) and a resonance (delocalization or *R*) effect on B—C bond properties. The *I*-effect (**20**), which will partly determine the polarity of the B—C bond, depends on the *lowering* of the effective electronegativity of the boron by its substituents R and is exerted along the B—C bond axis *independent* of the conformation of all substituents on boron in **21**. The *R*-effect (**22**), on the other hand, is favored by a *heightening* of the electronegativity on boron by its substituents R and is strongly *dependent* upon the conformation of the ligands on boron.

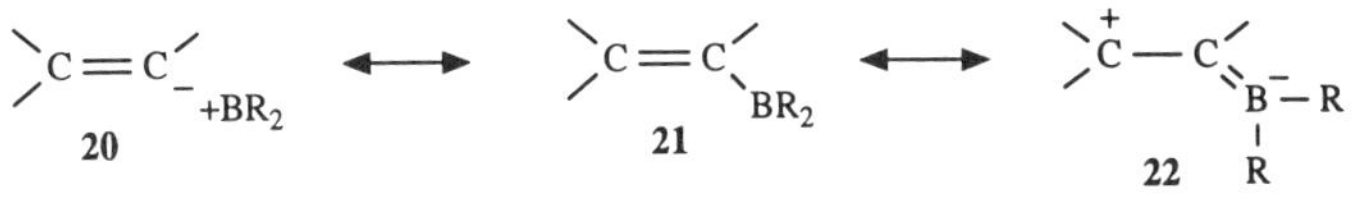

$$\gt C{=}C^{-}\ {}^{+}BR_2 \ (\mathbf{20}) \longleftrightarrow \gt C{=}C{-}BR_2 \ (\mathbf{21}) \longleftrightarrow \gt \overset{+}{C}{-}C{=}\overset{-}{B}R_2 \ (\mathbf{22})$$

For a given vinylic borane, therefore, a strong indication of an *R*-effect is to find some significant change in a structural parameter that is enhanced by electronegative R groups and is sensitive to conformation. Two such observable changes are the C—B bond distances and the C—B rotational barriers in vinylic boranes. Triarylborirene **23** and trivinylborane (**24**) have C—B bonds of 1.456 (av)[22] and 1.556 Å,[23] respectively; this difference is consistent with the strong dependence of the C—B bond separation on conformation. The boracyclopropene in **23** is of necessity planar and thus well-deposed for $2p_z$(B)–$2p_z$(C) overlap. Trivinylborane, on the other hand, has been shown by both IR and Raman spectroscopy to exist in freely interconverting planar and nonplanar conformers.[24]

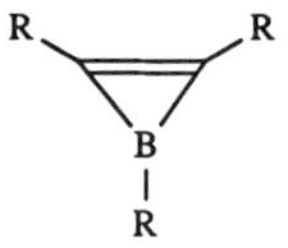

23 R = mesityl
R′ = 2,6-dimethylphenyl

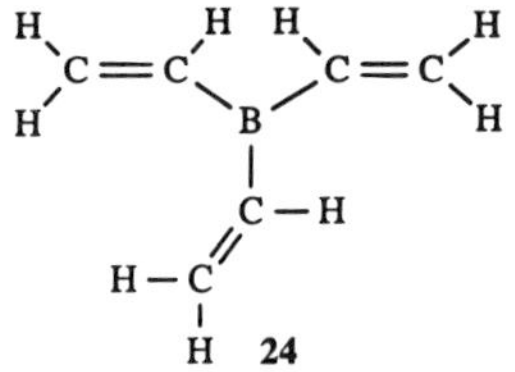

24

As to rotational barriers, that of **25** has been determined by microwave spectroscopy to be 17.4 kJ mol^{-1},[25] whereas the upper limit for **26** has been estimated at 4.2 kJ mol^{-1}.[11]

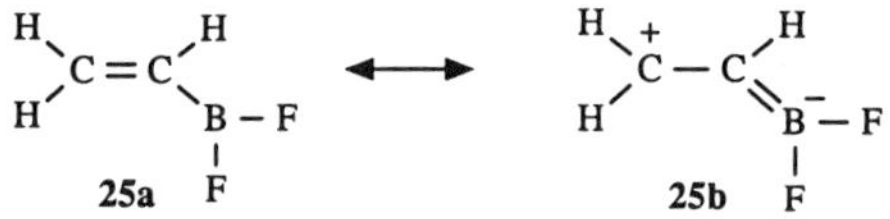

25a ⟷ **25b** 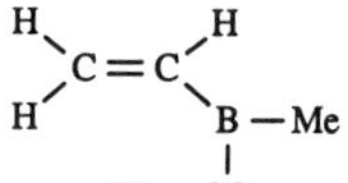   **26**

The higher barrier for **25a** is straightforwardly ascribable to the greater importance of **25b**. In further agreement with this is the shortening of the B—C bond in **25** to 1.536 Å[26] from its length in **24**, 1.556 Å.

Since the *R*-effect on boron–carbon π-bonding requires an sp^2- or tricoordinate boron for its operation, a second powerful experimental probe is to compare structural data for the tricoordinate borane $R^1R^2R^3B$ (**27**) with those for its Lewis complex $R^1R^2R^3B \cdot D$ (**28**), where D is some electron-pair donor, such as an amine R_3N or a carbanion R′. Any change in structural data between **27** and **28** can again arise from some combination of *I*- and *R*-effects (*cf*. **20–22**); but, taken together with the totality of other measurements on **27**, this change can then be attributed with considerable justice principally to the *R*-effect, or chiefly to the *I*-effect. Thus the change in the B—C bond length between Ph_3B (1.577 Å)[27] and that of the anionic complex Ph_4B^- (1.66 Å)[28] is most reasonably ascribed to the *I*-effect because the bond length of Ph_3B is almost the same as that of Me_3B (1.578 Å),[29] where π-bonding is not an obvious possibility. By contrast, the electronic spectrum of pentaphenylborole (**29**) displays a strong maximum at 607 nm, responsible for its deep blue color. Addition of weak or strong donors such as ethers, benzonitrile, or pyridine effects a marked hypsochromic shift in this maximum to 332 nm, resulting in a pale yellow color (**30**).[30] Here the preponderance of evidence points to the operation of an *R*-effect in **29**, leading to a Hückel antiaromatic delocalization (Eq. 6).

29a ⟷ **29b** —:D→ **30** (6)

Implicit in the foregoing discussion has been the assumption that structural data on bond angles and atom separations are the most important data for estimating boron–carbon π-bonding. The wide array of X-ray diffraction data for solids together with electron diffraction or microwave spectroscopic studies on volatile organoboranes in the gas phase offers the theoretician valuable data with which to evaluate various B—C bonding schemes. Next in utility are various kinds of vibrational (infrared and Raman), rotational (microwave), electronic (ultraviolet, visible, and photoelectron), and mass (electron bombardment and ion cyclotron resonance) spectroscopy, which permit assessment of fine structure details, such as bond rotational barriers, dipole moments, and excited states as well as parent, rearranged, and fragment ions.[10] Finally, multinuclear NMR data obtained by pulsed techniques are providing valuable data on ^{11}B, ^{13}C, ^{19}F, and ^{1}H nuclei in a variety of organoboranes.[31] Correlations of such data with boron's effective electronegativity, carbon hybridization indices, and the operation of possible *I*- and *R*-effects have been made; however, the utility of such data determining the extent of B—C π-bonding is neither straightforward nor clear.[32,33]

2. *Chemical Criteria*

The significance of B—C π-bonding in determining the course and the ease of chemical reaction with organoboranes is obviously a more difficult question for both the experimentalist and the theoretician. Again, both inductive and resonance effects can, in principle, play a role; but in addition, steric effects now become important, either for intramolecular reactions, as in Eq. (5), or for intermolecular reactions, as in Eqs. (7, 8).

$$CH_3-CH(R')(BR_2) \xleftarrow[\text{very minor}]{R_2BH} H_2C=CHR' \xrightarrow[\text{preponderant}]{R_2BH} R_2B\text{-}CH_2\text{-}CH_2R' \qquad (7)\ (8)$$

If the *R*-effect were operative in the ground state of a vinylic borane (**31a**), the effects would be to enhance the π-electron density of the α-carbon (**31b**) and to reduce C=C bond character (**31c**).

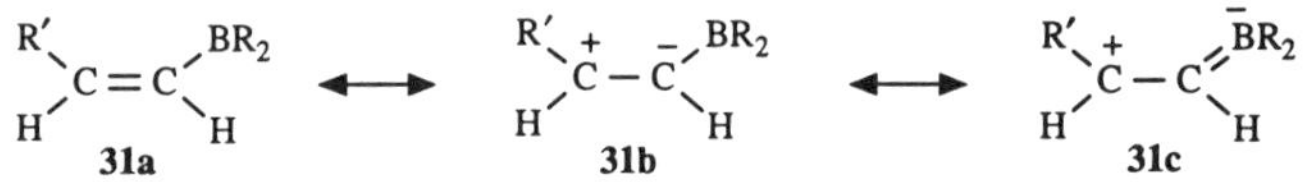

$$\mathrm{R'(H)C{=}CH(BR_2)}\ (\mathbf{31a}) \longleftrightarrow \mathrm{R'(H)\overset{+}{C}{-}\overset{-}{C}H(BR_2)}\ (\mathbf{31b}) \longleftrightarrow \mathrm{R'(H)\overset{+}{C}{-}CH{=}\overline{B}R_2}\ (\mathbf{31c})$$

The chemical effect would be to polarize the C=C bond and to foster addition of R_2BH to give **32**, an adduct formed regioselectively and *opposite* that expected on steric grounds. This mode of addition is indeed observed[34] (Eq. 9).

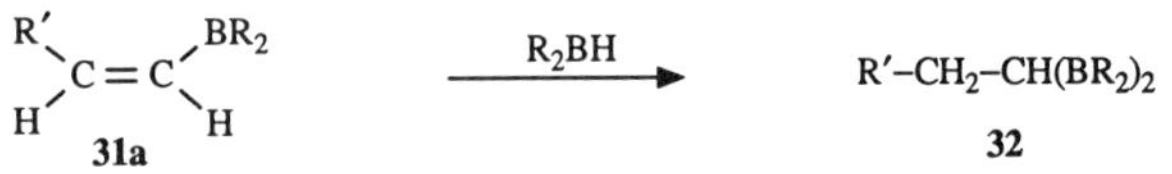

$$\underset{\mathbf{31a}}{\mathrm{R'(H)C{=}CH(BR_2)}} \xrightarrow{\mathrm{R_2BH}} \underset{\mathbf{32}}{\mathrm{R'{-}CH_2{-}CH(BR_2)_2}} \tag{9}$$

Furthermore, such boranes should undergo more facile cis, trans-isomerization to yield **33** as the contribution of **31c** becomes more important. To our knowledge, the ease of isomerizing (*Z*)-vinylboranes (**31a**) thermally to their *E*-isomers (**33**) has not yet been evaluated (Eq. 10), possibly because structures **31a** are not readily available.

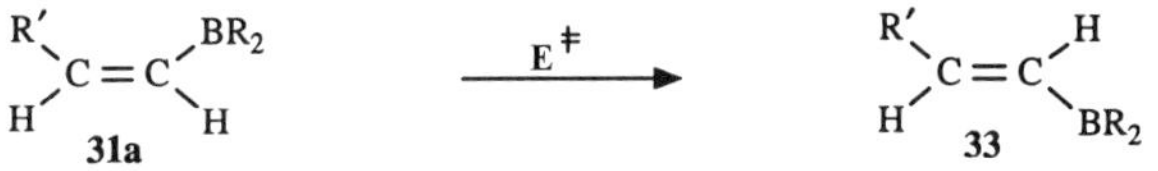

$$\underset{\mathbf{31a}}{\mathrm{R'(H)C{=}CH(BR_2)}} \xrightarrow{E^{\neq}} \underset{\mathbf{33}}{\mathrm{R'(H)C{=}C(H)BR_2}} \tag{10}$$

Thus, there is at least evidence that B—C π-bonding can influence both the pathway and the rapidity with which organoborane reactions (Eq. 9) take place.

The more challenging problem is to ascertain what quantitative effect such B—C π-bonding has on the activation energy ($E^{\neq}$) or enthalpy ($\Delta H^{\neq}$) of a reaction of an α,β-unsaturated organoborane. To answer this question, one has to assess the relative importance of inductive (*I*), resonance (*R*), and steric factors in both the ground state and the transition state of the reactants. Solution of the bimolecular case of two reactants seems to be beyond the realistic calculations of present *ab initio* methods. Hence, efforts in this area have been restricted to the unimolecular case, as in Eq. (5). When considering whether such π-bonding is reflected in bimolecular reactions of organoboranes, this review will offer explanations that are limited to qualitative arguments grounded in simple Hückel MO theory.

II

BORON–CARBON MULTIPLE BONDING IN OPEN-CHAIN UNSATURATED ORGANOBORANES

A. *Vinylboranes*

The detection of the structural effects of B—C π-bonding in unsaturated open-chain organoboranes should prove more difficult than in boracyclopolyenes for several reasons. First, as already mentioned, there is the conformational freedom of the open-chain boranes, which reduces $B(2p_z)$—$C(2p_z)$ overlap (*cf.* **17** vs. **19**). Second, there is the "linear" overlap of **34** versus the cyclic, Hückel aromatic overlap of **35**.

Third, substituents raising the effective electronegativity of boron will be expected to enhance B—C π-overlap (*cf.* **25a** and **26**). Finally, there can be competition between "linear" conjugation (**36**) and cross-conjugation (**24**).

Such cross-conjugation (where two of the vinyls are conjugated with each other through boron but not simultaneously with the third vinyl[35]) always lowers the total delocalization energy (DE) of the conjugated atomic set In simple Hückel MO theory, for example, (*E*)-1,3,5-hexatriene (**37**) has a DE of 0.988β, while 2-vinyl-1,3-butadiene (**38**) yields a DE of 0.900β.

The synergistic action of these four effects may explain why trivinylborane (**24**) shows no sign of B—C π-bonding but vinylboron fluorides (e.g., **25a**) do.

B. *1-Alkynylboranes*

For deciding on the importance of B—C π-bonding for alkyl-, 1-alkenyl, and 1-alkynylboranes, the most evident data of interest are the B—C and $C_1 \overset{...}{=} C_2$ bond separations exhibited by R_3B, $(RCH{=}CH)_3R$, and $(RC{\equiv}C)_3B$ systems. The decrease in the B—C bond separations in Me_3B (1.578 Å), $(H_2C{=}CH)_3B$ (1.558 Å), and $(mesityl)_2B{-}C{\equiv}C{-}mesityl$ (1.529 Å)[22] would seem to agree with an increase of π-bonding, as boron is successively bonded to an sp^3-, an sp^2- and finally to an sp-hybridized carbon center. The adjacent $C_1 \overset{...}{=} C_2$ bond does become longer in $(H_2C{=}CH)_3B$ (1.37 Å) but not noticeably longer in $(mesityl)_2B{-}C{\equiv}C{-}$ mesityl (1.217 Å) when compared with C=C and C≡C bonds in ordinary alkenes or alkynes. Thus, it can be concluded that the observed changes in the B—C bonds may be attributed to the *R*-effect in trivinylborane but not in dimesityl(mesitylethynyl)borane. In the latter, the shortened B—C bond may be induced by the *sp*-hybridization at C_1 but not by the *R*-effect. Similarly, an *I*-effect appears operative in shortening the B—C bonds (Å) in the following series: $MeBF_2$, 1.564; $H_2C{=}CHBF_2$, 1.532; and $H{-}C{\equiv}C{-}BF_2$, 1.513.[10] Thus, even though 1-alkynylboranes have no conformational restriction on B—C π-overlap, there seems to be no clear evidence of π-electron delocalization with the tricoordinate boron.

Even with the diborylacetylenes illustrated by structures **39a–c**, XRD data again show a shortening of the B—C bond as more electronegative groups are affixed to the boron centers but only a modest contraction of the C≡C bond[36]:

$$\underset{R'}{\overset{R}{>}}B-C\equiv C-B\underset{R'}{\overset{R}{<}}$$

39a–c

	a	**b**	**c**
	R,R′ = diisopropylamino	R = diisopropylamino R′ = t-butyl	R,R′ = o-phenylenedioxy
B—C (Å)	1.564(4)	1.557(4)	1.522(3)
C≡C (Å)	1.212(3)	1.212(5)	1.197(3)

These changes are more consistent with an *I*-effect, rather than an *R*-effect, on the B—C bond. That such diborylacetylenes have a diminished electron density at boron relative to alkyl- and vinylboranes is evident from their enhanced Lewis acidity: bis(diethylboryl)acetylene (**40**) forms a bis(tetrahydrofuran) complex, as is evident from the ^{11}B NMR spectrum[37] (Eq. 11). Alkyl- and vinylboranes exhibit no such shifts in their ^{11}B signals in THF solution.

$$Et_2B{-}C{\equiv}C{-}BEt_2 \ (\mathbf{40}) \xrightarrow{2\,THF} Et_2B{-}C{\equiv}C{-}BEt_2 \cdot 2\,THF \quad (11)$$

^{11}B NMR (C_6D_6), 72.2 ppm → ^{11}B NMR (THF), 36.7 ppm

C. *Arylboranes*

As already mentioned, B—C π-bonding in the ground state of Ph_3B is not reflected in X-ray diffraction (XRD) data, since the B—C bond length is essentially the same as that in Me_3B (1.578 Å). However, the ultraviolet spectra of triarylboranes (Ph_3B: maximum at 270 nm) compared with those of trialkylboranes (n-Bu_3B: maximum at 205 nm) have been interpreted as exhibiting bathochromic shifts due to charge-transfer (CT) transitions of the aromatic π-electrons into the boron $2p_z$-orbital.[38] A greater contribution of B—C π-bonding in the excited state of such arylboranes (**42** in Eq. 12)—over what exists in the ground state (**41**)—would reduce the energy of the CT transition and account for the bathochromic shift. As would be expected, addition of NH_3 to Ph_3B should remove the boron $2p_z$-orbital from conjugation and thus lead to a hypsochromic shift in the maximum ($Ph_3B \cdot NH_3$: 258 nm).[39]

$$\mathbf{41}\ (C_6H_5{-}BPh_2) \xrightarrow{h\nu} [\,(+)C_6H_5{=}\bar{B}Ph_2\,]^{*}\ \mathbf{42} \quad (12)$$

Such π-bonding is evident, however, in the ground state of triarylborane radical-anions. The Na^+ or K^+ salt of $[Ph_3B]^-$ (**43**) has the free spin densities shown in **43a**. Since fully 75% of the free spin is delocalized onto the phenyl groups, B—C π-overlap as expressed by structure **43b** is undoubtedly significant.[40]

43a: $[Ph_2\dot{B}{-}C_6H_5]^-$ (spin densities: B 7.84; ortho 1.99; meta 0.67; para 2.73) ⟷ **43b**: $[Ph_2B{=}C_6H_5\cdot]^-$

Such overlap requires a planarity between the phenyl ring and the trigonal, tricoordinate boron plane. Where that planarity is significantly disrupted by ortho substituents, as is evident in the XRD structure of lithium(12-crown-4)$_2$ cation–trimesitylboron anion,[41] only 65% of the free spin is delocalized onto the rings.[42]

The triphenylborane radical-anion itself has long been known to exist as a dimer (**44**),[43] but the structure of Krause's dimer, as it has come to be known, has been elucidated only recently.[44] A particularly stabilizing feature of **44**, it turns out, involves B—C π-bonding, for the union of two units of **43** forms an orange solid whose ^{1}H-NMR spectrum displays a telltale methine H at 4.03 ppm (* in **44**; Eq. 13).

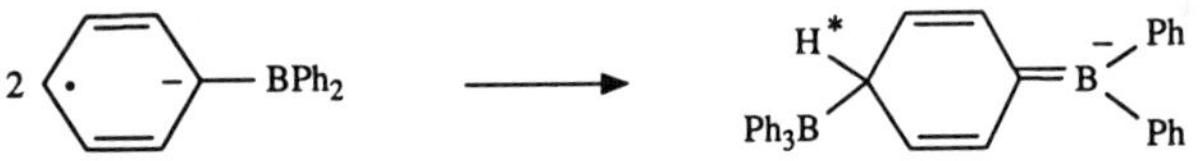

2 **43c** (BPh$_2$) ⟶ **44** (H*, Ph$_3$B, B, Ph, Ph) (13)

D. *Allyl- and Benzylboranes*

The intramolecular rearrangement of allylic boranes (Eq. 5) clearly involves a multiple boron–carbon bond in the transition state (**45**), as the boron $2p_z$-orbital interacts with the π-bonding MO.

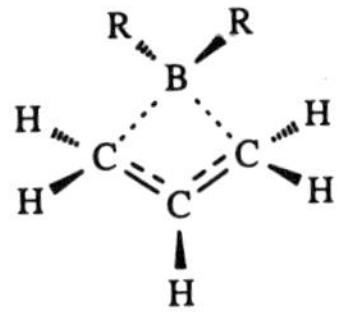

45

Lewis basic substituents on boron, themselves capable of π-bonding with boron, such as MeO and Me_2N, drastically slow down the rearrangement rate[45] by increasing the activation energy to attain $[\mathbf{45}]^{\ddagger}$. An *ab initio* calculation has estimated a $\Delta E^{\neq}$ of 9.2 kcal, which falls at the lower limit of experimental values (10–15 kcal/mol).[21]

Evidence of such homoallylic delocalization has already been gleaned from the spectral properties and the facile allylic rearrangements of 7-borabicyclo[2.2.1]heptadiene (**46**).[46] The most telling evidence supporting the boron $2p_z$-orbital's interaction with at least one of the C═C ring bonds is the hypsochromic shift in the UV spectrum of **46a** upon complexation with pyridine (Eq. 14) to form **47**. As in the UV spectrum of Ph_3B (Eq. 12), the absorption at 318 nm can be attributed to a CT transition from the C═C bond to the boron $2p_z$-orbital to form an excited state (**46b**), whose multiple B—C π-bonding has stabilized this state and lowered the CT transition energy (Eq. 15).

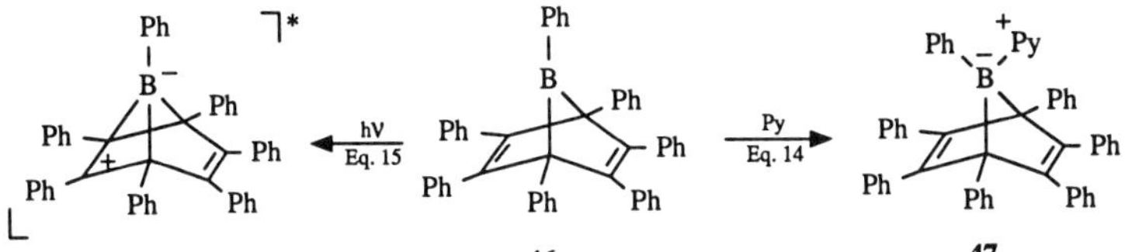

46b ⟵(hν, Eq. 15)— 46a —(Py, Eq. 14)⟶ 47 (14) (15)

46a: $\lambda_{max} = 318$ nm   47: $\lambda_{sh} = 273$ nm

In further corroboration of the enhanced electron density at boron in **46** due to the interaction of the B $2p_z$-orbital and the C=C bonds, is the ^{11}B NMR signal at −5.0 ppm. This is a huge upfield shift from the ^{11}B signal in Ph_3B (60 ppm) and is even more upfield from the signal of the tetracoordinate complex **47**, which occurs at 2.8 ppm. Such shielding is consistent with a boron $2p_z$-orbital interaction with both π-bonding MOs simultaneously without inclination of the Ph—B bridge toward either C=C bond (**46c**).

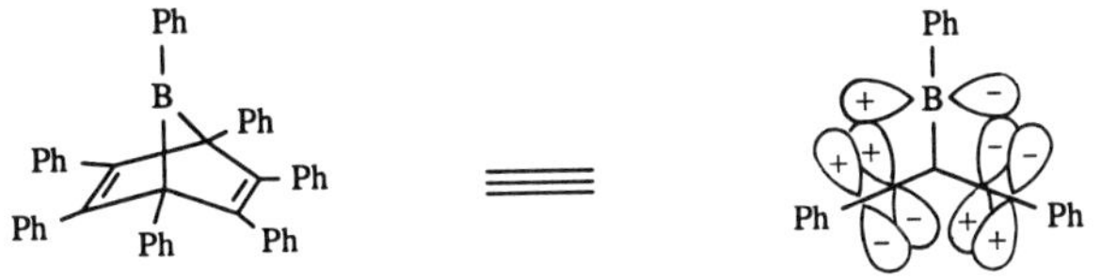

46a ≡ 46c

A similar situation obtains in a series of para-substituted tribenzylboranes (**48**) bearing X = H, F, Me, and MeO groups[47]; the UV spectra of **48** exhibit a CT band of medium intensity in the region of 240–285 nm. Such an assignment is supported by the existence of a linear correlation of the CT transition energies of the boranes with the ionization potentials of Ph—X (Eq. 16).

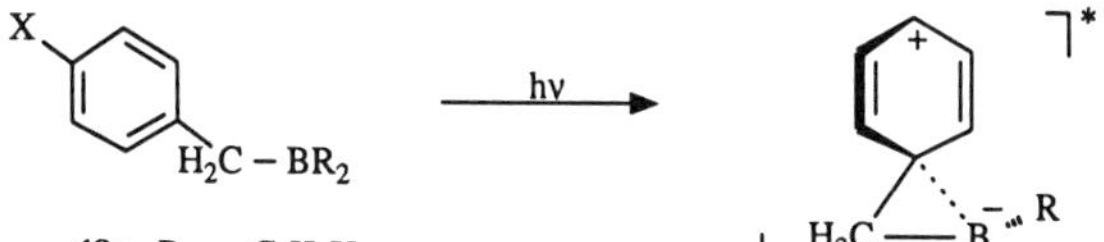

48a, R = p-C_6H_4X —(hν)⟶ 48b (16)

Consistent with the involvement of the boron $2p_z$-orbital in this transition is the almost total reduction of the long-wavelength absorption of $(PhCH_2)_3B$ above 250 nm by its complexation with ammonia. Strong chemical support for a multiply bonded boron in the excited state (**48b**) can be deduced from photolysis results with tribenzylborane (**49**).[48] While photolysis of

49 in ether under an inert atmosphere yielded 38% of toluene and 15% of bibenzyl, inclusion of methanol raised the yield of toluene to 82% with little change in the proportion of bibenzyl (10%). We suggest that the increased yield of toluene stems from the protonolysis of the zwitterionic, boratocyclopropane-like excited state **48b**, whose polar character and strained ring should provide for a high reactivity. Consider the ease with which even silacyclopropanes (**50**) are cleaved with methanol[49] (Eq. 17).

Mes, Mes, Me, Me, Si, Me, Me — **50** —MeOH, 20°C→ Me, Me, C, H, Me_2Si, OMe, Me, Me (17)

E. *Methyleneborates and Methyleneboranes*

1. *Methyleneborates*

The mental substitution of an olefinic carbon with tricoordinate boron would give rise to boraethenes or methyleneboranes (**51**); a similar substitution with a tetracoordinate borate anion would produce borataethenes or methyleneborates (**52a**), which are but a resonance representation of borylcarbanions (**52b**).

$>C=B-$ **51** $\quad >C=\bar{B}<$ **52a** $\longleftrightarrow$ $>\bar{C}-B<$ **52b**

A priori borataethenes would be expected to be much more stable than the corresponding boraethenes because, as Rundle presciently noted about organometallic structures,[50] "all low-lying atomic orbitals are involved in the bonding." In fact, the stabilization of **52** is the driving force for the acidity of C—H bonds of proper orientation α to tricoordinate boron. Such acidity was first detected by Rathke and Kow through the treatment of B-methyl-9-borabicyclo[3.3.1] nonane (**53**) with hindered lithium amides, such as lithium 2,2,6,6-tetramethylpiperidide[51] (Eq. 18):

CH_3, B, H*, H* — **53** —$LiNR_2$, $-HNR_2$→ H, H, $\bar{C}$, Li^+, B, H, H (18)

Note that by the investigators' design, the bridgehead protons (*) are not acidic, since any resulting carbanion would be orthogonal to the boron $2p_z$-orbital and thus incapable of π-bonding stabilization.

The acidity of such $R_2B—CH_2—R'$ systems has been utilized as the basis for developing the so-called boron-Wittig reaction for organic synthesis.[52–54] One of the distinct advantages of such reagents, the direct conversion of an aliphatic aldehyde to a ketone, is illustrated in Eq. (19).[55]

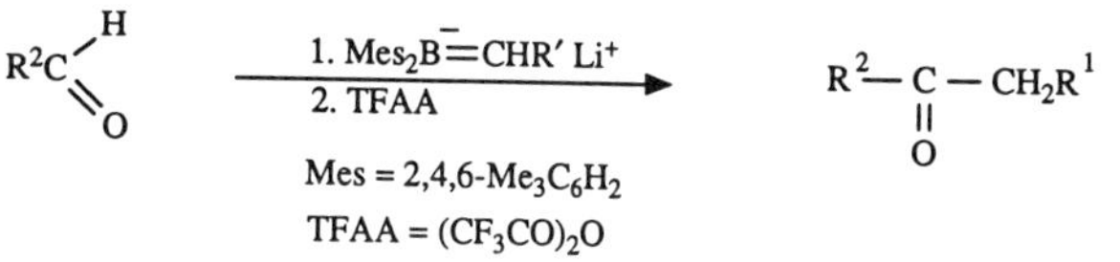

$$R^2C(H)=O \xrightarrow[2.\ TFAA]{1.\ Mes_2\bar{B}=CHR'\ Li^+} R^2—C(=O)—CH_2R^1 \quad (19)$$

$Mes = 2,4,6\text{-}Me_3C_6H_2$
$TFAA = (CF_3CO)_2O$

Any residual doubt about the actual π-stabilization of such carbanions, that is, the question concerning the importance of resonance structure **52a**, has been decisively banished by an XRD analysis of the lithium bis(12-crown-4) salt of dimesityl(methyl)borane (**54**).[56] Noteworthy are the slightly elongated mesityl C—B bonds at 1.617 Å (over those in Mes_2BMe at 1.586 Å) and the pronouncedly shorter methylene C—B bond at 1.444 Å. Moreover, the C_2BCH_2 core atoms lie essentially in a plane. All these data support the view of **54** as a borataethene, just as structure **52a** would imply.

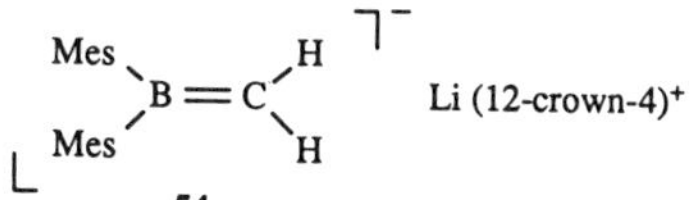

$$[Mes_2B=CH_2]^-\ Li(12\text{-crown-4})^+$$

54

2. *Methyleneboranes*

The heightened reactivity of these neutral boraethenes undoubtedly stems from the available boron $2p$-orbital and the greater electronegativity of the sp-hybridized, dicoordinate boron nucleus. The resulting increase in electrophilicity causes the boron center to accept electron density from external reagents or even from geometrically accessible groups within the boraethene molecule. Therefore, in order to synthesize and isolate such derivatives, chemists have had to provide for kinetic stabilization of such systems, either by affixing donor ligands to the system (e.g., **55a–55b**)[57,58] or by impeding disruptive intermolecular reactions by the tried-and-true strategem of employing bulky substituents (**56**).[59,60]

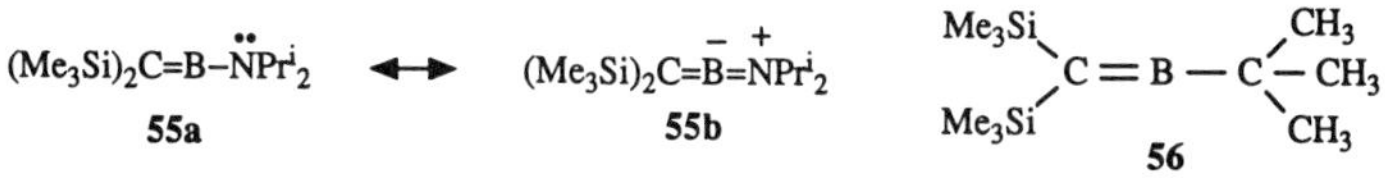

$$(Me_3Si)_2C=B-\ddot{N}Pr^i_2 \longleftrightarrow (Me_3Si)_2C=\overset{-}{B}=\overset{+}{N}Pr^i_2 \qquad (Me_3Si)_2C=B-C(CH_3)_3$$

55a **55b** **56**

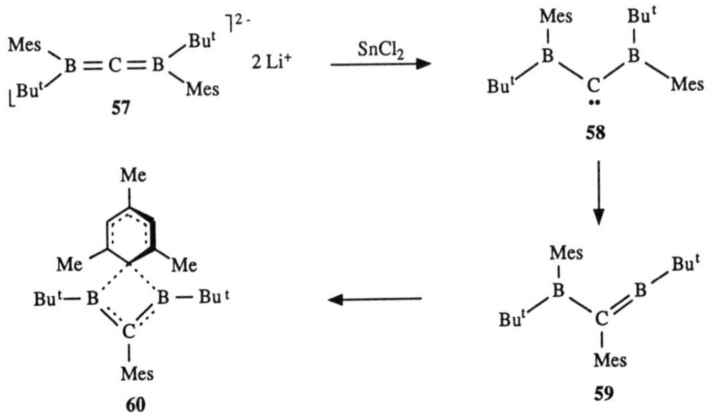

SCHEME 1

Even after the boron's electrophilicity has been arrested intermolecularly, intramolecular reorganization of the boraethene may still proceed. A remarkable instance is the structure **60** (by XRD) resulting from the oxidation of **57** by $SnCl_2$. Intermediates in this impressive transformation might well be carbene **58** and boraethene **59**. The borons in **59** are still sufficiently electrophilic to cause the mesityl group to bridge (Scheme 1).[61]

Such astonishing rearrangements as the foregoing have been found by Berndt and co-workers to abound both in synthetic routes to boraethenes and in reactions with a variety of substrates. To recount these fascinating observations here would take us rather far afield from our current focus on B—C multiple bonding. Fortunately, Berndt has recently presented an authoritative, thorough review of progress in this area, with an emphasis on the importance of σ–π delocalization in understanding the structural and chemical nature of so-called "nonclassical" methyleneboranes, of which **60** may be considered prototypical.[62]

What is of immediate interest are the various XRD determinations of B—C bond lengths in "classical" methyleneboranes, exemplified by **55** and **56**, which are 1.391 and 1.631 Å, respectively.[60] Furthermore, the C═B—C linkage is essentially linear (179.6(3) Å). The observed B—C bond length is only somewhat longer than the C═C bond in a trisubstituted ethene (1.35 Å). Thus there is no doubt that some boraethenes have a genuine B═C bond.

Closely related to boraethenes is the structure of carbanions derived from boraethene **61** (Eq. 20).

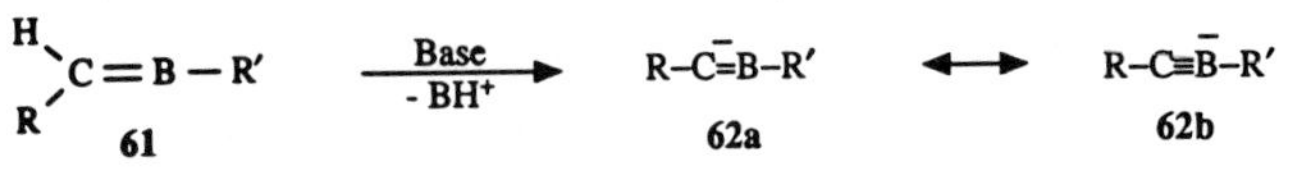

$$\underset{\mathbf{61}}{\text{H(R)C=B–R}'} \xrightarrow[\text{- BH}^+]{\text{Base}} \underset{\mathbf{62a}}{\text{R–}\bar{\text{C}}\text{=B–R}'} \longleftrightarrow \underset{\mathbf{62b}}{\text{R–C}\equiv\bar{\text{B}}\text{–R}'} \qquad (20)$$

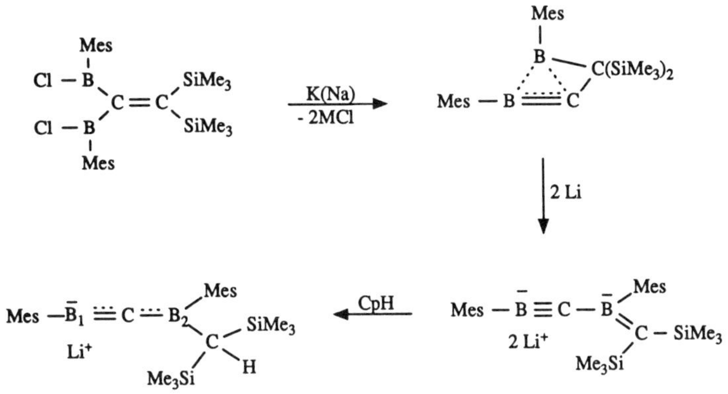

SCHEME 2

A derivative of **62** has been prepared, as shown in Scheme 2, and an XRD analysis of lithium salt **63** has been carried out.[63] The mesityl B_1—C bond length is 1.339 Å and the mesityl B_2—C bond is 1.492 Å, while the B_1—C—B_2 linkage is almost linear (1.765(5) Å). These findings support the conclusions that the B_1—C bond has considerable triple bond character, since it is 0.10 Å shorter than that in the $Mes_2B{=}CH_2$ anion (**54**) and the B_2—C bond has significant double bond charracter, being about 0.09 Å shorter than a B—C bond.

F. *Diborane(4) Derivatives*

Although π-bonding between two adjacent boron centers does not fall strictly within the present discussion, an inquiry into such bonding is prompted by our earlier treatment of boraethenes (**51**) and borataethynes (**62**). Before searching for derivatives of "diboraethene" (**63**) or "diborataethyne" (**64**), we should recall that derivatives of "diboraethane" (**65**), properly known as diborane(4), are well-known.[64]

H–B=B–H **63** $[R{-}B{\equiv}B{-}R]^{2-}$ **64** $H_2B{-}BH_2$ **65**

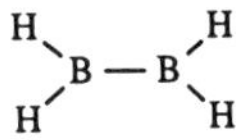

The availability of stable tetraorgano derivatives of **65**[65,66] has opened the way to synthesizing B—B multiple bonds. Treating 1,2,2-trimesityl-2-phenyldiborane(4) (**66**) with lithium metal in ether yielded **67** as dark red crystals (Eq. 21).

$$\text{Mes}_2\text{B—B(Mes)(Ph)}\ (\mathbf{66}) \xrightarrow[\text{Et}_2\text{D}]{2\ \text{Li}} 2\,[\text{Li}\cdot\text{Et}_2\text{O}]^+ \left[\text{Mes}_2\text{B}_1{=}\text{B}_2(\text{Mes})(\text{Ph})\right]^{2-}\ (\mathbf{67}) \tag{21}$$

A comparison of the XRD data for **66** and **67** reveals very significant structural differences: (1) The angle between the two planes containing, respectively, the mesityl-C—B_1—C-mesityl centers and the mesityl-C—B_2—C-phenyl centers changes from 79.1° in **66** to 7.3° in **67**; (2) the B—B bond distance of 1.706 Å in **66** is contracted to 1.636 Å in **67**; (3) the B—C bonds are lengthened in **67**; and (4) the C—B—C angle (6–11°) becomes smaller in **67**.[67] All such changes are consistent with a B—B π-bond that draws together the boron centers and makes the inner $C_2B{=}BC_2$ core of nuclei nearly coplanar. This planarity would increase the steric repulsion of the aryl groups, thereby leading to the observed changes in bond lengths and angles.

A worthwhile comparison can be made between **67** and the 2,3-diborata-butadiene dianion depicted in **68**.

$$\left[(\text{Me}_3\text{Si})_2\text{C}{=}\text{B(Mes)—B(Mes)}{=}\text{C}(\text{SiMe}_3)_2\right]^{2-}$$

68

In accord with the resonance structure drawn, there is little B=B bonding in this diborane(4) derivative and the B—B distance is found to be 1.859 Å,[68] which is considerably longer than even the normal expected value of 1.7 Å for a boron–boron single bond.[67] This is consistent with the normal repulsion of negative charges on adjacent atoms *not* stabilized by π-bonding.

III

BORON–CARBON MULTIPLE BONDING IN BORACYCLOPOLYENES

A. *The Question of Hyperconjugation in Boracycloalkanes*

The previously raised possibility of hyperconjugation for alkylboranes (**12a**↔**12b**) becomes of greater concern for the conformationally constrained rings of boracycloalkanes. The search for unambiguous physical evidence for hyperconjugation even in Me_3B (**12a**↔**12b**) has been inconclusive,[10] and theoretical treatments have both affirmed (extended Hückel)[69] and denied (*ab initio* and INDO)[70] its importance in describing the B—C bond. A further examination of available structural data for boriranes (**69**),[71,72] dihydroboretes (**70**),[73] 1,2-diboretanes (**71**),[74] and 1,3-diborolanes (**72**)[75] does not help to resolve this question.

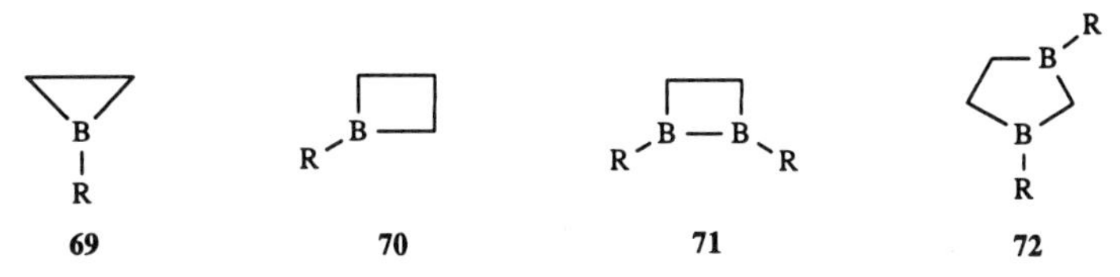

Perhaps the most convincing evidence favoring the operation of hyperconjugation with C—H bonds α to tricoordinate boron is chemical in nature: namely, the demonstrated acidity of B-methyl-9-borabicyclo[3.3.1]nonane (**53**, Eq. 18). At a minimum, this acidity requires that hyperconjugation be important in stabilizing the transition state (**73**). It would be reasonable to conclude that some degree of hyperconjugation must be important in the ground state of **53** as well, aiding the stretching of the C—H bond.

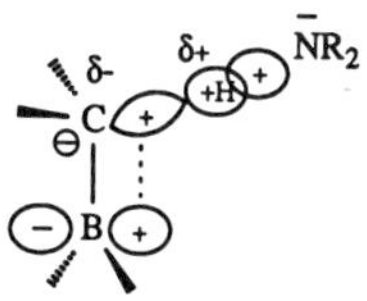

δ- δ+ $\bar{N}R_2$ C +H B

73

B. *Borirenes*

Over the last decade we have seen a variety of successful synthetic routes to the boron analog of the cyclopropenium ion, the boracyclopropene, or borirene nucleus (**74**). Illustrated examples are presented in Eqs. (22–25).[76–80]

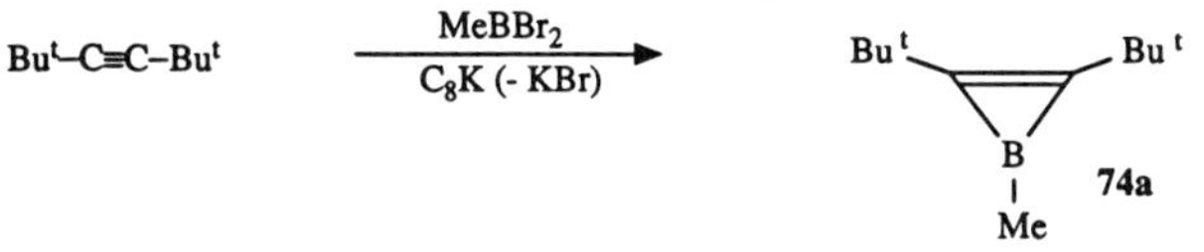

$$Bu^t{-}C{\equiv}C{-}Bu^t \xrightarrow[C_8K\ (-\ KBr)]{MeBBr_2} \textbf{74a}\ (Bu^t,\ Bu^t,\ B{-}Me) \tag{22}$$

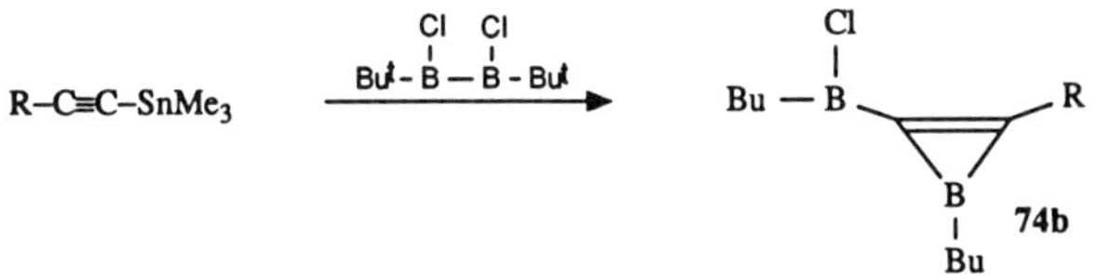

$$R{-}C{\equiv}C{-}SnMe_3 \xrightarrow{Bu^t{-}B(Cl){-}B(Cl){-}Bu^t} \textbf{74b}\ (Bu{-}B(Cl),\ R,\ B{-}Bu) \tag{23}$$

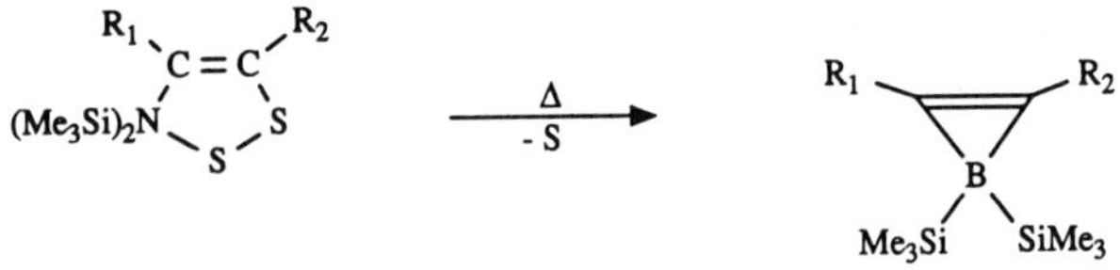

(24)

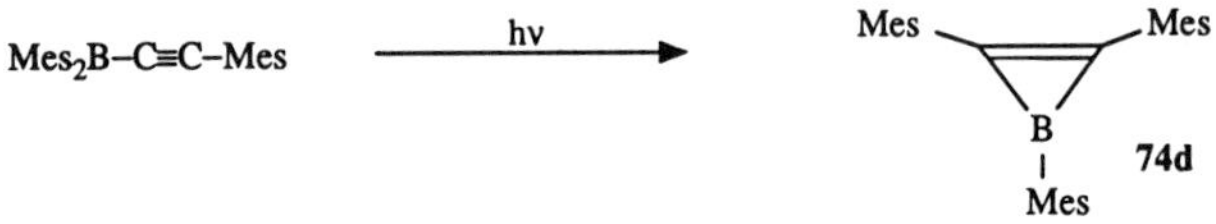

(25)

Despite appropriate nuclear spectroscopic and mass characterization of **74a–d**, the crucial question of the π-electron delocalization and possible aromatic character of the borirene ring remained unanswered. Although the photocyclized product **74d** produced well-formed crystals suitable for XRD analysis, crystal disorder arising from the high symmetry of the molecule prevented an unambiguous differentiation of boron and carbon atoms in the central ring. Hence, it could be concluded the C═C bond was lengthened and the B—C bond shortened, but not to what extent. Therefore, to disrupt such high symmetry, borirene **75** was then synthesized (Eq. 26).[81]

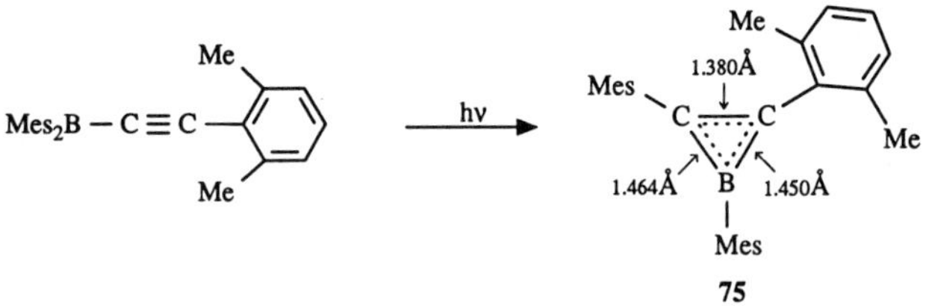

(26)

The resulting crystal proved amenable to a conclusive XRD analysis. As shown in Eq. (26), the C—C ring bond is lengthened over what it is in the structurally similar cyclopropene (1.304 Å) and the B—C bonds shortened relative to the electronically analogous bond in trivinylborane (1.558 Å). Thus, one can safely conclude that there is extensive π-electron delocalization and Hückel aromatic character in the borirene ring.

In view of the highly strained σ B—C bonds of **74–75**, there is experimental evidence that such π-bonding is actually responsible for the stability of the ring to cleavage. For example, addition of pyridine to colorless **74d** generates an intense yellow color and many new resonance signals in the ^{13}C-NMR spectrum consistent with isomeric zwitterions **76a**,**b**. Addition of *t*-BuOH (which does not itself react with **74d**) yields borinate ester **77** (Scheme 3). These results are consistent with the interpretation

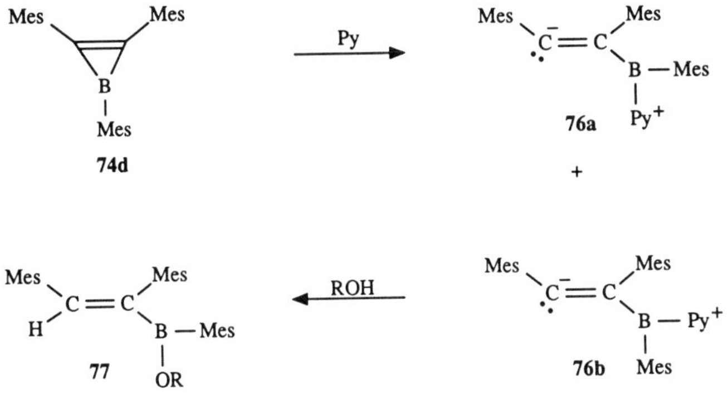

SCHEME 3

that coordination of pyridine with the boron in **74d** disrupts the stabilizing bonding in **74d** and causes ring opening to **76a** and **76b**. In fact, this interpretation helps explain previous puzzling observations made in this author's laboratory.

First, in an earlier attempt to prepare triphenylborirene (**81**), diphenylacetylene (**78**) was bromoborated with $PhBBr_2$ in high yield to form **79**. The structure of **79** was verified by protonolysis to **80**. Then **79** was treated with lithium metal at low temperature in the hope of producing **81**. Strong indication that **81** had indeed been formed came from treating the solution with DOAc and isolating a high yield of α,α'-dideutero-*cis*-stilbene (**82**) (Scheme 4).[82,83] But then pyridine was introduced in what we now recog-

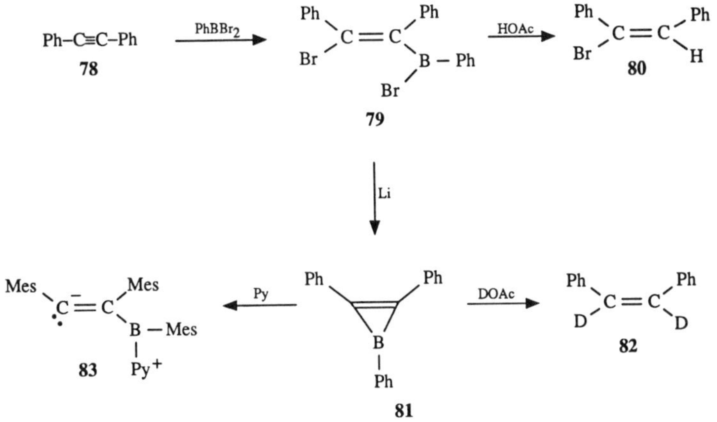

SCHEME 4

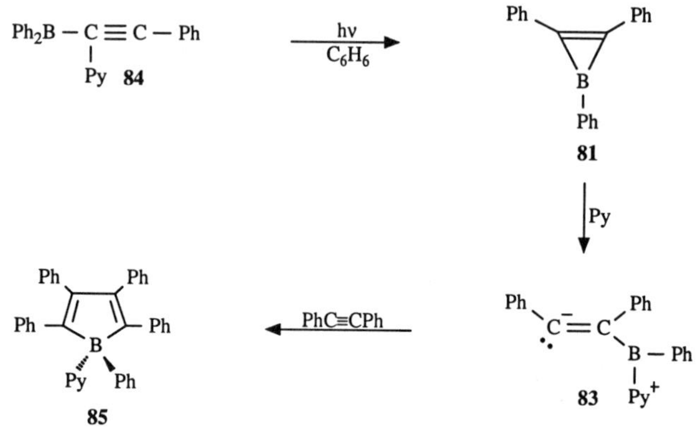

SCHEME 5

nize as a misguided attempt to isolate **81** as a pyridine complex. Instead, pyridine undoubtedly formed a mixture of zwitterions **83**, which could have formed oligomeric (PhC=CPh—BPh)$_x$.

A second observation stems from the photorearrangement of diphenyl (phenylethynyl)borane-pyridine (**84**). Again, treatment of the photolysate solution with DOAc gave high yields of **82**; but still, efforts to isolate **81** or its pyridine complex were unavailing. With the growing suspicion that the formation of **83** was the cause of our frustration, we attempted to trap **83**, a potential dipolarophile, with **78** in a 1,3-dipolar cycloaddition. Gratifyingly, the photolysis of **84** in the presence of **78** gave about a 35% yield of the expected cycloaddition product, pentaphenylborolepyridine (**85**)[84] (Scheme 5). There is, accordingly, no remaining doubt that even **81** ring-opens to a zwitterion in the presence of pyridine.

Electronically related to such borirenes are the salts of the dianion of 1,2-di-*tert*-butyl-3-[bis(trimethylsilyl)]methyl]-1,2-diborirane.[85] An XRD analysis of the dipotassium salt (**86a**,**b**) uncovered shortened B—B (1.58 Å) and B—C (1.50 Å) bonds, as suggested by resonance contributions **86a** and **86b**. Salt **86** can be considered to have largely a 2-π-electron Hückel aromatic ring.

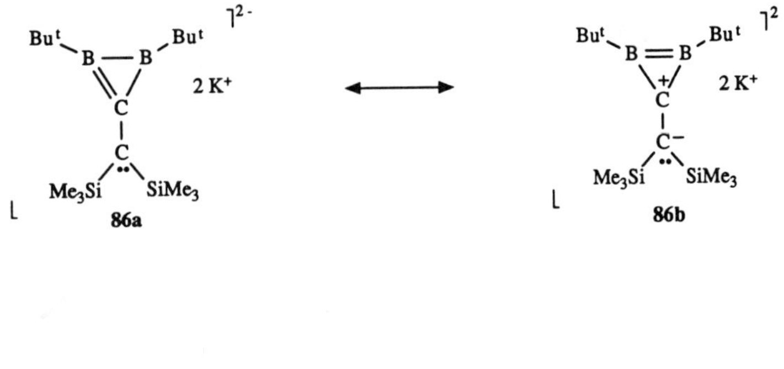

C. *Dihydroboretes and Dihydrodiboretes*

These systems are pertinent to the present review because, depending on the number and position of the boron centers, B—C multiple bonding can be involved. The 1,2-dihydroborete (**87**)[86] possesses a nonplanar borete ring with a fold angle of 148.8° along the B—C_3 diagonal, and its structure can fruitfully be compared with the planar 1,2-dihydro-1,2-diborete (**88**).[87]

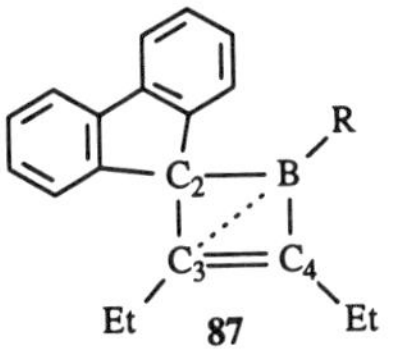

The C_3—C_4 distance in **87** is 0.064 Å longer than the C=C bond in **88** and the B—C_4 bond in **87** is shortened by 0.053 Å over what it is in **88**. These changes support a strong B—C_3 bond interaction that would justify viewing **87** as a homoborirene. Similar arguments applied to **88** itself would lead to the conclusion that there is little, if any, aromatic delocalization of the two π-electrons into the boron $2p_z$-orbitals. The C=C bond at 1.31 Å is even a little shorter than that of cyclobutene (1.342 Å). Apparently, electron donation to boron by nitrogen interferes with cyclic conjugation.

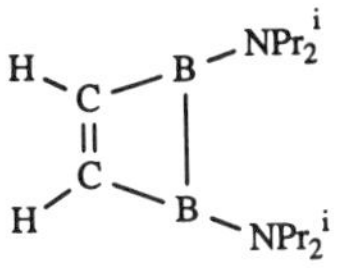

Attempts to synthesize other 1,2-dihydro-1,2-diboretes led to a remarkable rearrangement to 1,3-dihydro-1,3-diboretes, most likely via the 1,2-diborete (Eq. 27).[88]

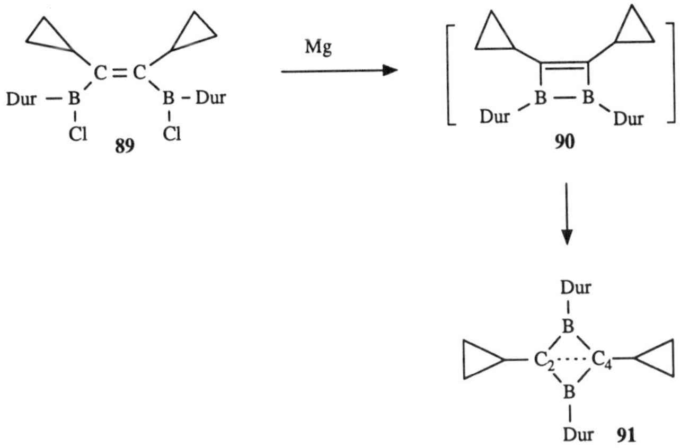

(27)

The resulting 1,3-dihydro-1,3-diborete **91** has a nonplanar structure with a 45° folding angle and an unusually long C_2—C_4 bond (1.965 Å). Such structural distortions are occasioned by delocalization of the two ring electrons among the four-atom nuclei. Both the instability of **90** and the structure of **91** were anticipated by *ab initio* calculations.[89] In keeping with π-electron delocalization in the ring, the B—C bonds are 1.504–1.515 Å in length. 1,2-Diboretes bearing diisopropylamino groups on boron show similar molecular dimensions.[90]

It might be amusing to conclude this section by noting the structure that a "1,2,3,4-tetrahydro-1,2,3,4-tetraborete" assumes. Treatment of *tert*-butylboron difluoride with sodium–potassium alloy (Eq. 28) gives a derivative (**92**) of such a system, whose XRD analysis reveals that the nuclei assume a maximally "folded" structure—namely a tetrahedron.[91]

$$4\ Bu^tBF_2 \xrightarrow{Na(K)} (Bu^tB)_4 \quad \mathbf{92} \tag{28}$$

D. *Boroles*

The first authenticated borole (**94**) was made by the exchange reaction of phenylboron dichloride with 1,1-dimethyl-2,3,4,5-tetraphenylstannole (**93**) (Eq. 29)[92] in toluene.

$$\mathbf{93} \xrightarrow[-\ Me_2SnCl_2]{PhBCl_2} \mathbf{94} \tag{29}$$

The resulting deep blue, air-sensitive solid was then fully characterized as pentaphenylborole (**94**) and stood in sharp contrast to the reported properties of pseudo-borole **95** as an air-stable, pale yellow solid,[93] prepared according to Eq. (30).

$$\underset{\text{(in ether)}}{\mathbf{96}} \xrightarrow[-\ LiCl]{PhBCl_2} \mathbf{97} \xrightarrow{-\ Et_2O} \mathbf{94} \tag{30}$$

Our repetition of this procedure with due provision during workup for the air- and moisture-sensitive nature of **94** also permitted us to isolate the same deep blue borole (**94**). It is pertinent that the initial reaction produced a pale yellow ether complex **97** and the blue **94** was formed after ether removal *in vacuo*.

However, the pseudo-borole **95** stemmed from another reaction of **96**. Unknown to the earlier researchers who reported the synthesis of **95**,[94] their attempted preparation of **96** instead gave partly **98**, a rearrangement observed previously,[95] whereupon the reaction of **98** with $PhBCl_2$ yielded phenyl(bis(2,3,4-triphenyl-1-naphthyl)borane, the actual identity of pseudo-borole **95** (Eq. 31).

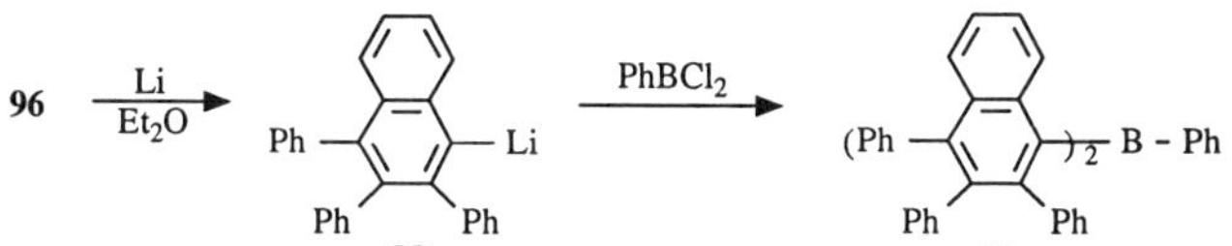

(31)

A third approach to the borole system bearing alkyl substituents (**99**) has recently been developed[93]; but because these substituents are smaller, the Diels–Alder dimer (**100**) is the actual product (Eq. 32).

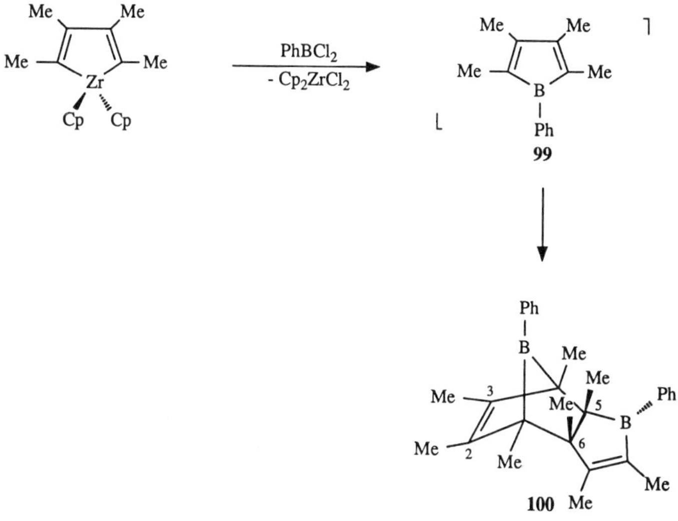

(32)

Of enormous interest to our present discussion would be XRD data on monomeric boroles in order to determine the structural consequences of π-electron delocalization in this four-electron, formally antiaromatic system.[30] However, many attempts to grow suitable crystals of **94** have failed, and prolonged storage of solutions has produced only yellow dimers, presumably similar in structure to **100**. Hence, in estimating the relative importance of structures **94a–c**, spectral data must be our guide.

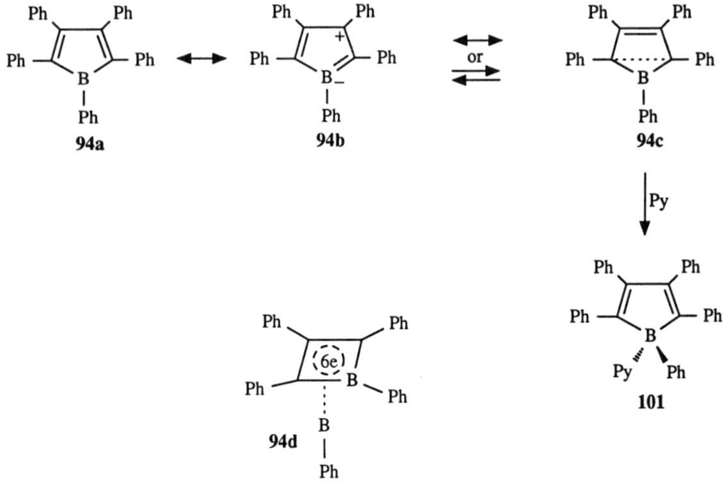

SCHEME 6

The two most striking spectral changes are those between **94** and its pyridine adduct **101** (Scheme 6): (1) **94** is deep blue while **101** is pale yellow; (2) the ^{11}B resonance at 55 ppm in **94** is shifted upfield to 5 ppm in **101**. Since triphenylborane (**102**) exhibits its ^{11}B signal at 60 ppm, it is reasonable to conclude that both have comparable π-electronic density on boron. Resonance structure **94b** should then have little importance. Structure **94c**, showing C_2—C_3 π-bonding, would be a resonance structure if the C_4B atoms lie in a plane. If the boron is out-of-plane, **94c** is isomeric with **94a**. In an extreme displacement **94c** could resemble a carborane or a cyclobutadiene–boron(I) π-complex (**94d**), where the boron would have an electron octet. Such an octet would also shift the ^{11}B signal upfield. Since the signal of **94** is not shifted, nonplanar isomers **94c** and **94d** are unlikely contributors to the ground state of **94** (Scheme 6).

But the visible spectral changes between **94** and **101** make it clear that the boron $2p_z$-orbital is involved in delocalization, except that such delocalization is destabilizing, thereby decreasing the HOMO–LUMO energy gap[30] (Scheme 7) by disrupting the degeneracy of the symmetric (S) and antisymmetric (A) MO levels of the isoelectronic cyclopentadienyl cation. The overall conclusion from these spectral considerations is that there is relatively little B—C multiple bonding in the ground state of boroles (small *R*-effect) and by its *I*-effect boron raises the cnergy of the A-MO somewhat and the S-MO markedly.

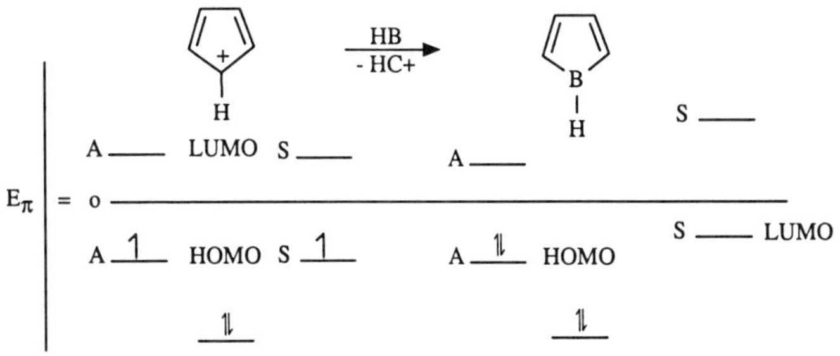

SCHEME 7

The chemical consequences of these MO changes is to activate the antiaromatic diene system for reaction. This activation is especially evident in Diels–Alder cyclizations, either with itself (Eq. 32) or with a variety of olefins and acetylenes.[93]

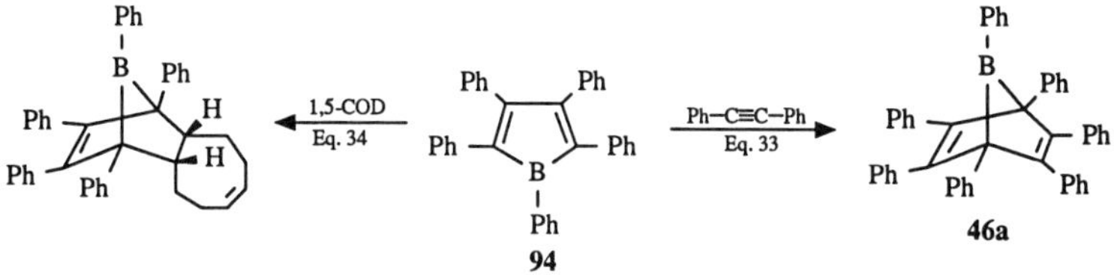

The relatively high-energy HOMO and the low-energy LUMO make **94** an eminently suitable Diels–Alder participant.

The cycloaddition product **46a** has already been considered in the foregoing discussion of B—C multiple bonding in allylic boranes (Section II.D; structures **46a**, **46b**, and **47**). It only remains here to comment on conclusions drawn from the XRD analysis of the structurally similar dimer **100**. Fagan and co-workers interpret the tilting of the bridging Ph-B group toward the C_2—C_3 double bond as evidence for the same kind of homoallylic π-interaction that is depicted in **46b**.[93] Unfortunately, no spectral or ^{11}B-NMR data are given to corroborate this interpretation. In their absence, one cannot dismiss the possibility that steric repulsion by the exomethyl groups at C_5 and C_6 are actually responsible for the bridge tilt.

The borole ring and various 1,3-diborolyl anions have been extensively employed as ligands to prepare a huge array of transition metal complexes and multidecker sandwich compounds.[96,97] Inevitably, the electronic character of the borole is profoundly changed upon complexation, so a study of such complexes can reveal nothing certain about B—C multiple bonding in the isolated ligand.

E. *Boron Derivatives of Six-Membered Rings*

Again, such boron-containing rings have often served as ligands in transition metal complexes, but we shall focus on what is known about the ligands themselves and their possible B—C π-bonding. Of interest first is the 1-phenylboratabenzene anion (**102**), prepared by Ashe and Shu[98] according to Scheme 8. The upfield shift of the ^{11}B resonance in **102** (27 ppm, compared with Me_3B at 86 ppm) and the relatively low-field proton signals of the C_5B ring (6.3–7.2 ppm) give evidence of substantial ring currents (electron delocalization) and heightened electron density on boron, both indicative of aromatic character.

The 1,5-dibora-2,5-cyclohexadiene system (**7**) is of unusual interest because such a planar array of atoms is predicted to be a Hückel antiaromatic system, similar to boroles themselves. Indeed, when **103** is generated at low temperatures, it readily rearrange into the *nido*-2,3,4,5-tetracarbahexaboranes[6] (**8**)[99] (Eq. 33).

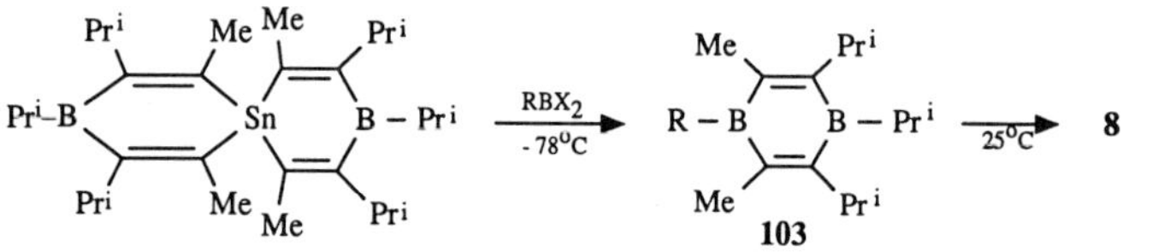

Pr^i Me Me Pr^i Pr^i–B Sn B – Pr^i Pr^i Me Me Pr^i $\xrightarrow[-78^\circ C]{RBX_2}$ Me Pr^i R – B B – Pr^i Me Pr^i **103** $\xrightarrow{25^\circ C}$ **8** (33)

Noteworthy is the ^{11}B resonance in **103**, which falls in the range 58–60 ppm, very little different from the signal of trivinylborane (55 ppm); therefore, any extensive π-electron delocalization and density on boron can be ruled out. Thus both in chemical and physical properties, such systems as **103** can be considered antiaromatic.

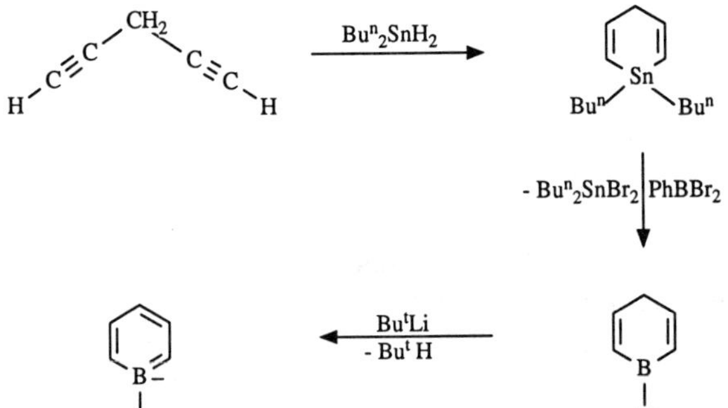

H–C≡C–CH_2–C≡C–H $\xrightarrow{Bu^n_2SnH_2}$ Sn(Bu^n)(Bu^n) ring

$\downarrow$ PhBBr$_2$, $- Bu^n_2SnBr_2$

B–Ph ring $\xrightarrow[- Bu^t H]{Bu^tLi}$ B$^-$–Ph **102**

SCHEME 8

The dianionic salts of the 1,2-diboraatabenzene (**104**) and the 2,3-diboratanaphthalene (**105**) systems have recently also been synthesized.[100,101] These systems appear to exhibit electron delocalization and aromatic stabilization similar to that shown by **102**.

104 **105**

One final fascinating example of an equilibrium set up between nonclassical and classical isomers of the C_4B_2 system is that resulting from the treatment of **106** with magnesium metal.[102] Depending upon whether or not THF is present, **107** or **108** is formed. The structures depicted are supported by XRD analysis (Scheme 9). The ^{11}B signals of **107** (9 and 47 ppm) appear to be outside the typical range of tetracarbahexaboranes[6] and are consistent with an equilibrium between **107** and **108** in solution.

F. *Borepins and Isomers of Borabicycloheptadiene Systems*

The first generation of the borepin nucleus without annulated substituents occurred during the heating of colorless heptaphenyl-7-borabicyclo[2.2.1]heptadiene (**46a**) into refluxing toluene.[103,104] The appearance

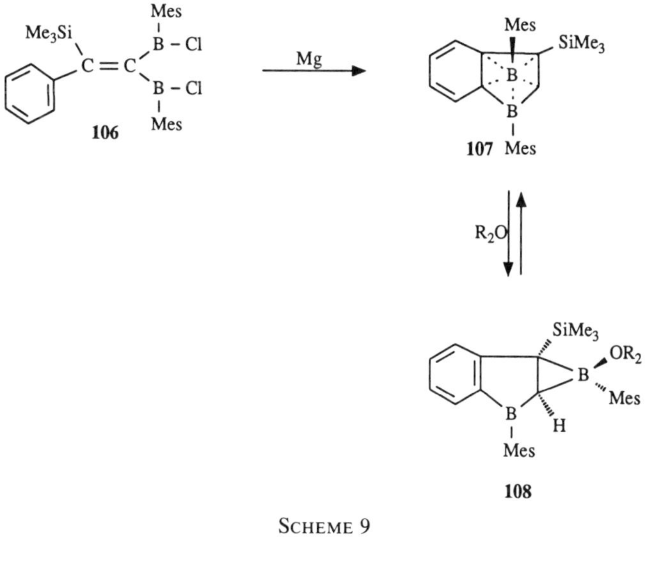

SCHEME 9

of the lime-green color signaled the formation of **109**, and its presence was supported by protolysis of the reaction mixture to produce the *cis*-1,2- and 1,4-dihydrobenzenes (**110** and **111**). Isomer **111** was thought to stem from **46a** and isomer **110** from **109**; **109** was viewed as cleaving to give the *cis*-hexatriene **112**, which underwent disrotatory ring closure to **111** (Scheme 10). Later studies revealed that the rearrangement leading to **109** was only a part of a whole series of pericyclic reorganizations in this C_6B array.[30] Such concatenated transformations prevented isolation of pure **109**.

Successful syntheses of monoannulated borepins were achieved by Ashe and co-workers by tin–boron exchanges with the tin heterocyclic precursors.[105,106] Particularly informative were the properties of **113**, which has no benzoannulation (Eq. 34):

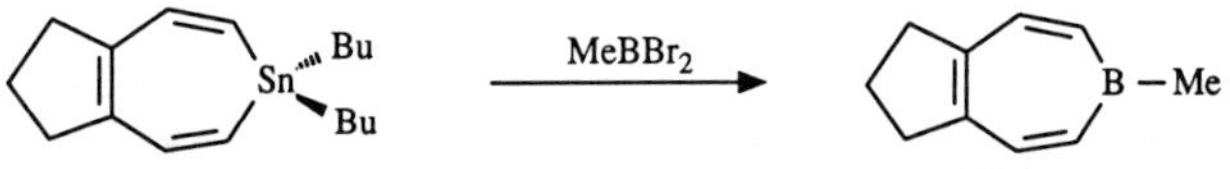

$$\text{(stannepin)}\ Sn(Bu)_2 \xrightarrow{MeBBr_2} \text{B–Me}\ (\mathbf{113}) \qquad (34)$$

The UV spectrum closely resembles that of tropone and exhibits a bathochromic shift over those of cycloheptatriene (unconjugated). Although the ^{11}B signal at 53.6 ppm is not much shifted over that in 1-phenyl-4,5-dihydroborepin (54.6 ppm), the borepin ring protons occur at markedly

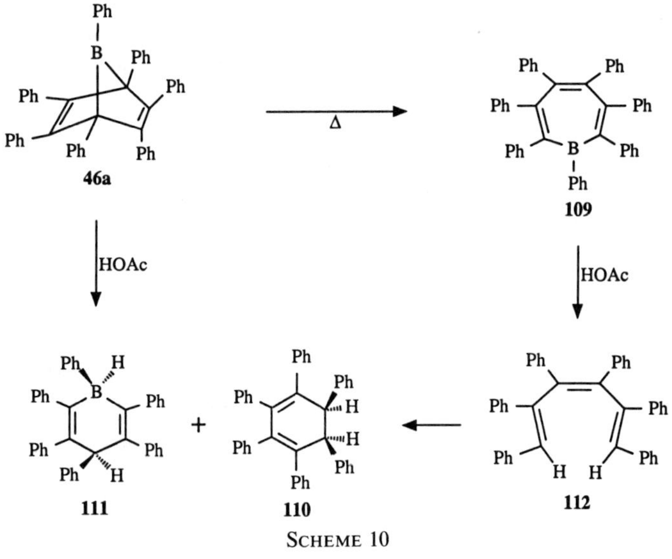

46a —Δ→ 109
46a —HOAc→ 111
109 —HOAc→ 112 → 110
111 + 110

SCHEME 10

low fields (7.49, 7.76 ppm), which is consistent with substantial ring current due to π-electron delocalization. The borepin nucleus can acccordingly be judged to be aromatic in character, but with relatively little π-bonding to the boron center.

This apparently moderate aromatic stabilization for the planar borepin (**114**) should therefore be only a small barrier for isomerization into the 7-borabicyclo[4.1.0]heptadiene system (**115**) (Eq. 35).

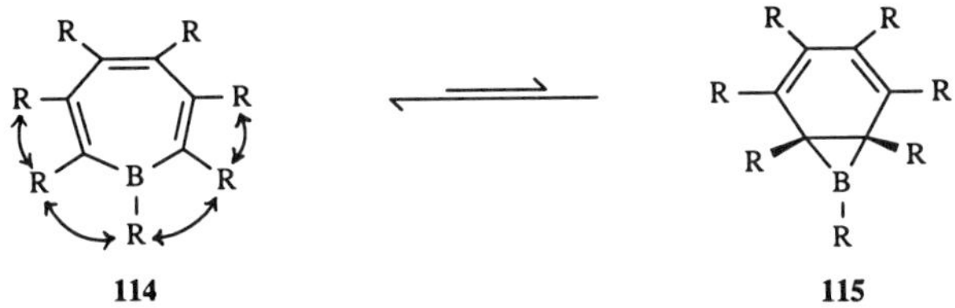

(35)

114 **115**

In fact, it would seem likely that the equilibrium could be shifted in favor of **115** by the in-plane steric repulsion of R groups in **114**. By forming **115**, such bulky adjacent R groups would no longer lie in the same plane and would be farther apart.

Exactly such steric repulsion seems to be responsible for the further rearrangements observed with heptaphenylborepin (**109** or **114** with R = Ph). The end result of a reasonable, but amazing sequence of pericyclic reactions is 1,2,3,3a,4,5-hexaphenyl-5-bora-3a,4-dihydro-5*H*-benz[*e*]indene (**119**), whose generation is initiated with **109** and continues through **116–118** (Scheme 11). The last step, **118–119**, is an impressive intramolecu-

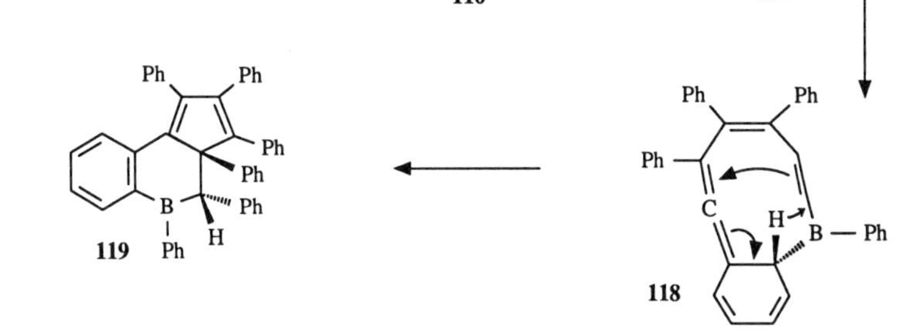

109 **116** **117** **118** **119**

SCHEME 11

lar ene reaction. Decisive in the ultimate formation of **119** is the predominance of releasing steric repulsion over any aromatic stabilization of the borepin nucleus in **109**.

IV

UNSOLVED PROBLEMS IN BORON–CARBON MULTIPLE BONDING AND REMAINING CHALLENGES

Although much progress has been made over the last 25 years in understanding the role of multiple bonding between carbon and neutral boron (III) centers, little is known about the importance of such bonding in subvalent or ionic organoboron intermediates as exemplified by boron cations (**120**),[107] boron anions that could exist as singlets (**121**) or triplets (**122**),[108–110] radicals (**123**),[111] and borenes (**124**).[83,112,113] Solving such problems, which are crucial in elucidating organoboron reaction mechanisms, will require the diligence and ingenuity of researchers for the next 25 years, at least.

$R\text{–}\overset{+}{B}\text{–}R$	$R_2B{:}^-$	$[R\text{–}\dot{B}\text{–}R]^{-}$ (with second dot below B)	$R_2B\cdot$	$R\text{–}B{:}$
120	**121**	**122**	**123**	**124**

ACKNOWLEDGMENTS

Our research in organoboron chemistry has been generously supported by the National Science Foundation, the U.S. Air Force Office of Scientific Research, and Akzo Corporate Research America.

REFERENCES

(1) Stock, A. E. *Hydrides of Boron and Silicon;* Cornell University Press: Ithaca, NY, 1933.
(2) Stock, A.; Pohland, E. *Ber. Dsch. Chem. Ges.* **1923,** *56*, 789.
(3) Niedenzu, K.; Dawson, J. W. In *The Chemistry of Boron and Its Compounds;* Muetterties, E. L., Ed.; Wiley: New York, 1967; pp. 410–422.
(4) Dewar, M. J. S. In *Progress in Boron Chemistry;* Steinberg, H.; McCloskey, A. L., Eds.; Pergamon: Oxford, 1964; Vol. 1.
(5) Lappert, M. F. In *The Chemistry of Boron and Its Compounds;* Muetterties, E. L., Ed.; Wiley: New York, 1967; pp. 485–495.
(6) Shapiro, I.; Good, C. D.; Williams, R. E. *J. Am. Chem. Soc.* **1962,** *84*, 3837.
(7) Grimes, R. N. *Carboranes;* Academic Press: New York, 1970.
(8) Coyle, T. D.; Stafford, S. L.; Stone, F. G. A. *J. Chem. Soc.* **1961,** 3103.
(9) Lappert, M. F. In *The Chemistry of Boron and Its Compounds;* Muetterties, E. L., Ed.; Wiley: New York, 1967; pp. 482–495.

(10) Odom, J. D. In *Comprehensive Organometallic Chemistry;* Wilkinson, G.; Stone, F. G. A.; Abel, E. W., Eds.; Pergamon: Oxford, 1982; Vol. 1, pp. 256–270.
(11) Odom, J. D.; Moore, T. F.; Johnston, S. A.; Durig, J. R. *J. Mol. Struct.* **1979,** *54*, 49.
(12) Matteson, D. S. *J. Am. Chem. Soc.* **1960,** *82*, 4228.
(13) Good, C. D.; Ritter, D. M. *J. Am. Chem. Soc.* **1962,** *84*, 1162.
(14) Hoffmann, R. *J. Chem. Phys.* **1963,** *39*, 1397.
(15) Armstrong, D. R.; Perkins, P. G. *Theor. Chim. Acta* **1966,** *4*, 352.
(16) Pople, J. A.; Segal, G. A. *J. Chem. Phys.* **1966,** *44*, 3289.
(17) Kuehnlenz, G.; Jaffee, H. H. *J. Chem. Phys.* **1973,** *55*, 2238.
(18) Allinger, N. L.; Siefert, J. H. *J. Am. Chem. Soc.* **1975,** *97*, 752.
(19) Krogh-Jespersen, K.; Cremer, D.; Dill, J. D.; Pople, J. A.; Schleyer, P. von R. *J. Am. Chem. Soc.* **1981,** *103*, 2589.
(20) Budzelaar, P. H. M.; Krogh-Jespersen, K.; Clark, T.; Schleyer, P. von R. *J. Am. Chem. Soc.* **1985,** *107*, 2773.
(21) Büll, M.; Schleyer, P. von R.; Ibrahim, M. A.; Clark, T. *J. Am. Chem. Soc.* **1991,** *113*, 2466.
(22) Eisch, J. J.; Shafii, B.; Odom, J. D.; Rheingold, A. L. *J. Am. Chem. Soc.* **1990,** *112*, 1847.
(23) Foord, A.; Beagley, B.; Reade, W.; Steer, I. A. *J. Mol. Struct.* **1975,** *24*, 131–137.
(24) Odom, J. D.; Hall, L. W.; Riethmiller, S.; Durig, J. R. *Inorg. Chem.* **1974,** *13*, 170.
(25) Durig, J. R.; Hall, L. W.; Carter, R. O.; Wurrey, C. J.; Kalasinsky, V. F.; Odom, J. D. *J. Phys. Chem.* **1974,** *80*, 1188.
(26) Durig, J. R.; Carter, R. O.; Odom, J. D. *Inorg. Chem.* **1974,** *13*, 701.
(27) Zettler, F.; Hausen, H. D.; Hess, H. *J. Organomet. Chem.* **1974,** *72*, 157.
(28) Kruger, K. J.; duPreez, A. L.; Haines, R. J. *J. Chem. Soc., Dalton Trans.* **1974,** 1302.
(29) Bartell, L. S.; Carroll, B. L. *J. Chem. Phys.* **1965,** *42*, 3076.
(30) Eisch, J. J.; Galle, J. E.; Kozima, S. *J. Am. Chem. Soc.* **1986,** *108*, 379.
(31) Nöth, H.; Wrackmeyer, B. *NMR: Basic Princ. Progr.* **1978,** *14*, 23.
(32) de Moor, J. E.; Vander Kelen, G. P. *J. Organomet. Chem.* **1966,** *6*, 235.
(33) Nöth, H.; Vahrenkamp, H. *J. Organomet. Chem.* **1968,** *12*, 23.
(34) Brown, H. C.; Zweifel, G. *J. Am. Chem. Soc.* **1961,** *83*, 3834.
(35) March, J. *Advanced Organic Chemistry,* 4th ed.; Wiley: New York, 1992; p. 34.
(36) Schulz, H.; Gabbert, G.; Pritzkow, H.; Siebert, W. *Chem. Ber.* **1993,** *126*, 1593.
(37) Eisch, J. J.; Kotowicz, B. unpublished studies.
(38) Ramsey, B. G. *J. Phys. Chem.* **1966,** *70*, 611.
(39) Onak, T. *Organoborane Chemistry;* Academic Press: New York, 1975; p. 13.
(40) Leffler, J. E.; Watts, G. B.; Tanigaki, T.; Dolan, E.; Miller, D. S. *J. Am. Chem. Soc.* **1970,** *92*, 6825.
(41) Olmstead, M. M.; Power, P. P. *J. Am. Chem. Soc.* **1986,** *108*, 4235.
(42) Griffin, R. G.; van Willigen, H. *J. Chem. Phys.* **1972,** *57*, 86.
(43) Krause, E. *Ber. Dtsch. Chem. Ges.* **1926,** *59*, 777.
(44) Eisch, J. J.; Dluzniewski, T.; Behrooz, M. *Heteroatom. Chem.* **1993,** *4*, 235.
(45) Hancock, K. G.; Kramer, J. D. *J. Organomet. Chem.* **1974,** *64*, C29.
(46) Eisch, J. J. In *Advances in Organometallic Chemistry;* West, R.; Stone, F. G. A., Eds.; Academic Press: New York, 1977; Vol. 16, pp. 67–109.
(47) Ramsey, B. G.; Das, N. K. *J. Am. Chem. Soc.* **1972,** *94*, 4227.
(48) Chung, V. V.; Inagaki, K.; Tokuda, M.; Itoh, M. *Chem. Lett.* **1976,** 209.
(49) Seyferth, D. *J. Organomet. Chem.* **1975,** *100*, 237.
(50) Rundle, R. E. *J. Phys. Chem.* **1957,** *61*, 45.
(51) Rathke, M. W.; Kow, R. *J. Am. Chem. Soc.* **1972,** *94*, 6854.

(52) Pelter, A.; Singaram, B.; Wilson, J. W. *Tetrahedron Lett.* **1983,** *24*, 643.
(53) Pelter, A.; Bus, D.; Colclough, E. *J. Chem. Soc., Chem. Commun.* **1987,** 297.
(54) Pelter, A.; Buss, D.; Pitchford, A. *Tetrahedron Lett.* **1985,** *26*, 5093.
(55) Pelter, A.; Smith, K.; Elgendy, S.; Rowlands, M. *Tetrahedron Lett.* **1989,** *30*, 5643.
(56) Olmstead, M. M.; Power, P. P.; Weese, K. J. *J. Am. Chem. Soc.* **1987,** *109*, 2541.
(57) Helm, S. W.; Linti, G.; Nöth, H.; Channareddy, A.; Hofmann, P. *Chem. Ber.* **1992,** *125*, 73.
(58) Tapper, A.; Schmitz, T.; Paetzold, P. *Chem. Ber.* **1989,** *122*, 595.
(59) Klusik, H.; Berndt, A. *Angew. Chem., Int. Ed. Engl.* **1984,** *23*, 825.
(60) Boese, R.; Paetzold, P.; Tapper, A.; Ziembinski, R. *Chem. Ber.* **1989,** *122*, 1057.
(61) Pilz, M.; Allwohn, J.; Massa, W.; Berndt, A. *Angew. Chem., Int. Ed. Engl.* **1990,** *29*, 399.
(62) Berndt, A. *Angew. Chem., Int. Ed. Engl.* **1993,** *32*, 985.
(63) Hunold, R.; Allwohn, J.; Baum, G.; Massa, W.; Berndt, A. *Angew. Chem., Int. Ed. Engl.* **1988,** *27*, 961.
(64) Malhotra, S. C. *Inorg. Chem.* **1964,** *3*, 862.
(65) Biffar, W.; Nöth, H.; Pommerening, H. *Angew. Chem., Int. Ed. Engl.* **1980,** *19*, 56.
(66) Schluter, K.; Berndt, A. *Angew. Chem., Int. Ed. Engl.* **1980,** *19*, 57.
(67) Moezzi, A.; Olmstead, M. M.; Power, P. P. *J. Am. Chem. Soc.* **1992,** *114*, 2715.
(68) Pilz, M.; Allwohn, J.; Willershausen, P.; Massa, W.; Berndt, A. *Angew. Chem., Int. Ed. Engl.* **1990,** *9*, 146.
(69) Ohkubo, K.; Shimada, H.; Okada, M. *Bull. Chem. Soc. Jpn.* **1971,** *44*, 2025.
(70) Guest, M. F.; Hillier, I. H.; Saunders, V. R. *J. Organomet. Chem.* **1972,** *44*, 59.
(71) Willershausen, P.; Höfner, A.; Allwohn, J.; Pilz, M.; Massa, W.; Berndt, A. *Z. Naturforsch., B: Chem. Sci.* **1992,** *47B*, 983.
(72) Denmark, S. E.; Nishide, K.; Faucher, A.-M. *J. Am. Chem. Soc.* **1991,** *113*, 6675.
(73) Schacht, W.; Kaufmann, D. *J. Organomet. Chem.* **1988,** *339*, 33.
(74) Littger, R.; Nöth, H.; Thomaun, M.; Wagner, M. *Angew. Chem., Int. Ed. Engl.* **1993,** *32*, 295.
(75) Köster, R.; Seidel, G.; Lutz, F.; Krüger, C.; Kehr, G.; Wrackmeyer, B. *Chem. Ber.* **1994,** *127*, 813.
(76) van der Kerk, S. M.; Budzelaar, P. H. M.; van der Kerk-van Hoof, A.; van der Kerk, G. J. M.; Schleyer, P. von R. *Angew. Chem., Int. Ed. Engl.* **1983,** *95*, 61.
(77) Pues, C.; Berndt, A. *Angew. Chem., Int. Ed. Engl.* **1984,** *23*, 313.
(78) Habben, C.; Meller, A. *Chem. Ber.* **1984,** *117*, 2531.
(79) Eisch, J. J.; Shafii, B.; Rheingold, A. L. *J. Am. Chem. Soc.* **1987,** *109*, 2526.
(80) West, R.; Pachaly, B. *Angew. Chem.* **1984,** *96*, 306.
(81) Eisch, J. J.; Shafii, B.; Odom, J. D.; Rheingold, A. L. *J. Am. Chem. Soc.* **1990,** *112*, 1847.
(82) Eisch, J. J.; Gonsior, L. J. *J. Organomet. Chem.* **1967,** *8*, 53.
(83) Eisch, J. J.; Becker, H. P. *J. Organomet. Chem.* **1979,** *197*, 141.
(84) Eisch, J. J.; Shen, F.; Tamao, K. *Heterocycles* **1982,** *18*, 245.
(85) Meyer, H.; Schmidt-Lukasch, G.; Baum, G.; Massa, W.; Berndt, A. *Z. Naturforsch., B: Chem. Sci.* **1988,** *43B*, 801.
(86) Pues, C.; Baum, G.; Massa, W.; Berndt, A. *Z. Naturforsch., B: Chem. Sci.* **1988,** *43B*, 275.
(87) Hildebrand, M.; Pturzkow, H.; Siebert, W. *Angew. Chem., Int. Ed. Engl.* **1985,** *24*, 759.
(88) Pilz, M.; Allwohn, J.; Bühl, M.; Schleyer, P. von R.; Berndt, A. *Z. Naturforsch., B: Chem. Sci.* **1991,** *46B*, 1085.
(89) Budzelaar, P. H. M.; Krogh-Jespersen, K.; Clark, T.; Schleyer, P. von R. *J. Am. Chem. Soc.* **1985,** *107*, 2773.

(90) Irngartinger, H.; Hauck, J.; Siebert, W.; Hildebrand, M. *Z. Naturforsch., B: Chem. Sci.* **1991,** *46B*, 1621.
(91) Mennekes, T.; Paetzold, P.; Boese, R.; Bläser, D. *Angew. Chem., Int. Ed. Engl.* **1991,** *30*, 173.
(92) Eisch, J. J.; Hota, N. K.; Kozima, S. *J. Am. Chem. Soc.* **1969,** *91*, 4575.
(93) Fagan, P. J.; Burns, E. G.; Calabrese, J. C. *J. Am. Chem. Soc.* **1988,** *110*, 2979.
(94) Braye, E. H.; Hübel, W.; Caplier, I. *J. Am. Chem. Soc.* **1961,** *83*, 4406.
(95) Bergmann, E.; Zwecker, O. *Justus Liebigs Ann. Chem.* **1931,** *487*, 155.
(96) Herberich, G. E.; Boveleth, W.; Hessner, B.; Köffer, D. P. J.; Negele, M.; Saive, R. *J. Organomet. Chem.* **1986,** *308*, 153.
(97) Siebert, W.; Bochmann, M.; Edwin, J.; Krüger, C.; Tsay, Y.-H. Z. *Z. Naturforsch., B: Anorg. Chem., Org. Chem.* **1978,** *33B*, 1410.
(98) Ashe, A. J., III; Shu, P. *J. Am. Chem. Soc.* **1971,** *93*, 1804.
(99) Wrackmeyer, B.; Kehr, G. *Polyhedron* **1991,** *10*, 1497.
(100) Herberich, G. E.; Hessner, B.; Hostalek, M. *J. Organomet. Chem.* **1988,** *355*, 473.
(101) Weinmann, W.; Pritzkow, H.; Siebert, W. *Chem. Ber.* **1994,** *127*, 611.
(102) Michel, H.; Steiner, D.; Wocadlo, S.; Allwohn, J.; Stamatis, N.; Massa, W.; Berndt, A. *Angew. Chem., Int. Ed. Engl.* **1992,** *31*, 607.
(103) Eisch, J. J.; Galle, J. E. *J. Am. Chem. Soc.* **1975,** *97*, 4436.
(104) Eisch, J. J.; Galle, J. E. *J. Organomet. Chem.* **1976,** *127*, C9.
(105) Ashe, A. J., III; Drone, F. J. *J. Am. Chem. Soc.* **1987,** *109*, 1879.
(106) Ashe, A. J., III; Kampf, J. W.; Kausch, C. M.; Konishi, H.; Kristen, M. O.; Kroker, J. *Organometallics* **1990,** *9*, 2944.
(107) Schneider, W. F.; Narula, C. K.; Nöth, H.; Bursten, B. E. *Inorg. Chem.* **1991,** *30*, 3919.
(108) Eisch, J. J.; Boleslawski, M. P.; Tamao, K. *J. Org. Chem.* **1989,** *54*, 1627.
(109) Eisch, J. J.; Shafii, B.; Boleslawski, M. P. *Pure Appl. Chem.* **1991,** *63*, 365.
(110) Eisch, J. J.; Shah, J. H.; Boleslawski, M. P. *J. Organomet. Chem.* **1994,** *464*, 11.
(111) Köster, R.; Benedikt, G.; Schrötter, H. W. *Angew. Chem., Int. Ed. Engl.* **1964,** *3*, 514.
(112) van der Kerk, S. M.; Budzelaar, P. H. M.; van Eekeren, A. L. M., van der Kerk, G. J. M. *Polyhedron* **1984,** *3*, 271.
(113) Bromm, D.; Seebold, U.; Noltemeyer, M.; Mellor, A. *Chem. Ber.* **1991,** *124*, 2645.

Index

A

C

D

E

K

L

M

T

U

V